T0310110

Geophysical Monograph Series

Including

IUGG Volumes

Maurice Ewing Volumes

Mineral Physics Volumes

156 **Particle Acceleration in Astrophysical Plasmas: Geospace and Beyond** *Dennis Gallagher, James Horwitz, Joseph Perez, Robert Preece, and John Quenby (Eds.)*

157 **Seismic Earth: Array Analysis of Broadband Seismograms** *Alan Levander and Guust Nolet (Eds.)*

158 **The Nordic Seas: An Integrated Perspective** *Helge Drange, Trond Dokken, Tore Furevik, Rüdiger Gerdes, and Wolfgang Berger (Eds.)*

159 **Inner Magnetosphere Interactions: New Perspectives From Imaging** *James Burch, Michael Schulz, and Harlan Spence (Eds.)*

160 **Earth's Deep Mantle: Structure, Composition, and Evolution** *Robert D. van der Hilst, Jay D. Bass, Jan Matas, and Jeannot Trampert (Eds.)*

161 **Circulation in the Gulf of Mexico: Observations and Models** *Wilton Sturges and Alexis Lugo-Fernandez (Eds.)*

162 **Dynamics of Fluids and Transport Through Fractured Rock** *Boris Faybishenko, Paul A. Witherspoon, and John Gale (Eds.)*

163 **Remote Sensing of Northern Hydrology: Measuring Environmental Change** *Claude R. Duguay and Alain Pietroniro (Eds.)*

164 **Archean Geodynamics and Environments** *Keith Benn, Jean-Claude Mareschal, and Kent C. Condie (Eds.)*

165 **Solar Eruptions and Energetic Particles** *Natchimuthukonar Gopalswamy, Richard Mewaldt, and Jarmo Torsti (Eds.)*

166 **Back-Arc Spreading Systems: Geological, Biological, Chemical, and Physical Interactions** *David M. Christie, Charles Fisher, Sang-Mook Lee, and Sharon Givens (Eds.)*

167 **Recurrent Magnetic Storms: Corotating Solar Wind Streams** *Bruce Tsurutani, Robert McPherron, Walter Gonzalez, Gang Lu, José H. A. Sobral, and Natchimuthukonar Gopalswamy (Eds.)*

168 **Earth's Deep Water Cycle** *Steven D. Jacobsen and Suzan van der Lee (Eds.)*

169 **Magnetospheric ULF Waves: Synthesis and New Directions** *Kazue Takahashi, Peter J. Chi, Richard E. Denton, and Robert L. Lysal (Eds.)*

170 **Earthquakes: Radiated Energy and the Physics of Faulting** *Rachel Abercrombie, Art McGarr, Hiroo Kanamori, and Giulio Di Toro (Eds.)*

171 **Subsurface Hydrology: Data Integration for Properties and Processes** *David W. Hyndman, Frederick D. Day-Lewis, and Kamini Singha (Eds.)*

172 **Volcanism and Subduction: The Kamchatka Region** *John Eichelberger, Evgenii Gordeev, Minoru Kasahara, Pavel Izbekov, and Johnathan Lees (Eds.)*

173 **Ocean Circulation: Mechanisms and Impacts—Past and Future Changes of Meridional Overturning** *Andreas Schmittner, John C. H. Chiang, and Sidney R. Hemming (Eds.)*

174 **Post-Perovskite: The Last Mantle Phase Transition** *Kei Hirose, John Brodholt, Thorne Lay, and David Yuen (Eds.)*

175 **A Continental Plate Boundary: Tectonics at South Island, New Zealand** *David Okaya, Tim Stem, and Fred Davey (Eds.)*

176 **Exploring Venus as a Terrestrial Planet** *Larry W. Esposito, Ellen R. Stofan, and Thomas E. Cravens (Eds.)*

177 **Ocean Modeling in an Eddying Regime** *Matthew Hecht and Hiroyasu Hasumi (Eds.)*

178 **Magma to Microbe: Modeling Hydrothermal Processes at Oceanic Spreading Centers** *Robert P. Lowell, Jeffrey S. Seewald, Anna Metaxas, and Michael R. Perfit (Eds.)*

179 **Active Tectonics and Seismic Potential of Alaska** *Jeffrey T. Freymueller, Peter J. Haeussler, Robert L. Wesson, and Göran Ekström (Eds.)*

180 **Arctic Sea Ice Decline: Observations, Projections, Mechanisms, and Implications** *Eric T. DeWeaver, Cecilia M. Bitz, and L.-Bruno Tremblay (Eds.)*

181 **Midlatitude Ionospheric Dynamics and Disturbances** *Paul M. Kintner, Jr., Anthea J. Coster, Tim Fuller-Rowell, Anthony J. Mannucci, Michael Mendillo, and Roderick Heelis (Eds.)*

182 **The Stromboli Volcano: An Integrated Study of the 2002–2003 Eruption** *Sonia Calvari, Salvatore Inguaggiato, Giuseppe Puglisi, Maurizio Ripepe, and Mauro Rosi (Eds.)*

183 **Carbon Sequestration and Its Role in the Global Carbon Cycle** *Brian J. McPherson and Eric T. Sundquist (Eds.)*

184 **Carbon Cycling in Northern Peatlands** *Andrew J. Baird, Lisa R. Belyea, Xavier Comas, A. S. Reeve, and Lee D. Slater (Eds.)*

185 **Indian Ocean Biogeochemical Processes and Ecological Variability** *Jerry D. Wiggert, Raleigh R. Hood, S. Wajih A. Naqvi, Kenneth H. Brink, and Sharon L. Smith (Eds.)*

186 **Amazonia and Global Change** *Michael Keller, Mercedes Bustamante, John Gash, and Pedro Silva Dias (Eds.)*

187 **Surface Ocean–Lower Atmosphere Processes** *Corinne Le Quèrè and Eric S. Saltzman (Eds.)*

188 **Diversity of Hydrothermal Systems on Slow Spreading Ocean Ridges** *Peter A. Rona, Colin W. Devey, Jérôme Dyment, and Bramley J. Murton (Eds.)*

189 **Climate Dynamics: Why Does Climate Vary?** *De-Zheng Sun and Frank Bryan (Eds.)*

190 **The Stratosphere: Dynamics, Transport, and Chemistry** *L. M. Polvani, A. H. Sobel, and D. W. Waugh (Eds.)*

Geophysical Monograph 191

Rainfall: State of the Science

Firat Y. Testik
Mekonnen Gebremichael
Editors

American Geophysical Union
Washington, DC

Library of Congress Cataloging-in-Publication Data

Rainfall : state of the science / Firat Y. Testik and Mekonnen Gebremichael, editors.
p. cm. — (Geophysical monograph, ISSN 0065-8448 ; 191)
Includes bibliographical references and index.
ISBN 978-0-87590-481-8 (alk. paper)
1. Rain and rainfall. 2. Rainfall probabilities. 3. Rain and rainfall—Measurement. I. Testik, Firat Y., 1977- II. Gebremichael, Mekonnen.
QC925.R24 2010
551.57'7—dc22

2010049230

ISBN: 978-0-87590-481-8
ISSN: 0065-8448

Cover Image: Raindrops on a window.

Printed in the United States of America.

CONTENTS

Preface
Firat Y. Testik and Mekonnen Gebremichael.....vii

Microphysics, Measurement, and Analyses of Rainfall
Mekonnen Gebremichael and Firat Y. Testik.....1

Section I: Rainfall Microphysics

Raindrop Morphodynamics
B. K. Jones, J. R. Saylor, and F. Y. Testik.....7

The Evolution of Raindrop Spectra: A Review of Microphysical Essentials
K. D. Beheng.....29

Raindrop Size Distribution and Evolution
Greg M. McFarquhar.....49

Section II: Rainfall Measurement and Estimation

Ground-Based Direct Measurement
Emad Habib, Gyuwon Lee, Dongsoo Kim, and Grzegorz J. Ciach.....61

Radar and Multisensor Rainfall Estimation for Hydrologic Applications
Dong-Jun Seo, Alan Seed, and Guy Delrieu.....79

Dual-Polarization Radar Rainfall Estimation
Robert Cifelli and V. Chandrasekar.....105

Quantitative Precipitation Estimation From Earth Observation Satellites
Chris Kidd, Vincenzo Levizzani, and Sante Laviola.....127

Section III: Statistical Analysis

Intensity-Duration-Frequency Curves
S. Rocky Durrans.....159

Frequency Analysis of Extreme Rainfall Events
Salaheddine El Adlouni and Taha B. M. J. Ouarda.....171

Methods and Data Sources for Spatial Prediction of Rainfall
T. Hengl, A. AghaKouchak, and M. Perčec Tadić.....189

Rainfall Generation
Ashish Sharma and Raj Mehrotra.....215

Radar-Rainfall Error Models and Ensemble Generators
Pradeep V. Mandapaka and Urs Germann.....247

Framework for Satellite Rainfall Product Evaluation
Mekonnen Gebremichael 265

AGU Category Index 277

Index 279

PREFACE

Rainfall, liquid precipitation, is a critical component of water and energy cycles. It is a critical source of fresh water, sustaining life on Earth, and an important process for energy exchanges between the atmosphere, ocean, and land, determining Earth's climate. It is central to water supply, agriculture, natural ecosystems, hydroelectric power, industry, drought, flood, and disease hazards for example. Therefore, rainfall is at the heart of social, economical, and political challenges in today's world. It is a high priority to use advancements in scientific knowledge of rainfall to develop solutions to the water-related challenges faced by society. The three main aspects of rainfall, "rainfall microphysics," "rainfall measurement and estimation," and "rainfall statistical analyses," have been widely studied as individual topics over the years. It is the goal of this book to synthesize all of these aspects to provide an integral picture of the state of the science of rainfall.

This book presents the state of the science of rainfall focusing on three areas: (1) rainfall microphysics, (2) rainfall measurement and estimation, and (3) rainfall statistical analyses. Each part consists of a number of self-contained chapters providing three forms of information: fundamental principles, detailed overview of current knowledge and description of existing methods, and emerging techniques and future research directions. Each book chapter is authored by preeminent researchers in their respective fields and has been reviewed by two renowned researchers within the same field for the scientific accuracy, quality, and completeness of the final content. The book is tailored to be an indispensable reference for researchers, practitioners, and graduate students who study any aspect of rainfall or utilize rainfall information in various science and engineering disciplines.

As editors of this book, we would like to express our utmost gratitude to everyone who has contributed in this publication. We are thankful to all the chapter authors for contributing their expertise and time. We are thankful also to all the reviewers who have selflessly served for the success of this project.

Firat Y. Testik
Clemson University

Mekonnen Gebremichael
University of Connecticut

Rainfall: State of the Science
Geophysical Monograph Series 191

10.1029/2010GM001026

Microphysics, Measurement, and Analyses of Rainfall

Mekonnen Gebremichael

Department of Civil and Environmental Engineering, University of Connecticut, Storrs, Connecticut, USA

Firat Y. Testik

Department of Civil Engineering, Clemson University, Clemson, South Carolina, USA

Rainfall, liquid precipitation, is a critical component of water and energy cycles. It is a critical source of water for water supply, agriculture, natural ecosystems, hydroelectric power, and industry and is central to issues of drought, flood, and disease hazards. The most desired characteristic of rainfall is the rainfall rate at Earth's surface. This book, "*Rainfall: State of the Science*," aims to synthesize the three main aspects (microphysics, measurement and estimation, and statistical analyses) of rainfall rate estimation efforts to provide an integral picture of this endeavor. In this introductory chapter, we present the issues that will be discussed in detail in the subsequent chapters.

1. MICROPHYSICS

Understanding the microphysics of rainfall is important to accurately estimate rainfall rate from microwave remote sensing and to model the rainfall process in process-based models. Rainfall microphysics, deals with the dynamical processes for individual and populated raindrops throughout their journey from cloud to surface. Rainfall and cloud microstructure is a broad topic, and there are several comprehensive books [*Pruppacher and Klett*, 1978; *Rogers and Yau*, 1989; *Mason*, 1971] devoted to this subject. For the purpose of rainfall rate retrievals, accurate information on the raindrop shape, fall velocity, and raindrop size distribution (DSD) are of particular interest. Therefore, main considerations on rainfall microphysics discussions in this book will be centered on these quantities with a perspective from rainfall rate retrievals. Aside from rainfall rate measurements, these quantities have important applications such as soil erosion studies [*Fox*, 2004; *Fornis et al.*, 2005], air pollution studies [*Mircea et al.*, 2000], and telecommunications [*Panagopoulos and Kanellopoulos*, 2002].

Raindrops demonstrate a variety of complex shapes and shape-altering oscillations under the action of a range of surface and body forces. Shapes and fall velocities of raindrops are closely coupled resulting in a dynamic interplay until equilibrium is reached. Raindrop shapes and fall velocities are important input parameters for extracting rainfall information via remote sensing. Polarimetric weather radars utilize vertical to horizontal chord ratios of "equilibrium" raindrop shapes and corresponding "terminal" fall velocities. Consequently, there have been a number of studies on raindrop morphodynamics (i.e., static and dynamic processes related to raindrop shape) over the years [*Laws*, 1941; *Spilhaus*, 1948; *Gunn*, 1949; *Gunn and Kinzer*, 1949; *Savic*, 1953; *McDonald*, 1954; *Pruppacher and Pitter*, 1971; *Green*, 1975; *Wang and Pruppacher*, 1977; *Beard*, 1977; *Beard and Chuang*, 1987; *Beard et al.*, 1989]. *Jones et al.* [this volume] review raindrop morphodynamics, providing a synthesis of information on raindrop shape and related physical processes, including forces shaping the raindrops, raindrop oscillations, and fall velocities.

Rainfall: State of the Science
Geophysical Monograph Series 191

10.1029/2010GM001025

Raindrops are formed within clouds through collisional interactions of cloud droplets. After formation, raindrops interact with each other via collisions throughout their journey from cloud to surface. These collisions may result in coalescence, breakup, and bounce of colliding drops. As a result of these collision outcomes, DSD continuously evolves with height. Accurate information on DSD at different heights is important for obtaining accurate rainfall information via remote sensing. Numerical simulations based on the stochastic coalescence/breakup equation and laboratory models of governing processes [*Low and List*, 1982a, 1982b] are used to obtain information on the DSD evolution with height [e.g., *Gillespie and List*, 1978; *List and McFarquhar*, 1990; *McFarquhar*, 2004]. *Beheng* [this volume] discusses the formation of raindrops from cloud droplets and collisional interactions of raindrops that result in an evolution of the raindrop size distribution. Environmental interactions are omitted.

Following the landmark study by *Marshall and Palmer* [1948], various distributions have been used to represent DSD, including exponential, Gaussian, and lognormal distributions. The assumed form of DSD plays a critical role in rainfall rate retrieval from both ground- and space-based systems. Various numerical simulations of DSD evolution under the action of collisional interactions have shown evidence for an equilibrium form of DSD after sufficient evolution time is given [e.g., *Valdez and Young*, 1985; *List et al.*, 1987; *List and McFarquhar*, 1990; *Hu and Srivastava*, 1995]. However, field observations have not verified the occurrence of an equilibrium DSD as predicted by these numerical simulations. *McFarquhar* [this volume] provides an overview of observed raindrop size distributions at both the ground and aloft as well as the evolution of the raindrop size distributions throughout the rain shaft from numerical simulations.

2. MEASUREMENT AND ESTIMATION

Information on rainfall properties can be obtained by means of different observing systems and associated algorithms: direct in situ sensors (i.e., rain gauges and disdrometers), ground-based remote sensors (i.e., weather radars), and space-based remote sensors (i.e., radars and infrared sensors). Each sensor has its own strengths and limitations.

Rain gauges and disdrometers provide direct in situ point measurements of rainfall properties at high temporal resolution. Rain gauge measurements of rainfall rate continue to be the main basis for "calibrating" rainfall remote sensing algorithms and for numerous research and operational applications that require rainfall data. However, even at their point measurement scales, rain gauge rainfall measurements are subject to systematic and random measurement errors, the most important of which are the following: the drift of rainfall particles due to wind field deformation around the gauge, losses caused by wetting of the inner walls of the gauge, evaporation of water accumulated in the gauge container, splashing of raindrops out or into the gauge, calibration-related errors, malfunctioning problems, poor location selection, and local random errors. Disdrometers measure DSD that describes the rainfall microstructure. The DSD data have been widely used in studying soil erosion and rainfall microphysical properties and, perhaps most importantly, derivation of radar rainfall retrieval algorithms. The DSD measurements can be affected by various sources of errors, which can be grouped into instrumental, sampling, and observational errors. *Habib et al.* [this volume] present an overview of the different types of rain gauges and disdrometers, discuss the major sources of uncertainties that contaminate measurements at the local point scale, and describe the recently developed methods for automatic quality control of the rain gauge data.

The availability of rainfall estimates from conventional ground-based scanning weather radars (i.e., single-polarization radar systems), such as the U.S. Weather Surveillance Radars-1988 Doppler (WSR-88D) radars, at high space-time resolutions and over large areas has greatly advanced our quantitative information on the space-time variability of rainfall. However, because radar measures volumetric reflectivity of hydrometeors aloft rather than rainfall near the ground, radar rainfall estimation is inherently subject to various sources of error. The major sources and possible practical consequences of these errors have been well recognized and discussed by many researchers [e.g., *Wilson and Brandes*, 1979; *Zawadzki*, 1982; *Austin*, 1987; *Krajewski and Smith*, 2002; *Jordan et al.*, 2003; *Krajewski et al.*, 2010]. Recent work on improving the accuracy of rainfall estimates from conventional weather radars consists of incorporating rain gauge measurements to remove the bias in the radar rainfall estimates and using multiple radar rainfall fields whenever possible. *Seo et al.* [this volume] describe the foundations of radar and multisensor rainfall estimation, recent advances and notable applications, and outstanding issues and areas of research that must be addressed to meet the needs of forecasting in various applications such as hydrology.

As part of the modernization of the WSR-88D, the U.S. National Weather Service and other agencies have decided to add a polarimetric capability to existing conventional single-polarization radars. Dual polarization provides additional information compared to single-polarization radar systems, which helps to significantly improve the accuracy of radar rainfall estimates. The three polarimetric variables that are often used in rainfall estimation are the radar reflectivity at horizontal polarization, the differential reflectivity (defined as the difference between reflectivities at horizontal and vertical polarizations), and the differential phase (defined as the difference between the phases of the radar signals at orthogonal polarizations). An overview of the methods of rainfall estimation from these variables is presented by *Cifelli and Chandrasekar* [this volume].

Remote sensing from a space platform provides a unique opportunity to obtain spatial fields of rainfall information over large areas of Earth. During the last two decades, satellite-based instruments have been designed to collect observations mainly at thermal infrared (IR) and microwave (MW) wavelengths that can be used to estimate rainfall rates. Observations in the IR band are available in passive modes from (near) polar orbiting (revisit times of 1–2 days) and geostationary orbits (revisit times of 15–30 min), while observations in the passive and active MW band are only available from the (near) polar-orbiting satellites. A number of algorithms have been developed to estimate rainfall rates by combining information from the more accurate (but less frequent) MW observations with the more frequent (but less accurate) IR observations to take advantage of the complementary strengths [*Sorooshian et al.*, 2000; *Scofield and Kuligowski*, 2003; *Joyce et al.*, 2004; *Turk and Miller*, 2005; *Huffman et al.*, 2007; *Ushio and Kachi*, 2010]. Satellite rainfall estimates are subject to a variety of error sources (gaps in revisiting times, poor direct relationship between MW cloud top measurements and rainfall rate, atmospheric effects that modify the radiation field, etc.). The errors increase with increasing space-time resolution and depend largely on the algorithm technique, type, and number of satellite sensors used and the study region [e.g., *Hong et al.*, 2004; *Gottschalck et al.*, 2005; *Brown*, 2006; *Ebert et al.*, 2007; *Tian et al.*, 2007; *Bitew and Gebremichael*, 2010; *Dinku et al.*, 2010; *Sapiano et al.*, 2010]. Ongoing research and development continues to address the accuracy and the resolution (temporal and spatial) of these estimates. *Kidd et al.* [this volume] cover the basis of the satellite systems used in the observation of rainfall and the processing of these measurements to generate rainfall estimates, discuss research challenges, and provide research and development recommendations.

3. STATISTICAL ANALYSES

Various statistical techniques are often applied to rainfall data depending on the application and source of data. Commonly employed statistical analyses are extreme event analysis, spatial interpolation of point rainfall, rainfall generation, and uncertainty analysis of remote sensing rainfall estimates.

Statistical analysis of extreme rainfall events is useful for a number of engineering applications including hydraulic design (culverts and storm sewers) and landslide hazard evaluations. Intensity-duration-frequency (IDF) curves, with areal reduction factors, are commonly used for design storm calculations. For any prescribed rainfall duration (which depends on the time of concentration for the watershed) and return period (which is often set as a standard value depending on the purpose and failure consequence of the hydraulic structure), the corresponding design rainfall intensity is obtained from regional IDF curves. The common method of developing IDF consists of the following steps: getting the annual maximum series of rainfall intensity for a given duration, using distributions (parametric or nonparametric) to find rainfall intensity for different return periods, and repeating the above steps for different durations. This method has recently come under criticism as it conveniently ignores the joint probability distribution among the rainfall characteristics: depth, intensity, and duration. The difficulty in modeling the joint distributions is that most parametric multivariate distributions are unable to handle rainfall because of the heavy-tailed distributions in the rainfall characteristics. The recently emerging technique of copula [*Sklar*, 1959] has shown promise in overcoming this difficulty because of its ability to model the dependence structure independently of the marginal distributions. Recent studies have successfully used the copula technique to model the joint distribution among rainfall depth, intensity, and duration [*De Michele and Salvadori*, 2003; *Salvadori and De Michele*, 2004a, 2004b; *Zhang and Singh*, 2007; *Kao and Govindaraju*, 2007, 2008; *Wang et al.*, 2010]. *Durrans* [this volume] presents an overview of the historical development of IDF, common methods for constructing IDF, and emerging new methods.

The frequency (or return period) analysis of extreme events is important to develop IDF curves and to test for any trends in the extremes. A number of parametric distributions (Gumbel, generalized extreme value, lognormal, log-Pearson type 3, Halphen, and generalized logistics) have been developed over the years to model the extreme rainfall events. There are two main statistical approaches to fit these distributions. The first approach applies to annual maxima of time series. The second approach looks at exceedances over high threshold, also known as the "peaks over threshold" approach. Prior to any statistical analyses, the data need to be checked for any outlier, dependence, and stationarity. Focusing on the first approach, *El Adlouni and Ouarda* [this volume] present detailed information on data preparation (detection and treatment of outliers, independence, and stationarity), the parametric distributions (and associated parameter estimation techniques) in the case of stationary time series, and modeling of nonstationary time series.

Spatial rainfall analysis is performed to estimate areal rainfall from point rain gauge data, or to estimate rainfall value at a site based on rainfall measured at another site and auxiliary information, or to generate a spatial pattern. *Hengl et al.* [this volume] present the spatial analysis techniques used in rainfall, with programming codes to help interested users apply the techniques. Stochastic rainfall generation is performed to generate long time series of rainfall data for a variety of applications including probabilistic failure assessment of natural or man-made systems where rain is an important input. *Sharma and Mehrotra* [this volume] present an overview of stochastic generation of rainfall, with a focus on daily and subdaily rainfall generation at point and multiple locations, for the current climate assuming climatic stationarity, as well as for future climates using exogenous inputs simulated using general circulation models under assumed greenhouse gas emission scenarios.

Remote sensing rainfall estimates are subject to systematic and random errors from various sources, some of which are inherent to the observation system and are unavoidable. Operational remote sensing rainfall products are deterministic and do not contain quantitative information on the level of the estimation errors. This has led to the current situation in which those who use remote sensing rainfall estimates know that there are significant errors in the estimates, but they have no quantitative information about the magnitudes of the estimation errors. Consequently, there are no mechanisms to account for the uncertainty of remote sensing rainfall estimates in applications and decision making. A possible solution to this major problem is to construct an error model that characterizes the conditional distribution of actual rainfall rate for any given remote sensing rainfall estimate. *Mandapaka and Germann* [this volume] present the advances in the area of weather radar rainfall error modeling that have taken place over the past decade. Compared to weather radar rainfall estimates, the satellite rainfall estimates are subject to additional error sources and therefore have higher estimation errors. *Gebremichael* [this volume] presents a recommended standard framework for quantifying errors in satellite rainfall estimates, reviews existing error models and presents emerging ones, and performs quantitative assessment of the utility of satellite rainfall estimates for hydrological applications in selected regions.

REFERENCES

Austin, P. M. (1987), Relation between measured radar reflectivity and surface rainfall, *Mon. Weather Rev.*, *115*, 1053–1071.

Beard, K. V. (1977), On the acceleration of large water drops to terminal velocity, *J. Appl. Meteorol.*, *16*, 1068–1071.

Beard, K. V., and C. Chuang (1987), A new model for the equilibrium shape of raindrops, *J. Atmos. Sci.*, *44*(11), 1509–1525.

Beard, K. V., H. T. Ochs, III, and R. J. Kubesh (1989), Natural oscillations of small raindrops, *Nature*, *342*, 408–410.

Beheng, K. D. (2010), The evolution of raindrop spectra: A review of microphysical essentials, in *Rainfall: State of the Science*, *Geophys. Monogr. Ser.*, doi:10.1029/2010GM000957, this volume.

Bitew, M. M., and M. Gebremichael (2010), Evaluation through independent measurements: Complex terrain and humid tropical region in Ethiopia, in *Satellite Rainfall Applications for Surface Hydrology*, edited by M. Gebremichael and F. Hossain, pp. 205–214, doi:10.1007/978-90-481-2915-7_12, Springer, New York.

Brown, J. E. M. (2006), An analysis of the performance of hybrid infrared and microwave satellite precipitation algorithms over India and adjacent regions, *Remote Sens. Environ.*, *101*, 63–81.

Cifelli, R., and V. Chandrasekar (2010), Dual-polarization radar rainfall estimation, in *Rainfall: State of the Science*, *Geophys. Monogr. Ser.*, doi:10.1029/2010GM000930, this volume.

De Michele, C., and G. Salvadori (2003), A generalized Pareto intensity-duration model of storm rainfall exploiting 2-copulas, *J. Geophys. Res.*, *108*(D2), 4067, doi:10.1029/2002JD002534.

Dinku, T., S. J. Connor, and P. Ceccato (2010), Comparison of CMORPH and TRMM-3B42 over mountainous regions of Africa and South America, in *Satellite Rainfall Applications for Surface Hydrology*, edited by M. Gebremichael and F. Hossain, pp. 193–204, doi:10.1007/978-90-481-2915-7_11, Springer, New York.

Durrans, S. R. (2010), Intensity-duration-frequency curves, in *Rainfall: State of the Science*, *Geophys. Monogr. Ser.*, doi: 10.1029/2009GM000919, this volume.

Ebert, E. E., J. E. Janowiak, and C. Kidd (2007), Comparison of near-real-time precipitation estimates from satellite observations and numerical models, *Bull. Am. Meteorol. Soc.*, *88*, 47–64.

El Adlouni, S., and T. B. M. J. Ouarda (2010), Frequency analysis of extreme rainfall events, in *Rainfall: State of the Science*, *Geophys. Monogr. Ser.*, doi:10.1029/2010GM000976, this volume.

Fornis, R. L., H. R. Vermeulen, and J. D. Nieuwenhuis (2005), Kinetic energy-rainfall intensity relationship for central Cebu, Philippines for soil erosion studies, *J. Hydrol.*, *300*, 20–32.

Fox, N. I. (2004), The representation of rainfall drop-size distribution and kinetic energy, *Hydrol. Earth Syst. Sci.*, *8*(5), 1001–1007.

Gebremichael, M. (2010), Framework for satellite rainfall product evaluation, in *Rainfall: State of the Science*, *Geophys. Monogr. Ser.*, doi:10.1029/2010GM000974, this volume.

Gillespie, J. R., and R. List (1978), Effects of collision-induced breakup on drop size evolution in steady state rainshafts, *Pure Appl. Geophys.*, *117*, 599–626.

Gottschalck, J., J. Meng, M. Rodell, and P. Houser (2005), Analysis of multiple precipitation products and preliminary assessment of their impact on global land data assimilation system land surface states, *J. Hydrometeorol.*, *6*, 573–598.

Green, A. E. (1975), An approximation for the shapes of large raindrops, *J. Appl. Meteorol.*, *14*, 1578–1583.

Gunn, R. (1949), Mechanical resonance in freely falling raindrops, *J. Geophys. Res.*, *54*, 383–385.

Gunn, R., and G. D. Kinzer (1949), The terminal velocity of fall for water droplets in stagnant air, *J. Meteorol.*, *6*, 243–248.

Habib, E., G. Lee, D. Kim, and G. J. Ciach (2010), Ground-based direct measurement, in *Rainfall: State of the Science*, *Geophys. Monogr. Ser.*, doi:10.1029/2010GM000953, this volume.

Hengl, T., A. AghaKouchak, and M. P. Tadić (2010), Methods and data sources for spatial prediction of rainfall, in *Rainfall: State of the Science*, *Geophys. Monogr. Ser.*, doi:10.1029/2010GM000999, this volume.

Hong, Y., K. L. Hsu, S. Sorooshian, and X. Gao (2004), Precipitation estimation from remotely sensed imagery using an artificial neural network cloud classification system, *J. Appl. Meteorol.*, *43*, 1834–1852.

Hu, Z., and R. C. Srivastava (1995), Evolution of raindrop size distribution by coalescence, breakup, and evaporation: Theory and observations, *J. Atmos. Sci.*, *52*(10), 1761–1783.

Huffman, G. J., R. F. Adler, D. T. Bolvin, G. Gu, E. J. Nelkin, K. P. Bowman, Y. Hong, E. F. Stocker, and D. B. Wolff (2007), The TRMM Multisatellite Precipitation Analysis (TMPA): Quasi-global, multilayer, combined-sensor, precipitation estimates at fine scale, *J. Hydrometeorol.*, *8*, 38–55.

Jones, B. K., J. R. Saylor, and F. Y. Testik (2010), Raindrop morphodynamics, in *Rainfall: State of the Science*, *Geophys. Monogr. Ser.*, doi:10.1029/2009GM000928, this volume.

Jordan, P. W., A. W. Seed, and P. E. Wienmann (2003), A stochastic model of radar measurement errors in rainfall accumulations at catchment scale, *J. Hydrometeorol.*, *4*, 841–855.

Joyce, R. J., J. E. Janowiak, P. A. Arkin, and P. Xie (2004), CMORPH: A method that produces global precipitation estimates from passive microwave and infrared data at high spatial and temporal resolution, *J. Hydrometeorol.*, *5*, 487–503.

Kao, S.-C., and R. S. Govindaraju (2007), A bivariate frequency analysis of extreme rainfall with implications for design, *J. Geophys. Res.*, *112*, D13119, doi:10.1029/2007JD008522.

Kao, S.-C., and R. S. Govindaraju (2008), Trivariate statistical analysis of extreme rainfall events via the Plackett family of copulas, *Water Resour. Res.*, *44*, W02415, doi:10.1029/2007WR006261.

Kidd, C., V. Levizzani, and S. Laviola (2010), Quantitative precipitation estimation from Earth observation satellites, in *Rainfall: State of the Science*, *Geophys. Monogr. Ser.*, doi: 10.1029/2009GM000920, this volume.

Krajewski, W. F., and J. A. Smith (2002), Radar hydrology: Rainfall estimation, *Adv. Water Resour.*, *25*, 1387–1394.

Krajewski, W. F., G. Villarini, and J. A. Smith (2010), Radar-rainfall uncertainties: Where are we after thirty years of effort?, *Bull. Am. Meteorol. Soc.*, *91*, 87–94.

Laws, J. O. (1941), Measurements of the fall-velocity of water-drops and rain drops, *Eos Trans. AGU*, *22*, 709–721.

List, R., and G. M. McFarquhar (1990), The evolution of three-peak raindrop size distributions in one-dimensional shaft models. Part I: Single-pulse rain, *J. Atmos. Sci.*, *47*(24), 2996–3006.

List, R., N. R. Donaldson, and R. E. Stewart (1987), Temporal evolution of drop spectra to collisional equilibrium in steady and pulsating rain, *J. Atmos. Sci.*, *44*(2), 362–372.

Low, T. B., and R. List (1982a), Collision, coalescence and breakup of raindrops. Part I: Experimentally established coalescence efficiencies and fragment size distributions in breakup, *J. Atmos. Sci.*, *39*(7), 1591–1606.

Low, T. B., and R. List (1982b), Collision, coalescence and breakup of raindrops. Part II: Parameterization of fragment size distributions, *J. Atmos. Sci.*, *39*(7), 1607–1618.

Mandapaka, P. V., and U. Germann (2010), Radar-rainfall error models and ensemble generators, in *Rainfall: State of the Science*, *Geophys. Monogr. Ser.*, doi:10.1029/2010GM001003, this volume.

Marshall, J. S., and W. McK. Palmer (1948), The distribution of raindrops with size, *J. Meteorol.*, *5*, 165–166.

Mason, B. J. (1971), *The Physics of Clouds*, Clarendon, Oxford, U. K.

McDonald, J. E. (1954), The shape and aerodynamics of large raindrops, *J. Meteorol.*, *11*, 478–494.

McFarquhar, G. M. (2004), A new representation of collision-induced breakup of raindrops and its implications for the shapes of raindrop size distributions, *J. Atmos. Sci.*, *61*(7), 777–794.

McFarquhar, G. M. (2010), Raindrop size distribution and evolution, in *Rainfall: State of the Science*, *Geophys. Monogr. Ser.*, doi: 10.1029/2010GM000971, this volume.

Mircea, M., S. Stefan, and S. Fuzzi (2000), Precipitation scavenging coefficient: Influence of measured aerosol and raindrop size distributions, *Atmos. Environ.*, *34*, 5169–5174.

Panagopoulos, A. D., and J. D. Kanellopoulos (2002), Adjacent satellite interference effects as applied to the outage performance of an Earth-space system located in a heavy rain climatic region, *Ann. Telecommun.*, *57*(9–10), 925–942.

Pruppacher, H. R., and J. D. Klett (1978), *Microphysics of Clouds and Precipitation*, Kluwer, Dordrecht, Netherlands.

Pruppacher, H. R., and R. L. Pitter (1971), A semi-empirical determination of the shape of cloud and rain drops, *J. Atmos. Sci.*, *28*(1), 86–94.

Rogers, R. R., and M. K. Yau (1989), *A Short Course in Cloud Physics*, Pergamon, New York.

Salvadori, G., and C. De Michele (2004a), Analytical calculation of storm volume statistics involving Pareto-like intensity–duration marginals, *Geophys. Res. Lett.*, *31*, L04502, doi:10.1029/2003GL018767.

Salvadori, G., and C. De Michele (2004b), Frequency analysis via copulas: Theoretical aspects and applications to hydrological events, *Water Resour. Res.*, *40*, W12511, doi:10.1029/2004WR003133.

Sapiano, M. R. P., J. E. Janowiak, W. Shi, R. W. Higgins, and V. B. S. Silva (2010), Regional evaluation through independent precipitation measurements: USA, in *Satellite Rainfall Applications for*

Surface Hydrology, edited by M. Gebremichael and F. Hossain, pp. 169–192, doi:10.1007/978-90-481-2915-7_10, Springer, New York.

Savic, P. (1953), Circulation and distortion of liquid drops falling through a viscous medium, *Rep. NRC-MT-22*, 50 pp., Natl. Res. Counc., Ottawa, Ont., Canada.

Scofield, R. A., and R. J. Kuligowski (2003), Status and outlook of operational satellite precipitation algorithms for extreme-precipitation events, *Weather Forecasting*, *18*, 1037–1051.

Seo, D.-J., A. Seed, and G. Delrieu (2010), Radar and multisensor rainfall estimation for hydrologic applications, in *Rainfall: State of the Science*, *Geophys. Monogr. Ser.*, doi:10.1029/2010GM000952, this volume.

Sharma, A., and R. Mehrotra (2010), Rainfall generation, in *Rainfall: State of the Science*, *Geophys. Monogr. Ser.*, doi:10.1029/2010GM000973, this volume.

Sklar, A. W. (1959), Fonctions de répartition à n dimension et leurs marges, *Publ. Inst. Stat. Univ. Paris*, *8*, 229–231.

Sorooshian, S., K. Hsu, X. Gao, H. V. Gupta, B. Imam, and D. Braithwaite (2000), Evaluation of PERSIANN system satellite-based estimates of tropical rainfall, *Bull. Am. Meteorol. Soc.*, *81*, 2035–2046.

Spilhaus, A. F. (1948), Raindrop size, shape, and falling speed, *J. Meteorol.*, *5*, 108–110.

Tian, Y., C. D. Peters-Lidard, B. J. Chaudhury, and M. Garcia (2007), Multitemporal analysis of TRMM-based satellite precipitation products for land data assimilation applications, *J. Hydrometeorol.*, *8*, 1165–1183.

Turk, F. J., and S. Miller (2005), Toward improving estimates of remotely-sensed precipitation with MODIS/AMSR-E blended data techniques, *IEEE Trans. Geosci. Remote Sens.*, *43*, 1059–1069.

Ushio, T., and M. Kachi (2010), Kalman filtering applications for global satellite mapping of precipitation (GSMaP), in *Satellite Rainfall Applications for Surface Hydrology*, edited by M. Gebremichael and F. Hossain, pp. 105–123, doi:10.1007/978-90-481-2915-7_7, Springer, New York.

Valdez, M. P., and K. C. Young (1985), Number fluxes in equilibrium raindrop populations: A Markov chain analysis, *J. Atmos. Sci.*, *42*(10), 1024–1036.

Wang, P. K., and H. R. Pruppacher (1977), Acceleration to terminal velocity of cloud and raindrops, *J. Appl. Meteorol.*, *16*, 276–280.

Wang, X., M. Gebremichael, and J. Yan (2010), Weighted likelihood copula modeling of extreme rainfall events in Connecticut, *J. Hydrol.*, *230*, 108–115.

Wilson, J. W., and E. A. Brandes (1979), Radar measurement of rainfall–A summary, *Bull. Am. Meteorol. Soc.*, *60*, 1048–1058.

Zawadzki, I. (1982), The quantitative interpretation of weather radar measurements, *Atmos. Ocean*, *20*, 158–180.

Zhang, L., and V. P. Singh (2007), Bivariate rainfall frequency distributions using Archimedean copulas, *J. Hydrol.*, *332*(1–2), 93–109.

M. Gebremichael, Department of Civil and Environmental Engineering, University of Connecticut, Storrs, CT 06269-2037, USA.

F. Y. Testik, Department of Civil Engineering, Clemson University, Clemson, SC 29634, USA. (ftestik@clemson.edu)

Raindrop Morphodynamics

B. K. Jones

Department of Mechanical Engineering, Columbia University, New York, New York, USA

J. R. Saylor

Department of Mechanical Engineering, Clemson University, Clemson, South Carolina, USA

F. Y. Testik

Department of Civil Engineering, Clemson University, Clemson, South Carolina, USA

In the absence of forces other than surface tension, a water drop will attain a perfectly spherical shape. Raindrops experience a range of forces, including those due to fluid flow (both inside and outside the drop), hydrostatic forces, and electrostatic forces. A falling raindrop deviates in shape from spherical, becoming a flattened oblate spheroid, a shape that becomes more prominent as the raindrop diameter increases. This shape is characterized by a chord ratio, which is the ratio of the height to the width of the raindrop. The drop shape is often variable, oscillating because of excitation of the natural frequencies of the drop by the flow of fluid around the drop and through interactions between the natural frequency of the drop and vortex shedding in the wake of the drop. These interactions make raindrop morphodynamics, the study of the dynamic and stable raindrop shape, an especially rich problem. Drop collisions also affect the transient behavior of drop shape. Polarimetric radar techniques have further motivated studies of raindrop morphodynamics, since knowledge of raindrop shape can be utilized to improve rain rate retrievals using these radars. Experimentation and analytical efforts have explored several facets of raindrop morphodynamical behavior, including raindrop fall speeds, nonoscillating and oscillating shapes, chord ratio versus diameter relationships, oscillation frequencies, and the preferred harmonic modes, for example. Herein, we provide a survey of the current state of knowledge of these aspects of raindrop morphodynamics.

1. INTRODUCTION

As water vapor in the atmosphere condenses, liquid droplets are initially sufficiently small to remain aloft, entrained in air currents. The motion of these cloud droplets causes them to collide with one another and form either permanent unions or smaller fragment droplets. In this manner, some drops increase in mass until the force of gravity exceeds the momentum available from the air motion, and they begin to fall. This collision and fragmentation occurs in falling drops as well. In fact, an upper size limit is determined as some falling drops coalesce until breakup invariably occurs due to hydrodynamic instability [*Pruppacher and*

Rainfall: State of the Science
Geophysical Monograph Series 191

10.1029/2009GM000928

Figure 1. Water drop levitated in a vertical wind tunnel illustrating the characteristic shape of a quiescent raindrop.

Pitter, 1971] or because of drop-drop collision. If they remain in the liquid phase, these hydrometeors are referred to as raindrops. Due to their fluid nature, raindrops assume a variety of complex shapes and shape-altering oscillations during free fall. The study of this static and dynamic behavior is referred to as raindrop "morphodynamics" and is based primarily in fluid mechanics. The characteristic shape of a nonoscillating (hereafter "equilibrium") raindrop is shown in Figure 1.

Voluminous work involving a variety of theoretical modeling and experimentation can be found on this subject in the literature (see *Testik and Barros* [2007] for an extensive review). Experimenters have utilized high-speed photography of natural rain [*Testik et al.*, 2006], for example, or devised creative methods to elucidate real raindrop behavior from water drops floating in wind tunnels [*Beard and Pruppacher*, 1969; *Pruppacher and Beard*, 1970; *Kamra et al.*, 1986; *Saylor and Jones*, 2005; *Szakall et al.*, 2009; *Jones and Saylor*, 2009] or falling from high stairwells, towers, or highway bridges [*Andsager et al.*, 1999; *Thurai and Bringi*, 2005]. Theoretical models of raindrop shape reflect the solution of complex differential equations that rely on prior empirical observations for boundary conditions and validation. The accuracy of these calculated shapes, when compared with observations of real raindrops, has chronologically increased as researchers have improved upon the assumptions, techniques, and errors of their predecessors.

Aside from scientific novelty, the study of raindrop morphodynamics is an important aspect of precipitation science, global hydrology, weather radar science [*Chandrasekar et al.*, 2008], and satellite and terrestrial communication techniques [*Thurai and Bringi*, 2005]. For example, accurate evaluation of dual-polarization (also "polarimetric") weather radar relies upon precise knowledge of raindrop shapes. Raindrops can also attenuate and disrupt wireless communication links operating at or above microwave frequencies, so correcting for these errors may be possible with more advanced knowledge of raindrop morphodynamic behavior [*Allnutt*, 1989].

The remainder of this chapter encompasses six sections. A brief background on the fluid dynamics of raindrops is provided in section 2. The equilibrium raindrop shape is introduced in section 3. Raindrop oscillations and resulting shape changes are discussed in section 4. The effect of raindrop shape on terminal fall velocity u_t is presented in section 5, and in section 6, the effect of electrical fields on raindrop shape is discussed. Experimental techniques used to study raindrops are presented in section 7.

2. BACKGROUND ON RAINDROP FLUID DYNAMICS

The airflow past a raindrop and the water flow inside a raindrop in free fall are governed by the continuity equation and the Navier-Stokes equations of motion, subject to the appropriate boundary conditions. However, being a nonlinear system of coupled partial differential equations, analytical solution of the Navier-Stokes equations is currently prohibitively complex unless simplifications can be introduced. Discussion of the Navier-Stokes equations and their theoretical treatment in the context of raindrops is given by *Pruppacher and Klett* [1997] and is not presented here. Further discussion of flows relevant to raindrops can be found in various graduate-level fluid mechanics textbooks [e.g., *Batchelor*, 1967; *Landau and Lifshitz*, 1959].

There are primarily three dimensionless parameters pertinent to the morphodynamics of raindrops in free fall: the Reynolds number *Re*, Weber number *We*, and Strouhal number *St*, defined as

$$Re = Ud/\nu, \quad (1)$$

$$We = \rho_a U^2 d/\sigma, \quad (2)$$

$$St = f_w d/U, \quad (3)$$

where U is the relative velocity (fall speed) between the airstream and the raindrop during free fall, d is the raindrop diameter, ν and ρ_a are the kinematic viscosity and density of air, respectively, σ is the surface tension of water, and f_w is the frequency of vortex shedding in the drop wake, relevant to larger raindrops (discussed below). Because of the variable

Table 1. Approximate Values of *Re* and *We* as a Function of Drop Diameter *d* for Raindrops

d (mm)	*Re*	*We*
0.2	10	$8.6(10^{-4})$
1.0	400	0.3
2.0	1380	1.4
3.0	2510	3.1
4.0	3670	5.0
5.0	4810	6.8
6.0	5920	8.6

nature of raindrop shape, the drop diameter d refers to the diameter of an equivalent volume sphere. The variation in morphodynamic behavior that raindrops exhibit can be correlated to these dimensionless parameters, since the ratios in equations (1)–(3) represent the relative magnitudes of underlying fluid forces. Specifically, Re is the ratio of inertial to viscous forces, while We is the ratio of inertial to surface tension forces. The Strouhal number St is the dimensionless frequency of periodic behavior in the raindrop wake, which arises for $d > 1$ mm. Based on similarity arguments [see, e.g., *Barenblatt*, 2003], by matching these parameters in fluid systems other than air/water, inferences can be made regarding raindrop behavior, an approach which can simplify experimentation (see below).

Because a drop of given diameter has a nominally fixed maximum (terminal) fall velocity (see section 5) (u_t, U in equations (1) and (2)) is a function of d. Hence, Re and We are essentially determined by d, except after drop collisions when U is readjusting to u_t. The values that Re and We attain at u_t for a span of d representative of raindrops is presented in Table 1, showing that the range of Re and We for raindrops spans nearly four orders of magnitude, an indication of the widely varying balance of forces.

The characteristics of the airflow around a falling raindrop vary significantly with d and thus fall speed and the governing dimensionless parameters. Laboratory visualizations of freely falling drops suggest that for raindrops with $Re \gtrsim 210–270$, a separated wake develops in the downstream region of the drop [*Margarvey and Bishop*, 1961]. The presence of this wake region alters the pressure distribution around the raindrop, inducing static and dynamic changes in the raindrop shape. Direct observation of this coupling between shape and wake behavior is a significant experimental challenge, and analogous fluid systems have been studied to elucidate the nature of the relationship for raindrops. For example, *Magarvey and MacLatchy* [1965] described the wakes of solid spheres falling through a liquid bulk, a fluid system which deviates from the raindrop case due to the rigidity of the sphere surface; for liquid drops, the deforming surface is a significant dissipator of energy. While these deviations somewhat complicate an exact comparison with raindrops, on the other hand, rigid sphere studies facilitate isolation of the wake formation mechanisms from the effects of the liquid drop free surface.

For solid spheres at $Re = 1–200$, *Magarvey and MacLatchy* [1965] reported a steady vortex trail in which vorticity is convected directly to the freestream. At $Re = 200–300$, an axisymmetric (refers to symmetry about the drop fall axis) near-wake develops immediately downstream of the sphere. This increasing vorticity is first dissipated by circulation within the wake, then convected to the freestream in two parallel vortex trails. Asymmetry develops in the range $Re = 300–450$, as the volume of the near-wake periodically varies at fixed Re in the following manner. With each circulation of the wake region, increasing amounts of freestream fluid are entrained until a portion of the near-wake detaches, forming an eddy that sheds downstream. The general pattern of this process is shown in Figure 2. Initially, at $Re \approx 300$, eddies detach from only one side of the sphere, but with increasing Re, they begin to detach from opposite sides similar to the Karman vortex street. For fixed Re, the periodicity of this process reaches a steady state and gives rise to an oscillatory wake, characterized by the Strouhal number St (equation (3)).

Margarvey and Bishop [1961] classified the wakes of dyed liquid drops falling in a liquid bulk at Re similar to raindrops falling at u_t and found the distinct regimes outlined in Table 2. This work is more detailed than the solid sphere observations reported by *Magarvey and MacLatchy* [1965] described above. They also described a steady, axisymmetric wake developing at $Re \approx 20$, appearing as a single thread downstream of the drop. This structure prevails for $d \leq$

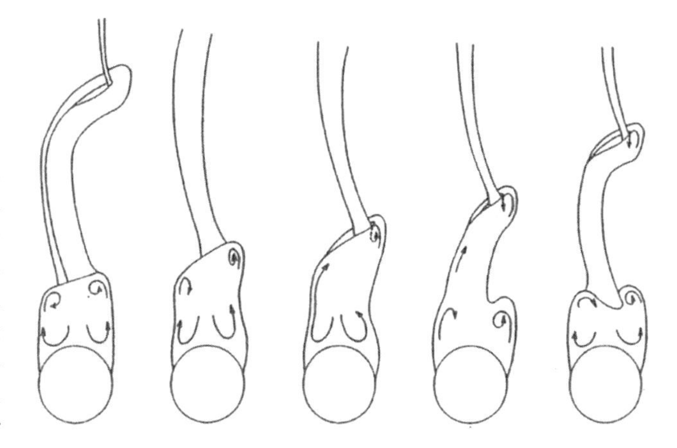

Figure 2. Temporal evolution (from left to right) of vortex shedding pattern in the wake of a solid sphere at $Re < 800$, as illustrated by *Sakamoto and Haniu* [1990]. From *Sakamoto and Haniu* [1990]. Copyright ASME.

Table 2. Classification of Wakes of Freely Falling Liquid Drops and Spheres[a]

Class	*Re* Range	*d* (mm)	Description of Wake
I	0–210	<0.9	Steady single thread
II	210–270	0.9–1.0	Asymmetric double thread
III	270–290	1.0–1.1	Double thread with waves
IV	290–410	1.1–1.3	Double row of vortex loops
V	290–700	1.1–1.8	Double row of vortex rings
VI	700–2500	1.8–4.2	Irregular vortex pattern

[a]The size (*d*) is given for raindrops falling in air. Class IV and V wakes simultaneously exist in the range *Re* = 290–410.

0.9 mm drops. With increasing *Re* and *d*, the circulating near-wake region grows until the vorticity generated can no longer be dissipated in this symmetric manner. Consequently, at $d \approx$ 0.9 mm, the point of detachment suddenly migrates from the fall axis to one side of the drop, and a double-threaded wake develops. The resulting lateral force on the drop produces a sideways drift, in agreement with the observations of *Gunn* [1949] describing the sideways and spiraling free fall of similar-sized drops in calm air, and the solid sphere observations of *Magarvey and MacLatchy* [1965]. At d = 1.0 mm, periodicity initiates at $Re = 270$ as the double-thread begins to oscillate. With further increases in *Re*, vortex loops and rings develop as outlined above and in Table 2, with the concomitant spiraling free fall seemingly caused by the precession of wake detachment points, also described by *Magarvey and MacLatchy* [1965].

3. DROP SHAPE

The shape of a falling raindrop is determined by the mechanical equilibrium of the liquid-gas interface defining its outer surface. During free fall, an aerodynamic pressure difference arises between the upper and lower poles and the equator of the raindrop, in addition to an internal circulation because of the no-slip boundary condition at the drop surface. The resulting forces, together with electrostatic forces, internal hydrostatic pressure and surface tension, balance to produce an equilibrium shape resembling a flattened sphere with a wide horizontal base and a smoothly curved upper surface. This shape varies with *d*, consequently small drops are essentially spherical while larger drops are more distorted. Figure 3 shows this effect, best characterized by the variation in the raindrop chord axis ratio, defined as the ratio of the vertical extent *a* to the horizontal extent *b* of the drop, or

$$\alpha = a/b. \quad (4)$$

Because of this flattening of raindrop shape with size, the ratio α decreases with increasing *d*. This trend persists until fragmentation occurs, typically around $d \approx$ 6–8 mm [*Pruppacher and Pitter*, 1971], although extraordinary instances of *d* larger than 8.8 mm [*Hobbs and Rangno*, 2004] and 10 mm [*Takahashi et al.*, 1995] have been observed in tropical clouds.

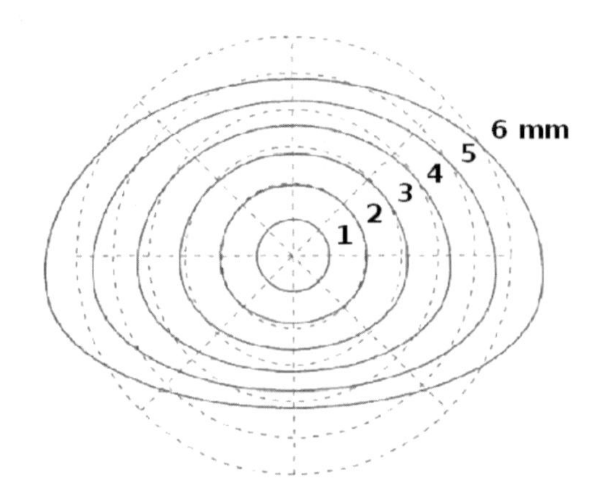

Figure 3. Calculated raindrop shapes from the numerical model due to *Beard and Chuang* [1987]. From *Beard and Chuang* [1987]. Copyright 1987 American Meteorological Society.

This predictable variation of α with *d* is the key principle behind polarimetric radar techniques [*Seliga and Bringi*, 1976]. Measurement of rain rate *R* using traditional single-polarization radar involves transmitting a microwave signal and measuring the intensity of the echo backscattered by raindrops. This intensity determines a reflectivity factor *Z*, which is used to estimate parameters of the drop size distribution. If the Rayleigh approximation is made for the backscattering cross-section of raindrops, the reflectivity factor *Z* and *R* are related to this drop size distribution by [*Doviak and Zrnić*, 1984]

$$Z = \int_0^\infty N(D) D^6 dD \quad (5)$$

$$R = \frac{\pi}{6}\int_0^\infty D^3 N(D) u_t(D) dD, \quad (6)$$

where *N(D)* is the drop size distribution (DSD), u_t is the terminal velocity, and *D* is the drop diameter (equivalent to *d*).

The DSD is typically modeled as the Marshall-Palmer spectrum

$$N(D) = N_0 e^{-\Lambda D}, \quad (7)$$

having two parameters Λ and N_0. Substituting equation (7) into equation (5) yields, after integration,

$$Z = N_0 (6!) \Lambda^{-7}. \quad (8)$$

Equation (8) reveals the primary obstacle to using single-polarization radar for the measurement of R: how to determine the two unknown parameters Λ and N_0 from the single measurement Z? The solution is to utilize dual-polarization radars. Because of the importance of drop shape to this measurement method, we now describe how dual-polarization radar measurement of rain is implemented.

In the most common implementation of this technique, the transmitted radar signal is repeatedly switched between a horizontal and vertical polarization so that two reflectivity factors, Z_H and Z_V, are measured by the receiver. The ratio of these factors gives the differential reflectivity Z_{DR}, defined as

$$Z_{DR} = 10\log(Z_H/Z_V) \quad \text{(dB)}. \tag{9}$$

This differential reflectivity varies with the specific drop sizes that are aloft during sensing, since different-sized drops exhibit distinctly different α values as shown in Figure 3. Following the treatment of *Ulbrich* [1986], R is related to Z_{DR} by equation (9) and the definitions of Z_H and Z_V, given by

$$Z_{H,V} = \frac{10^6\lambda^4}{\pi^5 K^2}\int_{D_{\min}}^{D_{\max}} \sigma_{H,V}(D)N(D)dD, \tag{10}$$

where $\sigma_{H,V}$ is [*Gans*, 1912]

$$\sigma_{H,V} = \frac{16\pi^7 D^6}{9\lambda^4}\left|\frac{\eta^2 - 1}{4\pi + (\eta^2 - 1)P_{H,V}}\right|^2 \tag{11}$$

$$P_V = \frac{4\pi}{e^2}\left[1 - \sqrt{(1-e^2)/e^2}\sin^{-1}e\right] = 4\pi - 2P_H. \tag{12}$$

Here λ is the radar signal wavelength, $|K|^2 = 0.93$ is the dielectric factor for water, η is the complex refractive index of water, and e is the drop eccentricity related directly to α by

$$e^2 = 1 - \alpha^2. \tag{13}$$

Estimates of R are obtained from measurements of Z_{DR} by determining N_0 and Λ (or the median volume diameter D_0, where $D_0 = 3.672/\Lambda$ after *Ulbrich* [1986]), which are then used in equations (6) and (7). Note that the parameter N_0 falls out of equation (9), since it is a constant that appears in the integrand for both Z_H and Z_V. Hence, by obtaining plots of Z_{DR} versus D_0 (viz. versus Λ) using the equations developed above, one can convert a measured Z_{DR} to a value of Λ after appropriate assumptions regarding the maximum and minimum drop sizes $D_{\max}$, $D_{\min}$ aloft during the rain event. Measurements of Z_H are then used along with this value in equation (10) to find N_0, which enters this equation in $N(D)$ (see equation (7)). This entire method is predicated on the variation of α with d, which enters into the retrieval of R in equation (13) and propagates through the other equations. This shows the critical nature of the α versus d relationship in the use of dual-polarization radar measurements of R [*Bringi and Chandrasekar*, 2001; *Goddard et al.*, 1994a].

Based upon laboratory observations, *Pruppacher and Pitter* [1971] described the variation with d of raindrop shapes as a continuum with three distinct diameter ranges: Class I ($d <$ 0.25 mm), Class II (0.25 mm $\leq d \leq$ 1 mm), and Class III ($d >$ 1 mm). Specifically, Class I drops exhibit no detectable distortion from sphericity. These shapes are dominated by surface tension, which effectively minimizes the energy and surface area of the drop, requiring a spherical shape. Class II drops exhibit a slight distortion with a discernible increase in radius of curvature of the lower hemisphere, a shape termed "oblate spheroidal." Class III category drops show further, marked distortion with increasing diameter, the oblate spheroid shapes exhibiting an increased flattening of the lower surface corresponding to a reduction in axis ratio. These observations are summarized in Table 3.

Early work on the development of a mathematical relationship between α and d focused on confirming and clarifying the relative roles of five physical factors: (1) surface tension, which forces a more spherical shape; (2) internal hydrostatic pressure, a vertical pressure gradient within the drop, acting outward against surface tension; (3) external aerodynamic pressure, which flattens the raindrop as it creates an increase in air pressure at the base and a decrement elsewhere; (4) internal circulation, which creates a toroidal-vortex flow within the drop, inducing complex effects on shape; and (5) electrostatic forces that may accentuate or suppress oblateness depending on drop electrical charge and field conditions [*Lenard*, 1904; *McDonald*, 1954a]. The dominance of the first three factors has been established; however, the role of internal circulation and electrostatic forces in controlling drop shape has yet to be fully understood [*Testik and Barros*, 2007]. This is because reported results on the amplitudes of internal circulation in raindrops falling at terminal velocity are rather contradictory [*Blanchard*, 1949; *McDonald*, 1954a; *Garner and Lane*, 1959; *Foote*, 1969; *Pruppacher and Beard*, 1970]. Additionally, electrostatic forces may have a strong, nonlinear effect on drop distortion in certain thunderstorm conditions [*Beard et al.*, 1989a; *Bhalwankar and Kamra*, 2007; *Beard et al.*, 2004].

Table 3. Classification of Drop Distortion With Size (d)

Class	d (mm)	Drop Shape
I	<0.25	No detectable distortion
II	0.25–1.0	Slightly aspherical
III	>1.0	Markedly oblate spheroidal

Two theoretical approaches have been used in raindrop shape models: "gravity" models and "perturbation" models. Gravity models derive an α versus d relationship from a balance of gravity and surface energy [*Beard*, 1984a], or surface tension and either external or internal pressure [*Green*, 1975; *Spilhaus*, 1948], often attaining considerable accuracy despite their relative simplicity. By comparison, the more rigorous perturbation models [*Imai*, 1950; *Savic*, 1953; *Pruppacher and Pitter*, 1971] utilize Laplace's pressure balance, which relates the curvature at each point on the drop surface to the internal and external pressures by

$$\sigma\left[\frac{1}{R_1}+\frac{1}{R_2}\right]=\Delta p, \qquad (14)$$

where R_1 and R_2 give the radii of curvature, and Δp gives the pressure difference across the drop surface (for derivation, see *Landau and Lifshitz* [1959]). The system of differential equations in equation (14) describes the complete drop silhouette. Typically, solutions to this system of equations are obtained numerically and incorporate empirical pressure measurements from wind tunnel data, a method first proposed by *Savic* [1953]. Pressure measurements from rigid spheres were used by *Pruppacher and Pitter* [1971] and others until more recently, when the technique was adapted by *Beard and Chuang* [1987] to account for an altered pressure field from drop distortions.

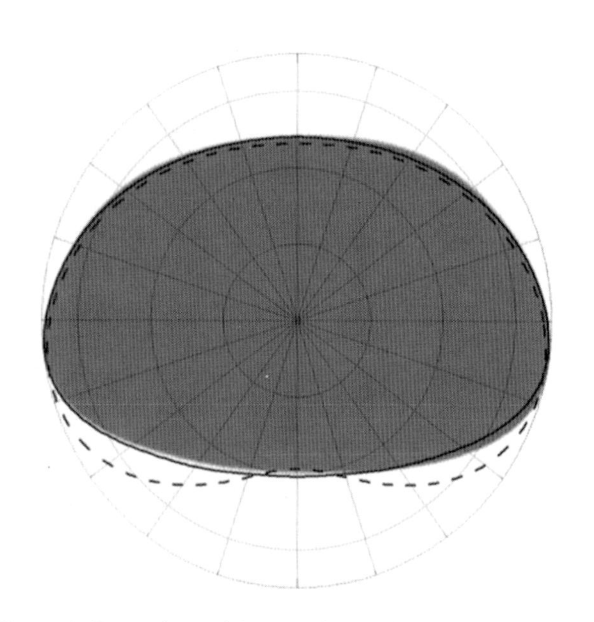

Figure 4. Comparison of the *Beard and Chuang* [1987] raindrop shape model (solid line) with the silhouette of a 6-mm wind tunnel drop (shadow) and the model of *Pruppacher and Pitter* [1971] (dashed line). From *Szakall et al.* [2009]. Copyright 2009 American Meteorological Society.

Their model, which compares well with field [*Chandrasekar et al.*, 1988; *Bringi et al.*, 1998] and laboratory [*Szakall et al.*, 2009; *Thurai et al.*, 2007] observations, is widely accepted as the most realistic for determining raindrop equilibrium shapes. A recent validation by *Szakall et al.* [2009] is shown in Figure 4.

4. DROP OSCILLATION

Schmidt [1913] was the first to observe oscillations in the shape of raindrops, and it is well known that $d \geq 1$ mm (Class III) drops may oscillate during free fall so that their shapes vary about the equilibrium shape [*Gunn*, 1949; *Jones*, 1959]. This complicates the α versus d relationship because instantaneous α measurements often scatter widely. Hence, the accurate interpretation of radar backscatter from oscillating drops requires precise knowledge of the "time-average" (mean) axis ratio as a function of drop diameter. Considerable research has been conducted to elucidate this behavior. Figure 5 shows a sequence of superposed images of a single oscillating drop, with the instants of maximum and minimum amplitude shown approximately at c and h, and f and j, respectively.

Rayleigh [1879] showed that drop oscillations occur at n discrete harmonics and with frequency f decreasing with d according to

$$f=[2n(n-1)(n+2)\sigma]^{1/2}[\pi^2\rho d^3]^{-1/2}, \qquad (15)$$

where ρ gives the bulk density of water. Rayleigh's solution assumes only axisymmetric motion of a spherical, inviscid drop oscillating with small amplitude ($A << r_0$, where A is the amplitude of drop oscillation, and r_0 is the unperturbed radius of the spherical drop); *Landau and Lifshitz* [1959] further generalized the problem and found that for each n harmonic frequency, there is one axisymmetric mode at m = 0 plus $m = n$ additional unique modes, differentiated hereafter by the ordered pair (n,m). The shapes are given for a spherical coordinate system by

$$r_{n,m}(t,\theta,\varphi)=r_0+A\sin\omega t P_{n,m}\cos m\varphi, \qquad (16)$$

where $\omega = 2\pi f$, and $P_{n,m}$ are the associated Legendre functions (see Appendix B). These oscillation modes are illustrated in Figure 6 superimposed on a sphere for the fundamental (n = 2) and first (n = 3) harmonic, the two deemed most realistic due to an incompressibility constraint for $n < 2$ and the role of viscous damping in diminishing the amplitudes of higher modes. It should be noted that while equation (15) gives the oscillation frequency as a function of n and d alone, calculations and empirical data show that the

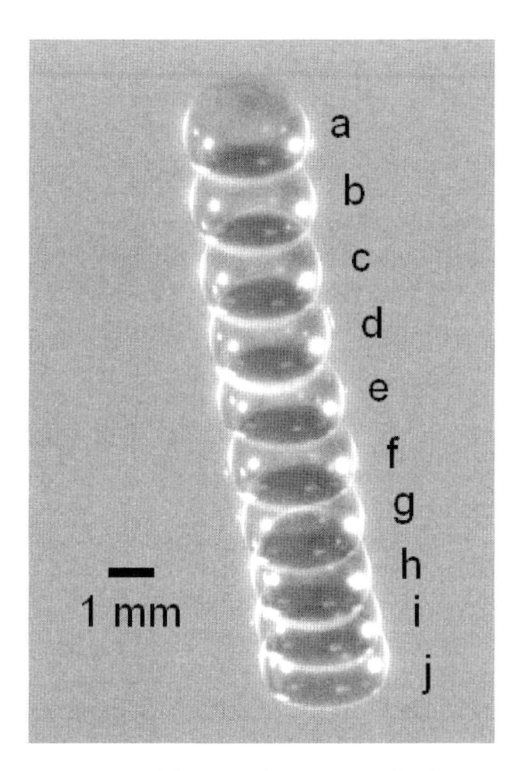

Figure 5. Superposed images of a single oscillating water drop slowly rising in the test section of a vertical wind tunnel ($d \approx 2.3$ mm, frame rate = 109 Hz).

frequency changes with mode m when the quiescent shape is distorted (i.e., nonspherical).

Observations of oscillating drops often reveal the higher-amplitude, simpler shapes of the fundamental more than the first harmonic. However, multiple modes may exist simultaneously. Moreover, modal preferences at a particular d seem to vary widely. From wind tunnel experiments, *Blanchard* [1948, 1950] reported observations of the fundamental axisymmetric (2,0) and fundamental horizontal (2,2) modes for large drops (d = 6–9 mm). *Brook and Latham* [1968] and *Nelson and*

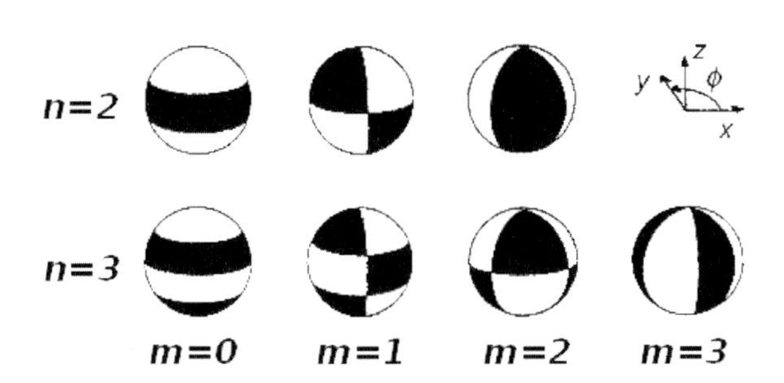

Figure 6. Orientation of perturbations given by spherical harmonic theory (equation (16)) for the fundamental (n = 2) and first (n = 3) harmonic. From *Beard and Kubesh* [1991]. Copyright 1991 American Meteorological Society.

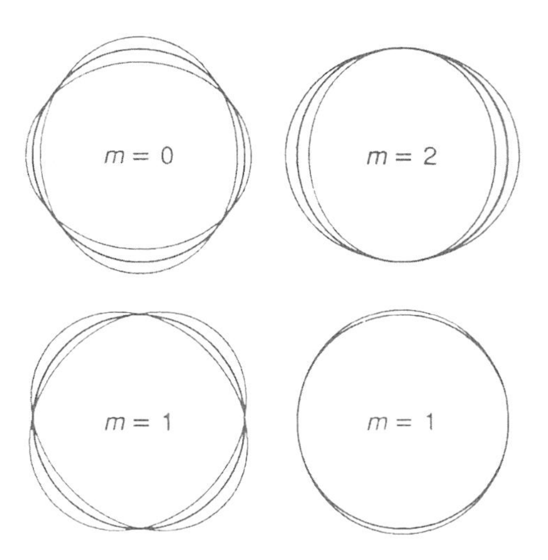

Figure 7. Views of the fundamental spherical harmonic (2,m). Two views of the fundamental transverse mode are shown, illustrating the dependence of α on the orientation ϕ of the drop. From *Beard et al.* [1989b]. Reprinted by permission from Macmillan Publishers Ltd. Copyright 1989.

Gokhale [1972] later described purely axisymmetric oscillations for d = 3.7–5.6 mm and d = 1–3 mm drops, respectively. However, *Nelson and Gokhale* [1972] noted the presence of additional modes for d = 4–7 mm drops, and *Beard* [1984b] observed evidence of the fundamental transverse (2,1) mode in the photographs of *Nelson and Gokhale* [1972] and *Musgrove and Brook* [1975]. These laboratory observations and the field results of *Jones* [1959] show a characteristic scatter in both instantaneous and mean α.

Beard and Kubesh [1991] examined the theoretical shapes shown in Figure 6 and identified the specific α variation for each mode. They determined that the transverse (2,1) mode gives a strictly positive variation in α toward unity. For this mode, the drop chords exhibit two variances, depending on the viewing angle ϕ: either together, giving no change in α (see bottom-left sketch in Figure 7), or independently with b remaining static and a varying positively (bottom-right sketch of Figure 7). In contrast, axisymmetric oscillations always produce a two-sided variation because both a and b vary in an opposing manner, independent of viewing angle. The horizontal mode (see top-right sketch in Figure 7) similarly produces a two-sided variation because of static a for this mode. Thus, with respect to the equilibrium α, the mean α of an oscillating drop may or may not shift, depending on the prevailing mode. As such, modal behavior can be inferred from the scatter and mean of α measurements of oscillating drops.

Because the shift can be significant and varying, a precise formulation of mean α versus d for radar and microwave scattering applications is important. *Seliga and Bringi* [1976] found polarimetric radar signals altered by 30% because of uncertainty in mean α, leading to erroneous estimates of drop size and rainfall rate [*Kubesh and Beard*, 1993].

Although the pioneering work of *Rayleigh* [1879] neglected any consideration of viscosity, *Lamb* [1881] showed that for small viscosity ($\nu/\omega r_0^2 \ll 1$, where $\omega = 2\pi f$), the main effects were a reduction in the oscillation amplitude, with higher-order modes dampening more quickly than the fundamental modes. Following this early work, most theoretical oscillation models [*Lamb*, 1932; *Foote*, 1973; *Tsamopoulos and Brown*, 1983; *Naterajan and Brown*, 1987] were based upon a spherical equilibrium shape in the absence of external fields. *Foote* [1973] introduced the first numerical model to study drop oscillations. This model used the finite-difference method to integrate the incompressible form of the Navier-Stokes equations, with surface tension effects incorporated through the use of Laplace's pressure balance (equation (14)) to define the drop surface curvature.

In an early work on the causes of oscillations, *Beard* [1984a] modeled the axisymmetric and horizontal mode for an ellipsoidal drop using a potential energy function that accounted for surface and energy due to gravity. Calculations of mean α, determined by assuming a steady state governed by a balance of collisional energy and viscous dissipation, showed a shift from equilibrium for larger drops ($d \geq 3$ mm), indicating the significant effect of collision-induced oscillations to an extent that increased with rainfall rate. *Beard* [1984b] compared this potential energy model to the available theory and found good agreement with the numerical result due to *Foote* [1973] with respect to time-varying α behavior. However, frequency calculations from this potential energy model compared more closely with experimental observations of the horizontal mode than with the axisymmetric mode, an error *Beard* [1984b] attributed to an inappropriate assumption regarding gravitational energy.

Feng and Beard [1991] described a rigorous multiple-parameter perturbation method that determined the characteristic frequencies for the specific fundamental modes ($n = 2$, $m = 0,1,2$). This eliminated the degeneracy whereby previously the spherical modes all had the same frequency (i.e., m independent; note the m-independence of equation (15)). The m-dependent equation they give,

$$f_{nm} = \frac{\omega_{nm}}{2\pi}\left(\frac{\sigma}{\rho r_0^3}\right)^{1/2}\left(1 - \frac{A_0^{\langle 2,1\rangle}(n,m)}{4\omega_{nm}^2}u_t^2\right), \qquad (17)$$

where u_t is the drop terminal velocity (see *Feng and Beard* [1991] for details), $A_0^{\langle 2,1\rangle}(n,m)(4\omega_{nm}^2)^{-1}$ has values -0.00804, 0.0241, and 0.121 for $m = 0$, 1 and 2, respectively, and

$$\omega_{nm}^2 = n(n-1)(n+2). \qquad (18)$$

Their result is shown in Figure 8 to be in considerable agreement with experimental data; specifically, the manner in which the bifurcation in the experimental data (indicative of m-dependence) aligns with their theory.

Although collisions are frequent enough in moderate to heavy rainfall to maintain oscillations for large raindrops ($d \geq 3$ mm), coupling with the unsteady drop wake is generally accepted as the primary physical mechanism for small drop oscillations [*Beard and Jameson*, 1983]. For smaller drop sizes, where collisions are more infrequent and viscous effects are more pronounced, coupling with the unsteady wake provides a sustained driving force to maintain oscillations against the time decay of viscous dissipation [*Johnson and Beard*, 1984]. First postulated by *Gunn and Kinzer* [1949], the mechanism is based on a match between vortex shedding and drop oscillation frequencies, as well as the simultaneous onset of both phenomena at $d \approx 1$ mm. The onset of oscillations at this drop diameter, combined with the diminished likelihood of collisional forcing as an oscillation mechanism for small

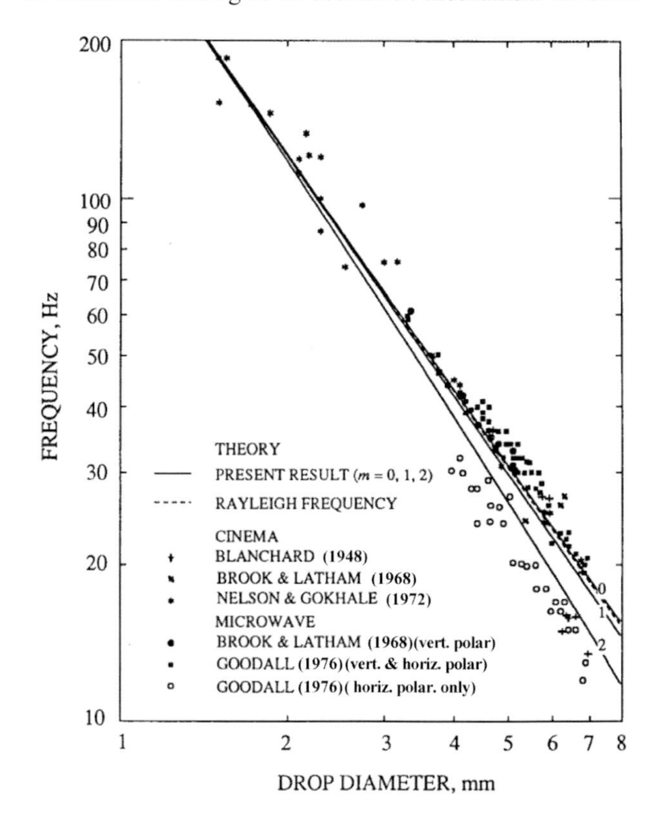

Figure 8. Oscillation frequency in Hz as a function of drop diameter in mm for the fundamental ($n = 2$) axisymmetric ($m = 0$), transverse ($m = 1$), and horizontal ($m = 2$) modes according to equation (17). Discrete points represent experimental data; lines show the m-dependent theoretical result due to *Feng and Beard* [1991]. From *Feng and Beard* [1991]. Copyright American Meteorological Society.

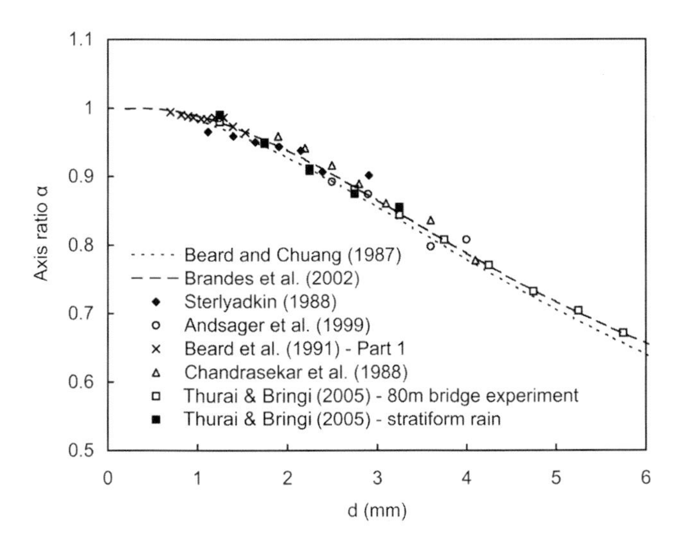

Figure 9. Empirical data for axis ratio α as a function of drop diameter d in mm, shown with the equilibrium raindrop shape model due to *Beard and Chuang* [1987] and the polynomial fit suggested by *Brandes et al.* [2002].

drop sizes, supports a causal relationship between vortex shedding and drop oscillations [*Beard and Jameson*, 1983]. With increasing d, the frequency match diverges, however, as the vortex-shedding frequency increases while the oscillation frequency decreases. Regardless, oscillations persist for larger d and increase in amplitude, as evidenced by the wide scatter in α. A transient effect reported by *Andsager et al.* [1999] suggests that the wide scatter may only occur after aerodynamic feedback reaches a steady state that is perhaps linked to eddy shedding or drag fluctuations. It is unknown exactly what resonant forcing mechanism lies behind this feedback or how and which specific oscillation modes are excited. However, the raindrop images of *Testik et al.* [2006] provide direct evidence of multimode oscillations and lateral drift. Combined with the long-speculated link between eddy shedding and lateral drift, these images suggest that the three may necessarily coexist.

A steady state combination of wake feedback, drop collisions, and turbulence seems to form the oscillation mechanism in raindrops. Because experiments have varied in their reproduction of these factors with regard to raindrop simulation, and oscillatory modal behavior has proven quite sensitive to these factors, it has been difficult to precisely characterize the mean α shift. This lack of consensus has made the determination of which α versus d relation to be used for radar calibration historically unclear. More recent work, however, seems to have identified boundaries for the mean α shift, with new data falling within range of the polynomial fits offered in the literature. A representative sampling of drop α data is presented in Figure 9. The recent data due to *Thurai and Bringi* [2005] is shown to agree reasonably well with the mean α curve due to *Brandes et al.* [2002].

A larger selection of the available α models in the literature is presented in Figure 10. Linear and nonlinear relationships are given, the former apparently being more appropriate for a particular polarimetric method of determining R [*Thurai and Bringi*, 2005] regardless of the inaccuracy due to linearization. The relationships shown in Figure 10 are summarized as follows (all units in millimeters unless specified):

1. *Pruppacher and Beard* [1970] derived a linear relationship from wind tunnel data for d = 1–9 mm, given by

$$\alpha = 1.03 - 0.062d \tag{19}$$

with d given in units of millimeters.

2. The calculated equilibrium shapes due to *Beard and Chuang* [1987] are given for d = 1–7 mm by

$$\alpha = 1.0048 + 5.7 \times 10^{-4}d - 2.628 \times 10^{-2}d^2 + 3.682 \times 10^{-3}d^3 - 1.677 \times 10^{-4}d^4, \tag{20}$$

with d given in units of millimeters.

3. *Andsager et al.* [1999] combined their data with those from *Chandrasekar et al.* [1988], *Beard et al.* [1991], and *Kubesh and Beard* [1993] to determine the following polynomial fit (in units of centimeters):

$$\alpha = 1.012 - 0.1445d - 1.03d^2, \tag{21}$$

valid for d = 1.1–4.4 mm, with equation (20) recommended for d outside this range.

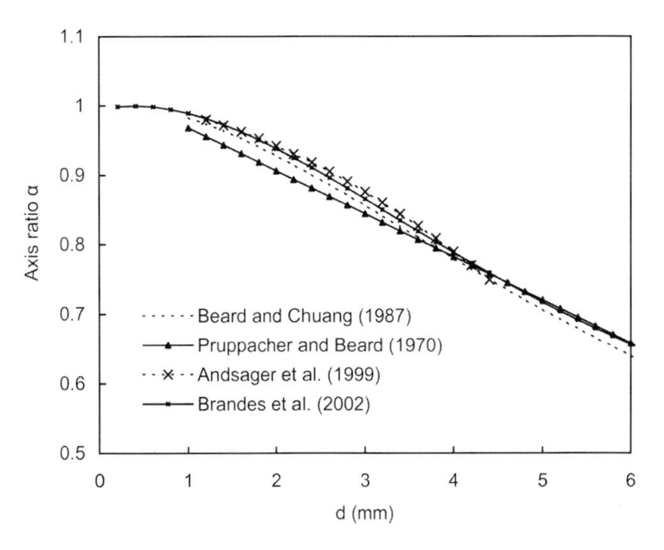

Figure 10. A selection of mean axis ratio formulations, given as a function of diameter d in mm.

4. *Brandes et al.* [2002] fit the following polynomial to the data due to *Pruppacher and Pitter* [1971], *Chandrasekar et al.* [1988], *Beard and Kubesh* [1991], and *Andsager et al.* [1999]:

$$\alpha = 0.9951 + 2.51 \times 10^{-2}d - 3.644 \times 10^{-2}d^2 + 5.303 \times 10^{-3}d^3 - 2.492 \times 10^{-4}d^4. \quad (22)$$

One method developed to address the α problem is the radar "self-consistency" principle, first described by *Gorgucci et al.* [1992], in which polarimetric radars are calibrated by calculating R from separate groups of measurables obtained by the same radar. The discrepancy arising between these R values is then correlated to raindrop shape and used to formulate an appropriate α versus d relationship. Versions of the method have been described by *Goddard et al.* [1994b], *Illingworth and Blackman* [2002], *Vivekanandan et al.* [2003], and *Ryzhkov et al.* [2005]. However, *Gourley et al.* [2009] recently utilized the method to evaluate the α models presented in Figure 10. They report minimal R error through use of either a hybrid relation consisting of equation (21) for small drops ($d = 0$–1.3 mm) and a formulation from *Goddard et al.* [1994b] for larger sizes ($d > 1.3$ mm) or equation (22).

Another method, first described by *Gorgucci et al.* [2000] and recently adapted by *Gorgucci et al.* [2008], utilizes a linear α model

$$\alpha = 1.03 - \beta d \quad (23)$$

and the self-consistency principle to optimize the variable slope parameter β (note that $\beta = 0.062$ mm^{-1} gives the *Pruppacher and Beard* [1970] α model, equation (19)). Although *Gourley et al.* [2009] determined that a similar linear model due to *Matrosov et al.* [2005] was quite sensitive to variability in α compared to a group of six other models, some radar practitioners may prefer the simultaneous DSD parameter retrieval that is a by-product of the variable β technique.

5. TERMINAL VELOCITY

The terminal velocity u_t of a raindrop is the velocity in still air achieved when the force due to gravity on the drop is exactly balanced by the (relatively small) buoyancy force and the aerodynamic drag force caused by airflow over the drop. While the aerodynamic drag increases roughly with the square of the drop diameter, the gravitational force on the drop increases with the drop mass and is proportional to the cube of the diameter. Hence, u_t increases with drop diameter approximately linearly for small d, where the drag coefficient is proportional to the inverse of the drop Reynolds number. However, for large d, the drag coefficient changes with Reynolds number in a more complicated fashion, and the relationship between the fall speed and d becomes nonlinear. The functional relationship between u_t and drop diameter has been a subject of research for some time, recently motivated by the necessity of knowing this function when converting the DSD obtained from radar scatter data into rain rates (see equation (6)). A variety of relationships giving u_t as a function of raindrop diameter d have been formulated. A selection of these are plotted in Figure 11 and listed in Appendix A. Velocities range up to 9 m s^{-1} for the largest raindrop sizes.

The terminal velocity is affected by several aspects of drop morphology, including drop shape, characteristics of the wake, and drop oscillation dynamics. Because of this dependence, a review of the literature on this subtopic is presented here.

Experimental studies of u_t for water drops began in the early twentieth century, with perhaps the earliest work being that due to *Lenard* [1904]. Other investigators of this period were *Schmidt* [1909], *Liznar* [1914], *Flower* [1928], and *Laws* [1941]. In 1949, *Gunn and Kinzer* [1949] conducted a careful investigation of raindrop u_t, taking great care to accurately measure the mass and speed of the drop. As noted by these authors, the earlier literature used relatively coarse methods for mass measurement, for example, relying on the size of the spot created by a drop after it impacted a specially treated piece of paper [*Schmidt*, 1909; *Lenard*, 1904] or by allowing drops to fall into fine flour which was subsequently measured [*Laws*, 1941]. *Gunn and Kinzer* [1949], on the other hand, measured drop masses in two ways. For large diameters, drops

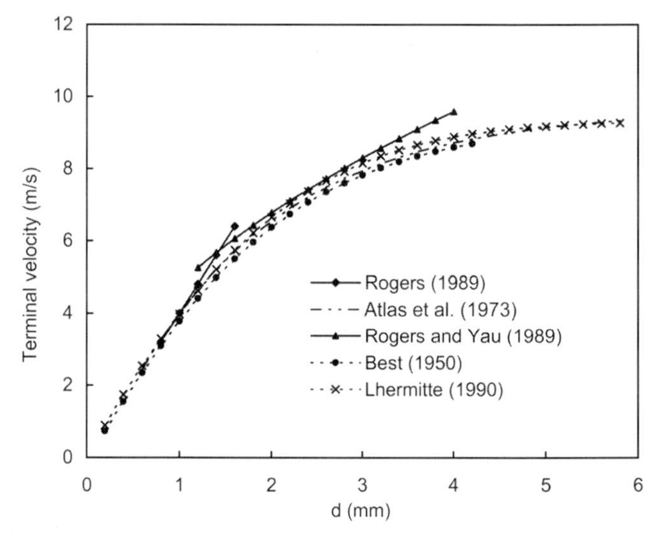

Figure 11. Terminal fall velocity u_t in m s^{-1} as a function of drop diameter d (mm) given by various authors (see Appendix A).

were carefully collected and their mass measured using a precise chemical balance. For small diameters, the drops were captured in a dish containing high quality vacuum-pump oil and the diameter then measured using a microscope. This latter method had the advantage of preventing any evaporation of the drop, since it was encased in an oil medium (an important advantage for small drops, where a small amount of evaporation could result in large errors). Although the data acquired by Gunn and Kinzer suffered from other problems (e.g., evaporation while falling through a 50% relative humidity environment), it is still of high quality, even by today's standards, and is frequently cited.

In the late 1960s, Beard and coworkers published several works pertaining to u_t. *Beard and Pruppacher* [1969] studied u_t for water drops in a special wind tunnel designed to levitate drops in an airflow of known speed, at a controlled temperature and relative humidity. These authors point out that the data of Gunn and Kinzer were obtained at a relative humidity of 50%. *Beard and Pruppacher* [1969] measured u_t for drop diameters ranging from 10 to 475 μm at a relative humidity of 100%, but found no significant deviation from Gunn and Kinzer's data (although they did observe very slight deviations, which they attributed to evaporation in Gunn and Kinzer's experiments). These authors also computed drag coefficients and showed that for a drop Reynolds number *Re* less than 200, the drag coefficient for drops deviated insignificantly from those for solid spheres. They concluded that for this range of *Re*, drops are essentially round, although a subsequent photographic study due to *Pruppacher and Beard* [1970] showed this conclusion to be strictly true only for $Re < 20$. *Pruppacher and Beard* [1970] also showed the velocity at the drop surface to be only about 1% of the terminal velocity. Hence, it is unlikely that the internal flow significantly affects the value u_t.

An interesting aspect of the variation of u_t with drop diameter concerns the behavior of drops in vertical updrafts during convective precipitation. A vertical updraft with velocity u_u will allow drops having a terminal velocity $u_t > u_u$ to fall, while those having smaller terminal velocities ($u_t < u_u$) will remain suspended, unable to fall. A model developed by *Srivastava and Atlas* [1969] shows that a convective cloud having an assumed linearly increasing updraft, capped by a linearly decreasing updraft, will cause a horizontal and vertical size-sorting of drops. Data obtained from aircraft support the results of this model [*Carbone and Nelson*, 1978; *Rauber et al.*, 1991; *Szumowski et al.*, 1998; *Atlas and Ulbrich*, 2000]. Measurements obtained from a Doppler radar further develops these ideas, showing how the relationship between the updraft and the variation in u_t with drop diameter results in a sorting of drop sizes in both the vertical and horizontal direction of the cloud structure [*Kollias et al.*, 2001].

Another interesting aspect of the literature on terminal velocity concerns the acceleration of drops to u_t. This acceleration can be characterized by the time and/or distance required for a drop to reach some fraction of u_t. This is an important topic, since raindrops are frequently perturbed from a stable fall velocity by a change in local air velocity as they fall and/or by collision with other drops. Knowing the time/distance needed to achieve u_t is also important in the design of drop towers.

Laws [1941] observed, in artificially generated drops, that the distance required for a drop to achieve 95% of its terminal fall velocity z_{95} did not increase monotonically with drop diameter, but rather achieved a maximum at a diameter of 4 mm (Laws presents z_{95} data for drop diameters d = 1, 2, 3, 4, 5, 6 mm, which have values of z_{95} = 2.2, 5.0, 7.2, 7.8, 7.6, 7.2 m). Laws explains this result using plots of drop velocity versus fall distance showing that for large drops (specifically d = 6.1 mm), the plot has a maximum, rather than asymptotically approaching u_t. That is, the drop exceeds u_t and then decelerates to u_t. This, in turn, he postulates, is due to a delay in the time required for the drop shape to attain an equilibrium (flattened) shape at a given instantaneous velocity. That is, near terminal velocity, the drop is closer to spherical than it should be for that given velocity, and hence accelerates further, achieving a velocity greater than terminal. The drop shape then becomes flatter, causing the drop to slow down and relax to its true terminal velocity. This phenomenon occurs only for large drops, since these exhibit significant deviations from spherical. Laws supports his argument with one presented by *Lenard* [1904], who similarly postulated that a time lag in achieving drop shape allowed drops to remain more spherical at a given instantaneous velocity than they would be at steady state, allowing them to accelerate to a velocity larger than u_t. Lenard furthermore argued that this phenomenon would be best explained if the drop shape was significantly affected by internal circulation within the drop, since the acceleration of the internal drop fluid would take a significant amount of time due to its mass. While a delay in attaining a steady state drop shape for a given velocity is certainly a factor in how a drop accelerates to its steady state velocity, Lenard's argument seems problematic to the present authors, since internal circulation in the drop results in a drop surface velocity, which reduces the net drop-to-air velocity, which would contribute to a continually decreasing drag force as the internal circulation flow ramps up, which one would expect to contribute to a monotonic approach to terminal velocity. *Beard* [1977a] also notes that the work of *Foote* [1969] shows that internal circulation results in less distortion, not more, further ruling out internal circulation as the result of Laws' and Lenard's observations. The work of *LeClair et al.* [1972] shows little effect of internal circulation on the drag on a drop for cases where the drop is nominally spherical.

Wang and Pruppacher [1977] revisit the problem of acceleration to terminal velocity through experiments in a drop tower and via the development of a theoretical method to compute the acceleration to terminal velocity. The theoretical and experimental results of *Wang and Pruppacher* [1977] do not reveal a maximum in their plots of instantaneous velocity versus fall distance. However, they do show a maximum in their plots of t_{99} (as well as z_{99}) versus drop size at a diameter of 3.2 mm, a pressure of 1000 mb, and a temperature of 20°C. This is similar to Laws' peak in t_{95} at a diameter of 4 mm. Considering the sparsity of Laws' data and the relatively crude methods by which it was attained, one should probably conclude agreement between his study and that of Wang and Pruppacher. However, the lack of a peak in the instantaneous velocity versus fall distance of *Wang and Pruppacher* [1977] requires a different explanation for the peak in t_{99} that they observe. Such an explanation of Wang and Pruppacher's results is probably best given by *Beard* [1977a] who notes that above a diameter of about 1 mm, drops begin to deviate significantly from sphericity, and as the diameter increases beyond 1 mm, this occurs earlier during the drop fall, thereby reducing t_{99} and z_{99} for $d > 1$ mm drops.

The studies described above focus almost exclusively on the terminal velocity of a single drop. In most analytical investigations and in most laboratory studies, the goal is to observe drops in the absence of air velocity fluctuations (turbulent or otherwise), collisions with other drops, and in the absence of evaporation or condensation. Of course, during actual rain, all of these effects are in play and can result in a drop velocity that differs from the terminal velocity predicted via analysis or from laboratory experiments. These factors will all affect drop velocity in that they perturb the drop from a nominally steady state condition into a transient one. It should be made clear that these effects do not "change" the value of terminal velocity. Rather, these effects simply cause a drop to undergo a transient and reapproach terminal velocity. This terminal velocity may be "new" only if the drop has changed somehow, e.g., it has increased in size due to collision with a smaller drop. There is some evidence to suggest that during heavy rain fall, many or perhaps even most drops do not achieve terminal velocity for any significant period of time. In other words, the number of drop collisions and the constantly changing velocity field of the air cause such a changing environment that drops are continually accelerating or decelerating to a new speed. This is suggested, for example in the work of *Montero-Martínez et al.* [2009] who show that for high rain rates ($R > 84$ mm h^{-1}), up to 50% of 0.44 mm diameter drops traveled at speeds greater than the predicted terminal velocity and that for a drop having a diameter of 0.24 mm, up to 80% of these drops are superterminal.

For very small drop diameters, mild deviations from the continuum flow assumption can result in errors in terminal velocity relations obtained for larger drops. As pointed out by *Beard* [1976], even though the terminal velocities of drops on the order of a micron are extremely small, they are needed to compute collision efficiencies, and this author computed a revised terminal velocity equation for these very small drops. Similar work was done by *Beard* [1977b] who accounted for noncontinuum effects and also presented an adjustment for sea-level terminal velocity equations that accounted for changes in terminal velocity due to temperature and pressure, permitting simple corrections to terminal velocity for drops over a range of altitude.

Ryan [1976] studied the effect of surfactants on the terminal velocity of drops using a drop levitation tunnel. Reduction in the drop surface tension by surfactants caused drops to deform (at equivalent drop volume) and flatten, reducing the terminal velocity. For example, using surfactants to reduce the surface tension to 17 dynes cm^{-1}, Ryan found that u_t for a drop having an equivalent spherical diameter of 3 mm dropped by more than 20%.

Due to the changes in air properties with altitude, the terminal velocity of a drop is a function of height above sea level. Several researchers have quantified this and developed equations for u_t in terms of air properties. These equations are sometimes reformatted to provided u_t in terms of altitude or barometric pressure. Examples of such studies include *Battan* [1964], *Cornford* [1965], *Foote and DuToit* [1969], *Berry and Pranger* [1974], *Beard* [1976], *Wang and Pruppacher* [1977], and *Beard* [1977b].

Note that the effect of freezing on u_t and velocities of partially or fully frozen drops is not considered here, and the effect of electrostatic forces on u_t is briefly discussed in the next section.

6. ELECTROSTATIC EFFECTS

Drops in many clouds are electrically charged at an early stage of the cloud life cycle when exposed to the external electric fields present in electrically charged clouds [*Rasmussen et al.*, 1985; *Despiau and Houngninou*, 1996]. Consequently, these drops are subject to electrostatic forces. The primary effects of electrostatic forces on the raindrop morphodynamics are threefold: (1) distortion, (2) fall speed alteration, and (3) disruption of raindrop stability. Due to scarce data on the behavior of charged drops falling steadily in an ambient field, our understanding of the effects of electrostatic forces on raindrop morphodynamics is limited mainly to simplified theoretical and numerical models

[*Coquillat and Chauzy*, 1993; *Beard et al.*, 1989a; *Chuang and Beard*, 1990; *Zrnić et al.*, 1984].

For an electrically distorted raindrop, the dependence of the drop shape on the electric field is highly nonlinear due to the coupling between the surface electrostatic stress and the aerodynamic distortion [*Chuang and Beard*, 1990]. The surface electrostatic stress caused by the electrical charge will tend to oppose surface tension, with the charge concentrated in regions of highest curvature. When the water drop is small and maintains its spherical shape, the electrical charge uniformly distributes over its surface [*Kamra et al.*, 1991]. However, since the maximum curvature of large, distorted drops is located just below the waist, these drops will be extended horizontally under the influence of electric charge. Thus, the effect of large surface charge is increased oblateness [*Chuang and Beard*, 1990]. On the other hand, an uncharged drop situated in an electric field will become elongated along the direction of the field [*Bhalwankar and Kamra*, 2007; *Coquillat et al.*, 2003]. Therefore, oblateness of a drop increases (decreases) in the presence of a horizontal (vertical) electric field compared to the equilibrium shape of the same-size drop in the absence of an electric field. Numerical simulations of *Coquillat et al.* [2003], later confirmed by experimental observations of *Bhalwankar and Kamra* [2007], showed that horizontal electric fields are more efficient than vertical ones in deforming the drop. This is because in a horizontal electric field, aerodynamic and electrostatic forces act together to distort the drop, whereas in a vertical field these forces counteract to suppress distortion effects.

In a recent vertical wind-tunnel study, *Bhalwankar and Kamra* [2009] investigated the effect of contaminants on the shapes of uncharged raindrops in a horizontal electric field, a simulation of contaminated raindrops in thunderstorms over large, polluted cities. *Bhalwankar and Kamra* [2009] observed that contaminated raindrops were more distorted (i.e., their oblateness increased) with respect to distilled water drops, and the observed difference in distortion increased with increasing electric field. They explained this observation as a consequence of the increased electrical forces acting on the drop due to the increase in the electrical conductivity of water when it is polluted. Based on their experimental results, *Bhalwankar and Kamra* [2009] provided brief qualitative discussions on the modification of raindrop size distribution and lightning activity in clouds formed over large cities. Further research is needed on electrostatic effects on contaminated raindrops, especially in vertical electric fields that are considered to be representative of thundercloud conditions, for quantitative conclusions.

When the drop is in motion, the combined effect of aerodynamic and electrostatic forces determines the raindrop shape. For example, *Chuang and Beard* [1990] reported that uncharged drops falling in strong vertical electric fields, representative of thundercloud conditions [*Rasmussen et al.*, 1985], show a pronounced extension of the upper pole and an enhanced flattening of the lower pole (i.e., triangular-like drop profiles) due to increased fall speed of electrostatically stretched drops. A summary of the calculations performed by *Chuang and Beard* [1990] are provided in Figure 12.

Electrostatic forces experienced by charged raindrops in a thundercloud may alter the force balance between the gravitational and aerodynamic forces acting on the raindrop. As a result, the terminal velocity of a raindrop under electrostatic effects may significantly deviate from the terminal velocity of a raindrop with the same diameter but isolated from electrostatic effects [*Coquillat and Chauzy*, 1993]. The raindrop terminal velocity is affected by the combined effects of the charge that the raindrop carries and

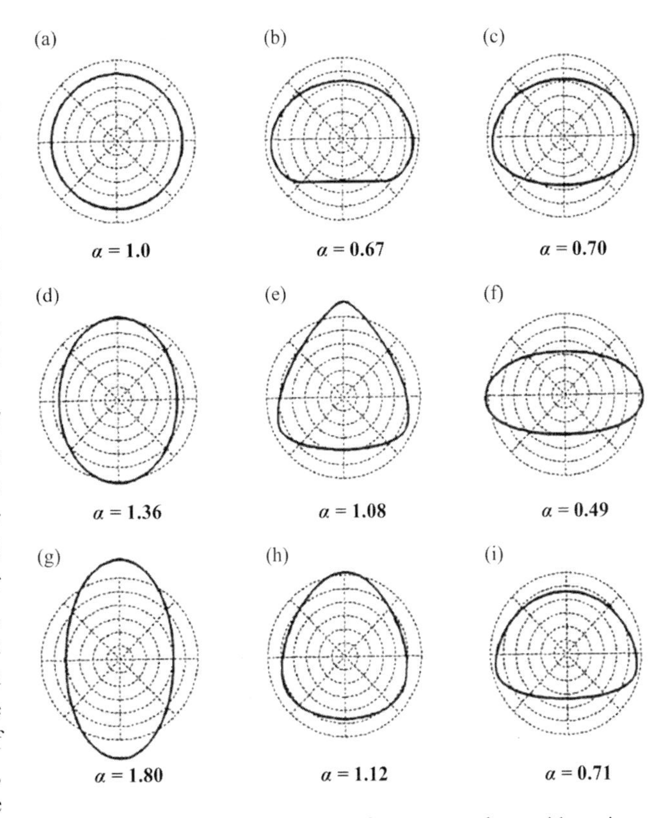

Figure 12. Shapes and axis ratios for a 5-mm drop with various distortion effects: (a) stationary drop (surface tension only); (b) sessile drop (surface tension and hydrostatic stress); (c) raindrop (surface tension, hydrostatic, and aerodynamic stresses); (d) stationary drop in vertical electric field; (e) uncharged raindrop in vertical electric field; (f) charged raindrop; (g) stationary drop in vertical electric field with largest possible distortion; (h) raindrop with maximum field charge combination with upward electric force; (i) same as Figure 12h with downward electric force. From *Chuang and Beard* [1990]. Copyright 1990 American Meteorological Society.

the electric field in which it moves. The effect of electrostatic forces may be to increase (e.g., positively charged raindrops in a negative vertical electric field) or to decrease (e.g., negatively charged raindrops in a negative vertical electric field) the raindrop terminal velocity depending upon the polarity of charge on the raindrop and the direction of the electric field [*Kamra*, 1975]. An important consequence of raindrop terminal velocity variations due to electrostatic effects is on the charge separation mechanism, which is assumed to be responsible for the negative electric fields that may cause lightning flashes in thunderclouds. The charge separation mechanism and the role of terminal velocity deviations of charged raindrops in this mechanism is out of the scope of this chapter, and the interested reader is kindly referred to *Kamra* [1970] and/or *Kamra* [1975]. It should be noted that induced velocity perturbations result in a change in the aerodynamic pressure around the raindrop, one of the primary factors governing the drop shape as discussed earlier. As a result, the simultaneous and accurate determination of the shape and the terminal velocity of a charged raindrop in an electric field are difficult because of the high interdependence of both parameters [*Coquillat and Chauzy*, 1993]. This aspect of raindrop dynamics is an open area of research.

Zrnić et al. [1984], extending the model of *Green* [1975] for describing the drop shape by including electrostatic effects, reported that commonly observed electric fields in clouds have only a modest effect on drop shape. *Rasmussen et al.* [1985] argued against this finding because of the discrepancy between Zrnić's predictions and wind-tunnel observations for the shape and disruption values of raindrops by *Dawson and Richards* [1970] and *Richards and Dawson* [1971]. Later, Zrnić's prediction was supported by model predictions of *Beard et al.* [1989a] and *Chuang and Beard* [1990], who added that somewhat stronger fields for highly charged raindrops can lead to instability/disruption. *Coquillat and Chauzy* [1993] predicted quite high field intensities for the disruption of uncharged raindrops, but they noted that the combination of field and net charge may lead to probable disruptions of charged raindrops in ambient fields of the order of those commonly measured in nature.

Accurate quantitative information on the electric field amplitude required for drop disruption remains incomplete. Moreover, the question of whether typical electrical fields in thunderclouds are sufficient to cause disruptions of drop stability is still a subject of debate. A classical reference for instability (i.e., disruption) of a drop acted upon by electrical forces is the work of *Taylor and Acrivos* [1964], which provided the following theoretical criterion for the onset of drop instability,

$$F\left(\frac{r_0}{\sigma}\right)^{1/2} = 1.63. \tag{24}$$

Here F is the value of critical electric field in esu, r_0 is the undistorted drop radius in cm, and σ is the surface tension in dynes cm^{-1}. Although Taylor's criterion agrees well with experimental observations reported by different authors, *Kamra et al.* [1993] reported a significantly lower critical electric field value for the onset of drop instability. Kamra et al. argued that the difference between their observations and those reported by others is because of long exposure of the freely suspended drops to the electric fields in their vertical wind-tunnel setup. This discrepancy between the observations of Kamra et al. and Taylor's criterion is later followed up by *Georgis et al.* [1997], who studied the onset of drop instability for free-falling drops at terminal velocity in a horizontal electric field. Georgis et al. reported a large discrepancy between their observations and those of Kamra et al., claiming they were possibly due to the experimental conditions of Kamra et al., specifically the high turbulence levels inherent to vertical wind tunnels. *Coquillat et al.* [2003] numerically studied the critical electric field for the onset of drop instability. Their simulation results showed good agreement with the experimental observations of *Georgis et al.* [1997]. However, the numerical simulations of Coquillat et al. did not take into account the effect of ambient turbulence and drop oscillations. As turbulence and drop oscillations are keys to the drop instability, Coquillat et al. noted that actual instability onset values in thundercloud conditions may be closer to the results by *Kamra et al.* [1993].

Bhalwankar and Kamra [2007] considered the stabilizing and destabilizing effects of ambient electric fields on uncharged raindrops. They suggested that vertical electric fields stabilize distorted drops by stretching along the drop vertical axis (reducing oblateness), whereas horizontal electric fields destabilize drops by stretching the drop along its horizontal axis, increasing oblateness. Based on this argument, Bhalwankar and Kamra discussed effects of electric field direction, which may differ between clouds and in different regions of the same cloud, on raindrop size distributions and raindrop growth rates.

Kamra et al. [1991] discussed the role of oscillations in the disruption of charged drops and postulated a destabilizing effect due to coupling between a surface charge density shift and drop shape distortion associated with oscillations. In the case of contaminated drops, such as those existing in rain over large cities, dissolved pollutants in the drop tend to reduce the amplitude of drop oscillations through an increase in surface tension. This effect of pollutants enhances drop stability against electric forces, which tend to break up the drops [*Bhalwankar and Kamra*, 2009]. Hence, a possible consequence of pollutant presence is to broaden the raindrop size distribution.

7. EXPERIMENTAL TECHNIQUES

Given the complex nature of the governing equations (see discussion in section 2), numerical and analytical studies on raindrop morphodynamics are challenging, thus experimental observations have played a critical role in studying raindrop morphodynamics. Experimental work on the shape, terminal velocity, and stability of raindrops began with *Lenard* [1887, 1904], who made nighttime flash photography observations and investigated the role of surface tension and internal circulation on drop shape using a wind tunnel. Prior work focused on pendant and sessile drops, likely owing to the experimental challenges of obtaining accurate measurements of unsupported, floating or falling drops. Later, *Flower* [1928] and *Laws* [1941] reported terminal velocity calculations, while the high-speed photographs of *Edgerton and Killian* [1939] and *Blanchard* [1950] further explored drop behavior. *McDonald* [1954b] outlined the available literature and also examined the photographs of *Magono* [1954] to deduce the distribution of aerodynamic pressure around the drop surface and show the distorting effects of flow separation. Figure 9 presents a representative selection of experimental data from the works outlined here, all of which depict the wide scatter indicative of oscillations, discussed in section 4.

Some aspects of raindrop morphodynamical behavior can be simulated in the laboratory, either using levitated or free-falling drops. Levitated drop studies are conducted using specially constructed vertical wind tunnels whereby drops are observed as they float in upward-oriented flow. Most wind tunnel designs condition the airflow using baffles or screens to reduce turbulence; however, more sophisticated tunnels may incorporate temperature and humidity controls [*Pruppacher and Beard*, 1970; *Mitra et al.*, 1992], electrical fields [*Kamra et al.*, 1986], or flow-visualization techniques [*Saylor and Jones*, 2005] in addition to the primary shape or frequency sensing method.

Data from the wind tunnel work of *Pruppacher and Beard* [1970] contributed to the important semiempirical drop shape model of *Pruppacher and Pitter* [1971], as well as an axis ratio relationship that was recently validated by *Thurai and Bringi* [2005]. *Beard and Pruppacher* [1969] also used wind tunnel measurements to develop a more sophisticated terminal velocity formula based on drag coefficients that improved upon the available literature, citing errors in the works of *Gunn and Kinzer* [1949], *Laws* [1941], and *Imai* [1950]. More recently, *Szakall et al.* [2009], using shape and axis ratio measurements of 2.5 mm $\leq d \leq$ 7.5 mm drops, found drop shapes to match the equilibrium model of *Beard and Chuang* [1987] in addition to oscillation behavior further outlined in section 4. *Jones and Saylor* [2009] also reported mean axis ratio measurements matching the *Beard and Chuang* [1987] equilibrium model, a curious result considering the shift in mean axis ratio usually reported from drop measurements [*Andsager et al.*, 1999]. *Bhalwankar and Kamra* [2009] and *Rasmussen et al.* [1985] have also utilized wind tunnels to study the effects of electrical fields and charging on drop shape and breakup behavior. Wind tunnels have also been utilized by *Blanchard* [1948], *Brook and Latham* [1968], *Nelson and Gokhale* [1972], *Musgrove and Brook* [1975], and *Goodall* [1976] to obtain quantitative observations of drop oscillation frequency from image measurements or microwave backscatter; some of these data are shown in Figure 8.

Free-falling drop studies are conducted using towers whereby drops fall from a height sufficient to approach terminal velocity. Drop towers consist of a drop production apparatus situated above cameras or sensors designed to measure shape or oscillation behavior. To accurately simulate real raindrop behavior, consideration should also be given to the fall height required to dampen anomalous surface dynamics resulting from drop generation [*Beard and Kubesh*, 1991] and for an oscillation steady state to fully develop [*Andsager et al.*, 1999]. Though outdoor experiments may provide additional height over enclosed arrangements, ambient winds must be monitored so that fall trajectories remain aligned with the instrument measurement volume or camera field of view.

Experimentation with falling drops has been an elemental technique in raindrop microphysics research, from the early studies mentioned above to the more recent work described here. For example, *Beard et al.* [1991] in their 4-m fall tower study of 0.70 mm $\leq d \leq$ 1.54 mm drops, largely determined what is known about drop oscillations, wakes, and fall behavior. The mean and scatter in their axis ratio measurements indicated distinct oscillation behavior at specific sizes, suggestive of the long-postulated causality between asymmetric oscillations and drop wake vortex shedding. *Andsager et al.* [1999] used a similar but taller 25-m arrangement to measure axis ratios of 2.5 mm $\leq d \leq$ 4 mm drops at discrete fall distances, reporting time development of oscillations even after drops had reached terminal velocity. *Thurai and Bringi* [2005] also obtained measurements of drop shape from an 80-m tower, finding good agreement with the equilibrium shapes due to *Beard and Chuang* [1987], implying an absence of the transverse oscillations evidenced in other data.

Measurements of oscillation frequency have been obtained from falling drop arrangements, wind tunnel experiments, and observations of real rain. *Brook and Latham* [1968] and *Goodall* [1976], for example, utilized microwave scattering to measure oscillation frequencies of drops floating in a wind tunnel. In a falling drop study, *Beard and Kubesh* [1991] described a specific lighting arrangement in which falling drops create rainbow streaks in photographs that can be measured to deduce oscillation frequency and amplitude. This technique has also been used in the falling drop studies of *Beard et al.* [1989b] and *Kubesh and Beard* [1993] and the rainfall observations of *Beard and Tokay* [1991] and *Tokay and Beard* [1996] to deduce modal behavior.

In many field and laboratory studies, the shape as well as the size of the drop is desired, and several methods exist to obtain drop shape. The most commonly used type of automated raindrop shape measurement in current use is the two-dimensional (2-D) video disdrometer (2DVD) manufactured by Joanneum Research [*Schönhuber et al.*, 1994, 1995, 1997]. The 2DVD disdrometer uses two line scan cameras to image perpendicular sections of a droplet falling through two parallel sheets of white light. The size and shape of the drop are determined by recording the portion of the light sheet occluded as the droplet falls passes through it. Hence, an instantaneous image is not acquired by a 2-D detector, but rather a pseudo-image of the raindrop constructed by stacking sequential line scans on top of one another.

The 2DVD system has been used by several researchers and is able to provide significant drop counts for shape and size measurements. However, it suffers from several problems. One of these is a simple lack of reliability. For example, *Tokay et al.* [2001, p. 2085] noted that the device "... is a delicate electronic optical device that requires frequent attendance," and that it frequently failed, due to overheating problems and an inability to cope with high humidity.

A more serious problem associated with the 2DVD concerns accuracy of the raindrop shape measurement. Since the raindrop image is acquired in planar sections as it falls through the light sheet, horizontal motion of the droplet results in a drop image which is more oblate and canted than the actual drop [*Schönhuber et al.*, 1994]. This is a significant problem, since drops frequently have a horizontal velocity due to the winds accompanying rain storms. Algorithms have been developed for the 2DVD to account for this error but have been limited in their success, making measurements of drop shape uncertain [*Schuur and Rhyzhkov*, 2001]. Furthermore, the number of scans obtained as the drop falls through the measurement volume is small (~10 scans for a 3-mm drop), creating a low quality pseudo-image of the drop.

Another significant concern regarding the 2DVD are problems due to drop splash during field measurements. The housing of the measurement volume is located above the line-scan cameras so splash drops often fall into the measurement volume, contaminating the results. The manufacturer provides a Velcro lining on the upper surface of the housing to eliminate this splashing; however, *Tokay et al.* [2001] report that this lining becomes saturated during heavy rainfall and splashing occurs nonetheless, resulting in spurious small drops. To eliminate these, some investigators have filtered out drops that are measured to have a velocity falling outside of some threshold above or below the theoretical terminal velocity [*Tokay et al.*, 2001; *Donnadieu*, 1980; *Hauser et al.*, 1984]. As many as 24% of the observations are discarded when this method is employed [*Hauser et al.*, 1984]. Recent results due to *Montero-Martínez et al.* [2009] suggest that such drops may not be spurious and that a significant number of raindrops may fall at speeds that deviate from terminal. This further complicates the processing of data from the 2DVD. Finally, during horizontal winds, small drops may simply pass over the 2DVD without falling into the measurement volume, resulting in an undercounting of small drops [*Nešpor et al.*, 2000].

Another automated device used particularly in many airborne experiments is the optical array probe (OAP), developed by *Knollenberg* [1970]. This device consists of a HeNe laser focused on a linear diode array [*Knollenberg*, 1970, 1981; *Cannon*, 1976]. As drops pass between the laser and the array, an occluded line is observed by the array, similar to the 2DVD mechanism. As the drop passes, sequential scans of the array are used to construct a pseudo-drop image [*Black and Hallett*, 1986]. Due to the reconstruction process, this device suffers some of the same problems as the 2DVD, namely, that the pseudo-drop images are excessively canted due to the velocity of the drop with respect to the probe. The OAP is typically mounted on an aircraft platform. *Chandrasekar et al.* [1988] have used this device and developed an algorithm for correcting canting in the resulting drop image. In many of the drop images, the entire image is composed of only about 10 scans (i.e., 10 columns of pixels). Through sophisticated data processing, *Chandrasekar et al.* [1988] were able to reduce uncertainties in their OAP measurements of axis ratio to less than 3% for drops greater than 2 mm in diameter.

We note in passing that other techniques employ a similar approach, but only obtain drop size. For example, the VIDIAZ spectropluviometer described by *Donnadieu* [1978] and *Donnadieu* [1980] obtains drop size based on occlusion of a light beam, and the self-evaluating disdrometer of *Stow and Jones* [1981] obtains drop size from a photomultiplier tube sensing scattered light from the drop. These and similar methods [*Hauser et al.*, 1984] use a photodetector to determine drop size based on occlusion or scattering by a drop, but do not provide drop shape.

Preferable to the methods described above are those that use a full frame imaging method so that the entire drop image is (effectively) captured at a single moment in time. This way, an image need not be constructed from successive scans, avoiding the concomitant uncertainties in shape that occur in so doing. Such methods have been employed by several researchers including the work of Saylor and coworkers [*Saylor et al.*, 2002; *Saylor and Jones*, 2005; *Jones et al.*, 2003; *Jones and Saylor*, 2009], *Testik et al.* [2006], *Testik* [2009], and *Thurai et al.* [2008].

While laboratory observations can be extended to infer the microphysical behavior of natural raindrops, discrepancies may arise due to the simulated environment. Differences in humidity and temperature, for example, may represent deviations from the environment of real raindrops sensed by weather radars, causing evaporation and uncertainty in determinations of drop size. *Tokay and Beard* [1996] concluded that although turbulence at high altitudes is probably too weak at the smaller scales needed to affect drop shape and oscillations, strong shear may sometimes alter axis ratios at the ground level. This observation may apply to some wind tunnel designs that utilize the shear induced by a drop-positioning apparatus to reposition floating drops in the horizontal plane of the tunnel test section [*Andsager et al.*, 1999]. Aircraft measurements of raindrops can also be affected by the accelerated, fluctuating air velocity, causing an uncertainty in axis ratio observations [*Beard and Jameson*, 1983; *Xu*, 1995]. The study of isolated drops may also present a discrepancy from natural rain because of the role of drop-drop interaction in rainfall. When multiplied over the vast number of raindrops sensed by radars in a typical rainfall event, these effects can become significant.

One primary difficulty encountered in field observations and raindrop simulations arises due to the nonvertical fall trajectories of some drops [*Beard et al.*, 1991] described in section 2. Some experimental techniques, for example, deduce drop size from vertical displacement using empirical terminal velocity relationships. Because it is now apparent that oscillatory behavior varies even with constant d, this may become a source of error when drop shape measurements are correlated to drop size. If transverse-oscillating drops necessarily spiral or drift laterally, as suggested by the observations of *Testik et al.* [2006], this behavior may introduce measurement bias by making these drops harder to observe and axisymmetric drop measurements more frequent. The effect is apparent, for example, in wind tunnel axis ratio data that apparently reflects an absence of steady state, fully developed transverse oscillations [*Jones and Saylor*, 2009; *Szakall et al.*, 2009], presumably due to the lateral drift and rejection of these drops from the tunnel test section. Falling drop experiments may also reflect this bias if drops produced directly above the camera measurement volume drift away and elude measurement. For example, in the work of *Thurai and Bringi* [2005], drops were generated by water released from a hose, 80 m above a 2DVD system. These authors observed no evidence of pure transverse oscillations. It is possible that such oscillations existed, but lateral drift caused these drops to impact far from the measuring system. Furthermore, the effect could be more pronounced than previously thought, as *Testik et al.* [2006] observed a lateral drift of 30% in oscillating raindrops, considerably larger than the 6% maximum reported by *Beard et al.* [1991]. These experimental anomalies may complicate attempts to characterize the varying nature of oscillatory behavior and the axis ratio shift with d.

APPENDIX A
RAINDROP TERMINAL VELOCITY FORMULAS

The terminal velocity formulas shown in Figure 11 are listed below. Equations (A4) and (A6) consider the dependence of u_t on altitude z by including the density ratio factor

$$\sqrt{\frac{\rho_{ref}}{\rho_{a,z}}}, \tag{A1}$$

where $\rho_{ref} = 1.2$ kg m^{-3} gives the reference density of dry air at 20°C and 101.3 kPa, and $\rho_{a,z}$ is the local density of moist air at altitude z. Figure 11 depicts equations (A4) and (A6) evaluated at 1 m altitude and $\rho_{a,z} = 1.176$ kg m^{-3}.

For d in millimeters over the range d = 0.08–1.5 mm, *Rogers* [1989] gives

$$u_t = 4000d \quad (\text{m s}^{-1}). \tag{A2}$$

For d in millimeters over the range d = 0.6–5.8 mm, *Atlas et al.* [1973] give

$$u_t = 9.65 - 10.3e^{-600d} \quad (\text{m s}^{-1}). \tag{A3}$$

For d in centimeters over the range d = 1.2–4.0 mm, *Rogers and Yau* [1989] give

$$u_t = 15\left(\frac{\rho_{ref}}{\rho_{a,z}}\right)^{1/2} d^{1/2} \quad (\text{m s}^{-1}). \tag{A4}$$

For d in millimeters over the range d = 0.3–6.0 mm, *Best* [1950] gives

$$u_t = 9.32e^{0.0405z}(1-e^{-(0.565d)^{1.147}}) \quad (\text{m s}^{-1}). \tag{A5}$$

where z is the altitude in kilometers.

For d in centimeters, *Lhermitte* [1990] gives

$$u_t = 9.23e\left(\frac{\rho_{ref}}{\rho_{a,z}}\right)^{1/2}(1-e^{(-6.8d^2-4.88d)}) \quad (\text{m s}^{-1}). \tag{A6}$$

APPENDIX B
LEGENDRE FUNCTIONS FOR SPHERICAL HARMONICS

Legendre functions for the two lowest frequencies (n = 2 and 3), from *Beard and Kubesh* [1991].

$$
\begin{aligned}
&P_{2,0} = \tfrac{1}{2}(3\cos^2\theta - 1) && P_{3,0} = \tfrac{1}{2}(5\cos^3\theta - 3\cos\theta) \\
&P_{2,1} = 3\cos\theta\sin\theta && P_{3,1} = \tfrac{3}{2}(5\cos^2\theta - 1)\sin\theta \\
&P_{2,2} = 3\sin^2\theta && \\
& && P_{3,2} = 15\cos\theta\sin^2\theta \\
& && P_{3,3} = 15\sin^3\theta.
\end{aligned}
\qquad \text{(B1)}
$$

REFERENCES

Allnutt, J. E. (1989), *Satellite-to-Ground Radiowave Propagation - Theory, Practice and System Impact at Frequencies Above 1GHz, IEE Electromagnetic Waves Series, vol. 29*, Peter Peregrinus, London, U. K.

Andsager, K., K. Beard, and N. Laird (1999), Laboratory measurements of axis ratios for large raindrops, *J. Atmos. Sci.*, *56*, 2673–2683.

Atlas, D., and C. W. Ulbrich (2000), An observationally based conceptual model of warm oceanic convective rain in the tropics, *J. Appl. Meteorol.*, *39*, 2165–2181.

Atlas, D., R. C. Srivastava, and R. S. Sekhon (1973), Doppler radar characteristics of precipitation at vertical incidence, *Rev. Geophys. Space Phys.*, *11*, 1–35.

Barenblatt, G. I. (2003), *Scaling*, 171 pp., Cambridge Univ. Press, New York.

Batchelor, G. K. (1967), *An Introduction to Fluid Dynamics*, 635 pp. Cambridge Univ. Press, New York.

Battan, L. J. (1964), Some observations of vertical velocities and precipitation sizes in a thunderstorm, *J. Appl. Meteorol.*, *3*, 415–420.

Beard, K. V. (1976), Terminal velocity and shape of cloud and precipitation drops aloft, *J. Atmos. Sci.*, *33*, 851–864.

Beard, K. V. (1977a), On the acceleration of large water drops to terminal velocity, *J. Appl. Meteorol.*, *16*, 1068–1071.

Beard, K. V. (1977b), Terminal velocity adjustment for cloud and precipitation drops aloft, *J. Atmos. Sci.*, *34*, 1293–1298.

Beard, K. V. (1984a), Oscillation models for predicting raindrop axis and backscatter ratios, *Radio Sci.*, *19*(1), 67–74.

Beard, K. V. (1984b), Raindrop oscillations: Evaluation of a potential flow model with gravity, *J. Atmos. Sci.*, *41*, 1765–1774.

Beard, K. V., and C. Chuang (1987), A new model for the equilibrium shape of raindrops, *J. Atmos. Sci.*, *44*(11), 1509–1524.

Beard, K. V., and A. R. Jameson (1983), Raindrop canting, *J. Atmos. Sci.*, *40*, 448–454.

Beard, K. V., and R. J. Kubesh (1991), Laboratory measurements of small raindrop distortion, part 2: Oscillation frequencies and modes, *J. Atmos. Sci.*, *48*(20), 2245–2264.

Beard, K. V., and H. R. Pruppacher (1969), A determination of the terminal velocity and drag of small water drops by means of a wind tunnel, *J. Atmos. Sci.*, *26*, 1066–1072.

Beard, K. V., and A. Tokay (1991), A field study of small raindrop oscillations, *Geophys. Res. Lett.*, *18*(12), 2257–2260.

Beard, K. V., J. Q. Feng, and C. Chuang (1989a), A simple perturbation model for the electrostatic shape of falling drops, *J. Atmos. Sci.*, *46*, 2404–2418.

Beard, K. V., H. T. Ochs, and R. J. Kubesh (1989b), Natural oscillations of small raindrops, *Nature*, *342*, 408–410.

Beard, K. V., R. J. Kubesh, and H. T. I. Ochs (1991), Laboratory measurements of small raindrop distortion, part 1: Axis ratios and fall behavior, *J. Atmos. Sci.*, *48*, 698–710.

Beard, K. V., H. T. Ochs III, and C. H. Twohy (2004), Aircraft measurements of high average charges on cloud drops in layer clouds, *Geophys. Res. Lett.*, *31*, L14111, doi:10.1029/2004GL020465.

Berry, E. X., and M. R. Pranger (1974), Equations for calculating the terminal velocities of water drops, *J. Appl. Meteorol.*, *13*, 108–113.

Best, A. C. (1950), Empirical formulae for the terminal velocity of water drops falling through the atmosphere, *Q. J. R. Meteorol. Soc.*, *76*, 302–311.

Bhalwankar, R. V., and A. K. Kamra (2007), A wind tunnel investigation of the deformation of water drops in the vertical and horizontal electric fields, *J. Geophys. Res.*, *112*, D10215, doi:10.1029/2006JD007863.

Bhalwankar, R. V., and A. K. Kamra (2009), A wind tunnel investigation of the distortion of polluted water drops in the horizontal electric fields, *J. Geophys. Res.*, *114*, D10205, doi:10.1029/2008JD011102.

Black, R. A., and J. Hallett (1986), Observations of the distribution of ice in hurricanes, *J. Atmos. Sci.*, *43*, 802–822.

Blanchard, D. C. (1948), Observations on the behavior of water drops at terminal velocity in air, *Occas. Rep. 7 Proj. Cirrus*, General Electr. Res. Lab., Schenectady, N. Y.

Blanchard, D. C. (1949), Experiments with water drops and the interaction between them at terminal velocity in air, *Occas. Rep. 17 Proj. Cirrus*, General Electr. Res. Lab., Schenectady, N. Y.

Blanchard, D. C. (1950), The behavior of water drops at terminal velocity in air, *Eos Trans. AGU*, *31*, 836–842.

Brandes, E. A., G. Zhang, and J. Vivekanandan (2002), Experiments in rainfall estimation with polarimetric radar in a subtropical environment, *J. Appl. Meteorol.*, *41*, 674–685.

Bringi, V. N., and V. Chandrasekar (2001), *Polarisation Doppler Weather Radar*, 636 pp., Cambridge Univ. Press, Cambridge, U. K.

Bringi, V. N., V. Chandrasekar, and R. Xiao (1998), Raindrop axis ratio and size distribution in Florida rainshaft: An assessment of multiparameter radar algorithms, *IEEE Trans. Geosci. Remote Sens.*, *36*, 703–715.

Brook, M., and D. J. Latham (1968), Fluctuating radar echo: Modulation by vibrating drops, *J. Geophys. Res.*, *73*(22), 7137–7144.

Cannon, T. W. (1976), Imaging devices, *Atmos. Technol.*, *8*, 32–37.

Carbone, R. E., and L. D. Nelson (1978), Evolution of raindrop spectra in warm-based convective storms as observed and numerically modeled, *J. Atmos. Sci.*, *35*, 2302–2314.

Chandrasekar, V., W. A. Cooper, and V. N. Bringi (1988), Axis ratios and oscillations of raindrops, *J. Atmos. Sci.*, *45*, 1323–1333.

Chandrasekar, V., A. Hou, E. Smith, V. N. Bringi, S. A. Rutledge, E. Gorgucci, W. A. Petersen, and G. S. Jackson (2008), Potential role of dual-polarization radar in the validation of satellite precipitation measurements: Rationale and opportunities, *Bull. Am. Meteorol. Soc.*, *89*(8), 1127–1145.

Chuang, C. C., and K. V. Beard (1990), A numerical model for the equilibrium shape of electrified raindrops, *J. Atmos. Sci.*, *47*(11), 1374–1389.

Coquillat, S., and S. Chauzy (1993), Behavior of precipitating water drops under the influence of electrical and aerodynamical forces, *J. Geophys. Res.*, *98*(D6), 10,319–10,329.

Coquillat, S., B. Combal, and S. Chauzy (2003), Corona emission from raindrops in strong electric fields as a possible discharge initiation: Comparison between horizontal and vertical field configurations, *J. Geophys. Res.*, *108*(D7), 4205, doi:10.1029/2002JD002714.

Cornford, S. G. (1965), Fall speeds of precipitation elements, *Q. J. R. Meteorol. Soc.*, *91*, 91–94.

Dawson, G. A., and C. N. Richards (1970), Discussion of the paper by J. Latham and V. Myers, 'Loss of charge and mass from raindrops falling in intense electric fields,' *J. Geophys. Res.*, *75*, 4589–4592.

Despiau, S., and E. Houngninou (1996), Raindrop charge, precipitation, and Maxwell currents under tropical storms and showers, *J. Geophys. Res.*, *101*(D10), 14,991–14,997.

Donnadieu, G. (1978), Mesure de la vitesse terminale des gouttes du pluie au sol à l'aide du spectropluviomètre VIDIAZ, *J. Rech. Atmos.*, *12*, 245–259.

Donnadieu, G. (1980), Comparison of results obtained with the VIDIAZ spectropluviometer and the Joss-Waldvogel rainfall disdrometer in a "rain of a thundery type", *J. Appl. Meteorol.*, *19*, 593–597.

Doviak, R. J., and D. Zrnić (1984), *Doppler Radar and Weather Observations*, 458 pp., Academic, New York.

Edgerton, H. E., and J. R. Killian (1939), *Flash! Seeing the Unseen by Ultra-High Speed Photography*, 203 pp., Hale, Cushman, and Flint, Boston

Feng, J. Q., and K. V. Beard (1991), A perturbation model of raindrop oscillation characteristics with aerodynamic effects, *J. Atmos. Sci.*, *18*, 1856–1868.

Flower, W. D. (1928), The terminal velocity of drops, *Proc. Phys. Soc. London*, *40*, 167–176.

Foote, G. B. (1969), On the internal circulation and shape of large raindrops, *J. Atmos. Sci.*, *26*, 179–181.

Foote, G. B. (1973), A numerical method for studying liquid drop behavior: Simple oscillation, *J. Comput. Phys.*, *11*, 507–530.

Foote, G. B., and P. S. DuToit (1969), Terminal velocity of raindrops aloft, *J. Appl. Meteorol.*, *8*, 249–253.

Gans, R. (1912), Uber die form ultramikroskopischer goldteilchen, *Ann. Phys.*, *37*, 881–900.

Garner, F. H., and J. J. Lane (1959), Mass transfer to drops of liquid suspended in a gas stream, *Trans. Inst. Chem. Eng.*, *37*, 167–172.

Georgis, J.-F., S. Coquillat, and S. Chauzy (1997), Onset of instability in precipitating water drops submitted to horizontal electric fields, *J. Geophys. Res.*, *102*(D14), 16,793–16,798.

Goddard, J. W. F., J. D. Eastment, and M. Thurai (1994a), The Chilbolton Advanced Meteorological Radar: A tool for multi-disciplinary atmospheric research, *IEEE Electron. Commun. Eng. J.*, *6*(2), 77–86.

Goddard, J. W. F., J. D. Eastment, and J. Tan (1994b), Self-consistent measurements of differential phase and differential reflectivity in rain, paper presented at 1994 International Geoscience and Remote Sensing Symposium, IEEE, Pasadena, Calif.

Goodall, F. (1976), Propagation through distorted water drops at 11 GHz, thesis, 205 pp., Univ. of Bradford, Bradford, U. K.

Gorgucci, E., G. Scarchilli, and V. Chandrasekar (1992), Calibration of radars using polarimetric techniques, *IEEE Trans. Geosci. Remote Sens.*, *30*, 853–858.

Gorgucci, E., G. Scarchilli, V. Chandrasekar, and V. N. Bringi (2000), Measurement of mean raindrop shape from polarimetric radar observations, *J. Atmos. Sci.*, *57*, 3406–3413.

Gorgucci, E., V. Chandrasekar, and L. Baldini (2008), Microphysical retrievals from dual-polarization radar measurements at x-band, *J. Atmos. Oceanic Technol.*, *25*, 729–741.

Gourley, J. J., A. J. Illingworth, and P. Tabary (2009), Absolute calibration of radar reflectivity using redundancy of the polarization observations and implied constraints on drop shapes, *J. Atmos. Oceanic Technol.*, *26*, 689–703.

Green, A. W. (1975), An approximation for the shape of large raindrops, *J. Appl. Meteorol.*, *14*, 1578–1583.

Gunn, R. (1949), Mechanical resonance in freely falling raindrops, *J. Geophys. Res.*, *54*, 383–385.

Gunn, R., and G. D. Kinzer (1949), The terminal velocity of fall for water droplets in stagnant air, *J. Meteorol.*, *6*, 243–248.

Hauser, D., P. Amayenc, and B. Nutten (1984), A new optical instrument for simultaneous measurement of raindrop diameter and fall speed distributions, *J. Atmos. Oceanic Technol.*, *1*, 256–269.

Hobbs, P. V., and A. L. Rangno (2004), Super-large raindrops, *Geophys. Res. Lett.*, *31*, L13102, doi:10.1029/2004GL020167.

Illingworth, A., and T. Blackman (2002), The need to represent raindrop size spectra as normalized gamma distributions for the interpretation of polarization radar observations, *J. Appl. Meteorol.*, *41*, 286–297.

Imai, I. (1950), On the velocity of falling raindrops, *Geophys. Mag. Tokyo*, *21*, 244–249.

Johnson, D. B., and K. V. Beard (1984), Oscillation energies of colliding raindrops, *J. Atmos. Sci.*, *41*, 1235–1241.

Jones, B. K., and J. R. Saylor (2009), Axis ratios of water drops levitated in a vertical wind tunnel, *J. Atmos. Oceanic Technol.*, *26*, 2413–2419.

Jones, B. K., J. R. Saylor, and L. F. Bliven (2003), Single-camera method to determine the optical axis position of ellipsoidal drops, *Appl. Opt.*, *42*(6), 972–978.

Jones, D. M. A. (1959), The shape of raindrops, *J. Meteorol.*, *16*, 504–510.

Kamra, A. K. (1970), Effect of electric field on charge separation by the falling precipitation mechanism in thunderclouds, *J. Atmos. Sci.*, *27*, 1182–1185.

Kamra, A. K. (1975), The role of electrical forces in charge separation by falling precipitation in thunderclouds, *J. Atmos. Sci.*, *32*(1), 143–157.

Kamra, A. K., A. B. Sathe, and D. V. Ahir (1986), A vertical wind tunnel for water drop studies, *Mausam*, *37*, 219–222.

Kamra, A. K., R. V. Bhalwankar, and A. B. Sathe (1991), Spontaneous breakup of charged and uncharged water drops freely suspended in a wind tunnel, *J. Geophys. Res.*, *96*, 17,159–17,168.

Kamra, A. K., R. V. Bhalwankar, and A. B. Sathe (1993), The onset of disintegration and corona in water drops falling at terminal velocity in horizontal electric fields, *J. Geophys. Res.*, *98*, 12,901–12,912.

Knollenberg, R. G. (1970), The optical array: An alternative to scattering or extinction for airborne particle size determination, *J. Appl. Meteorol.*, *9*, 86–103.

Knollenberg, R. G. (1981), Techniques for probing cloud microstructure, in *Clouds: Their Formation, Optical Properties, and Effects*, edited by P. V. Hobbs, and A. Deepak, pp. 15–89, Academic, New York

Kollias, P., B. A. Albrecht, and F. D. Marks Jr. (2001), Raindrop sorting induced by vertical drafts in convective clouds, *Geophys. Res. Lett.*, *28*, 2787–2790.

Kubesh, R. J., and K. V. Beard (1993), Laboratory measurements of spontaneous oscillations for moderate-size raindrops, *J. Atmos. Sci.*, *50*(8), 1089–1098.

Lamb, H. (1881), On the oscillations of a viscous spheroid, *Proc. London Math. Soc.*, *13*, 51–56.

Lamb, H. (1932), *Hydrodynamics*, 738 pp.Dover, New York.

Landau, L., and E. M. Lifshitz (1959), *Fluid Mechanics*, 536 pp., Pergamon, New York.

Laws, J. O. (1941), Measurements of the fall velocity of water and rain drops, *Eos Trans. AGU*, *22*, 709–721.

LeClair, B. P., A. E. Hamielec, H. R. Pruppacher, and W. D. Hall (1972), A theoretical and experimental study of the internal circulation in water drops falling at terminal velocity in air, *J. Atmos. Sci.*, *29*, 728–740.

Lenard, P. (1887), Über die schwingungen fallender tropfen, *Ann. Phys. Chem.*, *30*, 209–243.

Lenard, P. (1904), Über regen, *Meteorol. Z.*, *21*, 248–262.

Lhermitte, R. (1990), Attenuation and scattering of millimeter wavelength radiation by clouds and precipitation, *J. Atmos. Oceanic Technol.*, *7*, 464–479.

Liznar, J. (1914), Die fallgeschwindigheit der regenropfen, *Meteorol. Z.*, *31*, 339–347.

Magarvey, R. H., and C. S. MacLatchy (1965), Vortices in sphere wakes, *Can. J. Phys.*, *43*, 1649–1656.

Magono, C. (1954), On the shape of water drops falling in stagnant air, *J. Meteorol.*, *11*, 77–79.

Margarvey, R. H., and R. L. Bishop (1961), Transition ranges for three-dimensional wakes, *Can. J. Phys.*, *39*, 1418–1422.

Matrosov, S. Y., D. E. Kingsmill, B. E. Martner, and F. M. Ralph (2005), The utility of x-band radar for quantitative estimates of rainfall parameters, *J. Hydrometeorol.*, *6*, 248–262.

McDonald, J. E. (1954a), The shape and aerodynamics of large raindrops, *J. Meteorol.*, *11*, 478–494.

McDonald, J. E. (1954b), The shape of raindrops, *Sci. Am.*, *190*, 64–68.

Mitra, S. K., J. Brinkmann, and H. R. Pruppacher (1992), A wind tunnel study on the drop-to-particle conversion, *J. Aerosol Sci.*, *23*, 245–256.

Montero-Martínez, G., A. B. Kostinski, R. A. Shaw, and F. García-García (2009), Do all raindrops fall at terminal speed?, *Geophys. Res. Lett.*, *36*, L11818, doi:10.1029/2008GL037111.

Musgrove, C., and M. Brook (1975), Microwave echo fluctuations produced by vibrating water drops, *J. Atmos. Sci.*, *32*, 2001–2007.

Naterajan, R., and R. A. Brown (1987), Third-order resonance effects and the nonlinear stability of drop oscillations, *J. Fluid Mech.*, *183*, 95–121.

Nelson, A. R., and N. R. Gokhale (1972), Oscillation frequencies of freely suspended water drops, *J. Geophys. Res.*, *77*, 2724–2727.

Nešpor, V., W. F. Krajewski, and A. Kruger (2000), Wind-induced error of raindrop size distribution measurement using a two-dimensional video disdrometer, *J. Atmos. Oceanic Technol.*, *17*, 1483–1492.

Pruppacher, H. R., and K. V. Beard (1970), A wind tunnel investigation of the internal circulation and shape of water drops falling at terminal velocity in air, *Q. J. R. Meteorol. Soc.*, *96*, 247–256.

Pruppacher, H. R., and J. D. Klett (1997), *Microphysics of Clouds and Precipitation*, 2nd ed., 954 pp., Kluwer Acad., Dordrecht, Netherlands.

Pruppacher, H. R., and R. L. Pitter (1971), A semi-empirical determination of the shape of cloud and rain drops, *J. Atmos. Sci.*, *28*, 86–94.

Rasmussen, R., C. Walcek, H. R. Pruppacher, S. K. Mitra, J. Lew, V. Levizzani, P. K. Wang, and U. Barth (1985), A wind tunnel investigation of the effect of an external vertical electric-field on the shape of electrically uncharged rain drops, *J. Atmos. Sci.*, *42* (15), 1647–1652.

Rauber, R. M., K. V. Beard, and B. M. Andrews (1991), A mechanism for giant raindrop formation in warm, shallow convective clouds, *J. Atmos. Sci.*, *48*, 1791–1797.

Rayleigh, L. (1879), On the capillary phenomenon of jets, *Proc. R. Soc. London*, *21*, 71–97.

Richards, C. N., and G. A. Dawson (1971), The hydrodynamic instability of water drops falling at terminal velocity in vertical electric fields, *J. Geophys. Res.*, *76*, 3445–3455.

Rogers, R. R. (1989), Raindrop collision rates, *J. Atmos. Sci.*, *46*, 2469–2472.

Rogers, R. R., and M. K. Yau (1989), *A Short Course in Cloud Physics*, 293 pp.Elsevier, New York.

Ryan, R. T. (1976), The behavior of large, low-surface-tension water drops falling at terminal velocity in air, *J. Appl. Meteorol.*, *15*, 157–165.

Ryzhkov, A. V., S. E. Giangrande, V. M. Melnikov, and T. J. Schuur (2005), Calibration issues of dual-polarization radar measurements, *J. Atmos. Oceanic Technol.*, *22*, 1138–1155.

Sakamoto, H., and H. Haniu (1990), A study on vortex shedding from spheres in uniform flow, *Trans. ASME: J. Fluids Eng.*, *112*, 386–392.

Savic, P. (1953), Circulation and distortion of liquid drops falling through a viscous medium, *Rep. NRC-MT-22*, 55 pp., Natl. Res. Council, Canada, Ottawa, Ont., Canada.

Saylor, J. R., and B. K. Jones (2005), The existence of vortices in the wakes of simulated raindrops, *Physics of Fluids*, *17*(3), 031716.

Saylor, J. R., B. K. Jones, and L. F. Bliven (2002), A method for increasing depth of field during droplet imaging, *Rev. Sci. Instrum.*, *73*(6), 2422–2427.

Schmidt, W. (1909), Eine unmittelbare Bestimmung der Fallgeschwindigheit von Regentropfen, *Meteorol. Z.*, *26*, 183–184.

Schmidt, W. (1913), Die gestalt fallender regentropfen, *Meteorol. Z.*, *30*, 456–457.

Schönhuber, M., H. E. Urban, J. P. V. Poiares-Baptista, W. L. Randeu, and W. Riedler (1994), Measurements of precipitation characteristics by a new distrometer, paper presented at Atmospheric Physics and Dynamics in the Analysis and Prognosis of Precipitation Fields, SIMA, Rome.

Schönhuber, M., H. E. Urban, J. P. V. Poiares-Baptista, W. L. Randeu, and W. Riedler (1995), Weather radar versus 2d-video-distrometer data, paper presented at Third International Symposium on Hydrological Applications of Weather Radars, Lab. d'étude Transferts Hydrol. Environ., Sao Paulo, Brazil.

Schönhuber, M., H. E. Urban, J. P. V. Poaires Baptista, W. L. Randeu, and W. Riedler (1997), Weather radar versus 2d-video-distrometer data, in *Weather Radar Technology for Water Resources Management*, edited by B. Braga, Jr. and O. Massambani, pp. 159–171, UNESCO, Paris.

Schuur, T. J., and A. V. Rhyzhkov (2001), Drop size distributions measured by a 2D video disdrometer: Comparison with dual-polarization radar data, *J. Appl. Meteorol.*, *40*, 1019–1034.

Seliga, T. A., and V. N. Bringi (1976), Potential use of radar differential reflectivity measurements at orthogonal polarizations for measuring precipitation, *J. Appl. Meteorol.*, *15*, 69–76.

Spilhaus, A. F. (1948), Raindrop size, shape, and falling speed, *J. Meteorol.*, *5*, 108–110.

Srivastava, R. C., and D. Atlas (1969), Growth motion and concentration of precipitation particles in convective storms, *J. Atmos. Sci.*, *26*, 535–544.

Sterlyadkin, V. V. (1988), Field measurements of raindrop oscillations, *Izv. Acad. Sci. USSR Atmos. Oceanic Phys., Engl. Transl.*, *24*(6), 449–454.

Stow, C. D., and K. Jones (1981), A self-evaluating disdrometer for the measurement of raindrop size and charge at the ground, *J. Appl. Meteorol.*, *20*, 1160–1176.

Szakall, M., K. Diehl, and S. B. Mitra (2009), A wind tunnel study on the shape, oscillation, and internal circulation of large raindrops with sizes between 2.5 and 7.5 mm, *J. Atmos. Sci.*, *66*, 755–765.

Szumowski, M. J., R. M. Rauber, O. H. T. Ochs, III, and K. V. Beard (1998), The microphysical structure and evolution of Hawaiian rainband clouds. Part II: Aircraft measurements within rainbands containing high reflectivity cores, *J. Atmos. Sci.*, *55*, 208–226.

Takahashi, T., K. Suzuki, M. Orita, M. Tokuno, and R. Delamar (1995), Videosonde observations of precipitation processes in equatorial cloud clusters, *J. Meteorol. Soc. Jpn.*, *73*(2B), 267–290.

Taylor, T. D., and A. Acrivos (1964), On the deformation and drag of a falling viscous drop at low Reynolds number, *J. Fluid Mech.*, *18*, 466–476.

Testik, F. (2009), Outcome regimes of binary raindrop collisions, *Atmos. Res.*, *94*(3), 389–399.

Testik, F. Y., and A. P. Barros (2007), Toward elucidating the microstructure of warm rainfall: A survey, *Rev. Geophys.*, *45*, RG2003, doi:10.1029/2005RG000182.

Testik, F. Y., A. P. Barros, and L. F. Bliven (2006), Field observations of multimode raindrop oscillations by high-speed imaging, *J. Atmos. Sci.*, *63*, 2663–2668.

Thurai, M., and V. N. Bringi (2005), Drop axis ratios from a 2d video disdrometer, *J. Atmos. Oceanic Technol.*, *22*(7), 966–978.

Thurai, M., G. J. Huang, V. N. Bringi, W. L. Randeu, and M. Schönhuber (2007), Drop shapes, model comparisons, and calculations of polarimeteric radar parameters in rain, *J. Atmos. Oceanic Technol.*, *24*(6), 1019–1032.

Thurai, M., W. A. Peterson, V. N. Bringi, G. J. Huang, E. V. Johnson, and C. Schultz (2008), C-band polarimetric radar variables calculated using rain microstructure information from 2-d video disdrometer, paper presented at Fifth European Conference on Radar in Meteorology and Hydrology (ERAD 2008), Finn. Meteorol. Inst., Helsinki.

Tokay, A., and K. V. Beard (1996), A field study of raindrop oscillations. Part I: Observation of size spectra and evaluation of oscillation causes, *J. Appl. Meteorol.*, *35*, 1671–1687.

Tokay, A., A. Kruger, and W. F. Krajewski (2001), Comparison of drop size distribution measurements by impact and optical disdrometers, *J. Appl. Meteorol.*, *40*, 2083–2097.

Tsamopoulos, J. A., and R. A. Brown (1983), Nonlinear oscillations of inviscid drops and bubbles, *J. Fluid Mech.*, *127*, 519–537.

Ulbrich, C. W. (1986), A review of the differential reflectivity technique for measuring rainfall, *IEEE Trans. Geoscience and Remote Sensing*, *GE-24*(6), 955–965.

Vivekanandan, J., G. Zhang, S. M. Ellis, D. Rajopadhyaya, and S. K. Avery (2003), Radar reflectivity calibration using differential propagation phase measurement, *Radio Sci.*, *38*(3), 8049, doi:10.1029/2002RS002676.

Wang, P. K., and H. R. Pruppacher (1977), Acceleration to terminal velocity of cloud and raindrops, *J. Appl. Meteorol.*, *16*, 276–280.

Xu, X. (1995), Analyses of aircraft measurements of raindrop shape in cape, Master's thesis, Univ. of Illinois Urbana-Champaign, Urbana.

Zrnić, D. S., R. J. Doviak, and P. R. Mahapatra (1984), The effect of charge and electric field on the shape of raindrops, *Radio Sci.*, *19* (1), 75–80.

B. K. Jones, Department of Mechanical Engineering, Columbia University, 220 S. W. Mudd, 500 W. 120th Street, New York, NY 10027, USA. (bkj2106@columbia.edu)

J. R. Saylor, Department of Mechanical Engineering, Clemson University, Clemson, SC 29634, USA. (jrsaylor@ces.clemson.edu)

F. Y. Testik, Department of Civil Engineering, Clemson University, Clemson, SC 29634, USA. (ftestik@clemson.edu)

The Evolution of Raindrop Spectra: A Review of Microphysical Essentials

K. D. Beheng

Institute for Meteorology and Climate Research, Karlsruhe Institute of Technology, Karlsruhe, Germany

The evolution of raindrops from cloud droplets is presented on the theoretical basis of cloud microphysical principles in terms of the spectral balance equation. Also, interconnections of the spectral balance equation to some related bulk balance equations are shown. The process especially outlined in this study is collisional interaction of drops. This can result in coagulation (permanent unification) or disaggregation (production of fragment droplets by collision-induced breakup). In distinguishing between cloud droplets and raindrops, emphasis is laid on a stringent definition of autoconversion, accretion, and self-collection. The related conversion rates are strictly defined for both the size distribution functions, themselves, as well as for the number and mass densities of cloud droplets and raindrops. A numerical example shows the effects of these mechanisms. In case of collision-induced breakup, the time rates of change of the size distribution function and the total number and mass densities are formulated. Finally, collision kernels are discussed, which, on one hand, are valid for raindrops and, on the other hand, for large droplets/small raindrops as affected by turbulence.

1. INTRODUCTION

Raindrops reaching the ground are a manifestation of a number of processes taking place in clouds of different composition. In warm clouds, they are formed by a chain of microphysical processes beginning with heterogeneous nucleation, continuing with water vapor diffusion (condensation/evaporation) followed by collision/coalescence and breakup, and finally sedimenting to the surface. Originating from mixed-phase clouds, raindrops are melting products of ice hydrometeors. The temporal and spatial distribution of rain (drops) is thus intrinsically connected to the hydro- and thermodynamics of clouds embedded in small- and large-scale atmospheric structures, which initiate, evolve, and transport water mass. In terms of a mathematical-physical description, the consideration of rain-forming processes, as well as other processes related to condensed phases, leads to particular terms in some relevant basic equations, as, e.g., in the Navier-Stokes equations, for a mixture of humid air and hydrometeors, the liquid water drag appears, in the thermodynamic equation, phase transitions lead to temperature changes, and in the radiative transfer equation, cloud and rain particles are responsible for absorption and scattering effects.

2. BASICS

This section deals with general definitions and relations valid mostly for hydrometeors. These may be drops or ice particles. Where it seems appropriate, we specify the statements to a definite type of hydrometeor.

The formulation of cloud microphysical processes has to consider in its detailed description that, physically speaking, clouds are heterogeneous polydisperse systems, i.e., they consist of the three phases of water (identical to their aggregates' states) with their condensed elements (drops, ice particles) of different sizes or mass, respectively. However, some general definitions and relations which follow are applicable ceterum paribus also for particulate matter as to an aerosol.

Rainfall: State of the Science
Geophysical Monograph Series 191

10.1029/2010GM000957

With respect to polydispersity, all hydrometeor types are specified by distribution functions where, besides the space coordinates and time, an additional independent variable is introduced that can be mass m or, if mass and an other variable are bijectively related as in the case of spherical hydrometeors, radius r or diameter d. Such a function is mathematically given by $f(\mathbf{x},t;m) \geq 0$ with the independent space variables $\mathbf{x}$ = position vector and t = time. In a broader sense, the variables $(\mathbf{x},t)$ are denoted outer variables (or coordinates), whereas m (or r or d) is named an inner variable. For brevity, we abbreviate $f(\mathbf{x},t;m)$ in the following mostly by $f(m)$, where $f(m)\mathrm{d}m$ defines the number of hydrometeors with masses in the interval $[m, m + \mathrm{d}m]$ per unit volume.

In restricting to mass m as primary independent variable for now, $f(m)$ is, in kinetic terms, called the spectral number density function of drops and, in analogy to optics, also sometimes termed a drop number spectrum.

Changing the independent variable, here m, to a different one q, say, is then generally performed by obeying the relation $f(m)\mathrm{d}m = f(q)\mathrm{d}q$ with $m = m(q)$, and $q = q(m)$ vice versa, unique, differentiable and, in the cloud microphysical context, always strong monotonically increasing functions. Thus, the transformation condition is that the Jacobi determinant is $\mathrm{d}m/\mathrm{d}q > 0$ (for $q > 0$). So if we, for example, want to transform $f(m)$ to $f(r)$, we use the relation between mass and radius of a water sphere and arrive at $f(r) = 4\pi\bar{\rho}_w r^2 f(m)$ with $\bar{\rho}_w$ = material density of water.

For ice crystals, there is generally no unique relation between mass m and a "single" geometric dimension because even for same geometrical shapes as, e.g., hexagonal ice columns or hexagonal ice plates, the crystals may have different relations regarding their a and c axes lengths. This means, in turn, that the a and c axes relations may be used to define a specific habit of ice crystals. Thus, for each ice crystal habit, an additional relation between both axes has at least to be considered [cf. *Locatelli and Hobbs*, 1974].

It is emphasized that there is mathematically no limit for the number of inner variables, but one has to reason on the validity of large a number of additional inner variables in (cloud micro-) physical applications as the number of particles per unit volume and per unit of the inner variables has to be such large that a representative mean value can be calculated.

Integration of the spectral number density function $f(m)$ over its inner variable m or equivalently over r yields the total number density

$$N = \int_{m=0}^{\infty} f(m)\mathrm{d}m \equiv \int_{r=0}^{\infty} f(r)\mathrm{d}r. \tag{1}$$

This mathematical formulation may be different from an experimental point of view as every instrument has a lower and upper detection limit such that number densities calculated from observations are generally smaller than those computed by equation (1).

The total mass density L is another quantity inferred from the drop spectrum by integration. It is

$$\begin{aligned} L &= \int_{m=0}^{\infty} mf(m)\mathrm{d}m = \int_{m=0}^{\infty} g(m)\mathrm{d}m \\ &= (4\pi\bar{\rho}_w/3)\int_{r=0}^{\infty} r^3 f(r)\mathrm{d}r. \end{aligned} \tag{2}$$

The function $g(m) = mf(m)$ is called spectral mass density distribution function.

It is emphasized that throughout the text, the terms total number and mass density for N and L, respectively, are preferred compared to number concentration and liquid water content to be found in textbooks and journal publications. Although the latter denotations are commonly accepted and always correctly interpreted, we stick in this study mostly to the strict physical (synonymical) denotation of a quantity normalized by volume as a "density."

The expressions for N and L are, statistically speaking, the zeroth and first moments of the distribution functions, respectively, if mass is an independent variable. The second moment is proportional to the radar reflectivity factor

$$Z = \int_{m=0}^{\infty} m^2 f(m)\mathrm{d}m \propto \int_{r=0}^{\infty} r^6 f(r)\mathrm{d}r, \tag{3}$$

if for the radar cross-section, the Rayleigh approximation for dielectric spheres is made.

After these preparatory remarks and definitions, we now present a prognostic equation describing the spatiotemporal evolution of a drop spectrum. Such an equation is in ubiquitous use in science and technology and well-known as spectral-kinetic or population-dynamics equation. It roots in the physical description of gas kinetics as well as chemical reactions. The denotation "inner variables," referred to above, has been, to the author's knowledge first been proposed by *Prigogine and Mazur* [1953] who expand the description of intensive, i.e., mass-independent, quantities in terms of balance equations written solely with outer coordinates by some arbitrary inner coordinates. In analogy to balance equations for mass densities of single components of a gas mixture, they arrived at a formulation, which in the present context reads

$$\frac{\partial g(m)}{\partial t} + \nabla \cdot [g(m)\mathbf{v}(m)] + \frac{\partial [g(m)\dot{m}]}{\partial m} = \tilde{\gamma}, \tag{4}$$

with $\mathbf{v}(m)$ = effective velocity of a drop of mass m, $\dot{m} = \mathrm{d}m/\mathrm{d}t$ = "velocity" relative to the inner coordinate (here condensation/evaporation rate of a drop) and $\tilde{\gamma}$ = gain/loss terms comprising collisional interaction of drops and nucleation where the latter is almost always omitted in the following. This equation we call the spectral balance equation for the spectral mass density distribution function $g(m)$. Note that for clearness, we have again omitted the dependencies of $g(m)$ on $\mathbf{x}$ and t.

The effective velocity of a drop $\mathbf{v}(m)$ is defined by $\mathbf{v}(m) = \mathbf{v} + \mathbf{v}'(m)$ with $\mathbf{v}$ = barycentric velocity such that $\mathbf{v}'(m)$ is the drop's intrinsic velocity. In this form, $\mathbf{v}'(m)$ is assumed to be the same for same drop masses.

From a more general physical point of view, we may refer to equation (4) as the Boltzmann transport equation or equivalently as Reynolds transport theorem applied to the spectral mass density distribution function $g(m)$ with the configuration space spanned by the vector $(\mathbf{x},m)$.

We recognize that the local time rate of change of $g(m)$ is ruled, on the left-hand side, by the divergence of the effective spectral mass flux and a divergence expression in the space of the single inner coordinate m, i.e., movement of drops with mass m due to condensation/evaporation along this coordinate with $\mathbf{x}$ and t held fixed. The right-hand side (RHS) comprises, in this case, all contributions by nucleation, mutual collisions of drops, and breakup events. We will come back to this term later on.

If we write equation (4) for the spectral number density distribution function $f(m)$, we obtain a similar equation reading

$$\frac{\partial f(m)}{\partial t} + \nabla \cdot [f(m)\mathbf{v}(m)] + \frac{\partial [f(m)\dot{m}]}{\partial m} = \gamma. \tag{5}$$

The RHS of equation (4) relates to that of equation (5) by $\tilde{\gamma} = m\gamma + \dot{m}f(m)$.

To obtain equations for integrated values, we insert $\mathbf{v}(m) = \mathbf{v} + \mathbf{v}'(m)$ in equations (4) and (5) and integrate equation (5) over the inner coordinate m then yielding the time rate of change of the total number density (i.e., the zeroth moment of $f(m)$)

$$\frac{\partial N}{\partial t} + \nabla \cdot (N\mathbf{v}) + \nabla \cdot (N\overline{\mathbf{v}'}) = \int_0^\infty \gamma \mathrm{d}m \tag{6}$$

with $\overline{\mathbf{v}'} = \frac{1}{N}\int f(m)\mathbf{v}'(m)\mathrm{d}m$ arrived at by application of the mean-value theorem. Considering equation (4) leads equivalently to the time rate of change of the total mass density (i.e., the first moment of $f(m)$)

$$\frac{\partial L}{\partial t} + \nabla \cdot (L\mathbf{v}) + \nabla \cdot (L\overline{\overline{\mathbf{v}'}}) = \int_0^\infty m\gamma \mathrm{d}m + \int_0^\infty f(m)\dot{m}\mathrm{d}m = \int_0^\infty \tilde{\gamma}\mathrm{d}m, \tag{7}$$

with $\overline{\overline{\mathbf{v}'}} = \frac{1}{L}\int g(m)\mathbf{v}'(m)\mathrm{d}m = \frac{1}{L}\int mf(m)\mathbf{v}'(m)\mathrm{d}m$.

Note the difference between the mean intrinsic velocities in both integrated equations, since weighting is done in equation (6) by the total number density and in equation (7) by the total water mass density. Note also that the term $\int_0^\infty \{\partial[f(m)\dot{m}]/\partial m\}\mathrm{d}m$ is identical to zero as the total drop number does not change due to condensation or evaporation as long as no complete evaporation of drops occur or minute drops are nucleated. Changes due to collisional interactions can reduce or increase N but leave L constant as will be shown below.

Now, we turn to equations having included quantities, which are related to the existence of water and phase transitions, but take no notice on hydrometeors of different sizes such that we name them bulk quantities. In this sense is equation (6), a balance equation for the bulk quantity N and equation (7) for L.

The first equation is the balance equation for the partial density of water substance ρ_w given in usual notation by

$$\frac{\partial \rho_w}{\partial t} + \nabla \cdot (\rho_w \mathbf{v}) + \nabla \cdot \mathbf{J}_w = J_w, \tag{8}$$

with the bulk variables $\mathbf{J}_w = \rho_w(\mathbf{v}_w - \mathbf{v})$ = diffusion flux of water mass = difference between individual mass flux of water substance and the barycentric one and J_w = condensation/evaporation rate of water. Additionally, we have the thermodynamic equation (without consideration of friction, external forces and diffusion fluxes)

$$\rho c_p \frac{\mathrm{d}T}{\mathrm{d}t} + \nabla \cdot \mathbf{J}_q = \frac{\mathrm{d}p}{\mathrm{d}t} + \ell_{ce} J_w, \tag{9}$$

with c_p = isobaric specific heat capacity, p = pressure, T = temperature, $\mathbf{J}_q$ = heat flux and ℓ_{ce} = latent heat of condensation/evaporation. For deposition/sublimation as well as freezing/melting, different latent heats have to be considered. Contributions of water vapor and condensed phases to c_p are equally omitted here as was done for $\mathbf{J}_q$. In comparing equation (7) with equations (8) and (9), one recognizes the following identities:

$$\rho_w \equiv L, \quad \mathbf{J}_w \equiv L\overline{\overline{\mathbf{v}'}}, \quad J_w \equiv \int_0^\infty [f(m)\dot{m} + m\gamma]\mathrm{d}m. \tag{10}$$

The last equality comprises as first term contributions by condensation/evaporation and as second term of equation (10) loss/gain contributions by nucleation. Note that condensation/

evaporation ceases in equilibrium conditions, i.e., of $\dot{m} = 0$. It is also emphasized that contributions to the second term due to collision/coalescence and breakup disappear as in these cases $\int_0^\infty m\gamma dm = 0$ because these processes are mass conserving. Besides, we mention that $\int_0^\infty \gamma dm \gtrless 0$, i.e., the total number density N may change by collisional interactions of drops.

The vertical component of $\mathbf{J}_w$ defines the precipitation flux $\mathcal{P} = -\mathbf{J}_w \cdot \mathbf{k}$ with $\mathbf{k}$ = vertical unit vector in Cartesian coordinates. In this case, the appropriate intrinsic drop velocity $\mathbf{v}'(m)$ is the sedimentation velocity $\mathbf{v}_T(m)$, which is mostly assumed to be the stationary fall velocity in the Earth's gravity field, sometimes called terminal fall velocity or settling velocity. Consequently, we set $\mathbf{v}'(m) \equiv \mathbf{v}_T(m) = -\mathbf{v}_T(m)\mathbf{k}$ such that $\mathcal{P} = \int_0^\infty mf(m)v_T(m)dm$. Dividing $\mathcal{P}$ by the bulk water density $\bar{\rho}_w$, the precipitation intensity in units of, e.g., mm h^{-1}, results.

Since we concentrate in this study on processes important to the kinetics of raindrop "interactions" and not on the behavior of single drops, we do not further explore the description of their diffusional growth in this study. For details of such processes, the reader is referred to standard literature such as the comprehensive monograph of *Pruppacher and Klett* [1997] and the textbook of *Rogers and Yau* [1996].

Finally, we would like to stress that equations (4)–(7) are also valid if hydrometeors other than drops are accounted for. Clearly, then other mechanisms and processes have to be considered as deposition/sublimation of ice particles or accretional growth of ice crystals collecting supercooled droplets (riming). Furthermore, also combinations of different hydrometeor types as appearing in mixed-phase clouds or among habits of the ice phase (to the extent they are distinguished) can be described by those equations or a set of those equations. Such a set of equations is then the basis of so-called spectral (or bin) formulations [see, e.g., *Beheng*, 1982].

At last, we refer to an example of considering two inner coordinates in a size distribution function. This is the description of the uptake of aerosol particles by differently sized drops due to Brownian diffusion, phoretic forces, and gravity (particle scavenging [*Beheng and Herbert*, 1986]. The first inner coordinate is drop mass and the second the number of particles incorporated in drops. The in-drop particle number is then changed, except by scavenging itself, by mutual coagulation of drops having different numbers of incorporated particles.

3. PHYSICAL DESCRIPTION OF COLLISIONAL INTERACTION OF DROPS

As has been mentioned, the RHS of the balance equations (4) and (5) describe collisional interactions among drops. As in a diluted gaseous medium, only binary collisions are taken into account, since these are the most probable ones. The outcome of a collision between two drops can be twofold: One outcome could be a permanent coalescence of both drops after collision (subscript "CA") and the other is collision-induced breakup (subscript "BR"). The efficiency of both processes are measured by the coalescence efficiency $E_{CA}(m',m'')$ and by the breakup efficiency $E_{BR}(m',m'')$ yielding

$$E_{CA}(m',m'') + E_{BR}(m',m'') = 1, \qquad (11)$$

since a collision event leads exclusively to only one of these outcomes. $E_{CA}(m',m'')$ can attain unity as maximum value. But it should be noticed that E_{CA} may be $\ll 1$ depending on various factors such as the velocity of collision, the angle of impact, surface tension, and electrical forces associated with net charges on the drops.

It is emphasized that the efficiencies should be understood probabilistically as they are defined by the ratio of successful events to all possible events, i.e., they can be interpreted as probabilities.

In the following, another efficiency, the so-called collision efficiency $E_{CL}(m',m'')$ comes into play. It measures the effect, whereby the collision process of drops approaching one another is influenced by the ambient flow field or by the flow fields induced by the drops' own motion. In Figure 1, such a

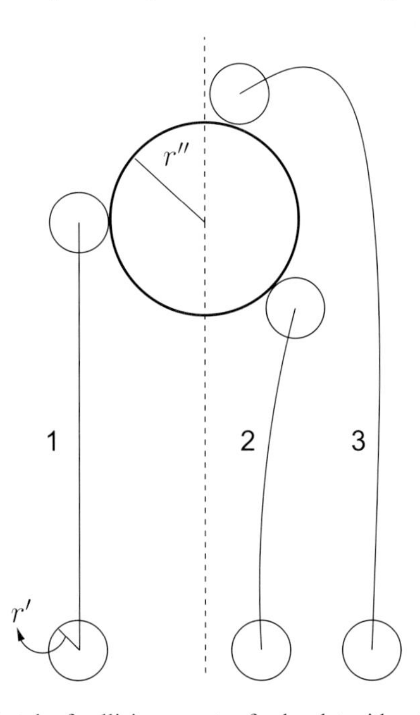

Figure 1. Sketch of collision events of a droplet with radius r' and a drop with radius r''. Numbers show the following: 1, grazing trajectory of a droplet unaffected by any flow field, i.e., the pure geometric case; 2, trajectory demonstrating the deflection of a smaller drop approaching a larger one by the superposition of their flow fields with an eventual collision event; and 3, trajectory indicating rear capture of the smaller drop in an assumed wake of the larger drop.

situation is depicted schematically. It shows on the left the grazing trajectory 1 of a droplet unaffected by any flow field, i.e., a pure geometric case. The maximum collision cross-section is then $\sigma_{geo} = \pi(r' + r'')^2$ with r', r'' the radii of the spherical assumed drops. On the right two trajectories are sketched: trajectory 2 demonstrating the effect if a smaller drop approaching a larger one is deflected by the superposition of their flow fields but collides eventually with the larger drop; trajectory 3 indicating rear capture of the smaller drop in an assumed wake of the larger drop, although the smaller drop was initially outside the area of the geometric collision cross-section thus resulting in $E_{CL} > 1$.

The collision efficiency E_{CL} can be defined as the ratio of the actual collision cross-section σ_{act} to the geometric one by

$$E_{CL} = \frac{\sigma_{act}}{\sigma_{geo}}. \tag{12}$$

It is emphasized that for reasons mentioned above for E_{CL}, no definite upper value can be given. Note that another factor that may affect the collision efficiency of drops is turbulence or fluctuations in their velocities, which may affect the outcome of a collision. We will come back to this issue in section 7.

Following *Gillespie and List* [1978/79], the result of a collision event can be described via the collisional interaction function $\chi(m;m',m'')$, giving the number of drops formed per collision and per mass interval $[m,m+\mathrm{d}m]$ by two initial drops with masses in the intervals $[m' + \mathrm{d}m']$ and $[m'' + \mathrm{d}m'']$, respectively, as the sum of two terms

$$\chi(m;m',m'') = \chi_{CA}(m;m',m'') + \chi_{BR}(m;m',m''), \tag{13}$$

where for coalescence

$$\chi_{CA}(m;m',m'') = E_{CA}(m',m'')\delta(m-m'-m'') - \delta(m-m') - \delta(m-m'') \tag{14a}$$

and for collision-induced breakup

$$\chi_{BR}(m;m',m'') = E_{BR}(m',m'')P(m;m',m''), \tag{14b}$$

with $P(m;m',m'')$ the fragment size distribution function defined below by equation (16).

Equation (14a) expresses by the first term on the RHS that one single drop of mass m is created by a coalescence of the two colliding drops with masses m' and m''. The Dirac δ-function ensures that the coalescing drops indeed yield a drop of mass m. The coalescence efficiency $E_{CA}(m',m'')$ is a weighting function here. The remaining two terms on the RHS of equation (14a) comprise the loss of a drop of mass m due to a collision with another drop, which necessarily produces a new drop whose mass is different from m.

Restricting ourselves momentarily to the case of coalescence only, we find from equation (14a) by integration and by taking into account the symmetry of the δ-function

$$\int_{m=0}^{\infty} \chi_{CA}(m;m',m'')\mathrm{d}m = \int_{m=0}^{\infty} [E_{CA}(m',m'')\delta(m-m'-m'') - \delta(m-m') - \delta(m-m'')]\mathrm{d}m = E_{CA}(m-m',m') - 2, \tag{15}$$

which yields for $E_{CA}(m - m',m') = 1$ a value of -1, i.e., the number of drops is, in the mean, reduced by one as expected.

The breakup term (equation (14b)) denotes the formation of drops with masses $[m,m + \mathrm{d}m]$ through the fragmentation or breakup of two drops of sizes m' and m'' that collide with one another. The function $P(m;m',m'')$ is the fragment size distribution function with

$$P(m;m',m'') \begin{cases} \geq 0 & \text{for } m \leq m' + m'' \\ = 0 & \text{for } m > m' + m'' \end{cases} \tag{16}$$

and

$$N_{BR} = \int_{m=0}^{m'+m''} P(m;m',m'')\mathrm{d}m \geq 2 \tag{17a}$$

$$m' + m'' = \int_{m=0}^{m'+m''} mP(m;m',m'')\mathrm{d}m. \tag{17b}$$

Equation (17a) describes that the total number of fragments N_{BR} is generally larger than 2. By equation (17b), we are simply expressing mass conservation, in that the mass of the sum of the fragments must equal the mass of the drops involved in the initial collision.

Since not every collision leads to a breakup event, the breakup term in equation (13) has to be weighted by the breakup efficiency $E_{BR}(m',m'') = 1 - E_{CA}(m',m'')$ (cf. equation (11)).

Next, we define the collision rate $\zeta(m',m'')$, which is the mean number of collisions between drops of masses $[m' + \mathrm{d}m']$ and $[m'' + \mathrm{d}m'']$ per unit time and unit volume, by

$$\zeta(m',m'') = K(m',m'')f(m')\mathrm{d}m'f(m'')\mathrm{d}m'', \tag{18}$$

with $f(m')\mathrm{d}m'$ and $f(m'')\mathrm{d}m''$ the spectral number densities of drops in the mass intervals $[m' + \mathrm{d}m']$ and $[m'' + \mathrm{d}m'']$, respectively, and the collision kernel $K(m',m'')$. The formulation of the collision kernel is deferred to section 7.

The time rate of change of the spectral number distribution function $f(m)$ by all collisional interactions follows from multiplying the collisional interaction function $\chi(m;m',m'')$

(equation (13)) by the collision rate $\zeta(m',m'')$ (equation (18)) and integrating over the entire mass intervals such that

$$\gamma \equiv \left.\frac{\partial f(m)}{\partial t}\right|_{\mathrm{CA+BR}} = \frac{1}{2}\int_{m'=0}^{\infty}\int_{m''=0}^{\infty} f(m')f(m'')K(m',m'') \times \chi(m;m',m'')\mathrm{d}m'\mathrm{d}m'' \quad (19)$$

The factor 1/2 avoids double counting.

4. COAGULATION

The theoretical basis for the formulation of binary collisions between particles resulting in a permanent coalescence, both processes together called coagulation, has been laid by *von Smoluchowski* [1916, 1917]. He investigated the coagulation of hard colloidal spheres, initially of same size, in a fluid to which an electrolyte had been added, such that larger particles are formed by attractive forces. This previous work has been extended later on by *Müller* [1928] to a system consisting of differently sized hard spheres at the beginning. *Telford* [1955] transformed this idea to cloud microphysics in accounting for the drops undergoing completely inelastic collisions. Thereafter, *Twomey* [1964, 1966] formulated the RHS of equation (5) by an integral expression analogous to that *Müller* [1928] arrived at. Later on *Warshaw* [1967] derived the same integral terms on similar statistical grounds.

In concentrating to coalescence only and denoting this contribution by γ_{CA}, the time rate of change of the spectral number density function $f(m)$ due to coagulation given by equation (19) can be written, with equation (14a) introduced and then in the loss term contributions only by coalescence via equation (11) considered, as

$$\gamma_{\mathrm{CA}} \equiv \left.\frac{\partial f(m)}{\partial t}\right|_{\mathrm{CA}} = \frac{1}{2}\int_{m'=0}^{\infty}\int_{m''=0}^{\infty} f(m')f(m'')C(m',m'')\delta(m-m'-m'')\mathrm{d}m''\mathrm{d}m' - \int_{m'=0}^{\infty}\int_{m''=0}^{\infty} f(m')f(m'')C(m',m'')\delta(m-m')\mathrm{d}m''\mathrm{d}m' \quad (20)$$

with the coagulation (or collection) kernel $C(m',m'') = K(m',m'')E_{\mathrm{CA}}(m',m'')$. In exploiting the properties of the δ-function, it yields

$$\gamma_{\mathrm{CA}} = \frac{1}{2}\int_{m'=0}^{m} f(m')f(m_c)C(m',m_c)\mathrm{d}m' - f(m) \times \int_{m'=0}^{\infty} f(m')C(m,m')\mathrm{d}m', \quad (21)$$

with $m_c = m - m'$. Because of the symmetry of the integrand, the factor 1/2 in front of the first integral can be avoided by changing the upper limit of this integral to $m/2$.

The first term on the RHS represents the increase of the number of drops with mass m being formed by two coagulating droplets with masses m' and $m - m'$. The second term describes the loss of drops with mass m due to collision and coalescence with any other drop.

This is the famous Smoluchowski equation [cf. also *Chandrasekhar*, 1943], which is used in several disciplines [see for example *Dohnanyi*, 1969, *van Dongen*, 1989, *Spahn et al.*, 2004]. Mathematical considerations have been presented by numerous authors, and we exemplarily refer to the comprehensive and rigorous reviews from *Aldous* [1999], *Norris* [1999], as well as *Fournier and Mischler* [2005].

Equation (21) is sometimes called the "stochastic collection equation" a term which can be misleading, for while it is indeed the mean field representation of a stochastic process, it is purely deterministic. For this reason, the author prefers to refer to equation (21) as the "kinetic coagulation equation." In comprehensively investigating stochasticity, emerging from the fact that spatial homogeneity of a drop population cannot be assumed on small scales in a cloud, *Bayewitz et al.* [1974] and *Gillespie* [1972, 1975a, 1975b] came, as *Warshaw* [1967], to the conclusion that equation (21) is valid only if large cloud volumes are under consideration with drop populations, which are always well mixed (named mean-field hypothesis in more general considerations).

For a detailed discussion of this issue including a review of past interpretations, the interested reader is referred to *Wang et al.* [2006] who, in their paper, also derived an equation, which they call the "true stochastic coalescence equation" (TSCE). An open question is if TSCE is the exact equation encapsulating all the effects impacting a drop population in a strongly turbulent (cloud) environment.

A review of the earlier literature related to the derivation of equation (21) is presented by *Drake* [1972].

4.1. Strict Definitions of Specific Coagulation Processes

In many applications concerning cloud microphysical processes and related parameterizations, one would like to discriminate between cloud droplets and raindrops. The first who considered such a partitioning in a parameterization context was *Kessler* [1969]. He expressed the mutual interaction of cloud droplets resulting in raindrops, which he called "autoconversion," solely in terms of the cloud water mass density. Thus, a definite influence of different shapes of drop spectra, comprising the same mass density, on autoconversion could not be included by this relation. Furthermore, the interaction between cloud droplets and raindrops *Kessler* [in the place cited] denoted "accretion" and arrived at a rate equation whose prerequisites were a monodisperse cloud droplet spectrum and, for raindrops, an exponential function previously formulated by *Marshall and Palmer* [1948].

In his seminal work, *Kessler* [in the place cited] did not consider a specific radius (or mass) of drops separating cloud droplets from raindrops.

With the advent of parameterizations of cloud microphysical processes, used at first in cloud models and later in cloud-resolving numerical weather prediction models, Kessler's parameterization has been tested and refined mostly by varying certain threshold values and by introducing different dependencies into the autoconversion parameterization. A different approach has been presented by *Berry* [1968] for autoconversion, derived from numerical solutions of equation (21). He related the time, at which the so-called predominant radius of the evolving spectrum reaches a value of 40 μm, to the initial number density of the drop spectrum, the total liquid water content, and the initial radius dispersion coefficient and described this as autoconversion. An analysis by *Beheng and Doms* [1986] has shown that Berry's relation has contributions not only from autoconversion but from accretion and cloud droplet self-collection as well, where the definition of "self-collection" is introduced below.

With respect to the description of specific cloud microphysical processes, where cloud droplets and raindrops should be distinguished, it becomes necessary to define a specific separation value for mass (or radius). This value is here denoted by m^* for a separation mass. Note that in the following, no specific numerical value for m^* is considered.

In so doing, two more specific coagulation processes appear besides autoconversion and accretion: self-collection of cloud droplets and of raindrops, following the terminology introduced by *Berry and Reinhardt* [1974]. Herewith, the mutual collisional interaction between, on one hand, cloud droplets is described where the end products have masses $m < m^*$, i.e., they remain cloud droplets and, on the other hand, between raindrops with the end products having masses $m \geq m^*$. A schematic of all specific coagulation processes is given in Figure 2.

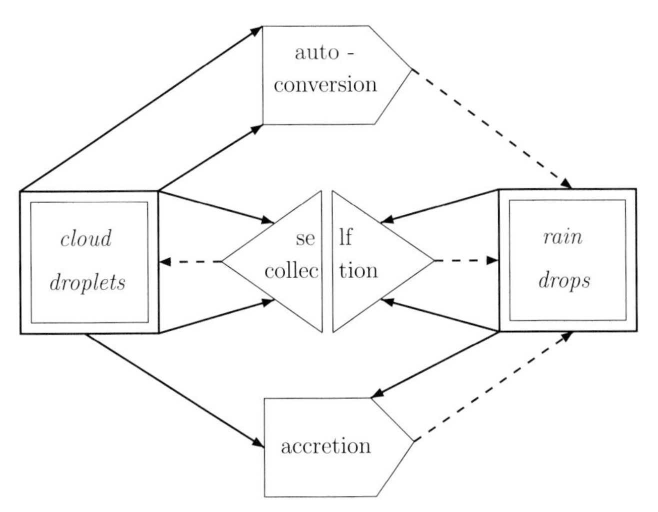

Figure 2. Sketch of specific coagulation processes. Solid arrows indicate partners before binary collision; dashed arrows indicate single partner after collision.

It should be clear already at this point that through the action of these specific coagulation processes, certain integral parameters related to the cloud droplet and raindrop part of the whole drop spectrum are changed as, for instance, mass and number densities. Thus, water mass densities of the component distribution are only affected by autoconversion and accretion, whereas the number densities are altered by self-collection as well.

According to the partitioning in cloud droplets and raindrops, we split the entire size distribution function $f(m)$ formally into two parts: $f_c(m)$ for cloud droplets and $f_r(m)$ for raindrops so that

$$f(m) = f_c(m) + f_r(m), \tag{22}$$

where

$$f_c(m) = \begin{cases} f(m), & m < m^* \\ 0, & m \geq m^* \end{cases} \tag{23a}$$

$$f_r(m) = \begin{cases} f(m), & m \geq m^* \\ 0, & m < m^* \end{cases} \tag{23b}$$

with m^* the separation mass.

With the definition of the specific coagulation processes in mind, the following special mass relations have to be taken into account:

Self-collection of cloud droplets (sc)

$$m', m'' < m^*,$$

where $m' + m'' < m^*$

Autoconversion (au)

$$m', m'' < m^*,$$

where $m' + m'' \geq m^*$

Accretion (ac)

$$m' \geq m^*, m'' < m^*$$

or

$$m' < m^*, m'' \geq m^*,$$

where $m + m'' \geq m^*$

Self-collection of raindrops (sc)

$$m', m'' \geq m^*,$$

where $m' + m'' \geq m^*$.

In the next paragraph, the specific coagulation rates related to the various coagulation processes are presented. They are derived by considering the above conditions and exploiting the properties of the δ-function in equation (20). Note that from now on, we summarize common terms in equation (20) by $H(m',m'') = f(m')f(m'')C(m',m'')$. The rates are termed spectral coagulation rates subsequently.

It should be noticed that the following discussion is taken from the works of *Beheng and Doms* [1986] as well as *Doms and Beheng* [1986]. A further refinement with respect to the cloud droplet range by assuming an additional separation mass $m^{\dagger} < m^*$ for better resolving the impact of autoconversion and self-collection has been presented by *Lüpkes et al.* [1989].

4.2. Spectral Rates of the Different Coagulation Processes

Considering at first the cloud droplet range of the spectrum, i.e., looking at $f(m) \equiv f_c(m)$ for masses $m < m^*$, the integration of equation (23) with respect to m'' yields for the contributions due to self-collection (sc), autoconversion (au), and accretion (ac)

$$\left.\frac{\partial f_c(m)}{\partial t}\right|_{sc} = \frac{1}{2}\int_{m'=0}^{m^*} H(m', m-m')\mathrm{d}m' - \int_{m'=0}^{m^*-m} H(m, m')\mathrm{d}m', \tag{24a}$$

$$\left.\frac{\partial f_c(m)}{\partial t}\right|_{au} = -\int_{m'=m^*-m}^{m^*} H(m, m')\mathrm{d}m', \tag{24b}$$

$$\left.\frac{\partial f_c(m)}{\partial t}\right|_{ac} = -\int_{m'=m^*}^{\infty} H(m, m')\mathrm{d}m'. \tag{24c}$$

The total rate of change of $f_c(m)$ is the sum of these contributions with the number of cloud droplets in any mass interval $[m, m+\mathrm{d}m]$ increasing or decreasing by self-collection, but only decreasing by autoconversion and accretion.

Second, we turn to the raindrop range of the spectrum, i.e., considering $f(m) \equiv f_r(m)$ for masses $m \geq m^*$. In integrating the first term of the RHS of equation (20) with respect to m'', one has to pay attention that the coagulation rates assume different forms if, on one hand, drops are considered for which $m^* \leq m \leq 2m^*$ or, on the other hand, for which $m > 2m^*$. For instance, the number of raindrops with masses $m^* \leq m \leq 2m^*$ cannot increase by self-collection, since coagulation of such drops results always in drops with masses $m > 2m^*$. In contrast, raindrops larger than $2m^*$ cannot be formed by autoconversion.

Thus, from performing integration with respect to m'' of equation (20), the following relations result for the mass interval $m^* \leq m \leq 2m^*$:

$$\left.\frac{\partial f_r(m)}{\partial t}\right|_{sc} = -\int_{m'=m^*}^{\infty} H(m, m')\mathrm{d}m', \tag{25a}$$

$$\left.\frac{\partial f_r(m)}{\partial t}\right|_{au} = \frac{1}{2}\int_{m'=m-m^*}^{m^*} H(m', m-m')\mathrm{d}m', \tag{25b}$$

$$\left.\frac{\partial f_r(m)}{\partial t}\right|_{ac} = \int_{m'=0}^{m-m^*} H(m', m-m')\mathrm{d}m' - \int_{m'=0}^{m^*} H(m, m')\mathrm{d}m', \tag{25c}$$

and for $m > 2m^*$:

$$\left.\frac{\partial f_r(m)}{\partial t}\right|_{sc} = \frac{1}{2}\int_{m'=m^*}^{m-m^*} H(m', m-m')\mathrm{d}m' - \int_{m'=m^*}^{\infty} H(m, m')\mathrm{d}m', \tag{26a}$$

$$\left.\frac{\partial f_r(m)}{\partial t}\right|_{au} = 0, \tag{26b}$$

$$\left.\frac{\partial f_r(m)}{\partial t}\right|_{ac} = \int_{m'=0}^{m^*} H(m', m-m')\mathrm{d}m' - \int_{m'=0}^{m^*} H(m, m')\mathrm{d}m'. \tag{26c}$$

Note that because H is a symmetric function of its arguments in equations (25a)–(25c), the following relations hold for masses $m^* \leq m \leq 2m^*$:

$$\frac{1}{2}\int_{m'=m-m^*}^{m^*} H(m', m-m')\mathrm{d}m' = \int_{m'=m-m^*}^{m/2} H(m', m-m')\mathrm{d}m' \tag{27a}$$

$$\int_{m'=0}^{m-m^*} H(m', m-m')\mathrm{d}m' = \int_{m'=m^*}^{m} H(m', m-m')\mathrm{d}m'. \quad (27b)$$

For an exemplary numerical evaluation of the spectral rates, we refer to section 5.

4.3. Integrals of the Specific Spectral Coagulation Rates

After having specified the contributions by self-collection, autoconversion, and accretion to the time rate of change of the cloud droplet and the raindrop size distribution function, we formulate the rates of change for certain integrals of these rates resulting from the different specific coagulation processes.

In detail, the following integrals are considered:

For cloud droplets only

$$M_c^k = \int_0^{m^*} m^k f(m)\mathrm{d}m = \int_0^{\infty} m^k f_c(m)\mathrm{d}m \quad (28a)$$

For raindrops only

$$M_r^k = \int_{m^*}^{\infty} m^k f(m)\mathrm{d}m = \int_0^{\infty} m^k f_r(m)\mathrm{d}m \quad (29a)$$

where the superscript k indicates an arbitrary non-negative real constant. For integer values of $k \geq 0$, these relations define the kth moments of $f_c(m)$ and $f_r(m)$. In particular, with $k = 0$ and $k = 1$, the number densities N_c and N_r as well as the mass densities (liquid water contents) L_c and L_r, respectively, result. From equations (28a)–(28b) we see

$$M_c^0 \equiv N_c \ , \ M_r^0 \equiv N_r \ , \ M_c^1 \equiv L_c \ , \ M_r^1 \equiv L_r. \quad (29)$$

Note that also non-integer values are permitted for k. For example, $M^{1/3}$ is proportional to the mean radius, and $M^{2/3}$ is related to the mean surface area as well as to the radius variance of the spectrum.

For completeness, we specify equation (5) in moments' form

$$\frac{\partial M^k}{\partial t} + \nabla \cdot (M^k \mathbf{v}) + \nabla \cdot \left[\int_0^{\infty} m^k f(m)\mathbf{v}'(m)\mathrm{d}m\right] = k\int_0^{\infty} f(m)m^{k-1}\,\dot{m}\ \mathrm{d}m + \int_0^{\infty} m^k \gamma\ \mathrm{d}m. \quad (30)$$

In the following, we only deal with number and mass densities and their time rates of change by collisional interaction processes.

Contributions to the evolution of kth moment of the droplet distribution by the various coagulation processes are then arrived at by multiplying equation (20) with m^k and then integrating with respect to m over the cloud droplet or raindrop range.

For the cloud droplets, this leads to

$$C_c^k = \int_{m=0}^{m^*} m^k \frac{\partial f(m)}{\partial t}\mathrm{d}m = \int_{m=0}^{\infty} m^k \frac{\partial f_c(m)}{\partial t}\mathrm{d}m. \quad (31)$$

For the raindrop range, an analogous expression results.

By applying some elementary mathematical operations, the following general integral coagulation rates result:
for cloud droplets

$$C_{c,sc}^k := \frac{\partial M_c^k}{\partial t}\bigg|_{sc} = \frac{1}{2}\int_{m'=0}^{m^*}\int_{m''=0}^{m^*-m'} H(m', m'') \times\left[(m'+m'')^k - 2(m')^k\right]\mathrm{d}m''\mathrm{d}m', \quad (32a)$$

$$C_{c,au}^k := \frac{\partial M_c^k}{\partial t}\bigg|_{au} = -\int_{m'=0}^{m^*}\int_{m''=m^*-m'}^{m^*} H(m', m'')(m')^k\mathrm{d}m''\mathrm{d}m', \quad (32b)$$

$$C_{c,ac}^k := \frac{\partial M_c^k}{\partial t}\bigg|_{ac} = -\int_{m'=0}^{m^*}\int_{m''=m^*}^{\infty} H(m', m'')(m')^k\mathrm{d}m''\mathrm{d}m', \quad (32c)$$

and for raindrops (where we have to take into account equations (25a)–(26a))

$$C_{r,sc}^k := \frac{\partial M_r^k}{\partial t}\bigg|_{sc} = \frac{1}{2}\int_{m'=m^*}^{\infty}\int_{m''=m^*}^{\infty} H(m', m'') \times\left[(m'+m'')^k - 2(m')^k\right]\mathrm{d}m''\mathrm{d}m', \quad (33a)$$

$$C_{r,au}^k := \frac{\partial M_r^k}{\partial t}\bigg|_{au} = \frac{1}{2}\int_{m'=0}^{m^*}\int_{m''=m^*-m'}^{m^*} H(m', m'') \times(m'+m'')^k\mathrm{d}m''\mathrm{d}m', \quad (33b)$$

$$C_{r,ac}^k := \frac{\partial M_r^k}{\partial t}\bigg|_{ac} = \int_{m'=0}^{m^*}\int_{m''=m^*}^{\infty} H(m', m'') \times\left[(m'+m'')^k - (m'')^k\right]\mathrm{d}m''\mathrm{d}m'. \quad (33c)$$

We now turn to specific moments by first setting $k = 0$ and thereafter $k = 1$. The case $k = 0$ refers to the number densities of cloud droplets and raindrops. Specifying $k = 0$ in equations (32a)–(33c) yields

For self-collection

$$\frac{\partial N_c}{\partial t}\bigg|_{sc} = -\frac{1}{2}\int_{m'=0}^{m^*}\int_{m''=0}^{m^*-m'} H(m', m'')\mathrm{d}m''\mathrm{d}m' \quad (34a)$$

$$\left.\frac{\partial N_r}{\partial t}\right|_{sc} = -\frac{1}{2}\int_{m'=m^*}^{\infty}\int_{m''=m^*}^{\infty} H(m',m'')\mathrm{d}m''\mathrm{d}m' \tag{34b}$$

Autoconversion

$$\left.\frac{\partial N_c}{\partial t}\right|_{au} = -\int_{m'=0}^{m^*}\int_{m''=m^*-m'}^{m^*} H(m',m'')\mathrm{d}m''\mathrm{d}m' \tag{35a}$$

$$\left.\frac{\partial N_r}{\partial t}\right|_{au} = -\frac{1}{2}\left.\frac{\partial N_c}{\partial t}\right|_{au} \tag{35b}$$

Accretion

$$\left.\frac{\partial N_c}{\partial t}\right|_{ac} = -\int_{m'=0}^{m^*}\int_{m''=m^*}^{\infty} H(m',m'')\mathrm{d}m''\mathrm{d}m' \tag{36a}$$

$$\left.\frac{\partial N_r}{\partial t}\right|_{ac} = 0. \tag{36b}$$

As expected, the number of cloud droplets tends to be reduced by all three specific coagulation processes. By equation (35b), it is expressed that two colliding cloud droplets result in one raindrop. Thus, autoconversion is the only mechanism whereby one raindrop is newly formed by two colliding cloud droplets. Since accretion does not change the number of raindrops (equation (36b)), self-collection of raindrops is the only process to reduce their number.

As a second case, we set $k = 1$ so that liquid water contents of cloud droplets and raindrops are addressed. According to its definition, self-collection must not change the mass densities, i.e.,

$$\left.\frac{\partial L_c}{\partial t}\right|_{sc} = \left.\frac{\partial L_r}{\partial t}\right|_{sc} = 0. \tag{37}$$

This results clearly with $k = 1$ in equations (32a) and (33a) because $H(m',m'')$ is symmetric.

For the autoconversion and accretion rates, the following relations appear

$$\left.\frac{\partial L_c}{\partial t}\right|_{au} = -\int_{m'=0}^{m^*}\int_{m''=m^*-m'}^{m^*} H(m',m'')m'\mathrm{d}m''\mathrm{d}m', \tag{38a}$$

$$\left.\frac{\partial L_c}{\partial t}\right|_{ac} = -\int_{m'=0}^{m^*}\int_{m''=m^*}^{\infty} H(m',m'')m'\mathrm{d}m''\mathrm{d}m', \tag{38b}$$

$$\left.\frac{\partial L_r}{\partial t}\right|_{au,ac} = -\left.\frac{\partial L_c}{\partial t}\right|_{au,ac}. \tag{39}$$

Summing up the terms of the latter equations expresses conservation of the total liquid water mass density $L = L_c + L_r$.

Given knowledge of the collision and coalescence efficiencies, the partial spectral as well as partial integral rates comprise a closed and complete set of equations allowing to investigate contributions of the different specific coagulation processes in detail (cf. *Beheng and Doms*, 1986, 1990 and *Doms and Beheng*, 1986). Furthermore, the separation mass m^*, not specified in the above derivations, has to be given an appropriate value. In accordance with *Berry* [1968] and led by numerical solutions of the kinetic coagulation equation applying standard collision efficiencies [cf. *Hall*, 1980] as justification, *Beheng and Doms* [1986] have chosen a separation radius $r^* = 40$ μm (cf. section 5 and Figure 4). For a further subdivision, *Lüpkes et al.* [1989] have selected $r^\dagger = 20$ μm as already mentioned above. Both radii r^* and $r^\dagger$ can be calculated from $m^\dagger$ and $m^\dagger$ by assuming spherical drops.

It is emphasized that these formulations can be used as benchmark for parameterizations with respect to the specific coagulation processes.

For rounding up this paragraph, it is noticed that the equations presented have been used to develop parameterizations of the different specific coagulation processes. *Beheng* [1994] as well as *Seifert and Beheng* [2001, 2006] developed a comprehensive scheme for two moments of the cloud droplet and raindrop size spectra, namely, the number and mass densities. There they considered a separation radius of $r^* =$ 40 μm. Also by a different approach, *Khairoutdinov and Kogan* [2000] arrived at parameterizations of the time rates of change of the liquid water contents and number densities of cloud droplets as well as raindrops, however, for autoconversion and accretion only. They applied a separation radius $r^* =$ 50 μm. Both parameterizations are currently widely used in cloud-resolving simulation models.

4.4. Integral Rates of the Total Number and Mass Densities

Next, we move to the time rate of change of the total number and mass densities, which are the sum of the integrated specific coagulation rates. However, the same results also appear if we proceed from the time rate of change of a general moment M^k. Multiplying equation (20) with m^k and then integrating with respect to m over the entire mass interval, we find

$$\int_{m=0}^{\infty} m^k \left.\frac{\partial f(m)}{\partial t}\right|_{CA}\mathrm{d}m = \left.\frac{\partial M^k}{\partial t}\right|_{CA} \equiv C^k$$
$$= \frac{1}{2}\int_{m'=0}^{\infty}\int_{m''=0}^{\infty} H(m',m'')\left[(m'+m'')^k - 2(m')^k\right]\mathrm{d}m''\mathrm{d}m'. \tag{40}$$

From this integral expression, the following relations are easily attained.

1. For coagulation only, for total number density N, i.e., $k = 0$, is

$$C^0\Big|_{CA} = \frac{\partial N}{\partial t}\Big|_{CA} = -\frac{1}{2}\int_{m'=0}^{\infty}\int_{m''=0}^{\infty} H(m', m'')\mathrm{d}m''\mathrm{d}m' < 0. \tag{41a}$$

For total liquid water content L, i.e., $k = 1$, is

$$C^1\Big|_{CA} = \frac{\partial L}{\partial t}\Big|_{CA} = 0. \tag{41b}$$

Thus, the total number density N decreases with time, and the total liquid water content L is conserved. By the way, we note that for $k = 2$

$$C^2\Big|_{CA} = \frac{\partial Z}{\partial t}\Big|_{CA} = \int_{m'=0}^{\infty}\int_{m''=0}^{\infty} H(m', m'')m'm''\mathrm{d}m''\mathrm{d}m' > 0, \tag{42}$$

i.e., the radar reflectivity Z increases with time.

2. For collision-induced breakup only, in anticipation, we refer to equation (48) given below, which is integrated in the same manner as above. We then find for total number density N, i.e., $k = 0$, is

$$C^0\Big|_{BR} = \frac{\partial N}{\partial t}\Big|_{BR} = \frac{1}{2}\int_{m'=0}^{\infty}\int_{m''=0}^{\infty} f(m')f(m'')B(m', m'') \times [N_{BR} - 2]\mathrm{d}m''\mathrm{d}m' \tag{43a}$$

with N_{BR} the average number of fragment drops as previously defined by equation (19). Note that if $N_{BR} = 2$, this is, strictly speaking, no breakup event as a binary collision results again into two final drops after an intermediate unification. However, since in that case two drops may eventually be created with masses very different from those of the original ones, also this process can lead to a modification of the drop size spectrum. As $N_{BR} > 2$ in the majority of cases, the time rate of change of the total drop number density N is mostly positive. for total liquid water content L, i.e., $k = 1$:

$$C^1\Big|_{BR} = 0. \tag{43b}$$

The proof of equation (43b), i.e., mass conservation by collision-induced breakup, is achieved by taking into account equation (17b) in the general moments' form of equation (48) and then setting $k = 1$.

5. A NUMERICAL CASE STUDY

An illustration and a short discussion of numerical results of a case study concerning the spectral rates and their integrals due to the different coagulation processes described so far is presented in this section. For a complete description and analysis, the interested reader is referred to the works of *Beheng and Doms* [1986, 1990].

The initial droplet size distribution function $f_0(r)$ is given by

$$f_0(r) = Ar^{\alpha-1}exp(-\beta r) \tag{44}$$

with

$$A = N_{c,0}\beta^{\alpha}/\Gamma(\alpha), \tag{45}$$

$$\beta = \{3\bar{x}_0/[4\pi\bar{\rho}_w(\alpha+2)(\alpha+1)\alpha]\}^{-1/3}, \tag{46}$$

$$\alpha^2 = \nu_0, \tag{47}$$

with $\bar{\rho}_w$ bulk density of water, $\bar{x}_0 = L_{c,0}/N_{c,0}$ mean initial drop mass, $\Gamma()$ gamma function, and ν_0 initial radius dispersion coefficient. The following parameters have been chosen: $L_{c,0} = 10^{-6}$ g cm^{-3}, $N_{c,0} = 100$ cm^{-3}, and $\nu_0 = 0.25$ such that the mean initial drop radius $\bar{r}_0$ that is proportional to $\bar{x}_0^{1/3}$ yields 13.4 μm. With these parameters, a moderate continental drop spectrum is prescribed, which is shown in Figure 3 in normalized form $f_{0,N}(r) = f_0(r)/N_{c,0}$. The function $f(r)$ can be transformed to $f(m)$ as given in section 2.

The collision kernel applied is given by equation (51), where detailed information on the terminal fall velocities and collision efficiency as well as on the numerical solution procedure are found in the works of *Beheng and Doms* [1986, 1990].

In Figure 4 the time evolution of the drop spectrum by numerically solving equation (21) is shown after 200 s model simulation time each, up to a maximum time of 2000 s. Note that breakup is neglected in these simulations. It is clearly seen that in the first hundred seconds, drops with increasing size are formed quickly. This can be addressed, as will be demonstrated later, to the combined action of self-collection of cloud droplets and autoconversion. With increasing time, accretion reduces more and more mass of cloud droplets in favor of the growth of larger raindrops. During the simulation, a distinct minimum evolves at about a radius of 40 μm, a feature which conforms to the work of *Berry* [1968]. The stationarity of this minimum at just this radius has been a justification of choosing this value as radius r^* separating (smaller) cloud droplets from (larger) raindrops (cf. section 4.3). At the end of the simulation, most drop mass is concentrated in raindrops.

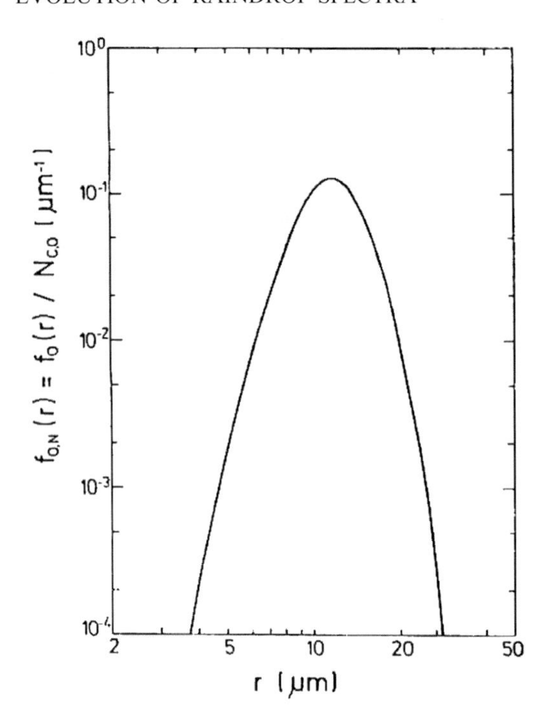

Figure 3. Normalized initial drop size distribution $f_{0,N}(r) = f_0(r)/N_{c,0}$ in μm^{-1} versus drop radius r in µm with parameters given in the text.

For a deeper understanding of the impact of the different coagulation processes, we first present number density conversion rates due to self-collection of cloud droplets and raindrops, respectively, (equations (34a), (34b)), autoconversion (equation (35a)) and accretion (equation (36a)) as a function of simulation time in Figure 5a. The formation rate of raindrops by autoconversion is omitted but can be deduced from the corresponding loss rate of cloud droplets (equation (35b)). One can see that in the beginning, the loss rate of cloud droplets by self-collection is the strongest by orders of magnitude compared to those by autoconversion and accretion. It is interesting to note that the number density loss rate due to autoconversion is small and remains small. However, it is emphasized that autoconversion is crucial because raindrops are created exclusively by this process. When after some hundreds of seconds raindrops are existent in appreciable number, accretion starts to become the dominant loss rate by efficiently collecting cloud droplets. Toward the end of the simulation, all loss rates decrease.

The effect of all number density loss rates on the time rate of change of the total cloud droplet and raindrop number densities is depicted in Figure 5b. The decrease of cloud droplets in number is at first only very slow and becomes significant not until raindrops have been formed in appreciable number. This feature can be traced back to accretion whose rate approaches its extreme value at about the same simulation time when the raindrop number attains its maximum.

Next, we turn to the mass density conversion rates due to autoconversion (equation (38a)) and accretion (equation (38b)) presented in Figure 6a. Compared to Figure 5b, we see a crude qualitative agreement between the number and mass density loss rates. The largest difference appears at the beginning of the simulation in that the loss rate by autoconversion is considerably stronger than that by accretion. In Figure 6b, the decrease of the cloud water (descending curve) is exactly compensated by the increase in rainwater mass (ascending curve), which is self-evident, since the total water mass density is conserved.

For a still closer look into the coagulation process, we analyze the spectral time rates of change of the mass density spectrum as affected by the various coagulation mechanisms. We begin with self-collection in the cloud droplet range (equation (24a)) presented in Figure 7. As expected, one sees a transport of smallest droplets in favor of the formation of larger ones. This rate is for droplet radii $r \leq 20$ µm, nearly the same for simulation times $t \lesssim 900$ s and thereafter decreases sharply. The reason for this behavior is found in the accretion mechanism by which for $t > 900$ s, small droplets are efficiently captured by raindrops thus leaving only a smaller number of cloud droplets for self-collection. For radii $r > 20$ µm and especially near $r^* = 40$ µm, a strong increase appears with time showing the importance of self-collection to create droplets with a size favorable to be converted to small raindrops by autoconversion.

The spectral dependence of autoconversion is shown in Figure 8. The primary maximum is always located near $r^* = 40$ µm, and the temporal sequence is a consequence of self-collection producing cloud droplets near this radius. Thus,

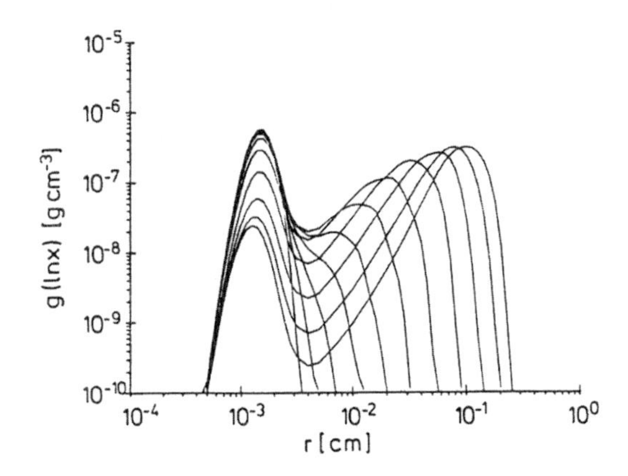

Figure 4. Mass density distribution function g (ln x) in g cm^{-3} as function of drop radius r in centimeters for simulation times $t = 0$, 200, 400, . . ., 2000 s. The variable x (in g (ln x)) is identical to mass m as used in the present study.

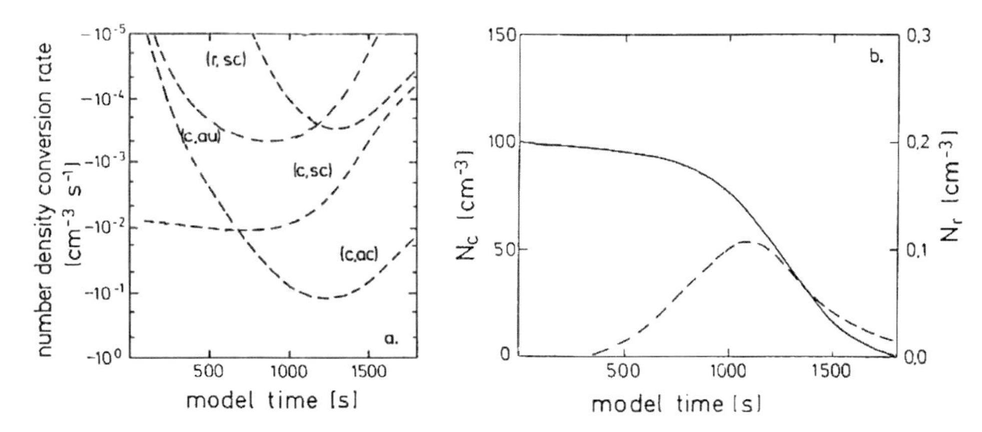

Figure 5. (a) Number density conversion rates in cm^{-3} s^{-1} as function of model time in s, specifically the loss rate of cloud droplet number by self-collection (indicated by (c,sc); equation (34a)), by autoconversion (c,au; equation (35a)), and by accretion (c,ac; equation (36a)) as well as loss rate of raindrop number by raindrop self-collection (r,sc; equation (34b)). (b) Cloud droplet number density N_c (solid line, left ordinate) and raindrop number density N_r (dashed line, right ordinate) in cm^{-3} as function of simulation time in s. Note different ordinate scales.

we recognize an intimate connection between autoconversion and self-collection with the effectivity of the last mechanism a necessary prerequisite of the former one.

Accretion clearly depletes the mass (and number) of cloud droplets after raindrops have formed (Figure 9). The shape of the spectral loss rates obviously reflects the shape of the

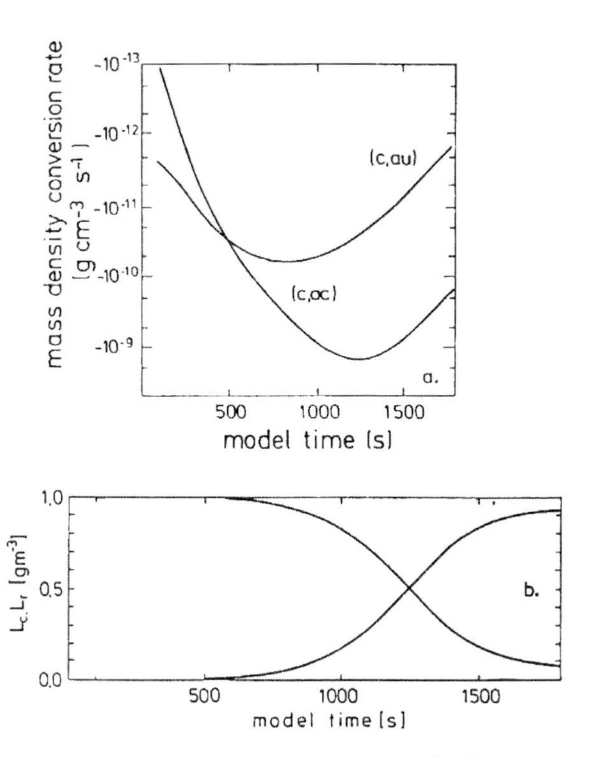

Figure 6. (a) Mass density conversion rates in g cm^{-3} s^{-1} as function of model time in s, specifically the loss rate of cloud water by autoconversion (c,au; equation (38a)) and by accretion (c,ac; equation (38b)). (b) Cloud water (L_c, descending curve) and rainwater (L_r, ascending curve) in g m^{-3} as function of model time in s.

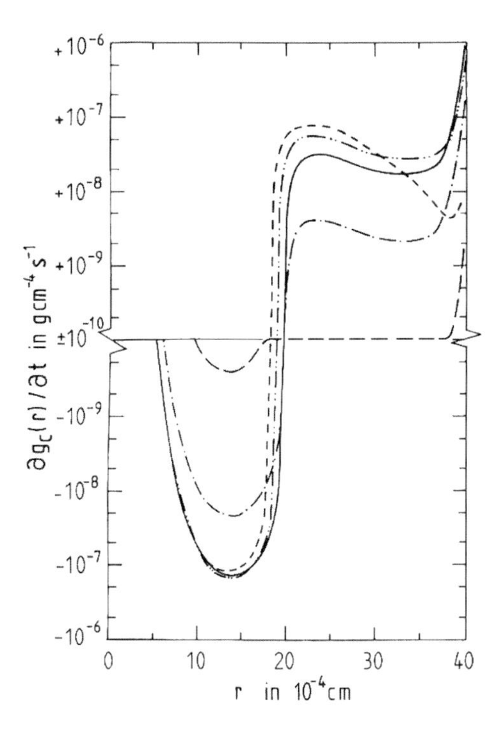

Figure 7. Time rate of change of the cloud droplet mass density spectrum $\partial g_c(r)/\partial t$ in g cm^{-4} s^{-1} due to self-collection (equation (30)) as a function of drop radius r in 10^{-4} cm for simulation times t = 100 s (short-dash line), 500 s (dash and double-dotted line), 900 s (solid line), 1300 s (dash and single-dotted line), and 1700 s (long-dash line).

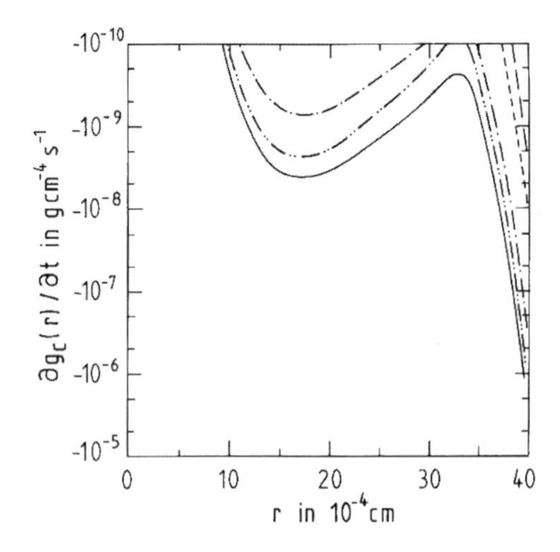

Figure 8. Same as Figure 7 but for autoconversion (equation (24b)).

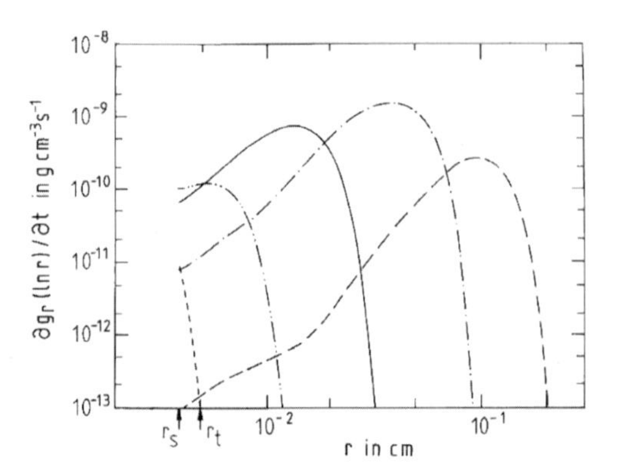

Figure 10. Time rate of change of the raindrop mass density spectrum $\partial g_r(\ln r)/\partial t$ in g cm^{-3} s^{-1} due to both autoconversion and accretion (equations (25b), (25c), and (26c)) as a function of drop radius r in cm for same simulation times as in Figure 7. The specific radius values $r_s \equiv r^* \propto (m^*)^{1/3} = 40$ μm and $r_t \equiv 2^{1/3}r^* \propto (2m^*)^{1/3} = 50$ μm that are referred to in the text are indicated.

cloud droplet part of the spectrum itself (cf. Figure 4), which is a kind of self-similarity. Reasons for this appearance are discussed by *Beheng and Doms* [1990]. The magnitude of this rate increases in this case study up to about 1300 s, which is coupled to the strong increase in rainwater content during this time (cf. Figure 6b).

Figure 10 is dedicated to the spectral rates due to the sum of both autoconversion and accretion in the raindrop range. In the figure, the range between radii $r_s \equiv r^*$ and $r_t \equiv 2^{1/3}r^* \propto (2m^*)^{1/3}$, where quite small raindrops are situated, is the only one where the autoconversion rate is positive with negligible negative contributions by self-collection and accretion. For $r > r_t$, where solely accretion contributes to Figure 10 (cf. equation (26b)), it has been found out [*Beheng and Doms*, 1990] that this growth mode behaves with high precision like a continuous growth mode as originally formulated by *Kessler* [1969]. However, in comparison to the numerical solution of equation (21), continuous (accretional) growth alone is not able to explain the accelerated growth of large raindrops, which is due to the self-collection mechanism of raindrops (see below). The spectral accretion rate is dominant up to simulation times of 1000 s. Thereafter, accretion is in competition with self-collection of raindrops whose spectral rate is shown in Figure 11. A certain qualitative similarity with

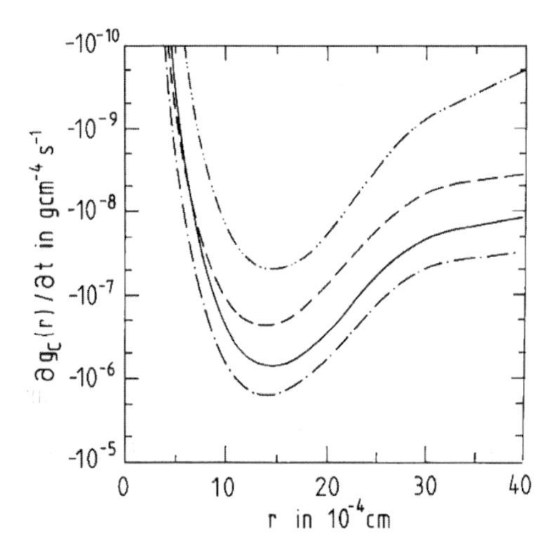

Figure 9. Same as Figure 7 but for accretion (equation (24c)) without the curve for $t = 100$ s.

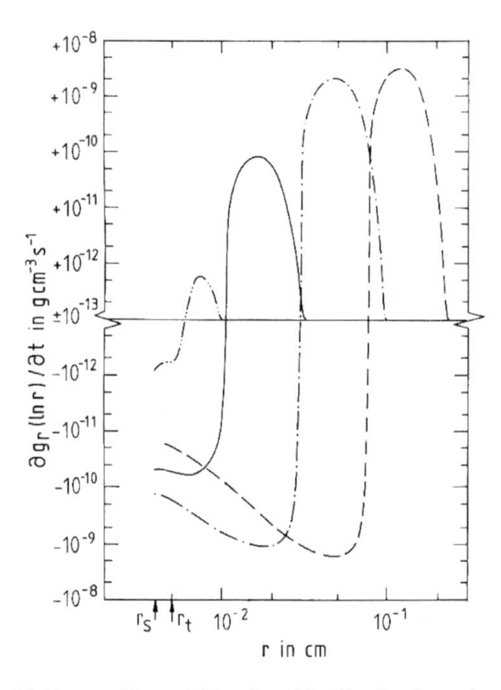

Figure 11. Same as Figure 10 but for self-collection (equations (25a), (26a)) without the curve for $t = 100$ s.

the spectral self-collection rate of cloud droplets (Figure 7) is striking and expresses again a transfer of smaller (rain)drops in favor of larger ones. However, in contrast to Figure 7, the zero of this rate is not stationary but moves toward larger radii with increasing simulation time. For simulation times larger than 1000 s, this rate is the largest of all coagulation mechanisms acting in the raindrop range.

We close this section by stating that this numerical case study has exemplarily demonstrated how the different coagulation mechanisms, self-collection of cloud droplets and raindrops, autoconversion, and accretion, interact and are partly prerequisites for processes becoming effective in later stages of coagulative drop growth.

6. BREAKUP

It is a well-known fact that raindrops are limited in size: Raindrops larger than about 6 mm in diameter are hardly observed at the surface. Larger raindrops are found in tropical clouds only [*Hobbs and Rangno*, 2004]. As responsible for this limitation, two mechanisms are usually assumed: the first involves the collision of two raindrops, which may temporarily unite but then fragmentate, i.e., collision-induced breakup (cf. *Pruppacher and Klett*, 1997); the other process is hydrodynamic breakup where a single very large drop becomes hydrodynamically unstable. In this case, the surface tension becomes insufficient to maintain a continuous surface of such a large drop, and the inflated surface breaks whereby the drop fragments into very many tiny droplets. Because this second process concerns very large drops only, which because of the first process are anyway very rare, it is not further discussed.

It should be noticed that large liquid drops may not only be formed by the warm rain process but also by melting of small hail (graupel) or by shedding from melting hailstones.

An extensive review on raindrop size distributions and their evolution is given by *McFarquhar* [2010]. Here we present only a brief description of the formulation of collision-induced breakup.

Breakup processes are formally included in the RHS of equation (5) and in equation (19). Introducing equation (14b) in equation (19) and indicating this contribution by γ_{BR}, collision-induced breakup can be formulated according to *List and Gillespie* [1976], *Gillespie and List* [1978/79], as well as *Hu and Srivastava* [1995] by

$$\gamma_{BR} \equiv \frac{\partial f(m)}{\partial t}\bigg|_{BR} = \int_{m'=0}^{\infty} \int_{m''=0}^{m'} f(m') f(m'') B(m', m'') P(m; m', m'') \mathrm{d}m'' \mathrm{d}m' - f(m) \int_{m'=0}^{\infty} f(m') B(m, m') \mathrm{d}m' \tag{48}$$

or equivalently by

$$\gamma_{BR} \equiv \frac{\partial f(m)}{\partial t}\bigg|_{BR} = \int_{m'=0}^{\infty} \int_{m''=0}^{m'} f(m') f(m'') B(m', m'') P(m; m', m'') \mathrm{d}m'' \mathrm{d}m' - \int_{m'=0}^{\infty} \frac{f(m) f(m') B(m, m')}{m + m'} \int_{m''=0}^{m+m'} m'' P(m''; m, m') \mathrm{d}m'' \mathrm{d}m' \tag{49}$$

with the breakup kernel

$$B(m, m') = K(m, m') E_{BR}(m, m'), \tag{50}$$

where $E_{BR}(m,m') = 1 - E_{CA}(m',m'')$ (cf. equation (11)), $K(m, m')$ the collision kernel (see section 7), and the fragment size distribution function $P(m:m',m'')$, which is defined roughly by equation (16).

In equation (49), the term describing mass conservation (cf. equation (17b)) has been introduced, i.e., all masses of the fragment drops sum up to the masses of the original two colliding drops. Clearly, this term can be omitted in equation (49), since it represents a multiplication of the last term of equation (48) by unity only. However, a numerical solution of this equation is facilitated if solved in the form of equation (49) as shown by *Gillespie and List* [1978/79], provided equation (17b) is exactly fulfilled [cf. *Feingold et al.*, 1988 as well as *Hu and Srivastava*, 1995].

For a comprehensive discussion of the fragment size distribution $P(m';m,m'')$, refer to the works of *Low and List* [1982a, 1982b], *McFarquhar* [2004], *Schlottke et al.* [2010], and *Straub et al.* [2010]. Furthermore, *McFarquhar* [2010] presents more details on collision-induced breakup with its parameterizations and on measurements of raindrop size spectra accompanied by a comparison of modeled and observed raindrop size distributions.

7. COLLISION KERNELS

As has been stated previously, the entire physics describing the collision of two drops is captured by the collision kernel. This function, thus, describes the condition that in a given time increment, two drops come into contact. Whether then a permanent unification of the drops appears is determined by the coalescence efficiency.

It is reasonable to discriminate the characteristics of collision process as a function of the forces acting on the respective partners and on the flow field wherein the drops are embedded.

If, on one hand, droplets are quite small (as fog droplets), their mutual collisions can be described in analogy to diffusion of small particles onto the surface of larger ones as for Brownian diffusion [cf. *Chandrasekhar*, 1943]. Then, the force due to gravity is negligible as is the effect of a laminar or turbulent flow field, since the small droplets follow immediately the ambient flow because of their small inertia. This process is relevant for small droplets with diameters roughly smaller than one micron, say. Diffusive coagulation is then proportional to the size-dependent diffusivity, which is a function of mobility (Einstein/von Smoluchowski relation), and mobility is in first approximation proportional to the drag exerted on a sphere with specific radius moving in a viscous medium. Since we are dealing in this review with raindrops, for which other forces become more and more dominant for mutual collisions, we do not discuss this situation further.

For larger droplets with diameters larger than about 100 μm and raindrops, a collision event between two differently sized larger hydrometeors, approaching each other, is related to the volume swept out by the two partners. This volume depends on the relative velocity of both drops. The (stationary) settling velocities of large drops result, in contrast to the diffusive motion of small droplets, from the equilibrium of the viscous force (drag), gravity, and buoyancy without any turbulence influence and is a function of drop diameter [*Pruppacher and Klett*, 1997]. The efficacy of such a collision event is usually measured by the collision kernel (see below for the definition).

In between these extremes, the motion of large droplets/drizzle drops with diameters of about some microns to 100 μm has to be considered as already affected considerably by gravity, but their relative smallness makes them susceptible to a turbulent flow field surrounding the collision partners.

We begin with the case of colliding raindrops, assumed spherical, such that any influence of turbulence can be neglected, i.e., assuming a laminar flow field. In assuming pure gravitational settling, the collision kernel (or the swept-out volume per unit time) is given by

$$\begin{aligned} K_g(m', m'') &= \sigma_{\text{act}} |v_T(m') - v_T(m'')| \\ &= \sigma_{\text{geo}} E_{\text{CL}}(m', m'') |v_T(m') - v_T(m'')| \\ &= \pi(r' + r'')^2 E_{\text{CL}}(r', r'') |v_T(r') - v_T(r'')| \\ &= K_g(r', r'') \end{aligned} \quad (51)$$

with σ_{act} and σ_{geo} the actual and geometric collision cross-sections, respectively, see equation (12). The collision efficiency $E_{\text{CL}}(m',m'')$ comprises all influences due to effects of the flow fields surrounding the approaching drops. Equation (51) is ceteris paribus equivalent to that used in kinetic gas theory for elastic scattering by single collision partners.

Note that the last expression in equation (51) is defined in terms of radii instead of masses. This is valid if there is an unambiguous relation between mass and radius as for drops with assumed spherical shape.

It is added that omitting E_{CL} in equation (51) defines the so-called "geometric" collision kernel

$$\begin{aligned} K_g^{\circ}(m', m'') &= \pi(r' + r'')^2 |v_T(r') - v_T(r'')| \\ &= K_g^{\circ}(r', r'') \end{aligned} \quad (52)$$

valid for drops moving in stagnant air with no mutual interactions of their flow fields.

Earliest attempts to deduce numerical values of the collision kernel defined by equation (51) relied on solving the Navier-Stokes equation for drops moving in a laminar flow field. Thereby, different mechanisms as deflection and wake capture could be taken into account. For details see the work of *Pruppacher and Klett* [1997].

An important aspect turns out if relatively large droplets (tens of microns in radius) undergo collisions. Owing to these collisions, the first raindrops are created, and then a turbulent flow field ubiquitous in clouds may affect a collision outcome considerably. This process is the subject of much current research [cf. *Pinsky et al.*, 2000, *Vaillancourt and Yau*, 2000, *Falkovich et al.*, 2002, *Shaw*, 2003, *Wang et al.*, 2005, *Khain et al.*, 2007, and others]. However, also in the past, scientists were aware of the effects due to the turbulent flow field surrounding cloud droplets and raindrops [*East and Marshall*, 1954; *Saffman and Turner*, 1955, 1988]. In reviewing this literature, it proves useful to categorize different types of contributions.

1. Theoretical descriptions are the goal of the early papers of *East and Marshall* [1954], *Saffman and Turner* [1955], as well as of some contributions in the more recent literature by *Sundaram and Collins* [1997], *Gosh et al.* [2005];

2. Numerical simulations that seek to quantify the effects of turbulence in flow fields with varying complexity are described in many recent papers [*Wang and Maxey*, 1993; *Pinsky and Khain*, 1997a, 1997b; *Pinsky et al.*, 2000; *Chun et al.*, 2005; *Franklin et al.*, 2005; *Pinsky et al.*, 2006; *Bourgoin et al.*, 2006; *Franklin et al.*, 2007; *Wang et al.*, 2008; *Grabowski and Wang*, 2009]; and

3. Observational attempts to investigate the prevalence and role of turbulence are described by *Vohl et al.* [1999]; *Siebert et al.* [2006b, 2006a]; *Lehmann et al.* [2007]; *Warhaft* [2009]; *Siebert et al.* [2010].

It is remarked that the numerical simulations, which are often used to explore such effects, the intensity of the turbulence can be measured by the Taylor microscale Reynolds number Re_λ. Because of computational limitations, simulations can only attain values of $Re_\lambda \leq 500$; however, in clouds, values up to 10^4 appear. Also, different numerical methods like LES and DNS have been and are being applied to these problems.

Related to turbulence, the subsequent mechanisms and effects have been hypothesized in literature to influence (and mostly enhance) collision of large cloud droplets [*Khain et al.*, 2007]: (1) enhanced relative motion following from differential acceleration and shear effects acting on droplets (turbulent transport effect), (2) increased average pair number densities resulting from enhanced local concentrations of droplets (preferential concentration effect), and (3) enhancement by variations of settling velocities due to turbulence modifying the collision efficiency (hydrodynamic interaction effect).

Analogously to equation (51), the turbulent collision kernel, which is here written in terms of drop radii instead of drop masses, is defined as based on kinematic pair statistics [*Sundaram and Collins*, 1997; *Grabowski and Wang*, 2009]

$$K_t(r',r'') = 2\pi R^2 \langle |w_r|(\Delta = R)\rangle g_{12} \times (\Delta = R) E_{CL}(r',r'')\eta_E \tag{53}$$

with $R = r' + r''$, w_r = radial relative velocity for two droplets separated by the center-to-center distance $\Delta = |\mathbf{\Delta}|$ and g_{12} = radial distribution function comprising the effect of preferential number densities on the pair number density at distance Δ, i.e., g_{12} is the ratio of actual pair number density to that in a uniform suspension and describes clustering. The angular brackets denote the average over all possible directions of $\mathbf{\Delta}$. The enhancement factor η_E is the ratio of the turbulent collision efficiency to the collision efficiency found for drops moving in quiescent air [*Wang et al.*, 2005].

If no turbulence effects are taken into account, and hydrodynamic forces are neglected by setting $E_{CL} = 1$, equation (53) reduces to the geometric collision kernel equation (52) [cf. *Saffman and Turner*, 1955] by

$$\begin{aligned} \langle |w_r|\rangle &\equiv \frac{1}{2}|v_T(r') - v_T(r'')| \quad , \\ g_{12} \equiv 1 \quad &\rightarrow \quad \overset{\circ}{K_g}(r',r'') = \pi R^2 \; |v_T(r') - v_T(r'')|, \end{aligned} \tag{54}$$

where $v_T(r')$,$v_T(r'')$ are the terminal fall velocities of the droplets with radii r',r'', respectively.

In the past and in ongoing research, great effort has been, and is being, devoted to the determination of the relative distribution function $\langle |w_r|\rangle$ and the preferential number densities g_{12}. Reliable numerical values of these functions for the conditions met in vigorous clouds are not at hand today. However, there are some hints that the application of turbulent collision kernels leads to an early broadening of drop size spectra as occasionally observed in clouds.

We close this section by a short remark on the breakup kernel $B(m,m') = K(m,m')E_{BR}(m,m')$ (cf. equation (50)), which are relevant ingredients for calculation of effects of collision-induced breakup on the temporal development of drop size distributions (cf. equations (48) or (49)). Details on both functions are presented in the works of *Low and List* [1982a, 1982b], *McFarquhar* [2004, 2010], *Schlottke et al.* [2010], and in *Straub et al.* [2010], wherein laboratory and numerical experiments are described and the corresponding data are evaluated with the aim to provide parameterizations of the breakup efficiency E_{BR} and the fragment size distribution function P.

8. SUMMARY

In this study, some elementary definitions and relations concerning the description of drop size distributions are presented. This comprises the derivation of the spectral balance equation for a size spectrum and related moments such as number and mass densities. Emphasis is given to the formulation of collisional interaction of drops, which can be twofold: on one hand, coagulation as a result of collision and subsequent coalescence (permanent unification) and, on the other hand, breakup arising from collision but subsequent disintegration of an interim coalesced system.

The process of raindrop formation, expressed by the kinetic coagulation equation, is then split into a number of partial mechanisms appearing if cloud droplets and raindrops are distinguished. These mechanisms are self-collection of cloud droplets and raindrops, autoconversion, and accretion for which stringent definitions are developed. They apply to spectral rates as well as integral rates for number and mass densities of cloud droplets and raindrops. Taking them all together, it is demonstrated that in case of pure coalescence, the total number density is decreasing and the total mass density conserved.

For collision-induced breakup of raindrops, spectral as well as integral rates of change are formulated for the total drop number and mass densities. Besides the expected result of mass conservation, it turns out that the number of cloud droplets and, under certain circumstances, small raindrops is increased by collision-induced breakup at the expense of the number of raindrops which decreases.

Finally, collision kernels comprising the intrinsic physics of binary collisions of drops are addressed. Depending on the sizes of the respective collision partners, the pure gravitational kernel for raindrops is relevant as well as a kernel concerning drop collisions in a turbulent environment. At the end, a remark is made on the breakup kernel and the fragment size distribution.

Acknowledgments. The text presented here is a compilation of both well-known relations but also of research pursued by the author since years. This has been made possible through collaboration with many esteemed colleagues, the prominent ones are F. Herbert, U. Wacker, and A. Seifert as well as G. Doms who died too young. Some of the research results shown have been supported by several grants of the German Research Foundation (Deutsche Forschungsgemeinschaft), which is gratefully acknowledged. Last but not the least, I am very thankful to an anonymous reviewer, to Bjorn Stevens for a constructive and extensive review as well as to Ulrich Blahak who all helped to improve the text considerably.

REFERENCES

Aldous, D. J. (1999), Deterministic and stochastic models for coalescence (aggregation and coagulation): A review of the mean-field-theory for probabilists, *Bernoulli*, *5*, 3–48.

Bayewitz, M. H., J. Yerushalmi, S. Katz, and K. Shinnar (1974), The extent of correlations in a stochastic coalescence process, *J. Atmos. Sci.*, *31*, 1604–1614.

Beheng, K. D. (1982), A numerical study on the combined action of droplet coagulation, ice particle riming and the splintering process concerning maritime cumuli, *Contrib. Atmos. Phys.*, *55*, 201–214.

Beheng, K. D. (1994), A parameterization of warm cloud microphysical conversion processes, *Atmos. Res.*, *33*, 193–206.

Beheng, K. D., and G. Doms (1986), A general formulation of collection rates of cloud and raindrops using the kinetic equation and comparison with parameterizations, *Contrib. Atmos. Phys.*, *59*, 66–84.

Beheng, K. D., and G. Doms (1990), The time evolution of a drop spectrum due to collision/coalescence: A numerical case study on the effects of selfcollection, autoconversion and accretion, *Meteorol. Rundsch.*, *42*, 52–61.

Beheng, K. D., and F. Herbert (1986), Mathematical studies on the aerosol concentration in drops changing due to particle scavenging and redistribution by coagulation, *Meteorol. Atmos. Phys.*, *35*, 212–219.

Berry, E. X. (1968), Modification of the warm rain process, *Proc. First Natl. Conf. Wea. Modification*, pp. 81–85, Am. Meteorol. Soc., Boston, Mass.

Berry, E. X., and R. L. Reinhardt (1974), An analysis of cloud drop growth by collection: Part I. Double distributions, *J. Atmos. Sci.*, *31*, 1814–1824.

Bourgoin, M., N. Ouelette, H. Xu, J. Berg, and E. Bodenschatz (2006), The role of pair dispersion in turbulent flow, *Science*, *311*, 835–838.

Chandrasekhar, S. (1943), Stochastic problems in physics and astronomy, *Rev. Mod. Phys.*, *15*, 1–89.

Chun, J., D. Koch, S. Rani, A. Ahluwalia, and L. Collins (2005), Clustering of aerosol particles in isotropic turbulence, *J. Fluid Mech.*, *536*, 219–251.

Dohnanyi, J. S. (1969), Collisional model of asteroids and their debris, *J. Geophys. Res.*, *74*, 2531–2554.

Doms, G., and K. D. Beheng (1986), Mathematical formulation of self-collection, autoconversion and accretion rates of cloud and raindrops, *Meteorol. Rundsch.*, *39*, 98–102.

Drake, R. (1972), A general mathematical survey of the coagulation equation, in *Topics in Current Aerosol Research*, edited by G. Hidy, and J. Brock, pp. 201–376, Pergamon, Oxford.

East, T. W. R., and J. S. Marshall (1954), Turbulence in clouds as a factor in precipitation, *Q. J. R. Meteorol. Soc.*, *80*, 26–47.

Falkovich, G., A. Fouxon, and M. G. Stepanov (2002), Acceleration of rain initiation by cloud turbulence, *Nature*, *419*, 151–154.

Feingold, G., S. Tzivion, and Z. Levin (1988), Evolution of raindrop spectra. Part I: Solution to the stochastic collection/breakup equation using the method of moments, *J. Atmos. Sci.*, *45*, 3387–3399.

Fournier, N., and S. Mischler (2005), A spatially homogeneous Boltzmann equation for elastic, inelastic and coalescing collisions, *J. Math. Pures Appl.*, *84*, 1173–1234.

Franklin, C., P. Vaillancourt, M. Yau, and P. Bartello (2005), Collision rates of cloud droplets in turbulent flow, *J. Atmos. Sci.*, *62*, 2451–2466.

Franklin, C., P. Vaillancourt, and M. Yau (2007), Statistics and parameterizations of the effect of turbulence on the geometric collision kernel of cloud droplets, *J. Atmos. Sci.*, *64*, 938–954.

Gillespie, D. T. (1972), The stochastic coalescence model for cloud droplet growth, *J. Atmos. Sci.*, *29*, 1496–1510.

Gillespie, D. T. (1975a), Three models for the coalescence growth of cloud drops, *J. Atmos. Sci.*, *32*, 600–607.

Gillespie, D. T. (1975b), An exact method for numerically simulating the stochastic coalescence process in a cloud, *J. Atmos. Sci.*, *32*, 1977–1989.

Gillespie, J. R., and R. List (1978/79), Effects of collision-induced breakup on drop size distributions in steady state rainshafts, *Pure Appl. Geophys.*, *117*, 599– 626.

Gosh, S., J. Dávila, J. Hunt, A. Srdic, H. Fernando, and P. Jonas (2005), How turbulence enhances coalescence of settling particles with applications to rain in clouds, *Proc. R. Soc. A.*, *461*, 3059–3088.

Grabowski, W. W., and L.-P. Wang (2009), Diffusional and accretional growth of water drops in a rising adiabatic parcel: Effects of the turbulent collision kernel, *Atmos. Chem. Phys.*, *9*, 2335–2353.

Hall, W. D. (1980), A detailed microphysical model within a two-dimensional dynamical framework: Model description and preliminary results, *J. Atmos. Sci.*, *37*, 2486–2507.

Hobbs, P. V., and A. L. Rangno (2004), Super-large raindrops, *Geophys. Res. Lett.*, *31*, L13102, doi:10.1029/2004GL020167.

Hu, Z., and R. Srivastava (1995), Evolution of raindrop size distribution by coalescence, breakup, and evaporation: Theory and observations, *J. Atmos. Sci.*, *52*, 1761–1783.

Kessler, E. (1969), *On the Distribution and Continuity of Water Substance in Atmospheric Circulations*, Meteor. Monogr., 32, 84 pp., Am. Meteorol. Soc., Boston, Mass.

Khain, A., M. Pinsky, T. Elperin, N. Kleeorin, I. Rogachechvskii, and A. Kostinski (2007), Critical comments to results of investigations of drop collisions in turbulent clouds, *Atmos. Res.*, *86*, 1–20.

Khairoutdinov, M., and Y. Kogan (2000), A new cloud physics parameterization in a large-eddy simulation model of marine stratocumulus, *Mon. Weather Rev.*, *128*, 229–243.

Lehmann, K., H. Siebert, M. Wendisch, and R. A. Shaw (2007), Evidence for inertial droplet clustering in weakly turbulent clouds, *Tellus, Ser. B*, *59*, 57–65.

List, R., and J. R. Gillespie (1976), Evolution of raindrop spectra with collision-induced breakup, *J. Atmos. Sci.*, *33*, 2007–2013.

Locatelli, J. D., and P. V. Hobbs (1974), Fall speeds and masses of solid precipitation particles, *J. Geophys. Res.*, *79*, 2185–2197.

Low, T., and R. List (1982a), Collision, coalescence and breakup of raindrops, Part I: Experimentally established coalescence efficiencies and fragment size distributions in breakup, *J. Atmos. Sci.*, *39*, 1591–1606.

Low, T., and R. List (1982b), Collision, coalescence and breakup of raindrops, Part II: Parameterization of fragment size distributions, *J. Atmos. Sci.*, *39*, 1607–1618.

Lüpkes, C., K. D. Beheng, and G. Doms (1989), A parameterization scheme for simulating collision/coalescence of water drops, *Contrib. Atmos. Phys.*, *62*, 289–306.

Marshall, J., and W. Palmer (1948), The distribution of raindrops with size, *J. Meteorol.*, *5*, 165–166.

McFarquhar, G. M. (2004), A new representation of collision-induced breakup of raindrops and its implications for the shape of raindrop size distributions, *J. Atmos. Sci.*, *61*, 777–794.

McFarquhar, G. M. (2010), Raindrop size distribution and evolution, in *Rainfall: State of the Science*, *Geophys. Monogr. Ser.*, doi:10.1029/2010GM000971, this volume.

Müller, H. (1928), Zur allgemeinen Theorie der raschen Koagulation, *Kolloidchem. Beihefte*, *27*, 223–250.

Norris, J. R. (1999), Smoluchowski's coagulation equation: Uniqueness, nonuniqueness and a hydrodynamic limit for the stochastic coalescent, *Ann. Appl. Probab.*, *9*, 78–109.

Pinsky, M., and A. Khain (1997a), Turbulence effects on the collision kernel: Part I: Formation of velocity deviations of drops falling within a turbulent three-dimensional flow, *Q. J. R. Meteorol. Soc.*, *123*, 1517–1542.

Pinsky, M., and A. Khain (1997b), Turbulence effects on the collision kernel: Part II: Increase of swept volume of colliding drops, *Q. J. R. Meteorol. Soc.*, *123*, 1543–1560.

Pinsky, M., A. Khain, and M. Shapiro (2000), Stochastic effects of cloud droplet hydrodynamic interaction in a turbulent flow, *Atmos. Res.*, *53*, 131–169.

Pinsky, M., A. Khain, B. Grits, and M. Shapiro (2006), Collisions of small drops in a turbulent flow. Part III: Relative droplet fluxes and swept volumes, *J. Atmos. Sci.*, *63*, 2131–2139.

Prigogine, I., and P. Mazur (1953), Sur l'extension de la thermodynamique aux phénomènes irréversibles liés aux degrés de liberté interne, *Physica*, *19*, 241–254.

Pruppacher, H., and J. Klett (1997), *Microphysics of Clouds and Precipitation*, 2nd ed., 954 pp., Kluwer Acad., Dordrecht, The Netherlands.

Rogers, R. R., and M. K. Yau (1996), *A Short Course in Cloud Physics*, 3rd ed., 304 pp., Butterworth-Heinemann, Woburn, Mass.

Saffman, P. G., and J. S. Turner (1955), On the collision of drops in turbulent clouds, *J. Fluid Mech.*, *1*, 16–30.

Saffman, P. G., and J. S. Turner (1988), Corrigendum: On the collision of drops in turbulent clouds, *J. Fluid Mech.*, *196*, 599.

Schlottke, J., W. Straub, K. D. Beheng, H. Gomaa, and B. Weigand (2010), Numerical investigation of collision-induced breakup of raindrops. Part I: Methodology and dependencies on collision energy and eccentricity, *J. Atmos. Sci.*, *67*, 557–575.

Seifert, A., and K. D. Beheng (2001), A double-moment parameterization for simulating autoconversion, accretion and selfcollection, *Atmos. Res.*, *59–60*, 265–281.

Seifert, A., and K. D. Beheng (2006), A two-moment cloud microphysics parameterization for mixed-phase clouds. Part 1: Model description, *Meteorol. Atmos. Phys.*, *92*, 45–66.

Shaw, R. A. (2003), Particle-turbulence interactions in atmospheric clouds, *Annu. Rev. Fluid Mech.*, *35*, 183–227.

Siebert, H., H. Franke, K. Lehmann, R. Maser, E. Saw, D. Schell, R. Shaw, and M. Wendisch (2006a), Probing fine scale dynamics and microphysics of clouds with helicopter-borne measurements, *Bull. Am. Meteorol. Soc.*, *87*, 1727–1738.

Siebert, H., K. Lehmann, and M. Wendisch (2006b), Observations of small-scale turbulence and energy dissipation rates in the cloudy boundary layer, *J. Atmos. Sci.*, *63*, 1415–1466.

Siebert, H., R. A. Shaw, and Z. Warhaft (2010), Statistics of small-scale velocity fluctuations and internal intermittency in marine stratocumulus clouds, *J. Atmos. Sci.*, *67*, 262–273.

Spahn, F., N. Albers, M. Sremčević, and C. Thornton (2004), Kinetic description in dilute granular particle ensembles, *Europhys. Lett.*, *67*, 545–551.

Straub, W., K. D. Beheng, A. Seifert, J. Schlottke, and B. Weigand (2010), Numerical investigation of collision-induced breakup. Part II: Parameterizations of coalescence efficiencies and fragment size distributions, *J. Atmos. Sci.*, *67*, 576–588.

Sundaram, S., and L. Collins (1997), Collision statistics in an isotropic particle-laden turbulent suspension. Part 1. Direct numerical simulations, *J. Fluid Mech.*, *335*, 75–109.

Telford, J. W. (1955), A new aspect of coalescence theory, *J. Meteorol.*, *12*, 436–444.

Twomey, S. (1964), Statistical effects in the evolution of a distribution of cloud droplets by coalescence, *J. Atmos. Sci.*, *21*, 553–557.

Twomey, S. (1966), Computation of rain formation by coalescence, *J. Atmos. Sci.*, *23*, 405–411.

Vaillancourt, P. A., and M. K. Yau (2000), Review of particle-turbulence interactions and consequences for cloud physics, *Bull. Am. Meteorol. Soc.*, *81*, 285–298.

van Dongen, P. G. J. (1989), Spatial fluctuations in reaction-limited aggregation, *J. Stat. Phys.*, *54*, 221–271.

Vohl, O., S. K. Mitra, S. C. Wurzler, and H. R. Pruppacher (1999), A wind tunnel study of the effects of turbulence on the growth of cloud drops by collision and coalescence, *J. Atmos. Sci.*, *56*, 4088–4099.

von Smoluchowski, M. (1916), Drei vorträge über diffusion, brownsche molekularbewegung und koagulation von kolloidteilchen, *Phys. Z.*, *17*, 557–599.

von Smoluchowski, M. (1917), Versuch einer mathematischen Theorie der Koagulationskinetik kolloidaler Lösungen, *Z. Phys. Chem.*, *92*, 129–168.

Wang, L.-P., and M. Maxey (1993), Settling velocity and concentration distribution of heavy particles in homogeneous isotropic turbulence, *J. Fluid Mech.*, *256*, 27–68.

Wang, L.-P., O. Ayala, S. Kasprzak, and W. Grabowski (2005), Theoretical formulation of collision rate and collision efficiency of hydrodynamically interacting cloud droplets in turbulent atmosphere, *J. Atmos. Sci.*, *62*, 2433–2450.

Wang, L.-P., Y. Xue, O. Ayala, and W. W. Grabowski (2006), Effects of stochastic coalescence and air turbulence on the size distribution of cloud droplets, *Atmos. Res.*, *82*, 416–432.

Wang, L.-P., O. Ayala, B. Rosa, and W. W. Grabowski (2008), Turbulent collision efficiency of heavy particles relevant to cloud droplets, *New J. Phys.*, *10*, 075013, doi:10.1088/1367-2630/10/7/075013.

Warhaft, Z. (2009), Laboratory studies of droplets in turbulence: Towards understanding the formation of clouds, *Fluid Dyn. Res.*, *41*, 011201, doi:10.1088/0169-5983/41/1/011201.

Warshaw, M. (1967), Cloud droplet coalescence: Statistical foundations and a one-dimensional sedimentation model, *J. Atmos. Sci.*, *24*, 278–286.

K. D. Beheng, Karlsruhe Institute of Technology, Institute for Meteorology and Climate Research, Kaiserstrasse 12, D-76131 Karlsruhe, Germany. (klaus.beheng@kit.edu)

Raindrop Size Distribution and Evolution

Greg M. McFarquhar

Department of Atmospheric Sciences, University of Illinois at Urbana-Champaign, Urbana, Illinois, USA

Collision-induced breakup of raindrops is assumed to be the main factor controlling the temporal evolution of raindrop size distributions (RSDs). Owing to this mechanism, fragment drops are produced, the size distributions of which have been determined from laboratory measurements of pairs of colliding drops conducted 30 years ago. From these measurements, different representations and parameterizations have been derived. When the earliest parameterizations, based on the results of laboratory collisions of raindrop pairs falling at their terminal velocities, were implemented in numerical models, stationary distributions with three peaks in number concentrations were realized. Subsequent studies, using a more physical basis for the parameterized relations or using relations based on computational fluid dynamics models, predicted distributions with two peaks from the opposing effects of breakup and coalescence or coagulation. Despite observations of peaks in RSDs in a variety of locations, there has been little systematic evidence for the occurrence of peaks at the specific diameters predicted by the modeling studies. Because factors other than collision-induced breakup, such as evaporation, size sorting, spontaneous breakup, updrafts, and mixing of rain shafts, also influence RSDs, peaks at specific diameters would not be expected to occur consistently. However, given sufficiently long averaging of observations acquired in heavy rain, some evidence of these peaks might be expected. Thus, there is the need for more observations under conditions of heavy rainfall and for studies to process heavy rain rate data acquired at a variety of locations in a consistent manner.

1. INTRODUCTION

Water is essential to all life on Earth, and rain is an important component of the hydrological cycle that governs its availability. Precipitation or the lack of it can have both beneficial and harmful effects. Knowledge and understanding of fundamental rain processes allows for better quantitative precipitation forecasts that help both policy makers and the public. Further, although weather modification and rain enhancement could benefit humankind, any effort on these first requires a thorough understanding of natural rain formation to identify potential opportunities for enhancement.

Rainfall: State of the Science
Geophysical Monograph Series 191

10.1029/2010GM000971

Rain forms through either the warm or cold rain process, where the latter involves the influence of ice and the former does not. Studies of both warm and cold rain require knowledge of processes that affect the temporal evolution of raindrop size distributions (hereafter RSDs). In addition, RSDs at the ground and throughout the atmosphere have many important applications both outside meteorology (e.g., soil erosion and telecommunications) and within. Assumptions about RSDs affect the derivation of rain rate from both ground-based radars and instruments on satellites. The dynamics of cloud systems are dependent on RSDs through their effect on the rate of evaporation that ultimately feeds back on the thermodynamics and dynamics.

Cloud droplets are formed from condensation of water vapor on cloud condensation nuclei, atmospheric aerosol particles that nucleate at specific supersaturations. By itself, condensation cannot explain the development of precipitation-sized particles within a reasonable time frame [*Reynolds*, 1877] because the rate of collision and coalescence of droplets with radii less than 20 μm is small [*Klett and Davis*, 1973]. There is a need for a few larger droplets with radii greater than 20 μm to give the broader cloud droplet size distributions necessary to initiate the growth of larger cloud droplets and raindrops (radii greater than 100 μm). There is considerable uncertainty in the exact mechanism by which the broader distributions are produced [*Beard and Ochs*, 1993]. Proposed explanations include the presence of ultragiant nuclei [e.g., *Woodcock*, 1953; *Johnson*, 1982], mixing favoring the growth of larger drops [e.g., *Baker and Latham*, 1979; *Cooper et al.*, 1986], entity-type entrainment mixing [*Telford*, 1995], turbulent mixing [*Jonas*, 1996; *Shaw*, 2003], the impact of film-forming compounds [*Feingold and Chuang*, 2002], and enhanced collision rates due to turbulence-induced spatial clustering of cloud droplets [*Bodenschatz et al.*, 2010].

Once large enough, drops fall at their terminal velocities modified by local updrafts or downdrafts and continue to grow by collision and coalescence. The fall speed might also be modified after a collision and depends upon the ratio between the raindrop relaxation time and the mean time interval between collisions [e.g., *Villermaux and Bossa*, 2009]. Other processes limit raindrop size. *Langmuir* [1948] hypothesized that aerodynamic breakup of raindrops occurred. On the other hand, *Magarvey and Geldart* [1962] concluded that collisions between raindrops resulting in breakup limited raindrop size. Subsequent studies showed that collision-induced breakup was more important than aerodynamic breakup in determining RSD evolution [*McTaggart-Cowan and List*, 1975; *Srivastava*, 1978] and further that drops with diameters less than 10 mm were theoretically stable and could be levitated in wind tunnels [*Pruppacher and Pitter*, 1971]. On the other hand, *Villermaux and Bossa* [2009] recently claimed that the breakup of larger drops could explain the exponential *Marshall and Palmer* [1948] size distribution of raindrops. However, their study, based on observations of drop breakup in an ascending stream of air, was not consistent with drops falling through air at their terminal velocities and did not adequately consider collision rates between raindrops of varying size [*Rogers*, 1989; *McFarquhar and List*, 1991b]. Thus, the role of collision-induced breakup and coalescence are still regarded as the main two factors that control RSDs.

In this review paper, the state of knowledge of RSDs and their evolution will be summarized. An overview of observed RSDs at both the ground and aloft by aircraft and a description of models used to simulate the evolution of RSDs are included. Laboratory studies and recent computational fluid dynamic studies on collision-induced breakup are also discussed. The remainder of this article is organized as follows. Section 2 describes laboratory studies of collision-induced breakup of raindrops and their implementation in numerical models designed to simulate the temporal evolution of RSDs. Section 3 discusses the instruments and techniques that are used to measure and retrieve RSDs, describes observations of RSDs at the ground, and compares them against the results of modeling studies. Section 4 summarizes the state of knowledge governing the evolution of RSDs and offers suggestions for future research.

2. REPRESENTATION OF COLLISION-INDUCED BREAKUP IN MODELS

In order to simulate the evolution of the size distribution of a population of raindrops as they fall through the atmosphere, accurate representations of the collision, coalescence, and breakup of raindrops are needed. *Beheng* [this volume] presents the spectral balance equation that must be solved in order to compute this temporal evolution, which includes the processes of nucleation, condensation, and the collision-induced interaction of drops. In this section, the representation of collision, coalescence, and breakup of raindrops is discussed.

The change in number distribution function of raindrops due to coalescence and breakup can be represented by

$$\frac{\partial f(m)}{\partial t} = \int_{x=m/2}^{\infty} \int_{y=m-x}^{x} G(m;x,y) f(x) f(y) dy dx, \quad (1)$$

where $f(m)$ is the number distribution function of raindrops in terms of mass m, $G(m; x, y)$ is the general "kernel" with $G(m; x, y)dm$ representing the probability a fragment with mass between m and $m + dm$ is produced or lost by a collision of a single drop of mass y swept out by a single drop of mass x, multiplied by the volume swept out. The kernel is given by

$$G(m;x,y) = \chi(m;x,y) \quad K(x,y) \quad (2)$$

with

$$\chi(m;x,y) = E_{\mathrm{CA}}(x,y)\delta(m-x-y) - \delta(m-x) - \delta(m-y) + E_{\mathrm{BR}}(x,y)P(m;x,y), \quad (3)$$

where E_{BR} is the breakup efficiency and E_{CA} is the coalescence efficiency for a collision of drops with masses x and y, $P(m; x, y)$ is the fragment distribution generated from a collision resulting in a breakup such that $P(m; x, y)$ dm gives the mean number of fragments with masses between m and $m+dm$, $K(x, y)$ the volume swept out by drop x for collisions with drop y per unit time, and $\delta(x)$ the Dirac-delta function used to describe the production of the coalesced drop and the disappearance of the original colliding drops. The volume $K(x,y)$ is given by

$$K(x,y) = \pi(r_x + r_y)^2 E_{CL}(x,y)|v_{x,T} - v_{y,T}|, \tag{4}$$

with r_x and r_y being the radii of drops with masses x and y, respectively, $v_{x,T}$ and $v_{y,T}$ being the relevant terminal velocities, and E_{CL} the collision efficiency.

In order to numerically solve equation (1), the dependence of $v_{x,T}$, $v_{y,T}$, E_{CL}, E_{BR}, and E_{CA} and especially in the present context $P(m; x, y)$ on the masses of the colliding drops must be known. It is emphasized that equation (1) is commonly interpreted to be valid by making the quasi-stochastic assumption. In a stochastic model, collisions between identically sized drops can produce different fragment distributions. For a quasi-stochastic model, the statistical nature of raindrop collisions is acknowledged, but all collisions between identical drops give the same result with $P(m; x, y)$ only a function of x and y. However, $P(m; x, y)$ is given as the average result of the interaction between drops of masses x and y, allowing fractional numbers of fragments to be generated in collisions. The use of the quasi-stochastic assumption is justified and gives similar results to fully stochastic models when raindrop collisions are sufficiently frequent [*Gillespie*, 1972] such that the average number of size-dependent fragments generated in a grid box time step by the stochastic model is similar to that predicted by the quasi-stochastic model.

The largest uncertainty in representing collision-induced breakup in models has been the representation of $P(m; x, y)$, the fragment distribution generated by a collision of drops with masses x and y. Initially, there was no theoretical basis for determining $P(m; x, y)$, so it was determined on the basis of laboratory experiments. Using a raindrop accelerator designed by *McTaggart-Cowan and List* [1975], *Low and List* [1982a] measured 1369 collisions between 10 specific pairs of colliding raindrops. More than 10,000 actual experiments were run, as only about 10% of the experiments resulted in collisions in order to preserve the random nature of raindrop collisions.

Based on this unique data set, *Low and List* [1982b] developed a parameterization that described the fragment size distribution that resulted from the collision of any pair of raindrops. Because each observed breakup event was classified as either filament (subscript f), sheet (subscript s), or disk (subscript d), according to the shape of the temporarily coalesced pair before breakup (Figure 1), the *Low and List* [1982b] parameterization described the average fragment distribution generated for each breakup type

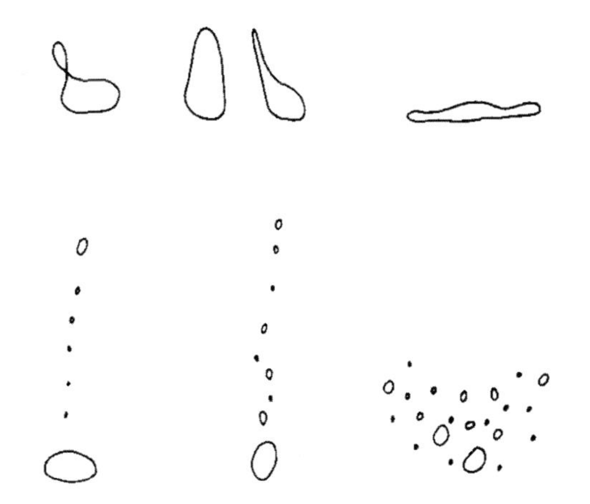

Figure 1. The three principal breakup types observed during the raindrop collision studies of *McTaggart-Cowan and List* [1975] and *Low and List* [1982a]. From left to right, the collision types are filament, sheet, and disk. Adapted from *McTaggart-Cowan and List* [1975].

($P_f(m; x, y)$, $P_s(m; x, y)$, and $P_d(m; x, y)$) and then determined $P(m; x, y)$ as

$$P(m;x,y) = R_f P_f(m;x,y) + R_s P_s(m;x,y) + R_d P_d(m;x,y), \tag{5}$$

where R_f, R_s, and R_d are the fractional occurrence of filament, sheet, and disk breakups, respectively, and are functions of x and y. The parameterization for each $P_i(m; x, y)$, where i is either f, s, or d, consists of combinations of Gaussian and lognormal functions, representing the remnants of the large and small (filament only) colliding drop, or the distribution of smaller satellite drops generated by the collision.

Before this collision-induced breakup kernel was available, earlier studies [*Srivastava*, 1971; *Young*, 1975; *List and Gillespie*, 1976] using simpler representations of breakup found that an "equilibrium distribution" was numerically approached for steady state conditions after sufficient evolution time from the opposing forces of coalescence and breakup. This distribution that is obtained from a numerical solution when coalescence equals breakup is called a "stationary distribution" hereafter because this more accurately reflects that this is a numerical solution, and as will be explained later, not an observationally or physically realized distribution in nature. The stationary distribution is also very useful for showing differences in numerical distributions that are derived from different representations of collision-coalescence-breakup.

Studies conducted using the *Low and List* [1982b] parameterization again showed that a stationary distribution was approached in cases of both steady [*Valdez and Young*, 1985; *Brown*, 1987; *Feingold et al.*, 1988; *List and McFarquhar*, 1990; *Chen and Lamb*, 1994] and nonsteady rain [*Tzivion et al.*, 1989; *McFarquhar and List*, 1991a]. Figure 2 shows an example of such a stationary distribution, which always had peaks at three specific diameters that did not vary significantly between studies. Thus, even though there were certain inherent difficulties, such as mass conservation errors, in the *Low and List* [1982b] parameterization, its use in models seemed robust because of the similarities in the stationary distributions.

Owing to uncertainties associated with the observational evidence for the three-peaked distributions (Section 3), *McFarquhar* [2004a] re-examined the foundation of the *Low and List* [1982b] parameterization. Earlier work [*Brown*, 1995] had examined the stability of the *Low and List* [1982b] parameterization by showing that hypothetical changes in the rate coefficients of up to 40% produced minor changes in the stationary solution. However, *McFarquhar*'s [2004a] work differed in that the original observations of *Low* [1977] were used as the starting point for the new parameterization, rather than merely modifying the *Low and List* [1982b] equations. Other improvements to the new parameterization included the fact that the new scheme ensured mass conservation, used adequate uncertainty analysis, and had more of a physical basis for deriving the parameterized relationships. Figure 3 shows that the stationary distribution produced from the *McFarquhar* [2004a] parameterization significantly

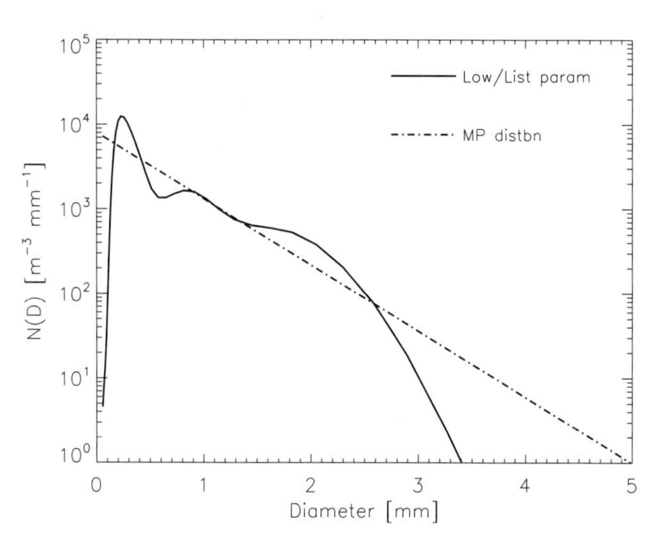

Figure 2. Number density function, $N(D)$, plotted against raindrop diameter, D, for stationary distribution derived using *Low and List* [1982a, 1982b] parameterization (Low/List param) derived from the shown *Marshall and Palmer* [1948] distribution (MP distbn) with a nominal rain rate of 50 mm h^{-1}.

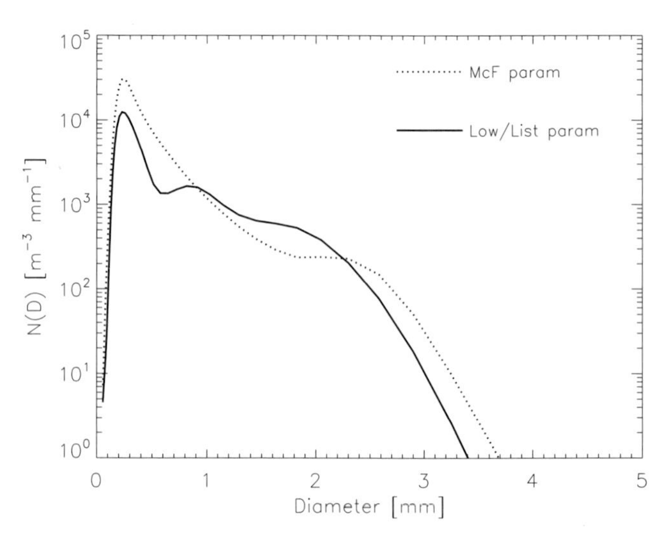

Figure 3. As in Figure 2, except that stationary distribution was derived from *McFarquhar* [2004a] parameterization (McF param) compared against that derived from *Low and List* [1982b] parameterization indicated by Low/List param.

differed from that obtained using the original *Low and List* [1982b] parameterization in that two peaks (produced from the opposing forces of coalescence and breakup) were present in the new stationary distribution which also had a factor of 2 larger total number concentration. A series of Monte Carlo simulations based on the uncertainty analysis in the parameterization showed that although the relative heights of the two peaks changed, the locations of the peaks were robust. Using a high-speed camera [*Testik et al.*, 2006; *Testik and Barros*, 2007], *Barros et al.* [2008] evaluated the results of a dynamic microphysics model [*Prat and Barros*, 2007a, 2007b] that used the *Low and List* [1982b] and *McFarquhar* [2004a] parameterizations. Most importantly, they found that the parameterizations underestimated fragment numbers for $D < 0.2$ mm, that size range that *McFarquhar* [2004a] had noted contained the most uncertainty in the collision-induced fragment distributions. Further, the sizes and fall speeds of these small raindrops are notoriously difficult to measure [e.g., *Montero-Martínez et al.*, 2009]. This suggested a need for more detailed understanding of the physical causes of the generated fragment distributions.

The next major breakthrough in the modeling of raindrop collisions was *Beheng et al.*'s [2006] use of a computational fluid dynamics program to directly model the results of collisions between raindrop pairs. This study not only provided more of a theoretical basis for the modeled size distributions, but also extended the database from the original 10 colliding pairs of *Low and List* [1982a, 1982b] to additional pairs that helped fill in the missing areas of the *x-y* drop

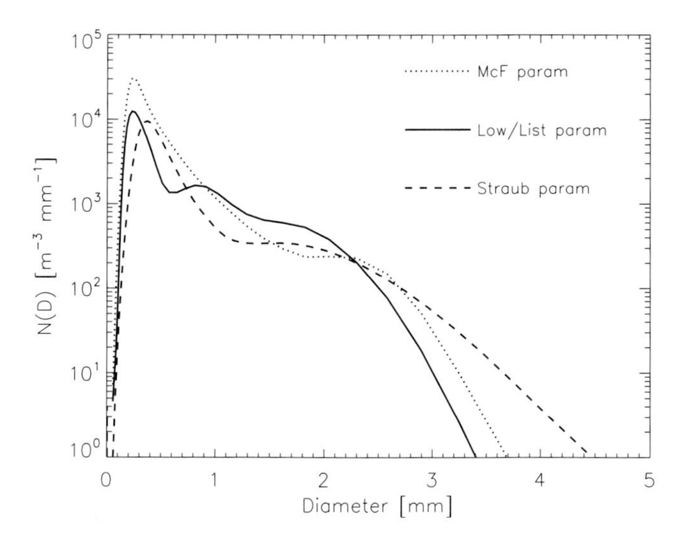

Figure 4. As in Figure 2, except that stationary distributions were derived from McF param and Low/List param compared against those derived from *Straub et al.* [2010] parameterization, indicated by Straub param.

collision space. Thereafter, *Schlottke et al.* [2010] numerically investigated collisions of 32 drop pairs, which were then used by *Straub et al.* [2010] to develop new parameterizations of the coalescence efficiencies and size distributions of breakup fragments of colliding raindrops.

Figure 4 compares the stationary distribution derived using the *Low and List* [1982b], the *McFarquhar* [2004a], and the *Straub et al.* [2010] parameterization. There is a striking similarity between that derived from the *McFarquhar* [2004a] and *Straub et al.* [2010] representations. Because the *McFarquhar* [2004a] parameterization was designed to reflect the best estimate of the laboratory collisions and the uncertainties associated with these representations, the similarity with the theoretical model suggests that that closure is being obtained on an optimal representation of collision-induced breakup of raindrops. Of course, the true test of any model is through comparison against observations. In this respect, Section 3 shows examples of observed RSDs both at the ground and in cloud.

3. MEASUREMENTS OF RSDS

3.1. Techniques for Measuring RSDs

Habib et al. [this volume] review the use of rain gauges and disdrometers for making direct measurements of rainfall at the surface. Here we review the use of filter papers, raindrop cameras, disdrometers, optical array probes, and dual-polarization and Doppler radars for measuring the properties of RSDs. *Marshall and Palmer* [1948], hereafter MP48, first measured RSDs by recording raindrop images on dyed filter papers. Thereafter, in the 1950s, the Illinois State Water Survey raindrop camera was used to measure RSDs in different locations [e.g., *Fujiwara*, 1965; *Stout and Mueller*, 1968; *Uijlenhoet et al.*, 2003]. A disdrometer developed by *Joss and Waldvogel* [1967] was then used to characterize relationships between radar reflectivity and rain rate. Although not specifically designed to measure RSDs, it does record RSDs. It consists of a thin sheet of aluminum 50 to 100 cm^2 in area surrounding a styrofoam cone. When a raindrop hits the surface, the cone and an attached coil are pushed down through a magnetic field, inducing an electrical pulse in the coil that is a function of the size of the impacting drop [*Kinnell*, 1976]. Integrating several such pulses and assuming a relationship between fall velocity and raindrop diameter determines the RSD because the Joss-Waldvogel disdrometer measures raindrop flux and not size distribution. Impact disdrometers have been used in a variety of studies investigating RSDs, including those examining Malaysian rain showers [*McFarquhar et al.*, 1996], Sahelian squall lines [*Nzeukou et al.*, 2004; *Moumouni et al.*, 2008], and tropical cyclones [*Tokay et al.*, 2008].

In addition to the mechanical disdrometer, an optical device named the two-dimensional (2-D) video disdrometer [*Kruger and Krajewski*, 2002] exists for recording orthogonal image projections of raindrops as they cross a sample volume. It has been used in a variety of studies, including those evaluating remote-sensing retrievals of rainfall [e.g., *Zrnic et al.*, 2000; *Bringi et al.*, 2003; *Williams et al.*, 2005], those examining small-scale fluctuations in RSDs [e.g., *Jameson et al.*, 1999; *Jameson and Kostinski*, 2002], and those characterizing RSDs at different locations [e.g., *Tokay et al.*, 2001; *Chang et al.*, 2009]. One big improvement is that, unlike the Joss-Waldvogel disdrometer, the optical disdrometer does not have a rain rate-dependent low diameter cutoff [*McFarquhar et al.*, 1996]. However, there are still uncertainties associated with the 2-D video disdrometers, such as the impact of heavy winds on the RSD [e.g., *Tokay et al.*, 2001].

Other instruments for measuring RSDs include optical array probes developed by Particle Measurement Systems (PMS): both ground-based and airborne versions exist. The PMS 2-D precipitation probe consists of an array of photodiode detectors attached to fast-response electronics that records a 2-D picture of a hydrometeor. This probe has been used to measure RSDs on aircraft [e.g., *Willis*, 1984; *McFarquhar et al.*, 2007], on the roof of vans driving at highway speeds [e.g., *Garcia-Garcia and Gonzalez*, 2000], and on the ground [*Liu and You*, 1994]. Finally, the optical spectropluviometer [*Salles et al.*, 1998] and some other optical disdrometers [*Löffler-Mang and Joss*, 2000] have the advantage of measuring both the fall speeds and diameters of raindrops.

In addition to in situ techniques for measuring RSDs, there are methods for remotely retrieving RSDs. For example, polarimetric or dual-polarization radars send out horizontally and vertically polarized electromagnetic pulses, which spherical and nonspherical particles impact differently. The return of both pulses is equal for spherical particles, but different for nonspherical particles. Because raindrops are more oblate at larger sizes [*Pruppacher and Klett*, 1997; *Jones et al.*, this volume], bigger drops give a larger difference between horizontal and vertical reflectivity [e.g., *Bringi et al.*, 1990]. *Anagnostou et al.* [2009] compared two different methods for retrieving parameters of normalized gamma RSDs using dual-polarization measurements from X-band radar, finding reasonable agreement with disdrometer measurements for the constrained method.

Dual-frequency radars send out two electromagnetic waves at different frequencies. As for polarimetric radar, differential attenuation is the key in deducing parameters that describe the RSDs. *Eccles and Mueller* [1971] first used 3 and 10 GHz radars to estimate rain rate using this technique. Recent studies have shown that the equations for estimating the RSD parameters are nearly equivalent for dual-polarization and dual-wavelength retrievals in the presence of attenuation [*Meneghini and Liao*, 2007]. Data from such techniques have been used to examine the characteristics of RSDs in various regimes. For example, *Kirankumar et al.* [2008] used data from dual-frequency wind profilers to examine how the characteristics of RSDs in stratiform and convective rain varied as a function of height in the Indian southwest monsoon season.

Wind profilers can also be used to retrieve information about RSDs. Here vertical profiles of Doppler spectra are measured and converted into RSDs using drop diameter, fall speed relations that have been established for raindrops. *Gossard* [1988] describes the fundamentals of this technique that has been used to characterize RSDs in tropical squall lines [*May and Rajopadhyaya*, 1996], mesoscale convective systems [*Cifelli et al.*, 2000], and in convective cells [*Lucas et al.*, 2004]. *Löffler-Mang et al.* [1999] also describe the use of a vertically pointing K-band Doppler radar for measuring RSDs. Nevertheless, returns from two wavelengths, two polarizations, or wind profilers can only provide a couple of parameters to describe the RSDs and do not provide the same detailed information about the RSD that in situ measurements can. In fact, in situ data also provide critical relationships between RSD parameters [e.g., *Chu and Su*, 2008] or assumptions about RSD parameters that are used in remote sensing schemes [e.g., *Munchak and Tokay*, 2008; *Zhang et al.*, 2008]. However, Doppler and dual-polarization radars do provide relevant information on the spatial variability of RSDs that is not provided by in situ probes, which is still critical for proper quantitative understanding of rainfall retrieval algorithms [*Uijlenhoet et al.*, 2006; *Schleiss et al.*, 2009]. Therefore, background knowledge of RSDs is needed for both process-oriented studies and for evaluation and development of remote retrieval schemes.

3.2. Representation of RSDs

Marshall and Palmer [1948] first hypothesized that the number density distribution function of raindrops in terms of maximum dimension, $N(D)$, was best characterized by an exponential function

$$N(D) = N_0 e^{-\lambda D}, \tag{6}$$

where N_0 is the intercept, λ the slope parameter, and D the raindrop diameter. The N_0 was assumed to be constant, whereas λ depended on rain rate. The use of an exponential function in cloud models is convenient because its exponential nature allows for easy integration and differentiation of growth and latent heating equations. However, single RSDs generally deviate from this shape, as the *Marshall and Palmer* [1948] distribution overestimates the numbers of small drops and underestimates the number of large drops compared to most observations [*Rogers and Yau*, 1989]. Some studies have compensated for this problem by using exponential distributions where both N_0 and λ depend upon the rain rate [e.g., *Markovitz*, 1976; *Ulbrich*, 1983].

In addition to the exponential distribution, other distributions have been used to characterize $N(D)$ including Gaussian distributions [*Maguire and Avery*, 1995], lognormal distributions [*Feingold and Levin*, 1986; *Sauvageot and Lacaux*, 1995], and exponential distributions, where both the constant and slope depend on precipitation rate [*Markovitz*, 1976]. This exponential distribution represents a special case of the scaling law for the description of RSDs originally proposed by *Sempere-Torres et al.* [1994] and applied by *Uijlenhoet et al.* [2003] and others to estimate rainfall. The gamma distribution, first used by *Ulbrich* [1983] and given by

$$N(D) = N_0 D^{\mu} e^{-\lambda D}, \tag{7}$$

where μ is the shape factor and N_0 and λ are still the intercept and slope respectively, has also been frequently used to characterize RSDs (and cloud droplet size distributions). The use of this distribution is still convenient for numerical models, as it is easily integrable, matches observations better, and can be used with newer parameterization schemes that predict two or three moments of the size distribution [*Milbrandt and Yau*, 2005; *Seifert and*

Beheng, 2006]. With three free parameters, it can also account for the presence of fewer small raindrops than represented by an exponential distribution. The gamma distribution has also been used in radar meteorology studies [e.g., *Kozu and Nakamura*, 2001; *Brandes et al.*, 2004] and for the development of relationships between radar reflectivity (Z) and rain rate (R).

3.3. Comparison of Observed and Modeled RSDs

In situ measurements of RSDs have been used to assess the RSDs that are predicted from numerical models. Because of the opposing effects of coalescence and breakup, such studies have shown that a stationary distribution is numerically approached given sufficient evolution time [e.g., *Srivastava*, 1971; *Young*, 1975; *List and Gillespie*, 1976]. Although many initial studies using the *Low and List* [1982b] breakup kernel showed that a stationary distribution with peaks at three specific diameters was approached for conditions of both steady [*Valdez and Young*, 1985; *Brown*, 1987; *List et al.*, 1987; *Feingold et al.*, 1988; *Chen and Lamb*, 1994; *Hu and Srivastava*, 1995] and nonsteady rain [*Levin et al.*, 1988; *McFarquhar and List*, 1991a, 1991b], later studies have shown that the stationary distribution has only two peaks when using kernels that have a more physical basis and complete uncertainty analysis [*McFarquhar*, 2004a]. Although the stationary distribution produced numerically is not dependent on the initial conditions, the time required to approach the distribution does depend on the initial conditions [*Brown*, 1987; *Prat and Barros*, 2007a] and rate of raindrop collisions [*Rogers*, 1989; *McFarquhar and List*, 1991a, 1991b].

The presence of a unique stationary distribution has not been obtained in observations. Some earlier studies [e.g., *Steiner and Waldvogel*, 1987; *Asselin de Beauville et al.*, 1988; *Zawadzki and de Agosthino Antonio*, 1988; *List et al.*, 1991] noted RSDs with peaks, whose locations were similar to those of the stationary distribution obtained from the *Low and List* [1982a, 1982b] breakup kernel. All these observations were obtained with a Joss-Waldvogel disdrometer. However, *Sheppard* [1990] examined the effects of irregularities in the diameter classification of the Joss-Waldvogel disdrometer and concluded that it produced artificial peaks at the locations where the peaks had been reported. Thereafter, *McFarquhar and List* [1993] determined that the magnitudes of the instrument-related peaks were similar to the magnitudes of the observed peaks, negating some of the previous evidence of stationary distributions.

Nevertheless, there are other reports of peaks in RSDs from techniques other than Joss-Waldvogel disdrometers [e.g., *Willis*, 1984; *Garcia-Garcia and Gonzalez*, 2000; *Sauvageout and Koffi*, 2000; *Radhakrishna and Rao*, 2009]. *Sauvageout and Koffi* [2000] explained the number of modes in the RSDs as a result of the overlapping of rain shafts. On the other hand, *Radhakrishna and Rao* [2009] found the occurrence of peaks at consistent diameters in the range of 0.45–0.65 and 0.9–1.3 mm, consistent with some of the earlier modeling studies [e.g., *Valdez and Young*, 1985] but not the later studies using the more physically based representation of the parameterization coefficients [*McFarquhar*, 2004a].

Although raindrop collision frequencies are certainly not frequent enough to produce any evidence of the peaks associated with stationary distributions for low rain rates [e.g., *Zawadzki et al.*, 1994], the frequency of raindrop collisions could be frequent enough to produce the peaks or stationary distributions for very heavy rain rates [*McFarquhar and List*, 1991a, 1991b; *McFarquhar*, 2004b]. Although evidence of such peaks has been noted in some observational studies [e.g., *Willis*, 1984; *McFarquhar et al.*, 1996; *Garcia-Garcia and Gonzalez*, 2000], these peaks of stationary distributions are not consistently seen in observations. Thus, the discrepancy between the models and observations must be explored.

McFarquhar [2004b] hypothesized that the large spread in fragment distributions generated by specific pairs of colliding drops would cause substantial variations in RSDs that were sampled at specific locations and times in space. This then would explain why stationary distributions are not observed in nature, and why peaks at the same specific locations are not seen in heavy rain, even when considering the possible role of raindrop clustering in enhancing collision interaction rates [*McFarquhar*, 2004b]. Further, the time required to approach a stationary distribution varies significantly depending on the initial RSD [e.g., *Prat and Barros*, 2007a], and there are other factors, such as evaporation [*Brown*, 1993; *Hu and Srivastava*, 1995] and sorting by updrafts [*Kollias et al.*, 2001], whose impact on the RSD depends on local conditions. Thus, there seems to be no reason to expect stationary distributions to consistently occur in nature when individual rain events are sampled.

However, the question still arises as to whether some evidence for the existence of the peaks would be expected when integrating over many rain events over a long time period, especially given that the small peak is produced primarily by filament breakups and the large peak by coalescence events. *McFarquhar* [2004b] estimated that it would take approximately 6 h of data from ground-based disdrometers, integrated over several events in conditions with rain rates greater than 50 mm h^{-1}, for comparing against results of models, given the volumes needed to justify the quasi-stochastic assumption in the model. Although such data may exist if data from a variety of field campaigns are combined, to the best of our knowledge, an integrated analysis of such data has not been presented. Future research efforts should concentrate not only on analysis of data from single field campaigns, but also on integrating the analysis of data collected from a variety of programs.

4. SUMMARY AND FUTURE WORK

This review article has summarized past modeling and observational efforts concerning the temporal and spatial evolution of RSDs. Because collision-induced breakup has the biggest influence on the temporal evolution of RSDs, studies have concentrated on developing improved representations of it. Initially, the results of laboratory collisions of raindrops falling at their terminal velocities [*Low and List*, 1982a] were used to develop parameterization schemes [*Low and List*, 1982b] that were subsequently used in models to predict the evolution of RSDs in conditions of both steady and nonsteady rain.

Despite shortcomings in the original parameterization scheme that different authors accounted for in different ways, all studies found that stationary distributions of similar shapes were approached given sufficiently high rain rates and evolution times. Some field evidence of stationary distributions was seen, as some observed RSDs had concentration peaks at diameters similar to those predicted by models. However, when many observed peaks were found to be instrument related, the basis of the peaks in the modeled distributions was questioned.

Thus, *McFarquhar* [2004a] developed a more physically based parameterization of the fragment-size distributions generated by colliding raindrops whose use gave a stationary distribution with two peaks, one generated by breakup and the other by coalescence. Thereafter, *Beheng et al.* [2006] used a computational fluid dynamics program to directly model the results of collisions between raindrop pairs. *Schlottke et al.* [2010] used the model to compute the fragment distributions generated by collisions of 32 drop pairs, which were then used by *Straub et al.* [2010] to develop a new parameterization for the fragment distributions of colliding raindrops. The stationary distributions generated by the *Straub et al.* [2010] and *McFarquhar* [2004a] parameterizations are similar, suggesting that theoretical and laboratory-based parameterizations are approaching closure.

However, despite a plethora of observations in heavy rain by a variety of authors in different locations with varying instruments, there has been little evidence of the consistent appearance of peaks at diameters predicted by the models. No doubt, effects such as evaporation, overlapping rain shafts, gravitational sorting, updrafts, downdrafts, and variability in fragment distributions, generated by specific pairs of colliding drops, suggest that a local stationary distribution would not be expected to be observed. Nevertheless, when integrating over long time periods in heavy rain, one might expect evidence of the peaks. Despite occasional evidence, there has yet to be convincing evidence that peaks at the specific diameters predicted by the models frequently occur. Because large amounts of data at heavy rain rates are required for an investigation, it is recommended that there be continuing efforts at acquiring such data at heavy rain rates in a variety of locations and that data from a variety of locations be combined and processed in a common analysis strategy. Such efforts will be critical for assessing reasons for variations in RSDs and associated properties, such as relations between radar reflectivity and rain rate in different regions, because rain rates estimated from a single value of radar reflectivity can vary by well over a factor of two depending on the RSD [*List*, 1988].

Acknowledgments. The assistance of Junshik Um and Fiona Weingartner in the preparation of this manuscript is appreciated. The review of Klaus Beheng and an anonymous reviewer considerably improved the clarity of the presentation. We thank Firat Testik for his support in the writing of this manuscript. The preparation of this manuscript was partially supported by the Richard and Margaret Romano scholar fund at the University of Illinois.

REFERENCES

Anagnostou, M. N., E. N. Anagnostou, J. Vivekanandan, and F. L. Ogden (2009), Comparison of two raindrop size distribution retrieval algorithms for X-band dual polarization observations, *J. Hydrometeorol.*, *9*, 589–600.

Asselin de Beauville, C. A., R. Petit, G. Marion, and J. P. Lacaux (1988), Evolution of peaks in the spectral distribution of raindrops from warm isolated maritime clouds, *J. Atmos. Sci.*, *45*, 3320–3332.

Baker, M. B., and J. Latham (1979), The evolution of droplet spectra and rate of production of embryonic raindrops in small cumulus clouds, *J. Atmos. Sci.*, *36*, 1612–1615.

Barros, A. P., O. P. Prat, P. Shrestha, F. Y. Testik, and L. F. Bliven (2008), Revisiting Low and List (1982): Evaluation of raindrop collision parameterizations using laboratory observations and modeling, *J. Atmos. Sci.*, *65*, 2983–2994.

Beard, K. V., and H. T. Ochs, III (1993), Warm-rain initiation: An overview of microphysical mechanisms, *J. Appl. Meteorol. Climatol.*, *32*, 608–625.

Beheng, K. D. (2010), The evolution of raindrop spectra: A review of microphysical essentials, in *Rainfall: State of the Science*, *Geophys. Monogr. Ser.*, doi:10.1029/2010GM000957, this volume.

Beheng, K. D., K. Jellinghaus, W. Sander, W. Roth, and B. Weigand (2006), Investigation of collision-induced breakup of raindrops by numerical simulations: First results, *Geophys. Res. Lett.*, *33*, L10811, doi:10.1029/2005GL025519.

Bodenschatz, E., S. P. Malinowski, R. A. Shaw, and F. Stratmann (2010), Can we understand clouds without turbulence?, *Science*, *327*, 970–971.

Brandes, E. A., G. Zhang, and J. Vivekanandan (2004), Drop size distribution retrieval with polarimetric radar: Model and application, *J. Appl. Meteorol.*, *43*, 461–475.

Bringi, V. N., V. Chandrasekar, V. Balakrishnan, and V. Zrnic (1990), An examination of propagation effects in rainfall on radar measurements at microwave frequencies, *J. Atmos. Oceanic Technol.*, *7*, 829–840.

Bringi, V. N., V. Chandrasekar, J. Hubbert, E. Gorgucci, W. L. Randeu, and M. Schoenhuber (2003), Raindrop size distribution in different climatic regimes from disdrometer and dual-polarized radar analysis, *J. Atmos. Sci.*, *60*, 354–365.

Brown, Jr., P. S. (1987), Parameterization of drop-spectrum evolution due to coalescence and breakup, *J. Atmos. Sci.*, *44*, 242–249.

Brown, Jr., P. S. (1993), Analysis and parameterization of the combined coalescence, breakup, and evaporation processes, *J. Atmos. Sci.*, *50*, 2940–2952.

Brown, Jr., P. S. (1995), Structural stability of the coalescence/breakup equation, *J. Atmos. Sci.*, *52*, 3857–3865.

Chang, W.-Y., T.-C. Chen Wang, and P.-L. Lim (2009), Characteristics of the raindrop size distribution and drop shape relation in typhoon systems in the Western Pacific from the 2D video disdrometer and NCU C-band polarimetric radar, *J. Atmos. Oceanic Technol.*, *26*, 1973–1993.

Chen, J.-P., and D. Lamb (1994), Simulation of cloud microphysical and chemical processes using a multicomponent framework, Part I: Description of the microphysical model, *J. Atmos. Sci.*, *51*, 2613–2630.

Chu, Y.-S., and C.-L. Su (2008), An investigation of the slope-shape relation for gamma raindrop size distribution, *J. Appl. Meteorol. Climatol.*, *47*, 2531–2544.

Cifelli, R., C. R. Williams, D. K. Rajopadhyaya, S. K. Avery, K. S. Gage, and W. L. Ecklund (2000), Drop-size distribution characteristics in tropical mesoscale convective systems, *J. Appl. Meteorol.*, *39*, 760–777.

Cooper, W. A., D. Baumgardner, and J. E. Dye (1986), Evolution of the droplet spectra in Hawaiian orographic clouds, in *9th Conference on Cloud Physics*, pp. 52–55, Am. Meteorol. Soc., Snowmass, Colo.

Eccles, P. J., and E. A. Mueller (1971), X-band attenuation and liquid water content estimation by a dual-wavelength radar, *J. Appl. Meteorol.*, *10*, 1252–1259.

Feingold, G., and P. Y. Chuang (2002), Analysis of the influence of film-forming compounds on droplet growth: Implications for cloud microphysical processes and climate, *J. Atmos. Sci.*, *59*, 2006–2018.

Feingold, G., and Z. Levin (1986), The lognormal fit to raindrop spectra from frontal convective clouds in Israel, *J. Clim. Appl. Meteorol.*, *25*, 1346–1363.

Feingold, G., S. Tzivion, and Z. Levin (1988), Evolution of raindrop spectra. Part I: Solution to the stochastic collection/breakup equation using the method of moments, *J. Atmos. Sci.*, *45*, 3387–3399.

Fujiwara, M. (1965), Raindrop-size distribution from individual storms, *J. Atmos. Sci.*, *22*, 585–591.

Garcia-Garcia, F., and J. E. Gonzalez (2000), Raindrop spectra observations from convective showers in the valley of Mexico, paper presented at 13th International Conference on Clouds and Precipitation, Int. Comm. on Cloud Phys., Reno, Nev.

Gillespie, D. T. (1972), The stochastic coalescence model for cloud droplet growth, *J. Atmos. Sci.*, *29*, 1496–1510.

Gossard, E. E. (1988), Measuring drop size distributions in clouds with a clear air sensing Doppler radar, *J. Atmos. Oceanic Technol.*, *5*, 640–649.

Habib, E., G. Lee, D. Kim, and G. J. Ciach (2010), Ground-based direct measurement, in *Rainfall: State of the Science*, *Geophys. Monogr. Ser.*, doi: 10.1029/2010GM000953, this volume.

Hu, Z., and R. C. Srivastava (1995), Evaporation of raindrop size distribution by coalescence, breakup and evaporation, *J. Atmos. Sci.*, *52*, 1761–1783.

Jameson, A. R., and A. B. Kostinski (2002), When is rain steady? *J. Appl. Meteorol.*, *41*, 81–90.

Jameson, A. R., A. B. Kostinski, and A. Kruger (1999), Fluctuation properties of precipitation. Part IV: Finescale clustering of drops in variable rain, *J. Atmos. Sci.*, *56*, 82–91.

Johnson, D. B. (1982), The role of giant and ultragiant aerosol particles in warm rain initiation, *J. Atmos. Sci.*, *39*, 448–460.

Jonas, P. R. (1996), Turbulence and cloud microphysics, *Atmos. Res.*, *40*, 283–306.

Jones, B. K., J. R. Saylor, and F. Y. Testik (2010), Raindrop morphodynamics, in *Rainfall: State of the Science*, *Geophys. Monogr. Ser.*, doi: 10.1029/2009GM000928, this volume.

Joss, J., and A. Waldvogel (1967), Ein spectrograph fur Niederschlagstropfen mit automatischer Auswertung, *Pure Appl. Geophys.*, *68*, 240–246.

Kinnell, P. I. A. (1976), Some observations on the Joss-Waldvogel disdrometer, *J. Appl. Meteorol.*, *15*, 499–502.

Kirankumar, N. V. P., T. N. Rao, B. Radhakrishna, and D. N. Rao (2008), Statistical characteristics of raindrop size distribution in southwest monsoon season, *J. Appl. Meteorol. Climatol.*, *47*, 576–590.

Klett, J. D., and M. H. Davis (1973), Theoretical collision efficiencies of cloud droplets at small Reynolds number, *J. Atmos. Sci.*, *30*, 107–117.

Kollias, P., B. A. Albrecht, and F. D. Marks Jr. (2001), Raindrop sorting induced by vertical drafts in convective clouds, *Geophys. Res. Lett.*, *28*, 2787–2790.

Kozu, T., and K. Nakamura (2001), Rainfall parameter estimation from dual-radar measurements combining reflectivity profile and path-integrated attenuation, *J. Atmos. Oceanic Technol.*, *8*, 259–270.

Kruger, A., and W. F. Krajewski (2002), Two-dimensional video disdrometer: A description, *J. Atmos. Oceanic Technol.*, *19*, 602–617.

Langmuir, I. (1948), The production of rain by a chain reaction in cumulus clouds at temperatures above freezing, *J. Atmos. Sci.*, *5*, 175–192.

Levin, Z., G. Feingold, and L. Tzivion (1988), The evolution of raindrop spectra in a rainshaft through collection/breakup and evaporation for steady and pulsating rain, in *Report of the 2nd International Cloud Modeling Workshop, Toulouse, France*, report, pp. 145–154, World Meteorol. Organ., Geneva, Switzerland.

List, R. (1988), A linear radar reflectivity-rainrate relationship for steady tropical rain, *J. Atmos. Sci.*, *45*, 3564–3572.

List, R., and J. R. Gillespie (1976), Evolution of raindrop spectra with collision-induced breakup, *J. Atmos. Sci.*, *33*, 2007–2013.

List, R., and G. M. McFarquhar (1990), The role of breakup and coalescence in the three-peak equilibrium distribution of raindrops, *J. Atmos. Sci.*, *47*, 2274–2292.

List, R., N. R. Donaldson, and R. E. Stewart (1987), Temporal evolution of drop spectra to collisional equilibrium in steady and pulsating rain, *J. Atmos. Sci.*, *44*, 362–372.

List, R., R. Nissen, G. M. McFarquhar, D. Hudak, N. P. Tung, T. S. Kang, and S. K. Soo (1991), Properties of tropical rain, paper presented at 25th International Conference on Radar Meteorology, Am. Meteorol. Soc., Paris.

Liu, Y., and L. You (1994), Error analysis of GBPP-100 probe, *Atmos. Res.*, *34*, 379–387.

Löffler-Mang, M., and J. Joss (2000), An optical disdrometer for measuring size and velocity of hydrometeors, *J. Atmos. Oceanic Technol.*, *17*, 130–139.

Löffler-Mang, M., M. Kunz, and W. Schmid (1999), On the performance of a low-cost K-band Doppler radar for quantitative rain measurements, *J. Atmos. Oceanic Technol.*, *16*, 379–387.

Low, T. B. (1977), Products of interacting raindrops, experiments and parameterizations., Ph.D. thesis, 230 pp., Dep. of Phys., Univ. of Toronto, Toronto, Ont., Canada.

Low, T. B., and R. List (1982a), Collision, coalescence and breakup of raindrops. Part I: Experimentally established coalescence efficiencies and fragment size distributions in breakup, *J. Atmos. Sci.*, *39*, 1591–1606.

Low, T. B., and R. List (1982b), Collision, coalescence and breakup of raindrops. Part II: Parameterizations of fragment size distributions, *J. Atmos. Sci.*, *39*, 1607–1618.

Lucas, C., A. D. MacKinnon, R. A. Vincent, and P. T. May (2004), Raindrop size distribution retrievals from a VHF boundary layer profiler, *J. Atmos. Oceanic Technol.*, *21*, 45–60.

Magarvey, R. H., and J. W. Geldart (1962), Drop collisions under conditions of free fall, *J. Atmos. Sci.*, *19*, 107–113.

Maguire, II, W. B., and S. K. Avery (1995), Retrieval of raindrop size distributions using two Doppler wind profilers: Model sensitivity testing, *J. Appl. Meteorol.*, *33*, 1623–1635.

Markovitz, A. H. (1976), Raindrop size distribution expressions, *J. Appl. Meteorol.*, *15*, 1029–1031.

Marshall, J. S., and W. M. Palmer (1948), The distribution of raindrops with size, *J. Meteorol.*, *5*, 165–166.

May, P. T., and D. K. Rajopadhyaya (1996), Wind profiler observations of vertical motion and precipitation microphysics of a tropical squall line, *Mon. Weather Rev.*, *124*, 621–633.

McFarquhar, G. M. (2004a), A new representation of collision-induced breakup of raindrops and its implication for the shape of raindrop size distributions, *J. Atmos. Sci.*, *61*, 777–794.

McFarquhar, G. M. (2004b), The effect of raindrop clustering on the collision-induced break-up of raindrops, *Q. J. R. Meteorol. Soc.*, *130*, 2169–2190.

McFarquhar, G. M., and R. List (1991a), The evolution of three-peak raindrop size distributions in one-dimensional shaft models. Part II: Multiple pulse rain, *J. Atmos. Sci.*, *48*, 1587–1595.

McFarquhar, G. M., and R. List (1991b), The raindrop mean free path and collision rate dependence on rainrate for three-peak equilibrium and Marshall-Palmer distributions, *J. Atmos. Sci.*, *48*, 1999–2004.

McFarquhar, G. M., and R. List (1993), The effect of curve fits for the disdrometer calibration on raindrop spectra, rainfall rate and radar reflectivity, *J. Appl. Meteorol.*, *32*, 774–782.

McFarquhar, G. M., R. List, D. R. Hudak, R. P. Nissen, J. S. Dobbie, N. P. Tung, and T. S. Kang (1996), Flux measurements of pulsating rain packages with disdrometers and Doppler radar made during Phase II of the Joint Tropical Rain Experiment in Penang, Malaysia, *J. Appl. Meteorol.*, *35*, 859–874.

McFarquhar, G. M., M. S. Timlin, R. M. Rauber, B. F. Jewett, J. A. Grim, and D. P. Jorgensen (2007), Vertical variability of cloud hydrometeors in the stratiform region of mesoscale convective systems and bow echoes, *Mon. Weather Rev.*, *135*, 3405–3428.

McTaggart-Cowan, J. D., and R. List (1975), An acceleration system for water drops, *J. Atmos. Sci.*, *32*, 1395–1400.

Meneghini, R., and L. Liao (2007), On the equivalence of dual-wavelength and dual-polarization equations for estimation of raindrop size distribution, *J. Atmos. Oceanic Technol.*, *24*, 806–820.

Milbrandt, J. A., and M. K. Yau (2005), A multimoment bulk microphysics parameterization. Part I: Analysis of the role of the spectral shape parameter, *J. Atmos. Sci.*, *62*, 3051–3064.

Montero-Martínez, G., A. B. Kostinski, R. A. Shaw, and F. Garcia-Garcia (2009), Do all raindrops fall at terminal fall speed?, *Geophys. Res. Lett.*, *36*, L11818, doi:10.1029/2008GL037111.

Moumouni, S., M. Gosset, and E. Houngninou (2008), Main features of rain drop size distributions observed in Benin, West Africa, with optical disdrometers, *Geophys. Res. Lett.*, *35*, L23807, doi:10.1029/2008GL035755.

Munchak, S. J., and A. Tokay (2008), Retrieval of raindrop size distribution from simulated dual-frequency radar measurements, *J. Appl. Meteorol. Climatol.*, *47*, 223–239.

Nzeukou, A., H. Sauvageot, A. D. Ochou, and C. M. F. Kebe (2004), Raindrop size distribution and radar parameters at Cape Verde, *J. Appl. Meteorol.*, *43*, 90–105.

Prat, O. P., and A. P. Barros (2007a), Exploring the use of a column model for the characterization of microphysical processes in warm rain: Results from a homogeneous shaft model, *Adv. Geosci.*, *10*, 145–152.

Prat, O. P., and A. P. Barros (2007b), A robust numerical solution of the stochastic collection-breakup equation for warm rain, *J. Appl. Meteorol. Climatol.*, *46*, 1480–1498.

Pruppacher, J. R., and J. D. Klett (1997), *Microphysics of Clouds and Precipitation*, 736 pp., Academic, Dordrecht.

Pruppacher, H. R., and R. L. Pitter (1971), A semi-empirical determination of the shape of cloud and raindrops, *J. Atmos. Sci.*, *28*, 86–94.

Radhakrisha, B., and T. N. Rao (2009), Statistical characteristics of multipeak raindrop size distributions at the surface and aloft in different rain regimes, *Mon. Weather Rev.*, *137*, 3501–3518.

Reynolds, O. (1877), On the manner in which raindrops and hailstones are formed, *Proc. Lit. Phil. Soc. Manchester*, *16*, 23–33.

Rogers, R. R. (1989), Raindrop collision rates, *J. Atmos. Sci.*, *46*, 2469–2472.

Rogers, R. R., and M. K. Yau (1989), *A Short Course in Cloud Physics*, 3rd ed., 304 pp., Reed-Elsevier, Oxford, U. K.

Salles, C., J.-D. Creutin, and D. Sempere-Torres (1998), The optical spectropluviometer revisited, *J. Atmos. Oceanic Technol.*, *15*, 1215–1222.

Sauvageot, H., and M. Koffi (2000), Multimodal raindrop size distributions, *J. Atmos. Sci.*, *57*, 2480–2492.

Sauvegeout, H., and J. P. Lacaux (1995), The shape of averaged drop size distributions, *J. Atmos. Sci.*, *52*, 1070–1083.

Schleiss, M. A., A. Berne, and R. Uijlenhoet (2009), Geostatistical simulation of two-dimensional fields of raindrop size distributions at the meso-γ scale, *Water Resour. Res.*, *45*, W07415, doi:10.1029/2008WR007545.

Schlottke, J., W. Straub, K. D. Beheng, H. Gomaa, and B. Weigand (2010), Numerical investigation of collision-induced breakup of raindrops. Part I: Methodology and dependencies on collision energy and eccentricity, *J. Atmos. Sci.*, *67*, 557–575.

Seifert, A., and K. D. Beheng (2006), A two-moment cloud microphysics parameterization for mixed-phase clouds. Part 1: Model description, *Meteorol. Atmos. Phys.*, *92*, 45–66.

Sempere-Torres, D., J. M. Porr, and J.-D. Creutin (1994), A general formulation for raindrop size distributions, *J. Appl. Meteorol.*, *33*, 1494–1502.

Shaw, R. A. (2003), Particle turbulence interactions in atmospheric clouds, *Annu. Rev. Fluid Mech.*, *35*, 183–227.

Sheppard, B. E. (1990), Effect of irregularities in the diameter classification of raindrops by the Joss-Waldvogel disdrometer, *J. Atmos. Oceanic Technol.*, *7*, 180–183.

Srivastava, R. C. (1971), Size distribution of raindrops generated by their breakup and coalescence, *J. Atmos. Sci.*, *28*, 410–415.

Srivastava, R. C. (1978), Parameterization of raindrop size distributions, *J. Atmos. Sci.*, *35*, 108–117.

Steiner, M., and A. Waldvogel (1987), Peaks in raindrop size distributions, *J. Atmos. Sci.*, *44*, 3127–3133.

Stout, G. E., and E. A. Mueller (1968), Survey of relationships between rainfall rate and radar reflectivity in the measurement of precipitation, *J. Appl. Meteorol.*, *7*, 465–474.

Straub, W., K. D. Beheng, A. Seifert, J. Schlottke, and B. Weigand (2010), Numerical investigation of collision-induced breakup of raindrops. Part II: Parameterizations of coalescence efficiencies and fragment size distributions, *J. Atmos. Sci.*, *67*, 576–588.

Telford, J. W. (1995), Comments on "Warm-rain initiation: An overview of microphysical mechanisms", *J. Appl. Meteorol.*, *34*, 2098–2099.

Testik, F. Y., and A. P. Barros (2007), Toward elucidating the microstructure of warm rainfall: A survey, *Rev. Geophys.*, *45*, RG2003, doi:10.1029/2005RG000182.

Testik, F. Y., A. P. Barros, and L. F. Bliven (2006), Field observations of multi-mode raindrop oscillations by high-speed imaging, *J. Atmos. Sci.*, *63*, 2663–2668.

Tokay, A., A. Kruger, and W. F. Krajewski (2001), Comparison of drop size distribution measurements by optical and video disdrometers, *J. Appl. Meteorol.*, *40*, 2083–2097.

Tokay, A., P. G. Bashor, E. Habib, and T. Kasparis (2008), Raindrop size distribution measurements in tropical cyclones, *Mon. Weather Rev.*, *136*, 1669–1686.

Tzivion, S., G. Feingold, and Z. Levin (1989), The evaporation of raindrop spectra. Part II: Collisional collection/breakup and evaporation in a rainshaft, *J. Atmos. Sci.*, *46*, 3312–3327.

Uijlenhoet, R., J. A. Smith, and M. Steiner (2003), The microphysical structure of extreme precipitation as inferred from ground-based raindrop spectra, *J. Atmos. Sci.*, *60*, 1220–1238.

Uijlenhoet, R., J. M. Porr, D. Sempere-Torres, and J.-D. Creutin (2006), Analytical solutions to sampling effects in drop size distribution measurements during stationary rainfall: Estimation of bulk rainfall variables, *J. Hydrol.*, *328*, 65–82.

Ulbrich, C. W. (1983), Natural variations in the analytical form of the raindrop size distribution, *J. Clim. Appl. Meteorol.*, *22*, 1764–1775.

Valdez, M. P., and K. C. Young (1985), Number fluxes in equilibrium raindrop populations: A Markov chain analysis, *J. Atmos. Sci.*, *42*, 1024–1036.

Villermaux, E., and B. Bossa (2009), Single drop fragmentation determines size distribution of raindrops, *Nat. Phys.*, *5*, 697–702, doi:10:103.8/nphys1340.

Williams, C. R., K. S. Gage, W. Clark, and P. Kucera (2005), Monitoring the reflectivity calibration of a scanning radar using a profiling radar and a disdrometer, *J. Atmos. Oceanic Technol.*, *22*, 1004–1018.

Willis, P. T. (1984), Functional fits of observed drop size distributions and parameterization of rain, *J. Atmos. Sci.*, *41*, 1648–1661.

Woodcock, A. H. (1953), Salt nuclei in marine air as a function of altitude and wind force, *J. Appl. Meteorol.*, *10*, 362–371.

Young, K. C. (1975), The evolution of drop-spectra due to condensation, coalescence, and breakup, *J. Atmos. Sci.*, *32*, 965–973.

Zawadzki, I., and M. de Agostinho Antonio (1988), Equilibrium raindrop size distributions in tropical rain, *J. Atmos. Sci.*, *45*, 3542–3549.

Zawadzki, I., E. Monteiro, and F. Fabry (1994), The development of drop size distributions in light rain, *J. Atmos Sci.*, *51*, 1100–1114.

Zhang, G., M. Xue, Q. Cao, and D. Dawson (2008), Diagnosing the intercept parameter for exponential raindrop size distribution based on video disdrometer observations: Model development, *J. Appl. Meteorol. Climatol.*, *47*, 2983–2992.

Zrnic, D. S., T. D. Keenan, L. D. Carey, and P. T. May (2000), Sensitivity analysis of polarimetric variables at a 5-cm wavelength, *J. Appl. Meteorol.*, *39*, 1514–1526.

G. M. McFarquhar, Department of Atmospheric Sciences, University of Illinois at Urbana-Champaign, 105 S. Gregory Street, Urbana, IL 61801-3070, U.S.A. (mcfarq@atmos.uiuc.edu)

Ground-Based Direct Measurement

Emad Habib,[1] Gyuwon Lee,[2] Dongsoo Kim,[3] and Grzegorz J. Ciach[1]

Rain gauges and disdrometers provide direct in situ measurements of rainfall properties at relatively high temporal resolutions. In this chapter, the most common types of these instruments are described. While rain gauges measure rainfall accumulations and intensities, disdrometers provide data on the drop size distribution that describes the rainfall microphysical structure. In addition, some disdrometers can also measure the fall velocities of the drops. The most significant error sources, problems of automated data quality control of rain gauge data and example schemes for correcting their systematic errors, as well as selected issues of disdrometer data representation are presented here.

1. INTRODUCTION

Rain gauges provide the most direct tools for measuring surface rainfall rates and accumulations. Despite the advances in remote sensing, rain gauge measurements continue to be the main basis for numerous research and operational applications that require quantitative data about the surface rainfall amounts (e.g., agricultural, water-resource monitoring, calibration, and assessment of remotely sensed rainfall estimates). Although rain gauges are relatively inexpensive and simple to operate, they can provide fairly accurate data on the rainfall volumes and intensities over their collection areas. On the other hand, disdrometers can measure more detailed properties of the raindrops. Therefore, disdrometer observations can have more specialized applications such as studies of soil erosion, rainfall microphysics research, and development of radar and satellite estimation algorithms, for example.

[1]Department of Civil Engineering, University of Louisiana at Lafayette, Lafayette, Louisiana, USA.
[2]Department of Astronomy and Atmospheric Sciences, Kyungpook National University, Daegu, South Korea.
[3]NOAA/NESDIS/NCDC, Ashville, North Carolina, USA.

Rainfall: State of the Science
Geophysical Monograph Series 191

10.1029/2010GM000953

This chapter provides an overview of the different types of rain gauges and disdrometers. The measuring principles of different types of these instruments are briefly presented in sections 2.1 and 3.2. The major sources of the measurement errors are discussed in sections 2.2 and 3.3. The vital issues and example schemes of automated quality control (QC) of rain gauge data are described in more detail in section 2.3. Finally, section 4 summarizes and concludes this review.

This presentation is limited to measurements in liquid precipitation only. Also, the uncertainty aspects that are related to spatial rainfall variability are beyond the scope of this review.

2. RAIN GAUGES

2.1. *Types of Rain Gauges*

There are several types of rain gauges that can be classified according to their principle of operation. Detailed technical description of each gauge type, its operational mechanism, construction details, and maintenance requirements are provided in several publications [e.g., *Nystuen*, 1996; *Strangeways*, 2007; *WMO*, 2008]. Short descriptions and color pictures of rain gauges can be also found online in Wikipedia (http://en.wikipedia.org/wiki/Rain_gauge). Thus, only brief technical outlines of the most commonly used types are provided below. Instead, more space in this chapter is devoted to the problems of measurement errors and QC of the collected rain gauge data.

2.1.1. Nonrecording storage gauges. These gauges are based on manual measurements of cumulative rainfall amounts over given periods of time (e.g., a day, a month, or storm duration). Different shapes and designs of manual gauges have been used worldwide where most of them consist of a collector that funnels the accumulated rainwater into a container. The stored water is measured using a graduated stick or a graduated glass cylinder. Depending on the cross-sectional area of the funnel and the measuring cylinder, the measurement accuracy can be as high as 0.01 in. (0.25 mm). The collected water can also be poured from the collecting container into an outside container and weighted so that more accurate measurements can be obtained. In the operational regime, the nonrecording rain gauges are usually read once a day, with some locations providing only monthly or seasonal readings.

2.1.2. Recording gauges. Recording gauges automatically record the amount of rain collected as a function of time. According to their principle of operation, the most common types of these instruments are the following.

2.1.2.1. Float gauges. In this type of gauge design, the rain from the funnel passes into a chamber containing a light float. The movement of the float with the rise of the water level is recorded by a pen on a chart or a digital transducer. A siphoning mechanism is usually used for automatic emptying of the float chamber whenever it becomes full to allow the chart pen to return to zero.

2.1.2.2. Weighing gauges. A weighing gauge consists of a collection container that sits on top of a scale where the weight of water collected is measured automatically as a function of time. Rainfall depths and rates can be inferred from in the weight changes of the collected water. The weighting mechanism is typically mechanical by using springs or a system balance weights. The readings are recorded automatically by different mechanisms such as a pen trace on graph paper mounted on a rotating drum, punched tape recorder, voltage output from a tensometric sensor, or other electronic device for weighing the collected water. The resolution of these gauges depends on the mechanism of recordings and is typically 0.1 in. (2.5 mm) like in the U.S. National Weather Service Universal gauge, but can also be as low as 0.4 in. (1 mm) like in the U.S. National Weather Service Fischer and Porter gauge [*Higgins et al.*, 1996, and references therein].

In the United States, the National Weather Service (NWS) has been operating more than 2000 Fischer and Porter weighing gauges that record the measurements by punching a paper tape every 15 min. The punched paper is retrieved at the end of the month and a tape reader converts the data into digital records for further processing. Unfortunately, damaging the paper tape at any stage of its usage can make the whole month of data unusable. NWS is currently in the process of upgrading the paper tapes to digital memory card, which circumvent the tape punching, storing, and reading steps. This will improve considerably both the data availability [*U.S. Department of Commerce*, 2009] and any further data processing [*Wilson et al.*, 2010].

The U.S. Climate Reference Network (USCRN) consists of 114 stations whose rain gauges are equipped with the weighting sensors of the vibrating wire type. The precipitation depth is calibrated as a function of the resonance frequencies of three vibrating wires, on which the collecting container is suspended. This results in high accuracy of the rainfall measurements (http://www.ncdc.noaa.gov/crn/instrdoc.html). The technical details about the Geonor T-200B rain gauge used by USCRN can be found online in the following document: ftp.ncdc.noaa.gov/pub/data/uscrn/documentation/program/technotes/TN04001GeonorHeater.pdf.

The recent upgrade of some Automatic Surface Observing System (ASOS) precipitation gauges from heated tipping bucket to All-Weather Precipitation Accumulation Gauge (AWPAG) belongs to this type of gauge. The AWPAG upgrade with windshields is implemented on about 350 sites out of total 948 ASOS stations as of April 2010.

2.1.2.3. Tipping bucket gauges. Tipping bucket rain gauge is, by far, the most common type of the rainfall measuring instrument that is used by both the research and operational organizations. A typical tipping bucket gauge is composed of a funnel that drains the water into a pair of joined buckets that are balanced in unstable equilibrium on a horizontal axis. When a prespecified amount of rain has accumulated in the upper bucket, the buckets become unstable and tip over so that the other bucket starts filling. The movement of both buckets as they tip over excites an electronic switch to record the timing of each tip. The rain rate and accumulation can be determined from the timing and the number of the recorded tips.

2.1.2.4. Optical (noncollecting) gauges. These gauges do not physically collect the rainwater. Instead, they are based on measuring the scintillations in an optical beam resulting from raindrops as they fall between a light source and an optical receiver. The measured scintillations are analyzed by an algorithm to estimate instantaneous rain rates, which can then be integrated into cumulative amounts of rainfall. Because of its operational mechanism, optical rain gauges come at a high cost compared to other types of gauges. Because of their measurement principle, the rain amounts from optical rain gauges can be subject to inaccuracies that depend on the rainfall microphysical structure. Any disdrometer can also be used as another type of optical rain gauge. A separate section of this chapter is dedicated to the disdrometers and their principles of operations.

2.2. *Errors in Rain Gauge Measurements and Correction Methods*

Like most meteorological sensors, rain gauge measurements are subject to various sources of errors. In this context, errors are defined as the deviations of a measurement from the corresponding true value of rainfall that one attempts to determine, falling on the rain gauge collection area. We call these errors "local" to distinguish them from the errors related to the area-point differences in rainfall, which are beyond the scope of this chapter. A schematic classification of local errors is presented in Figure 1. While the relative significance of different error sources may depend on the gauge type, the following discussion applies to most types of the rain gauges. Special emphasis is placed on the errors in tipping bucket gauges, since this gauge type can provide fairly accurate measurements at a relatively low cost. Therefore, they have been increasingly used in a wide range of operational and research applications.

2.2.1. Local systematic errors. Local systematic errors include wind-induced errors, losses due to evaporation, wetting and splashing, and errors caused by lack of proper calibration.

2.2.1.1. Wind effects. Undercatchment in rain gauge measurements due to wind effects has been the subject of numerous studies [see *Sevruk and Hamon*, 1984 and *Sevruk and Lapin*, 1993, and references therein], and various correction formulas for this effect have been developed. Wind-effect correction methods can be classified into two main approaches: empirical formulas based on field intercomparison studies and numerical modeling simulations. In the first approach, observations from gauges elevated above the ground are compared to those from ground level pit gauges and used to estimate correction factors (CF) as a function of wind speed (u). Typical examples of such formulas are those developed by *Legates and DeLiberty* [1993] for correction of monthly rainfall measurements in the United States (equation (1a)) and *Yang et al.* [1998] for correction of daily rainfall measurements collected by the NWS 8-in. standard (storage) gauges (equation (1b)):

$$\mathrm{CF}(u) = \frac{100}{100-2.12u} \tag{1a}$$

$$\mathrm{CF}(u) = 100\mathrm{e}^{(-4.606+0.041u^{0.69})} \tag{1b}$$

where u is the monthly (equation (1a)) and daily (equation (1b)) average wind speed in m s^{-1}. Once calculated, the CF are then multiplied by the measured rainfall amounts to yield the desired corrected rainfall amounts.

It is noted that the application of the above formulas and other similar ones is possible only at coarse time scales (monthly, daily, or event scale at best), which may not be sufficient for hydrological analyses that require finer temporal resolutions. The convenience of using a single input parameter (average wind speed) is also offset by the significant degree of scatter that was reported with the fitting of these formulas. For example, fitting (equation (1b)) to field intercomparison data was associated with the determination coefficient, R^2, being less than 0.3.

In the second and more recent approach, computational fluid dynamics (CFD) techniques are applied to simulate the

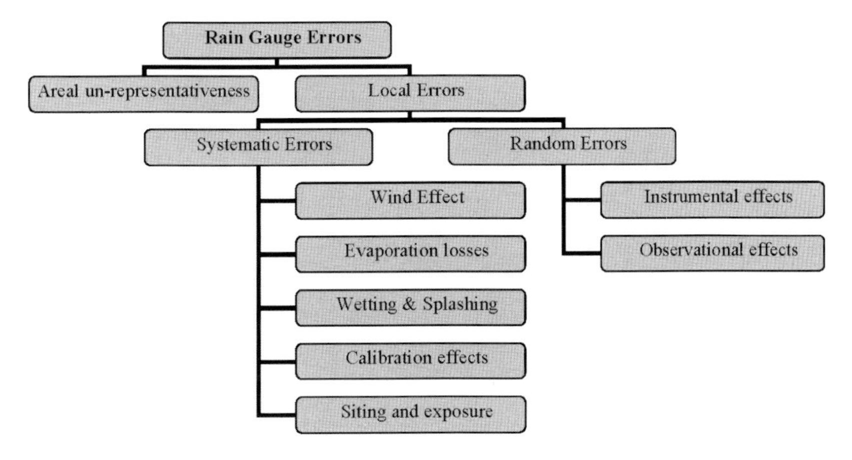

Figure 1. Classification of rain gauge measurement errors.

wind flow field around the rain gauge and to track the trajectories of rain droplets as they approach the gauge orifice. Following this approach, *Nespor and Sevruk* [1999] developed an approximate formula to estimate wind-induced errors as a function of wind velocity, rainfall intensity, and raindrop size distribution:

$$\mathrm{CF}(u,R,k) = \frac{1}{1-e(u,R,k)}, \tag{2}$$

where u is the wind speed in m s^{-1}, R is the rainfall intensity in mm h^{-1}, k is a rain-type parameter that characterizes the raindrop sizes (e.g., $k = -1$ for orographic rains, $k = 1$ for thunderstorms rain and $k = 4$ for showers), and e is a function derived by integrating partial errors of individual drops over the full DSD spectrum. An example graphical illustration of (equation (2)) is shown in Figure 2 for a fixed rain rate of 1 mm h^{-1} and different values of the parameter k. The importance of including information about rain rate and microphysical regime to correct for wind-related errors has been emphasized by both numerical [*Habib et al.*, 1999] and field [*Duchon and Essenberg*, 2001] studies. However, such information is not always available in operational situations. An important difference between empirical and numerical formulas is the time scale at which the corrections can be applied. In formulas (1a) and (1b), the corrections are done at monthly, daily, or individual rainfall event scales. On the other hand, because of their numerical-based derivation, formulas such as equation (2) can be applied to the finest possible temporal scale at which the rainfall rates R can be measured. In a comparative analysis, *Habib et al.* [1999] showed that performing the correction on relatively large time scales (daily and higher) would lead to an overestimation of the wind-induced error by a factor of as high as five times more than if the correction was made at hourly time scale. Based on the results of the computed corrections, *Habib et al.* [1999] recommended that, whenever possible, an hourly scale or finer is used for correction of the rainfall measurements for the wind undercatchment.

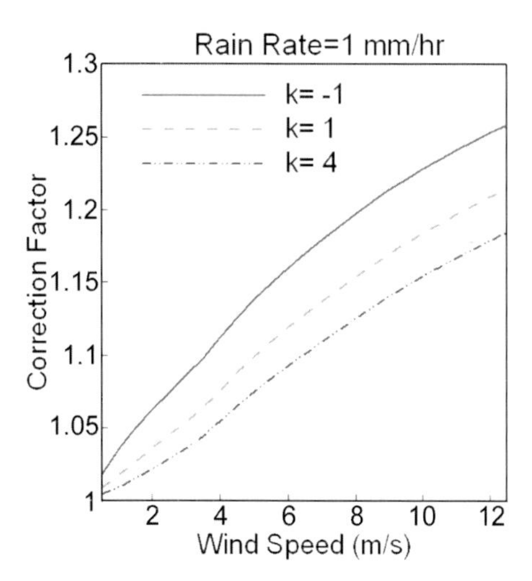

Figure 2. Correction factor of wind-induced error based on numerical simulation-derived formula as a function of wind speed, rain rate, and a drop size distribution (DSD) parameter k. From *Habib et al.* [1999].

2.2.1.2. Calibration-related errors. Because of their simple operation mechanisms, nonrecording manual gauges may not require extensive calibration. Basically, calibration of such gauges includes checking the diameter of the orifice and ensuring that the measuring cylinder or stick is consistent with the chosen size of the collector. Weighing-recording gauges usually do not require calibration because they have few moving parts. On the other hand, tipping bucket gauges require more calibration effort. These gauges are typically regulated by the manufacturer to adjust the tipping bucket volume to its nominal specification. However, experience shows that this regulation might not hold at the time of the gauge field deployment and after a longer period of its operation. Besides, due to the measurement principle of the tipping bucket gauges, some of the rainfall water is missed between successive flips of the bucket, which results in an underestimation of actual rainfall volumes and intensities. This dynamic underestimation effect is negligible at very small rainfall rates, but grows nonlinearly with the increasing rainfall rates [*Niemczynowicz*, 1986; *Humphrey et al.*, 1997; *Luyckx and Berlamont*, 2001]. If this dynamic effect is not accounted for, it results in significant underestimation of high rain rates. To correct for it, a dynamic calibration procedure should be performed under controlled laboratory conditions. This calibration involves passing known amounts of water through the tipping bucket for a fixed number of tips at several fixed rates to determine the effective volume of passing water per one tip at each rain rate. As the flow rate increases, this effective volume also increases due to the dynamic losses during the bucket flips. The following formula can be adjusted based on the dynamic calibration procedure to provide corrected rainfall intensity as a function of the measured nominal rainfall intensity:

$$R_{\mathrm{corrected}} = C_s R_{\mathrm{measured}}(1 + C_d R_{\mathrm{measured}}), \tag{3}$$

where C_s and C_d are the gauge static and dynamic calibration coefficients, respectively. Typical values of these two coefficients obtained for 24 newly purchased tipping bucket rain gauges used in an experimental network near Lafayette, Louisiana, U.S., ranged from 0.99 to 1.04 for C_s and from 0.0006 to 0.0008 for C_d (with R expressed in mm h^{-1}).

Thus, relying on the factory specification would lead to the overall rainfall underestimation of up to 4%. On the other hand, the lack of dynamic correction would result in considerable additional underestimation of high rainfall intensities (e.g., up to 8% for $R = 100$ mm h^{-1} and 24% at $R = 300$ mm h^{-1}).

2.2.1.3. Wetting, evaporation, and splashing losses. The cause and magnitude of these losses vary with the type of the rain gauge. In storage-type gauges, wetting losses occur on the internal walls of the gauge collector and every time when the container is emptied, and therefore, its magnitude will depend on how many times the gauge is actually emptied. According to *WMO* [2008] and studies therein, average wetting losses in storage gauges can be up to 0.2 mm per observation in rainfall. In weighing-type gauges, wetting losses are not a concern because all water falling into its container is weighted. On the other hand, in tipping bucket gauges, water may adhere to the walls of the funnel and to the bucket resulting in water residue and delay in the tipping action. While considered small in comparison to other systematic errors, the wetting losses can still be reduced by using a smooth coating in the container of storage gauges and waxed surface coating for the tipping bucket gauges.

Evaporation losses are important in hot climates and they also depend on the rain gauge type. For storage gauges, losses over 0.8 mm d^{-1} have been reported in warm months [*WMO*, 2008]. These losses can be kept low in well-designed gauges where only a small amount of water is exposed with minimum inside ventilation and by using reflective outer surfaces of the rain gauges. In tipping bucket gauges, the exposed water surface in the gauge bucket is rather large with respect to its volume; this can be problematic during light rain where buckets remain partially filled for durations long enough to cause significant evaporation especially under high-temperature conditions. In weighing-type gauges, evaporation effects can be observed as the loss of weight in the between-rain periods. This information can be then used to reduce the evaporation losses during the rainy periods. Therefore, weighing gauges may perform better, if long-term seasonal rainfall volumes are of interest. Using other gauge types for such purposes can result in trace rainfall being assigned zero values and thus underestimating cumulative rainfall volumes.

Significant splashing losses occur with large drops and during heavy rain. These can be minimized when the gauge has deep vertical side walls around the collecting funnel and when the funnel has steep sloping sides. While such gauge design can minimize the outsplashing of raindrops, it increases evaporation and wetting losses. It can also cause more deformation to the airflow around the gauge, thus leading to larger wind-induced errors. Overall, the splashing losses are rather small and can be in the order of 1% to 2% of the total collected rainfall volume.

2.2.2. Local random errors. Local random errors are mostly attributed to several instrumental factors. They are best recognized for the tipping bucket rain gauges. Therefore, their description in this chapter is limited only to this type of gauges. The discrete sampling mechanism of the tipping bucket gauges results in random quantization errors, which are significant especially during light rain events. It also leads to ambiguities in defining start and end of rainfall events. Additional uncertainties are caused by the instabilities in the water flow in the collecting funnel and oscillations of the water surface in the collecting buckets. These hydrodynamic effects result in weakening the strict correspondence between the intertip time and the fixed amount of water falling into the rain gauge orifice. The local random errors are strongly affected by gauge data recording strategy (e.g., recording the number of tips in prespecified intervals versus recording the exact time of each tip occurrence) and the interpolation schemes used to convert discrete tips into continuous records of rainfall rates [*Ciach*, 2003]. Quantitative assessments of the local random errors in tipping bucket rain gauges were attempted using both simulation and experimental approaches. Using data-driven simulation of the measurements, *Habib et al.* [1999] analyzed the effects of discrete sampling. They demonstrated considerable levels of these errors at short time scales (less than 10 min) and for low rain rates. The magnitude of the errors was strongly dependent on the data-recording strategy, the sampling interval, and the data postprocessing scheme. Using a more direct, fully empirical approach based on the data from a cluster of 15 collocated rain gauges, *Ciach* [2003] was able to provide direct estimates of the local random errors. The errors (ε_l), defined as the differences between single-gauge measurements and the 15-gauge averages normalized by these averages, were estimated for different averaging time scales (T) and interval-averaged rainfall intensities (R). The local random errors defined this way are free of conditional biases. For each time scale, the standard deviation of ε_l as a function of interval-averaged rainfall intensity was approximated using the following formula:

$$\sigma_{\varepsilon_l}(R) = \sigma_0 + \frac{R_0}{R}, \tag{4}$$

where σ_0 are R_0 empirical parameters that depend on the time scale T and the data-processing scheme. Two schemes were examined: one based on linear interpolation between the tip-times, and another based on counting the number of tips in each interval (the second method is common in current operational practices). The values of the parameters σ_0 and R_0 as functions of T are shown in Figure 3 for both processing methods. It is evident that the tip-counting scheme results in much larger errors, especially for shorter time scales. Further reduction of the quantization errors in tipping bucket rain gauges can be achieved using more sophisticated schemes for processing the tip-time information [e.g., *Wang et al.*, 2008].

The studies by *Ciach* [2003] and *Habib et al.* [1999] provide the examples of simple yet practical tools that can be used to statistically characterize the magnitude of the local random error of any type of rain gauge measurements for a given time scale and a rainfall intensity. However, different models might be required with other gauge types than those described in *Ciach* [2003] and *Habib et al.* [1999]. For most of the currently available operational collections of rainfall data, the local random errors are unknown and are usually ignored by the users. While such a practice can be justified when users are working with the rather coarse time scales of the collected rainfall data (e.g., 1 h or larger), the results discussed above indicate that local random error can be significant when high-resolution data (5–15 min) are desired. Rainfall data in such short time scales are increasingly demanded, especially with the recent advances in hydrologic and engineering modeling and analysis tools that can digest rainfall data with high resolutions. The practical implications of the local random errors depend on the particular application. For example, *Habib et al.* [2001] showed that

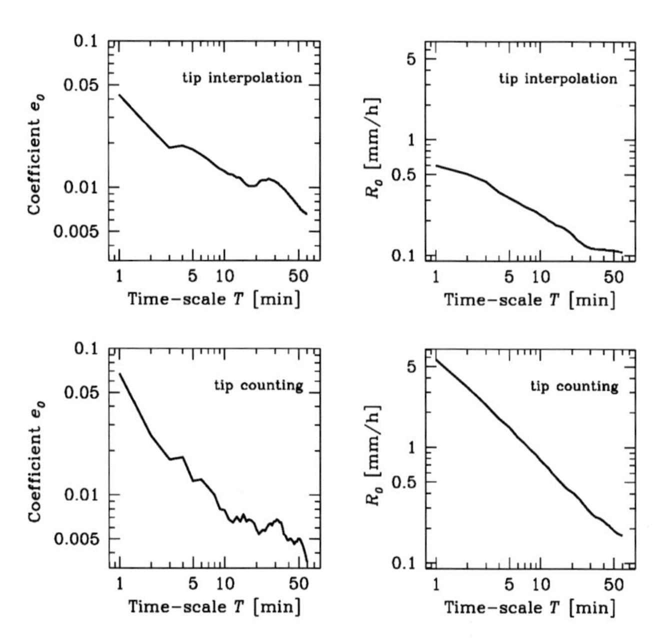

Figure 3. Estimates of two coefficients in the analytical model of the local random error of the tipping bucket gauge defined by equation (4) as a function of the time scale for two data processing schemes (interpolation and tip counting). From *Ciach* [2003]. © American Meteorological Society, reprinted with permission.

the effect of such errors on rainfall-runoff simulations was negligible. On the other hand, the large local random errors associated with low rainfall intensities can lead to difficulties in proper calibration and validation of radar-rainfall products.

2.2.3. Other errors. Other errors that may affect rain gauge measurements can be caused by mechanical and electrical malfunctioning and improper gauge exposure and siting.

2.2.3.1. Malfunctioning problems. Rain gauges also suffer from measurement errors due to simple malfunctioning that is usually caused by mechanical and electrical problems. For example, any gauge can provide incorrect measurements due to partial or complete clogging of the funnel that collects the water. The situations of partial clogging lead to specific distortions of the data that are often undetected by the QC procedures. Frequent maintenance visits (preferably everyday) at the measurement sites could protect the users from being misled by such erroneous data. However, a more reasonable and less expensive alternative to the time-consuming maintenance visits is using more than one collocated rain gauges in each site. The clogging practically never occurs the same way in the collocated gauges, which allows easy detection of such situations. Other common malfunctioning problems include data logger transmission interruptions and temporary power failures. These types of errors are unpredictable, rather difficult to detect, but can be quite serious and may undermine the integrity of the data, especially when they go unnoticed. From our own experience, the multiple-gauge setups, where two or three gauges are located side-by-side at each measuring station, is about the most efficient way to early detect and account for many malfunctioning problems.

2.2.3.2. Bad location. The location and exposure of rain gauges have a significant effect on the reliability of the collected measurements. The surrounding objects that are too close cause shadowing and turbulence deflections of the incident rain that can considerably distort the rain gauge measurements. Some useful location rules indicate that surrounding objects (e.g., a tree or a building) should be located no closer from the gauge than 4 to 10 times of their height above the gauge [e.g., *Brakensiek et al.*, 1979].

2.3. QC of Rain Gauge Measurements

Significant concerns about the quality of rain gauge data have always been recognized by both the operating agencies and the data users. Low quality of the collected data is caused by various factors such as gauge frequent malfunctioning, data storage and transmission problems especially in remote unattended sites, human errors, and others. Such problems are usually manifested in the form of data gaps, physically impossible values, and unlikely high or low values [*Einfalt et al.*, 2006]. We note that the characteristics of these problems are quite different from those that we discussed earlier (e.g., wind-induced errors, calibration errors, etc.) and therefore need special attention. In this section, we focus on reviewing and discussing QC methods and tools that have been developed to help detect and isolate suspect data before they propagate into subsequent applications.

2.3.1. Automated real-time QC methods. Manual data checks by experienced data collectors and users are probably the best means for QC of rain gauge data. However, thorough manual QC analysis is quite laborious and time-consuming and cannot be afforded in operational applications as well as in most research projects. Therefore, recent efforts by the scientific and operational communities have focused on developing automated or semiautomated QC methods that are more appropriate in a real-time operational environment. By "operational environment" we mean an environment where users deal with real-time data around the clock (e.g., operational agencies in charge of flood forecasting or fire-threat monitoring). In fact, in such operational environments, one may argue that fast delivery of time-sensitive observations is more critical than late delivery of better quality of observations. A particular example of an operational environment that we refer to in our discussion is the U.S. National Weather Service (NWS) River Forecast Center (RFC). The different RFC offices rely on a satellite-based real-time data acquisition and distribution system known as the Hydrologic Automated Data System (HADS). The HADS system supports the NWS mission of warning of high impact weather and flood events in the RFC respective service areas. Due to the vast amount of data transmitted every hour through the HADS system (e.g., data from a few thousands of gauges), the NWS developed a suite of automated and semiautomated QC tools [*Kondragunta and Shrestha*, 2006], which is now in-lined as a module within the NWS Multisensor Precipitation Estimator (MPE) algorithm. This algorithm is composed of a series of hierarchical checks that start with the simplest checks performed on a single observation in a given location and at a given observation time. At this level, gross errors (e.g., negative values; clear transmission interruptions) are first detected and flagged as invalid, unless it can be fixed through the user's intervention. The second level of checks involves a single observation and compares it against observations from other gauges that lie within certain spatiotemporal limits that are usually determined based on climatological information. The next level of analysis is more advanced and involves two types of checks: spatial consistency checks and temporal consistency checks. The spatial consistency checks are used to identify outliers in the data that are not consistent with neighboring gauges within a predefined area (e.g., $1° \times 1°$). The occurrence of lightning strikes within a 10-km radius is checked to avoid false identification of extreme rain rates during localized convective rain situations. Similarly, temporal consistency checks are used to detect situations when gauges stopped working during given periods of time. Other operational examples with similar features include the MeteoSwiss QC system [*Musa et al.*, 2003], which consists of different modules that include physical and climatological tests, temporal variability tests, interparameter consistency tests and spatial consistency tests.

2.3.2. QC using multisensor observations. While most of the existing QC automated methods rely solely on the gauge data itself, recent efforts have explored the incorporation of parallel information from other sensors and data sources to perform the QC checks. Such additional information includes rainfall estimates from weather radars and satellites and the outcomes of numerical weather prediction models. Within the NWS-HADS QC system, a multisensor check is designed to help detect suspicious zero rainfall reports by checking the gauge records against the corresponding radar estimates above the gauge site and in the surrounding nine radar pixels. A radar-based temporal consistency check is also designed to compare cumulative gauge values versus radar estimates. More emphasis has been placed on the use of multisensor checks in another recent QC algorithm developed by the National Severe Storms Laboratory (NSSL) in the United States to QC gauge data that are used to bias-adjust the NSSL multisensor precipitation products (NMQ-Q2). In this algorithm [*Zhang et al.*, 2009], radar-gauge differences at a certain gauge are compared to differences at other surrounding gauges within a radius of 10 km. An iterative cross-validation procedure with increasing levels of difference thresholds is followed to identify and eliminate suspect gauges from the NMQ-Q2 bias adjustment procedure.

2.3.3. QC of historical rain gauge data. While the development of automated QC algorithms primarily supports real-time operational applications, they can also be used to reprocess large volumes of historical data. Reprocessing of historical data can additionally take advantage of accumulated human experience by data experts at operational agencies. For example, forecasters at the Lower Mississippi River Forecast Center (LMRFC) and Southeast River Forecast Center (SERFC) had recently begun to archive their results of quality flags with error type and decisions on hourly precipitation data of every gauge station within the forecasting service area that includes thousands of stations

[*Caldwell and Palmer*, 2009]. Preliminary analyses revealed that the most frequently encountered error type is "light precipitation" where no radar and/or satellite images indicate any rainfall occurrence. They are followed by error type of "zero precipitation" when there is an evidence of rainfall as judged by forecasters [*Kim et al.*, 2009a]. Similar experience on more than 20,000 gauge years of data [*Einfalt et al.*, 2006] showed that most quality problems were associated with data gaps and data inconsistencies. *Kim et al.* [2009b] demonstrated quality improvement of reprocessed HADS hourly precipitations using two quality metrics, the fractional of missing values, and the fraction of top-of-the-hour data by month. Figure 4 displays significant reduction of the missing values by the reprocessing recovery algorithm and more top-of-the-hour data recovered by using the original subhourly measurements. Besides improving the existing records of historical data for the purposes of climatological and scientific analyses, documentation of field QC procedures and

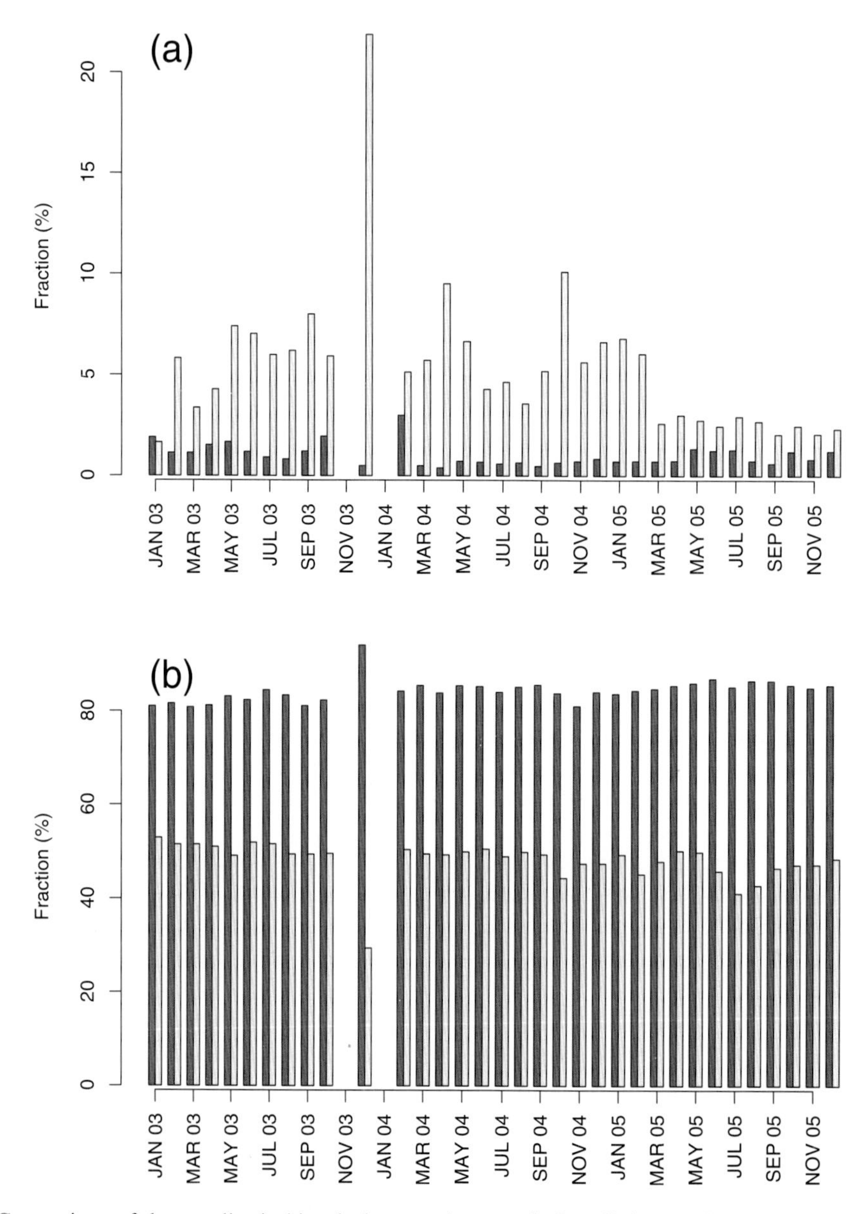

Figure 4. Comparison of data quality in historical gauge data sets before (light gray bars) and after (dark gray bars) applying a retrospective quality control reprocessing analysis: (a) fractional missing values (the smaller the better) and (b) percentage of top-of-the-hour observations (the larger the better). From *Kim et al.* [2009b]. © American Meteorological Society, reprinted with permission.

maintaining archives of associated quality flags is quite important in testing and benchmarking newly proposed real-time QC algorithms.

3. DISDROMETERS

3.1. Basic Definitions

The DSD is defined as the number of raindrops, $N(D)$, with diameter D, per unit volume and unit interval of drop diameter. The two units that are typically used to it are m^{-3} mm^{-1} and cm^{-4}. Mathematically, the DSD can be expressed by a product of the two factors:

$$N(D) = N_T p(D), \tag{5}$$

where N_T is the number of all drops within a unit volume and $p(D)$ is the probability density function of the drop diameters.

We can define various physical quantities of interest from the DSDs (see Table 1 for a summary of physical quantities). The nth moment of DSDs can be calculated by the following integral equation:

$$M_n = \int_{D_{min}}^{D_{max}} D^n N(D) dD \approx \sum_{D_i = D_{min}}^{D_{max}} D_i^n N(D_i) \Delta D_i. \tag{6}$$

Here D_{min} and D_{max} are the minimum and maximum diameters. When the units of $N(D)$, D, and ΔD are m^{-3} mm^{-1}, mm, and mm are used respectively, we can express M_n in m^{-3} mm^{-n}. Due to the nth power of the diameter, the lower moments weigh the smaller diameter and vice versa for the higher moments. For example, the 0th moment does not depend on diameters (that is, 0th power to the diameter) but depends on number concentration.

$$M_0 = N_T = \int_{D_{min}}^{D_{max}} N(D) dD. \tag{7}$$

Thus, M_0 corresponds to the total number of drops per unit volume N_T [m^{-3}]. The radar reflectivity factor can be calculated from DSDs:

$$z[mm^6 m^{-3}] = M_6 = \int_{D_{min}}^{D_{max}} D^6 N(D) dD. \tag{8}$$

By taking "log" of the above equation, we express the radar reflectivity in dBZ units.

$$Z[dBZ] = 10\log_{10}\left(\frac{z}{1 mm^6 m^{-3}}\right). \tag{9}$$

This logarithmic representation of the radar reflectivity factor has benefit for compressing the wide range of radar reflectivity factor in linear scale especially for extreme values. In addition, equation (9) assumes the Rayleigh scattering for raindrops. The Z values are different from the one measured by actual radars, since the Rayleigh scattering assumption may not be valid all the time. Thus, we use "equivalent" radar reflectivity to separate the radar observed radar reflectivity from disdrometric radar reflectivity derived from equations (8) and (9). However, when the scattering medium is composed of raindrops, and the wavelength is much larger than the diameter of raindrops, the Rayleigh assumption is valid (that is, $D < 0.16\ \lambda$) [see *Battan*, 1973].

Table 1. Integral Quantities That Are Defined by Drop Size Distributions[a]

Quantity Symbol	Formula	Unit
Total number concentration M_0	$\int_{D_{min}}^{D_{max}} N(D) dD$	m^{-3}
Optical extinction coefficient S	$\frac{\pi}{2} 10^{-3} \int_{D_{min}}^{D_{max}} N(D) D^2 dD$	km^{-1}
Liquid water content LWC	$\frac{\pi}{6} \rho_w \int_{D_{min}}^{D_{max}} N(D) D^3 dD$	$g\ m^{-3}$
Rainfall intensity R	$6\pi \times 10^{-4} \int_{D_{min}}^{D_{max}} v(D) D^3 N(D) dD$	$mm\ h^{-1}$
Radar reflectivity factor z	$\int_{D_{min}}^{D_{max}} D^6 N(D) dD$	$mm^6 m^{-3}$

[a]The ρ_w represents the density of water.

The rainfall intensity is the mass flux that is the mass of precipitation fallen on a unit area per unit time. Thus, the terminal fall speed of raindrops (precipitation particles) should be included in the equation that calculates the rainfall intensity from DSDs. The following equation is commonly used to calculate the rainfall intensity in $mm\ h^{-1}$ or $m\ s^{-1}$.

$$R = 6\pi \times 10^{-4} \int_{D_{min}}^{D_{max}} v(D) D^3 N(D) dD. \tag{10}$$

The $v(D)$ is the terminal fall velocity of raindrops. Based on the data of the terminal fall velocity of *Gunn and Kinzer* [1949], *Atlas and Ulbrich* [1977] derived a relationship where $v(D)$ is proportional to $D^{0.67}$. When this relationship is applied in equation (10), the rainfall intensity can be approximated by 3.67th moment of DSDs. Thus, the rainfall intensity is sensitive to the number concentration at relatively small and medium size drops rather than larger size. However, bigger drops will have more impact on the calculations of Z. Besides these integral parameters, many different parameters such as liquid water content, kinetic energy, etc. can be derived.

There were several attempts to parameterize DSDs in functional forms such as the exponential, gamma, and lognormal function. Certainly, the functional description of DSD has various advantages: (1) the variation of DSDs can be explained by a few characteristic parameters such as intercept, shape, and/or slope parameters, (2) the integral parameters (e.g., R and Z) can be easily expressed with these characteristic parameters. The most widely used functional forms of DSDs are the exponential distribution [*Marshall and Palmer*, 1948; *Sekhon and Srivastava*, 1971; *Joss and Gori*, 1978], the gamma distribution [*Ulbrich*, 1983; *Ulbrich and Atlas*, 1998], and the lognormal distribution [*Bradley and Stow*, 1974]. The generalized gamma distribution is used as well [*Auf der Maur*, 2001; *Szyrmer et al.*, 2005]. The reader is referred to the work of *McFarquhar* [this volume] for a description of some such DSDs models that are used in the community.

Figure 6. A photo of Joss-Waldvogel disdrometer.

3.2. Types of Disdrometers

3.2.1. Impact disdrometers. In the late 1800s, W. A. Bentley tried to observe DSDs using fine sifted flour [*Blanchard*, 1970]. He measured dough pellets formed by the raindrops fallen in the flour. The filter paper technique was commonly used in many countries during the mid-twenties and lead to many early works on rainfall DSD. The filter paper was dyed with gentian violet and exposed to rain for a short time period. Figure 5 shows an example of the marks made by raindrops on the Whatman #1 filter paper. The size of these marks is larger than the actual size of the drops, and conversion formulas were derived based on laboratory experiments with water drops of known sizes. The first automatic disdrometer was developed by *Joss and Waldvogel* [1969, 1970] and is commonly referred to as the Joss-Waldvogel disdrometer (JWD) (Figure 6). The sensor of the JWD (circular part in Figure 6) has an area of 50 cm^2. This JWD converts the impacts of raindrops falling on the sensor into electrical signals from which the sizes of drops are calculated. The JWD has several limitations. While the sensor responses to a big drop and, during this response time, a smaller drop hits it, the sensor cannot properly receive the signal from the smaller drop. This causes underestimation of the number of small drops, which can be statistically corrected only to a certain extent [*Sheppard and Joe*, 1994; *Sauvageot and Lacaux*, 1995]. Therefore, the DSDs measured during high rainfall intensities can be strongly deformed. Wind and acoustic noise from the surroundings can also affect the JWD measurements. The JWD electronic circuitry continuously sets the noise threshold to avoid counting of false drops due to acoustic noise. Signals from raindrops that are below the noise threshold are lost [*Sauvageot and Lacaux*, 1995]. Thus, a careful consideration of instrument location, such as the middle of grassy area or on a carpet, and away from noise sources is essential for obtaining reliable data from the JWD. Nevertheless, the JWD is still commonly used in the hydrometeorological community due to its relative simplicity and the fact that it has been extensively evaluated over the last few decades, since its inception [e.g., *Tokay et al.*, 2005, and references therein].

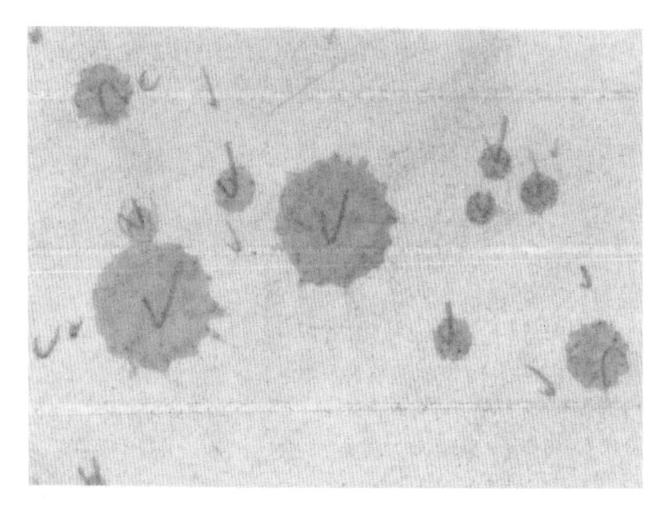

Figure 5. An example of measurement of raindrops on the Whatman #1 filter paper. After *Palmer* [1949].

3.2.2. Optical disdrometers. In these instruments, a laser (or another source of light) and a light detector (e.g., a photodiode) are used to measure the signals that are generated by precipitation particles passing through the measuring area. Several designs of optical disdrometers have been developed. In Figure 7, few examples are presented for a one-dimensional (1-D) optical disdrometer (Parsivel) [*Löffler-Mang and Joss*,

Figure 7. (a) Two-dimensional video disdrometer, (b) Parsivel, and (c) hydrometeor velocity and shape detector (courtesy of R. Schefold).

2000], a 2-D video disdrometer (2DVD) [*Kruger and Krajewski*, 2002], a hydrometeor velocity and shape detector (HVSD) [*Barthazy et al.*, 2004], and a snow video imager (SVI) [*Newman et al.*, 2009].

Depending on the light detector, the optical disdrometer can be divided into 1- and 2-D. The 1-D detectors (e.g., Parsivel, 2DVD, and HVSD) measure the temporal changes of the light intensity. The reduction in the intensity of transmitted light is measured using a photodiode as particles pass through the measurement plane. The horizontal size of particles is indirectly obtained from the maximum reduction of the light intensity. Thus, the shape of particles is not obtained. The falling velocity of particles is derived from the duration of signal reduction based on an assumption on particle axis ratio. Therefore, the accuracy of the fall velocity estimates depends on the validity of this assumption. Nevertheless, the design of Parsivel is relatively simpler than other disdrometers, leading to more stable operation and less field maintenance.

The 2DVD and HVSD use two-line CCDs (charge coupled devices) where each line performs consecutive scans of particle images. The shape of particles is derived by combining information from these consecutive scans. The derived shapes of particles from each line CCD are matched to obtain the falling velocity and to correct distortion of shape by winds. The accuracy of the derived falling velocity is determined by the matching quality. The 2DVD shows some difficulty in matching particles from two perpendicular planes in which particle images can be different under conditions of strong winds especially with the case of anisotropic particles, thus causing significant errors in the derivation of fall velocities. The HVSD uses measurements from the same direction by two parallel light planes. Thus, the accuracy of the fall velocity from HVSD is higher than from 2DVD.

Unlike 1-D detector, the 2-D detectors (e.g., SVI) measure particles within a measurement volume that is defined by the field of view at the focal plane and the depth of field. The horizontal and vertical velocities have no effect on DSD measurements due to their volume measuring principle. A high shutter speed minimizes blurring of particles' images. Consecutive images obtained from very high speed cameras [*Testik et al.*, 2006] are used to measure falling velocities by matching individual particles. Strobe lighting can also be used with slow frame rates to obtain particle velocities. SVI uses a slower frame rate for observation during longer periods and provides 2-D gray scale images that can represent the texture of particles. Thus, these gray images can be very useful for classification of both liquid and solid particles.

Unlike Parsivel, which cannot identify two particles falling simultaneously, the 2DVD provides unique information on different views of particles at two perpendicular directions.

Although HVSD has the advantage of measuring falling velocity, it does have one view only. While this may not be relevant when assuming isotropy in measuring raindrops, it can be rather critical for measuring snow particles. Although the optical disdrometers have significant advantages against the impact disdrometers, proper alignment of optics is usually difficult and can be time-consuming.

3.2.3. Radar-based "disdrometers". In the absence of the vertical air velocity, the Doppler radial velocity from a vertically pointing radar is identical to the terminal fall velocity of precipitation particles. Thus, the power spectrum measured from a vertically pointing radar contains information on DSDs. The Doppler radial velocity can be converted into the drop size. The drop number concentration can be derived by dividing the power value per velocity interval by the power that a drop should have. This method is proposed by *Rogers* [1964] and is widely used in different instrumentations, particularly in wind profilers. The DSDs are retrieved rather than measured, and thus, these instruments are not considered as a direct disdrometer. Nevertheless, this technique provides excellent information on DSDs that cannot be shown in ground-based disdrometers.

There are several instrumentations that utilize this technique to retrieve DSDs (Figure 8): microrain radar (MRR) [*Peters et al.*, 2005], Precipitation Occurrence Sensor System (POSS) [*Sheppard*, 1990], pluviometro-disdrometro in X band (Pludix) [*Caracciolo et al.*, 2006], vertically pointing X band radar (VertiX) [*Zawadzki et al.*, 2001], wind profilers [*Gage et al.*, 2004; *Nikolopoulos et al.*, 2008; *Williams and May*, 2008]. MRR is susceptible to the precipitation attenuation since MRR uses the K band wavelength. The attenuation is less critical for Pludix, POSS, and VertiX. The wind profiler at long wavelength (~ meter) has signals from precipitation and clear air.

An advantage of the radar-based "disdrometer" is its large sampling volumes due to the measuring principles. In particular, POSS has the highest sample volume by utilizing the bistatic principle of radars. The large sampling volume reduces the sampling uncertainty in DSDs measurement. The

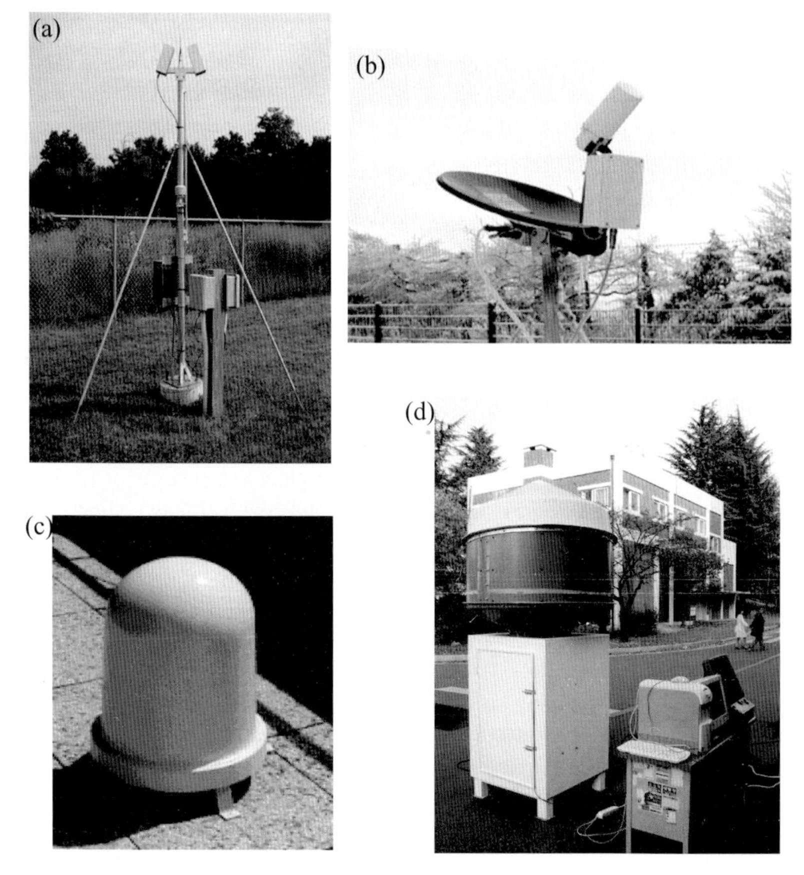

Figure 8. (a) Precipitation Occurrence Sensor System (POSS), (b) microrain radar (MRR), (c) pluviometro-disdrometro in X band (Pludix) [*Caracciolo et al.*, 2006], and (d) vertically pointing X band radar (VertiX).

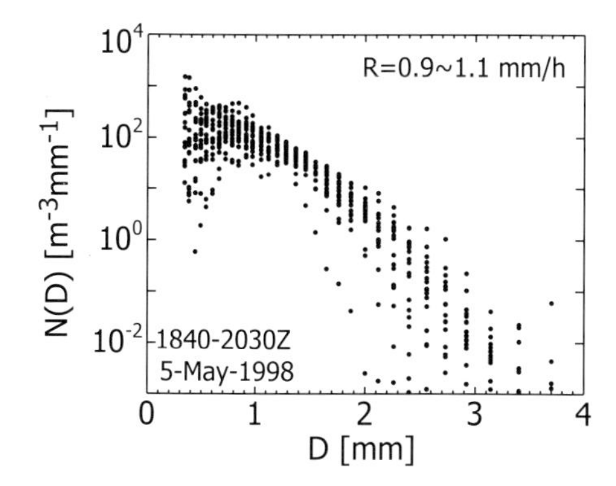

Figure 9. Observed DSDs for rainfall intensity of 0.9~1.1 mm h^{-1} during the quasi-homogeneous microphysical processes by POSS.

advantage is that DSD is retrieved at high temporal resolution as function of heights. Thus, the radar based "disdrometer" is an excellent tool to study the evolution of DSDs. However, the accuracy of retrieved DSDs depends on the degree of the vertical air velocity.

3.3. Measurement Errors

The DSD measurements by disdrometers are affected by various sources of uncertainty, which can be classified into three types: (1) instrumental errors, (2) sampling errors, and (3) observational errors [*Lee and Zawadzki*, 2005]. Observations of DSDs in Figure 9 show a significant degree of scatter within the rainfall intensity of 0.9–1.1 mm h^{-1} during quasi-homogeneous microphysical processes. Since these observations were collected under similar rainfall rate conditions and with no significant changes in the underlying microphysical processes, it is reasonable to attribute such scatter to the three uncertainty sources listed herein.

3.3.1. Instrumental uncertainty. The instrumental uncertainty is linked to the measuring principles and occurs during the data-processing stage. For examples, the accuracy of DSDs from the radar-based "disdrometer" depends on the degree of vertical air velocity. In addition, the accuracy of the Doppler radial velocity also contributes to the error in DSD retrieval. The winds and "dead time' is the main source of the instrumental errors in JWD. The error related to the dead time depends on DSDs and rainfall intensity. The wind effects are subject to the installation of JWD and, of course, the intensity of winds. The accuracy of optical alignment, intensity of light and threshold to detect signals, and distortion by winds are the main contributing factors that degrade the accuracy of measured DSDs in optical disdrometers. The identification and reduction of the instrumental error can be done through laboratory experiments or simulation studies. This is not an easy task but is essential to improve the accuracy. Furthermore, the instrumental uncertainty can be partly quantified by intercomparison of DSDs from the same type disdrometers located next to each other.

When the solid particles are under consideration, the definition of physical size and measured size further contributes to this uncertainty. Parsivel measures the largest horizontal dimension. However, HVSD can provide various measurements of sizes (maximum horizontal dimension, equivalent diameter, maximum occupied diameter, etc.) from one side view. 2DVD can provide similar sizes to HVSD from two side views. Under the condition of strong winds, the matching quality and subsequently derived falling velocity controls the accuracy of these sizes. Thus, the derived falling velocity and horizontal displacement is applied to correct the measured shapes and sizes.

The falling velocity of individual particles determines the accuracy of sampling volumes and hence DSDs. The relationship between falling velocity and size are well defined in raindrops but variable in solid particles [*Barthazy et al.*, 2004; *Brandes et al.*, 2008]. Thus, a unique relationship cannot be assumed. The measured or assumed falling velocity should be used to determine sampling volumes. Thus, the accuracy of falling velocity-size relationship is another factor to determine DSDs in solid particles. The DSDs from volume measuring disdrometers (i.e., SVI) is immune to the relationship.

3.3.2. Sampling uncertainty. The sampling uncertainty is originated from the small sampling volume of disdrometers. When drops distribute uniformly in a given volume as in Figure 10, the measured DSDs depend on the size of sampling volume. For example, in the case of measuring DSDs with two disdrometers that have small but equivalent

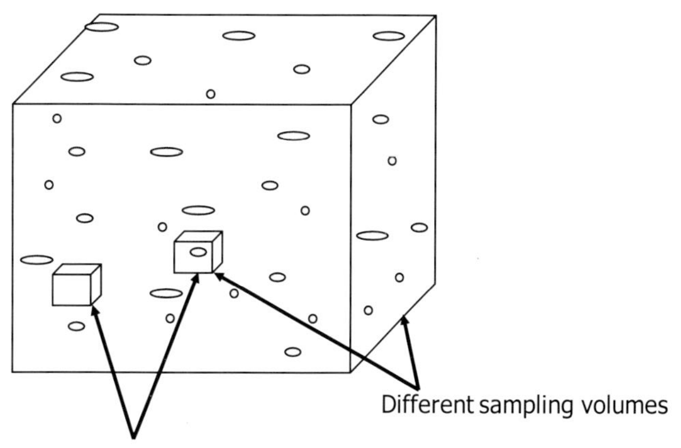

Figure 10. Illustration of drop sampling with different volumes and same size of the small volume.

sampling volumes, one disdrometer detects the drop, but the other does not. On the other hand, two disdrometers of different sampling volumes will show different DSDs, since the disdrometer with the small sampling volume detects one drop, while the other disdrometer detects many drops with different sizes. Thus, although the drops distribute uniformly in the space, the measured DSDs vary with the size of sampling volume. We call the error related to this discrepancy as the sampling uncertainty. The sampling error decreases with increasing sampling volume. The radar-based "disdrometers" has the smallest sampling error, while the sampling uncertainty is significant in the optical and impact disdrometers. In addition, the sampling error reduces by time integration but at the expense of smoothing out the natural variation in the DSDs.

3.3.3. Observational uncertainty. The terminal fall velocity depends on the size of drops, That is, the bigger the sizes, the larger the terminal fall velocity. When the two drops of different sizes start to fall from the same location, the larger drop will fall faster than the smaller one. If the precipitation system moves, the two drops fall at completely different places. We call this phenomenon as the drop sorting. In the perspective of the ground-based disdrometer, the measured drops at different sizes are actually originated from the different parts of storm. For the case of two drops (0.5 and 5 mm with fall velocities of 2 and 9 m s^{-1}, respectively) falling from the height of 1 km, there is a time difference of about 6.5 min when they both arrive at the ground surface. Assuming the precipitation system moves at a speed of about 11 m s^{-1}, the time difference translates into 4.3 km in horizontal distance. Thus, when detected at the ground, these two drops may actually have originated from the different parts of storm that are 4.3 km apart. The convective precipitation system can have a horizontal gradient of reflectivity of about 7~8 dBZ km^{-1}. Thus, measured DSDs at the ground may come from different microphysical processes. In such case, the measured DSDs at the ground are far from the Marshall-Palmer (MP) DSDs, although the DSDs at 1 km altitude were MP distributions (see the simulation by *Lee and Zawadzki* [2005]). This variation of DSDs is not related to the accuracy or measuring principles of the disdrometers and increases as the horizontal gradient of radar reflectivity gets larger.

Various filtering techniques are developed to minimize the above three uncertainties. The original works of *Marshall and Palmer* [1948] is an example of applying the filtering. The filtering should depend on the characteristics of uncertainty. For example, to minimize the observational error, mixing and averaging of DSDs that may have originated from different parts of storm with different horizontal gradient of radar reflectivity are desired. For detailed discussions, see the works of *Joss and Gori* [1978] and *Lee and Zawadzki* [2005].

4. CONCLUDING REMARKS

In this chapter, we presented an overview of the different types of rain gauges and disdrometers, discussed the major sources of uncertainties that contaminate their measurements at the local spatial scale, and described the recently developed methods for automatic QC of the rain gauge data.

Well-maintained rain gauges provide the most accurate measurements of the true surface rainfall accumulations and intensities, at least at single points. Organized in spatial networks, they can provide accurate data on the area-averaged rainfall. The accuracy of such measurements can be increased without limits by increasing the network densities, assuring careful selection of the locations of the measurement points, using multiple-gauge setups of the stations, and providing thorough maintenance of the instruments. Therefore, despite the steady increase of using radars and satellites for rainfall estimation, rain gauges are expected to remain an important source of information for a variety of applications that require relatively accurate rainfall data. For example, the rain gauges are expected to remain indispensable as a source of reference data for meaningful evaluation, tangible improvement, and reliable validation of the rainfall estimation algorithms based on any remote-sensing technique. While rain gauge observations are subject to various local errors, they can be minimized using careful locations and multiple-gauge designs of the measurement stations, and applying error correction schemes based on thorough knowledge of the problems.

In the authors' view, the most alarming factor about the existing rain gauge measurements is the fact that they are notoriously prone to poor quality issues that eventually affect the integrity of large data sets and their utility for further applications. Low-quality rain gauge measurements often go undetected by the end users who must rely on the received rainfall data. Thorough QC of the data can also be an overwhelming task, especially when dealing with large amounts of historical data, and does not guarantee detection of all possible distortions in the collected measurements. The problems increase in the operational applications that rely on real-time streams of rain gauge observations. Therefore, development of rigorous and experience-based automatic QC procedures for the data is a necessity in such situations. Also, a simple upgrade of the measurement stations from single- to double-gauge setups would be the most efficient way to improve considerably on the current situation.

Disdrometers provide data on rainfall microphysical structure in the form of rain DSD. In addition, some disdrometer types can also measure the raindrop falling velocities. Several integral quantities can be derived from these raindrop measurements. Rainfall intensity is one of these quantities, which allows using any disdrometer as a specific type of a rain gauge. Besides, disdrometers have more broad applications including studies of soil erosion and rainfall microphysical processes, developing new DSD parameterization schemes in numerical models of precipitation, and derivation of rainfall retrieval algorithms based on radar and satellite remote sensing, etc. The first automated impact disdrometer, JWD, advanced these research areas in an unprecedented way. Currently, the broad availability of relatively inexpensive optical disdrometers (e.g., the latest models of PARSIVEL) provides the opportunity to expand the precipitation research in several ways, as well as to improve considerably on the previously derived results. Using local clusters of many disdrometers is the simplest way to minimize the effects of their sampling errors, which can lead to new discoveries concerning the actual physical relationships between DSD parameters. Deploying dense networks of such clusters would allow serious research on the spatiotemporal structure of precipitation microphysics based on accurate and direct measurements. Therefore, further development of less expensive and more accurate versions of these instruments is of utmost importance for the future of an important part of the Earth sciences.

REFERENCES

Atlas, D., and C. W. Ulbrich (1977), Path- and area-integrated rainfall measurement by microwave attenuation in the 1–3 cm band, *J. Appl. Meteorol.*, *16*, 1322–1331.

Auf der Maur, A. N. (2001), Statistical tools for drop size distribution: Moments and generalized gamma, *J. Atmos. Sci.*, *58*, 407–418.

Barthazy, E., S. Göke, R. Schefold, and D. Högl (2004), An optical array instrument for shape and fall velocity measurements of hydrometeors, *J. Atmos. Oceanic Technol.*, *21*, 1400–1416.

Battan, L. J. (1973), *Radar Observation of the Atmosphere*, 324 pp., Univ. of Chicago Press, Chicago.

Blanchard, D. C. (1970), Wilson Bentley, the snowflake man, *Weatherwise*, *23*(6), 260–269.

Bradley, S., and C. Stow (1974), The measurement of charge and size of raindrops: Part II. Results and analysis at ground level, *J. Appl. Meteorol.*, *13*, 131–147.

Brakensiek, D. L., H. B. Osborn, and W. J. Rawls (1979), *Field Manual for Research in Agricultural Hydrology, Agriculture Handbook*, *224*, Science and Education Administration, U.S. Department of Agriculture, Beltsville, Md.

Brandes, E. A., K. Ikeda, G. Thompson, and M. Schönhuber (2008), Aggregate terminal velocity/temperature relations, *J. Appl. Meteorol. Climatol.*, *47*, 2729–2736.

Caldwell, R. J., and J. M. Palmer (2009), LOUZIE: An operational quality control procedure at the Lower Mississippi River Forecast Center, Tech memo. (Available online from http://www.srh.noaa.gov/images/lmrfc/tech/louzie.pdf).

Caracciolo, C., F. Prodi, and R. Uijlenhoet (2006), Comparison between Pludix and impact/optical disdrometers during rainfall measurement campaigns, *Atmos. Res.*, *82*, 137–163.

Ciach, G. J. (2003), Local random errors in tipping-bucket rain gauge measurements, *J. Atmos. Oceanic Technol.*, *20*, 752–759.

Duchon, C. E., and G. R. Essenberg (2001), Comparative rainfall observations from pit and aboveground rain gauges with and without wind shields, *Water Resour. Res.*, *37*, 3253–3263.

Einfalt, T., M. Jessen, and M. Qurimbach (2006), Can we check raingauge data automatically?. Proceedings 7th International Workshop on Precipitation in Urban Areas, St. Moritz, Switzerland, 7–10 December.

Gage, K. S., W. L. Clark, C. Williams, and A. Tokay (2004), Determining reflectivity measurement error from serial measurement using paired disdrometers and profiles, *Geophys. Res. Lett.*, *31*, L23107, doi:10.1029/2004GL020591.

Gunn, R., and G. D. Kinzer (1949), The terminal velocity of fall for water drops in stagnant air, *J. Meteorol.*, *15*, 452–461.

Habib, E., W. F. Krajewski, V. Nespor, and A. Kruger (1999), Numerical simulation studies of rain-gage data correction due to wind effect, *J. Geophys. Res.*, *104*, 19,723–19,733.

Habib, E., W. F. Krajewski, and A. Kruger (2001), Sampling errors of tipping-bucket rain gauge measurements, *J. Hydrol. Eng.*, *13*, 159–166.

Higgins, R. W., J. E. Janowiak, and Y.-P. Yao (1996), A gridded hourly precipitation data base for the United States (1963–1993), in *NCEP/Climate Prediction Center Atlas No. 1*, 47 pp., National Weather Service, NOAA, U.S. Department of Commerce, Camp Springs, Md.

Humphrey, M. D., J. D. Istok, J. Y. Lee, J. A. Hevesi, and A. L. Flint (1997), A new method for automated dynamic calibration of tipping-bucket rain gauges, *J. Atmos. Oceanic Technol.*, *14*, 1513–1519.

Joss, J., and E. G. Gori (1978), Shapes of raindrop size distribution, *J. Appl. Meteorol.*, *17*, 1054–1061.

Joss, J., and A. Waldvogel (1969), Raindrop size distribution and sampling size errors, *J. Atmos. Sci.*, *26*, 566–569.

Joss, J., and A. Waldvogel (1970), A method to improve the accuracy of radar measured amounts of precipitation, *14th Conf. on Radar Meteor., Tucson*, pp. 237–238, Am. Meteorol. Soc., Boston, Mass.

Kim, D., E. Tollerud, S. Vasiloff, and J. Caldwell (2009a), Comparison of manual and automated quality control of operational hourly precipitation data of the National Weather Service, Preprints 23rd AMS Conf. on Hydrology, 11–15 January 2009, Phoenix, Ariz.

Kim, D., B. Nelson, and D. J. Seo (2009b), Characteristics of reprocessed Hydrometeorological Automated Data System (HADS) hourly precipitation data, *Weather Forecast.*, *25*, 1287–1296.

Kondragunta, C., and K. Shrestha (2006), Automated real-time operational rain gauge quality controls in NWS hydrologic operations. Preprints, 20th AMS Conf. on Hydrology, 29 Jan.–2 Feb. 2006, Atlanta, Ga.

Kruger, A., and W. F. Krajewski (2002), Two-dimensional video disdrometer: A description, *J. Atmos. Oceanic Technol.*, *19*, 602–617.

Lee, G. W., and I. Zawadzki (2005), Variability of drop size distributions: Noise and noise filtering in disdrometric data, *J. Appl. Meteorol.*, *44*, 634–652.

Legates, D. R., and T. L. DeLiberty (1993), Precipitation measurement biases in the United States, *Water Resour. Bull.*, *29*, 855–861.

Löffler-Mang, M., and J. Joss (2000), An optical disdrometer for measuring size and velocity of hydrometeors, *J. Atmos. Oceanic Technol.*, *17*, 130–139.

Luyckx, G., and J. Berlamont (2001), Simplified method to correct rainfall measurements from tipping bucket rain gauges, in Proc. Urban Drainage Modelling (UDM) Symposium, part of the World Water Resources & Environmental Resources Congress, Orlando, Florida, 20–24 May 2001, edited by R. W. Brashear and C. Maksimovic, ASCE Publications, Reston, Va.

Marshall, J. S., and W. M. Palmer (1948), The distribution of raindrops with size, *J. Meteorol.*, *5*, 165–166.

McFarquhar, G. M. (2010), Raindrop size distribution and evolution, in *Rainfall: State of the Science*, *Geophys. Monogr. Ser.*, doi: 10.1029/2010GM000971, this volume.

Musa, M., E. Grüter, M. Abbt, C. Häberli, E. Häller, U. Küng, T. Konzelmann, and R. Dössegger (2003), Quality control tools for meteorological data in the MeteoSwiss Data Warehouse System. Proc. ICAM/MAP 2003, Brig, Switzerland, May 19th–23rd 2003.

Nespor, V., and B. Sevruk (1999), Estimation of wind-induced error of rainfall gauge measurements using a numerical simulation, *J. Atmos. Oceanic Technol.*, *16*, 450–464.

Niemczynowicz, J. (1986), The dynamic calibration of tipping-bucket raingauges, *Nord. Hydrol.*, *17*, 203–214.

Nikolopoulos, E. I., A. Kruge, W. F. Krajewski, C. R. Williams, and K. S. Gage (2008), Comparative rainfall data analysis from two vertically pointing radars, an optical disdrometer, and a raingauge, *Nonlin. Processes Geophys.*, *15*, 987–997.

Nystuen, J. A., et al. (1996), A comparison of automatic rain gauges, *J. Atmos. Oceanic Technol.*, *13*, 62–73.

Newman, A. J., P. A. Kucera, and L. F. Bliven (2009), Presenting the snowflake video imager (SVI), *J. Atmos. Oceanic Technol.*, *26*, 167–179.

Palmer, W. M. (1949), Studies of continuous precipitation, Ph.D. thesis, 123 pp., McGill Univ., Montreal, Que., Canada.

Peters, G., B. Fischer, H. Münster, M. Clemens, and A. Wagner (2005), Profiles of raindrop size distributions as retrieved by microrain radars, *J. Appl. Meteorol.*, *44*, 1930–1949.

Rogers, R. R. (1964), An extension of the Z-R relation for Doppler radar, *11th Conference on Radar Meteor., Boulder*, pp. 158–161, Am. Meteorol. Soc., Boston, Mass.

Sauvageot, H., and J.-P. Lacaux (1995), The shape of averaged drop size distributions, *J. Atmos. Sci.*, *52*, 1070–1083.

Sekhon, R. S., and R. C. Srivastava (1971), Doppler radar observations of drop-size distributions in a thunderstorm, *J. Atmos. Sci.*, *28*, 983–994.

Sevruk, B., and W. R. Hamon (1984), International comparison of national precipitation gauges with a reference pit gauge, *Instruments and Observing Methods Rep. 17*, World Meteorological Organization, Geneva, Switzerland.

Sevruk, B., and M. Lapin (Eds.) (1993), Precipitation measurement and quality control, Proceedings of the International Symposium on Precipitation and Evaporation, vol. 1, Slovak Hydrometeorological Institute, Bratislava.

Sheppard, B. E. (1990), Measurement of raindrop size distribution using a small Doppler radar, *J. Atmos. Oceanic Technol.*, *7*, 255–268.

Sheppard, B. E., and P. I. Joe (1994), Comparison of raindrop size distribution measurements by a Joss–Waldvogel disdrometer, a PMS 2DG spectrometer, and a POSS Doppler radar, *J. Atmos. Oceanic Technol.*, *11*, 874–887.

Strangeways, I. (2007), *Precipitation: Theory, Measurement and Distribution*, 290 pp., Cambridge Univ. Press, New York.

Szyrmer, W., S. Laroche, and I. Zawadzki (2005), A microphysical bulk formulation based on scaling normalization of the particle size distribution. Part I: Description, *J. Atmos. Sci.*, *62*, 4206–4221.

Testik, F. Y., A. P. Barros, and L. F. Bliven (2006), Field observations of multimode raindrop oscillations by high-speed imaging, *J. Atmos. Sci.*, *63*, 2663–2668.

Tokay, A., P. G. Bashor, and K. R. Wolff (2005), Error characteristics of rainfall measurements by collocated Joss–Waldvogel disdrometers, *J. Atmos. Oceanic Technol.*, *22*, 513–527.

Ulbrich, C. W. (1983), Natural variations in the analytical form of the raindrop size distribution, *J. Appl. Meteorol.*, *22*, 1764–1775.

Ulbrich, C. W., and D. Atlas (1998), Rainfall microphysics and radar properties: Analysis methods for drop size spectra, *J. Appl. Meteorol.*, *37*, 912–923.

U.S. Department of Commerce (2009), *NWSREP's Fischer-Porter Rebuild (FPR-D) Operations Manual*, 87 pp., NOAA, NWS, Silver Spring, Md. (Available at http://www.nws.noaa.gov/ops2/Surface/documents/0AFPR_ObserverGuide21July2009.pdf).

Wang, J., B. L. Fisher, and D. B. Wolff (2008), Estimating rain rates from tipping-bucket rain gauge measurements, *J. Atmos. Oceanic Technol.*, *25*, 43–56.

Williams, C. R., and P. T. May (2008), Uncertainties in profiler and polarimetric DSD estimates and their relation to rainfall uncertainties, *J. Atmos. Oceanic Technol.*, *25*, 1881–1887.

Wilson, A. M., S. Hinson, D. Manns, R. Ray, and J. Lawrimore (2010), Hourly precipitation data processing changes at NCDC. 15th Symposium on Meteorological Observation and Instrumentation, Am. Meteorol. Soc. Atlanta, Georgia, Jan. 2010.

World Meteorological Organization (2008), Guide to Meteorological Instruments and Methods of Observation, 7, WMO No. 8, WMO, Geneva.

Yang, D., B. E. Goodisson, and J. R. Metcalfe (1998), Accuracy of NWS 8″ standard nonrecording precipitation gauge: Results and application of WMO intercomparison, *J. Atmos. Oceanic Technol.*, *15*, 54–67.

Zawadzki, I., F. Fabry, and W. Szyrmer (2001), Observations of supercooled water and secondary ice generation by a vertically pointing X-band Doppler radar, *Atmos. Res.*, *59–60*, 343–359.

Zhang, J., et al. (2009), National mosaic and QPE (NMQ) system—description, results and future plan. 34th Conference on Radar Meteor, Williamsburg, Virginia.

G. J. Ciach and E. Habib, Department of Civil Engineering, University of Louisiana at Lafayette, Lafayette, LA 70504, USA. (habib@louisiana.edu)

D. Kim, National Climate Data Center, NESDIS, NOAA, Ashville, NC 28801, USA.

G. W. Lee, Department of Astronomy and Atmospheric Sciences, Kyungpook National University, Buk-gu-Daegu 702-701, South Korea.

Radar and Multisensor Rainfall Estimation for Hydrologic Applications

Dong-Jun Seo,[1,2,3] Alan Seed,[4] and Guy Delrieu[5]

In the last 20 years, ground-based scanning weather radars have revolutionized quantitative rainfall estimation around the world. Today, in many places they serve as the primary observing system for rainfall estimation in various applications. One can only expect that the use of radar will spread more widely and play an even more important role in quantitative rainfall estimation and related applications. Because radar measures volumetric reflectivity of hydrometeors aloft rather than rainfall near the ground, radar rainfall estimation is inherently subject to various sources of error. For improved rainfall analysis, in particular for quantitative hydrologic applications, it is generally necessary to use rain gauge observations to bias correct and to merge with radar rainfall data. Today, such multisensor rainfall analysis routinely accompanies radar-only rainfall estimation as part of radar-based rainfall estimation. While tremendous advances have been made in the science and technology of radar and multisensor rainfall estimation and the use of radar-based rainfall data and products, a number of large challenges remain for the scientific, operational, and user communities to capitalize on and realize fully the potential of radar-based rainfall estimation for a wide range of applications. The purpose of this chapter is to describe the foundations of radar and multisensor rainfall estimation, recent advances and real-world applications, and outstanding issues and emerging areas of research aiming toward fully realizing their promises, particularly from the perspective of operational hydrologic forecasting.

1. INTRODUCTION

Rainfall is a critical variable in terrestrial and atmospheric mass and energy balance over a wide range of spatiotemporal scales. As such, accurate quantitative estimation of rainfall is extremely important to a wide range of applications in Earth, environmental, agricultural and other science and engineering disciplines. In particular, rainfall is arguably the most important input in many hydrology and water resources applications. For this reason, accurate estimation, or analysis, of rainfall has been one of the long-standing and outstanding challenges in hydrology [*Chow et al.*, 1988; *Hudlow*, 1988]. Owing to extremely large space-time variability of rainfall, point observations based on in situ sensors such as rain gauges may provide only very limited information about the spatiotemporal distribution of rainfall, depending on the spatiotemporal scale at which the analysis is desired. To attain

[1]Formerly at Office of Hydrologic Development, National Weather Service, NOAA, Silver Spring, Maryland, USA.

[2]Formerly at University Corporation for Atmospheric Research, Boulder, Colorado, USA.

[3]Now at Department of Civil Engineering, University of Texas at Arlington, Arlington, Texas, USA

[4]Australian Bureau of Meteorology, Melbourne, Victoria, Australia.

[5]Laboratoire d'étude des Transferts en Hydrologie en Environnement, Grenoble, France.

Rainfall: State of the Science
Geophysical Monograph Series 191

10.1029/2010GM000952

even the slightest improvement, numerous rainfall mapping techniques have been developed over the years [see, e.g., *Creutin and Obled*, 1982; *Tabios and Salas*, 1985, and references therein]. Radar has changed the above picture dramatically by providing spatially continuous estimates of rainfall at small temporal sampling intervals, thereby filling the observation gap of rain gauges in space and time. However, radar does not measure surface rainfall but the backscattered power from the hydrometeors aloft. As such, radar rainfall estimation is inherently subject to various sources of error. To improve the quality of radar-based rainfall estimates, it is therefore necessary to understand, assess, reduce, quantify and account for these errors. The purpose of this chapter is to describe the foundations of radar and multisensor rainfall estimation, recent advances and notable real-world applications, and outstanding issues and areas of research that must be addressed to meet the needs of quantitative hydrologic forecasting. This chapter builds on the previous review papers by *Wilson and Brandes* [1979], *Joss and Waldvogel* [1990], *Krajewski and Smith* [2002], *Delrieu et al.* [2009a], *Krajewski et al.* [2010], and *Villarini and Krajewski* [2010], and text books by *Battan* [1973], *Sauvageot* [1982] and *Doviak and Zrnic* [1993].

Hydrometeors aloft may be liquid and/or solid and may precipitate to the surface as rainfall or snowfall depending on the temperature in the subcloud layer and other hydrometeorological and microphysical factors. Because single-polarization radar measurement of returned power does not distinguish hydrometeor type, it is not possible to separate cleanly radar rainfall estimation from radar snowfall estimation. The latter is subject to substantially larger uncertainties because of the complex nature of the scattering of electromagnetic waves with snowflakes and ice crystals, and to the fact that accurate measurement of snowfall rate, necessary for establishing ground truth, is rather difficult. Radar estimation of snow water equivalent (SWE) is also subject to large uncertainty due to large variability in snow density [*Super and Holroyd*, 1998; *Judson and Doesken*, 2000; *Dubé*, 2003, and references therein; *Ware et al.*, 2006]. For these reasons, radar snowfall estimation [*Rasmussen et al.*, 2003] warrants separate attention and is not dealt with in this chapter. Also, we focus on rainfall estimation using single-polarimetric radar in this chapter; dual-polarimetric radar rainfall estimation is described by *Cifelli and Chandrasekar* [this volume]. The organization of this chapter is as follows. Sections 2 through 4 describe the foundations and major sources of error, real-world applications, and advances and challenges in radar rainfall estimation. Sections 5 and 6 describe multisensor rainfall estimation and challenges therein. Section 7 summarizes the main conclusions.

2. FOUNDATIONS AND MAJOR SOURCES OF ERROR IN RADAR RAINFALL ESTIMATION

A ground-based scanning weather radar transmits electromagnetic waves of near-constant power in very short pulses that are concentrated into a narrow beam at predetermined elevation and radial angles. Between outgoing pulses, the radar measures the returned power of the electromagnetic waves backscattered from targets within the sampling volume as the pulse travels away from the radar. By translating to distance the difference between the transmission and reception times of the outgoing and incoming waves, respectively, one may obtain a map of the returned power within the three-dimensional space comprising all sampling volumes.

The average received power $\bar{P}_r$ (W) is related to the transmitter power P_t (W) via the following radar equation [*Battan*, 1973; *Doviak and Zrnic*, 1993]:

$$\bar{P}_r = 10^{-20}\frac{P_t G^2 \theta^2 \pi^3 h |K|^2 Z}{1024 \ln 2 \lambda^2 r^2}, \tag{1}$$

where G denotes the antenna gain (dimensionless), θ denotes the half power beam width (rad), h denotes the pulse width (m), K denotes the complex dielectric factor of the targets (dimensionless), Z denotes the reflectivity factor (mm^6 m^{-3}), λ denotes the wavelength of the radar (cm), and r denotes the distance to the target (km). In arriving at equation (1), it is assumed that the beam is uniformly filled with targets, multiple scattering may be ignored, the total average power equals the sum of the scattered power from individual particles, the beam is Gaussian-shaped, and the targets are Rayleigh scatterers. If these assumptions are met and the targets are known (water, ice, etc.) so as to specify accurately the complex dielectric factor, one may accurately evaluate the radar reflectivity factor, Z, via equation (1). Radars operating at X and, to a lesser extent, C bands are subject to attenuation, i.e., the reduction of a signal as it propagates through gases, particles and precipitation. The effects of attenuation may be incorporated into equation (1) by multiplying the attenuation factor f_a to Z:

$$f_a = exp(-0.46\int_0^r k\, dr), \tag{2}$$

where k denotes the specific attenuation (dB km^{-1}) which may be approximated by a power law function of Z. The reflectivity factor Z (mm^6 m^{-3}) is, by definition, proportional to the 6th moment of the diameter of the raindrop:

$$Z = \int_0^\infty N(D) D^6 dD, \tag{3}$$

where $N(D)$ (mm^{-1} m^{-3}) denotes the raindrop size distribution (DSD) in a unit volume (m^3) and D denotes the diameter of the raindrop (mm). Rain rate ($mm\ h^{-1}$), on the other hand, is given by

$$R = 6\pi 10^{-4}\int_0^{\infty} N(D)D^3 v_t(D)dD, \qquad (4)$$

where $v_t(D)$ denotes the terminal velocity ($m\ s^{-1}$) of a raindrop of diameter D, often approximated by a power law formula proposed by *Atlas and Ulbrich* [1977], $v_t(D) = 3.80\ D^{0.67}$. Hence, rain rate is approximately proportional to the 3.67th moment of the raindrop size.

Equations (3) and (4) indicate that the DSD plays a key role in radar rainfall estimation as attested by the very extensive literature on the DSD and the subsequent *Z-R* relationships generally approximated by power law functions $Z = AR^b$. There are two fundamental sources of uncertainty in the *Z-R* relationship for rainfall estimation using ground-based scanning radars. The first is that the DSD and the phase of the hydrometeors aloft, where the radar beam samples the atmosphere, are generally different from those near the ground, where rainfall estimates are desired, due to the various microphysical processes, such as growth, coalescence, collision, breakup, evaporation, etc., and possible phase changes (and hence changes in the dielectric constant) among rain, snow, ice crystals, graupel, hail, etc., that the hydrometeors aloft may undergo [*Austin*, 1987; *Pruppacher and Klett*, 1997]. Note also that, because the radar beam gets wider as the range, i.e., the distance from the radar, increases, the sampling volume at far ranges is much larger than that at close ranges. The second is that, even if the DSD aloft is the same as that near the surface, the DSD, and hence the *Z-R* relationship, may vary significantly from storm to storm and within a storm in both space and time depending on the precipitation type, such as convective, stratiform, tropical, etc., and the microphysical processes involved [*Joss and Waldvogel*, 1990; *Uijlenhoet et al.*, 2003; *Rosenfeld and Ulbrich*, 2003; *Lee and Zawadzki*, 2005]. The *Z-R* parameters are usually derived from DSD measurements taken at ground level with electromechanical or optical devices or from radar–rain gauge comparisons. In the latter case, the parameters represent "effective" *Z-R* relationships as they are influenced by other sources of error associated with radar and rain gauge observations of rainfall.

The closer the radar beam is to the ground, the more likely the DSD sampled by the radar beam is closer to that near the surface. The lowest-elevation angle in the scanning strategy of a ground-based radar on a flat terrain, however, is usually set above zero (approximately 0.5° for the Weather Surveillance Radar - 1988 Doppler (WSR-88D) version in the United States) to avoid returns from ground clutter and beam blockage by natural or man-made obstructions. For this reason and because of the curvature of the Earth, the height of the axis of the radar beam as well as the size of the beam increases with range. As such, if the precipitating cloud is shallow, the radar beam will begin to overshoot the cloud top at some range beyond which the radar can no longer detect precipitation. Within the maximum precipitation-detectable range, the reflectivity profile may vary greatly in the vertical, depending on the type of the precipitating cloud. Hence, even if beam overshooting does not occur, accuracy gradually decreases with range. Above the freezing level in stratiform clouds, the radar observes ice particles, which have much smaller effective backscattering cross section than raindrops of comparable size. As the ice particles fall past the freezing level, they begin to melt and, just below the freezing level before the ice melts completely, the radar observes water-coated ice particles, which produce significantly larger reflectivity than the resulting raindrops. The vertical extent of this layer of enhanced reflectivity, or the bright band, is quite small, typically less than 500 m [*Fabry and Zawadzki*, 1995]. The bright band, which shows up as a very sharp spike in the vertical profile of reflectivity (VPR), strongly affects the radar reflectivity measurements at ranges where the beam intercepts the band and the size of the beam is larger than the thickness of the melting layer. For convective clouds, the VPR is usually deeper and more uniform in the vertical. Plate 1 shows a very well developed squall line with trailing stratiform precipitation as observed by the WSR-88D at Twin Lakes, Oklahoma (KTLX). Also shown are the apparent mean VPRs in the convective front and in the stratiform region [*Vignal and Krajewski*, 2001]. If uncorrected, the combined effects of VPR and its integration into radar measurement of reflectivity through the beam propagation pattern may result in highly erroneous radar rainfall estimates even if the spatiotemporally varying *Z-R* relationship is perfectly known.

The above considerations of reflectivity morphology and the sampling geometry of the radar assume that the radar beam navigates over the Earth's surface through the standard atmosphere [*Battan*, 1973; *Doviak and Zrnic*, 1993]. The actual propagation, however, depends on the density variations in the atmosphere, and hence on the temperature and moisture profiles [*Bean and Dutton*, 1968; *Smith et al.*, 1996c]. When anomalous propagation (AP) such as super-refraction, or ducting, occurs because of a strong low-level inversion, the radar returns may not be from the atmosphere but from the ground, which usually results in irregular patterns of very high reflectivity, referred to as ground clutter. Ground clutter may also result from direct interactions of the radar beam and the side lobes with topographic relief and/or anthropogenic targets.

3. REAL-WORLD APPLICATIONS OF RADAR RAINFALL ESTIMATION

With the operational implementation of the network of 160 Weather Surveillance Radars - 1988 Doppler (WSR-88D) version in the early to mid-1990s, radar rainfall data have become an indispensable source of precipitation information in the United States for operational forecasting and a wide spectrum of other applications. The WSR-88D rainfall products [*Klazura and Imy*, 1993] are generated by the Precipitation Processing Subsystem (PPS) [*Fulton et al.*, 1998]. In the U.S. National Weather Service (NWS), radar rainfall products are used for flash flood forecasting as input to tools such as the Flash Flood Monitoring and Prediction System (FFMP) [*Smith et al.*, 2000], river and water resources forecasting using lumped and distributed hydrologic models [*Smith et al.*, 2004] (see *Cole and Moore* [2009] for similar applications in the United Kingdom), continental-scale land surface modeling for initialization of weather and climate models (see *Mitchell et al.* [2004] and also the companion papers in the "GEWEX Continental-Scale International Project, Part 3" special section in *Journal of Geophysical Research*, volume 109) and assimilation into numerical weather prediction (NWP) models [*Lin and Mitchell*, 2005]. Of these applications, flash flood forecasting often relies solely on radar-only rainfall estimates as timeliness is critical. For other applications, multisensor rainfall estimates are used, for which radar rainfall estimates are a major input. Plate 2 depicts how the radar rainfall data are used in NWS for various quantitative hydrologic applications. In Plate 2, the acronyms, WFO, RFC, NCEP, and stage IV, refer to the Weather Forecast Office, the River Forecast Center, the National Centers for Environmental Prediction, and the mosaic of the RFC-produced multisensor quantitative precipitation estimation (QPE) products over the contiguous United States (CONUS) (see section 5.1).

The UK Met Office provides operational radar rainfall estimates through the Nimrod system [*Golding*, 1998; *Harrison et al.*, 2000]. The system includes algorithms to mitigate errors that arise from anomalous propagation, variations in the VPR, occultation, attenuation, and variations in the DSD. Nimrod uses NWP forecasts, satellite imagery, and rain gauge data in real time to inform the various quality control algorithms of the meteorological situation. For the use of radar data in hydrological applications and in NWP models in the U.K., the reader is referred to *Inter-Agency Committee on the Hydrological Use of Weather Radar* [2007] and *Macpherson* [2001], respectively.

4. ADVANCES AND CHALLENGES IN RADAR RAINFALL ESTIMATION

In this section, we describe the major sources of error, recent advances, and outstanding and emerging areas of research in rainfall estimation using ground-based scanning radars. Owing to space limitations, it is not possible to cover all significant sources of error. For an exhaustive treatment, the reader is referred to *Villarini and Krajewski* [2010]. Here we focus on the major sources of error as identified, in particular, from the operational experience with the WSR-88D network in the United States for operational hydrologic forecasting. Extensive experience with and assessment of other radar processing algorithms elsewhere in other hydroclimatological regimes are available in the literature (*Germann et al.* [2006a], *Tabary* [2007], and *Tabary et al.* [2007], to name a few). The major error sources may be grouped into the following three headings: (1) sampling geometry and reflectivity morphology, (2) microphysics, and (3) hardware and beam propagation.

4.1. Sampling Geometry and Reflectivity Morphology

As noted above, the base elevation angle of radially scanning single-polarized weather radars is typically set to above zero to overcome ground clutter and beam shielding by the ground. Hence, the farther the range is, the higher the axis of the radar beam and the larger the sampling volume is. Because of this sampling geometry and the fact that the reflectivity profile of precipitating clouds is not uniform in the vertical (see Plate 1), radar rainfall estimates are subject to biases that are range-dependent (*Joss and Waldvogel* [1990], *Fabry et al.* [1992a], *Kitchen and Jackson* [1993], *Smith et al.* [1996b], and *Vignal et al.* [1999], just to name several). In addition, depending on the sampling geometry of the radar beam relative to the reflectivity morphology of the precipitating cloud and topography, the spatial extent to which radar can consistently see precipitation may vary significantly [*O'Bannon*, 1997; *Breidenbach et al.*, 1999; *Maddox et al.*, 2002; *Wood et al.*, 2003; *Germann et al.*, 2006a]. As such, for successful radar rainfall estimation it is necessary to delineate accurately the effective coverage of the radar. Such maps may be obtained from a combination of digital terrain model (DTM), radar beam models and climatological VPRs [*Pellarin et al.*, 2002; *Krajewski et al.*, 2006b]. The DTMs, however, may not account for man-made or biological obstructions such as tall buildings, water towers, trees, etc. If a long-term archive of radar data is available, more accurate maps may be obtained from long-term climatology of radar reflectivity or rainfall estimates (see section 6). For WSR-88D, a combination of the beam model and DTM is used to determine the lowest unobstructed tilt and to compensate for partial obstructions [*Fulton et al.*, 1998]. The effective coverage maps thus obtained are necessarily static and do not account for storm-to-storm or intrastorm variability of the effective coverage because of, e.g., changes

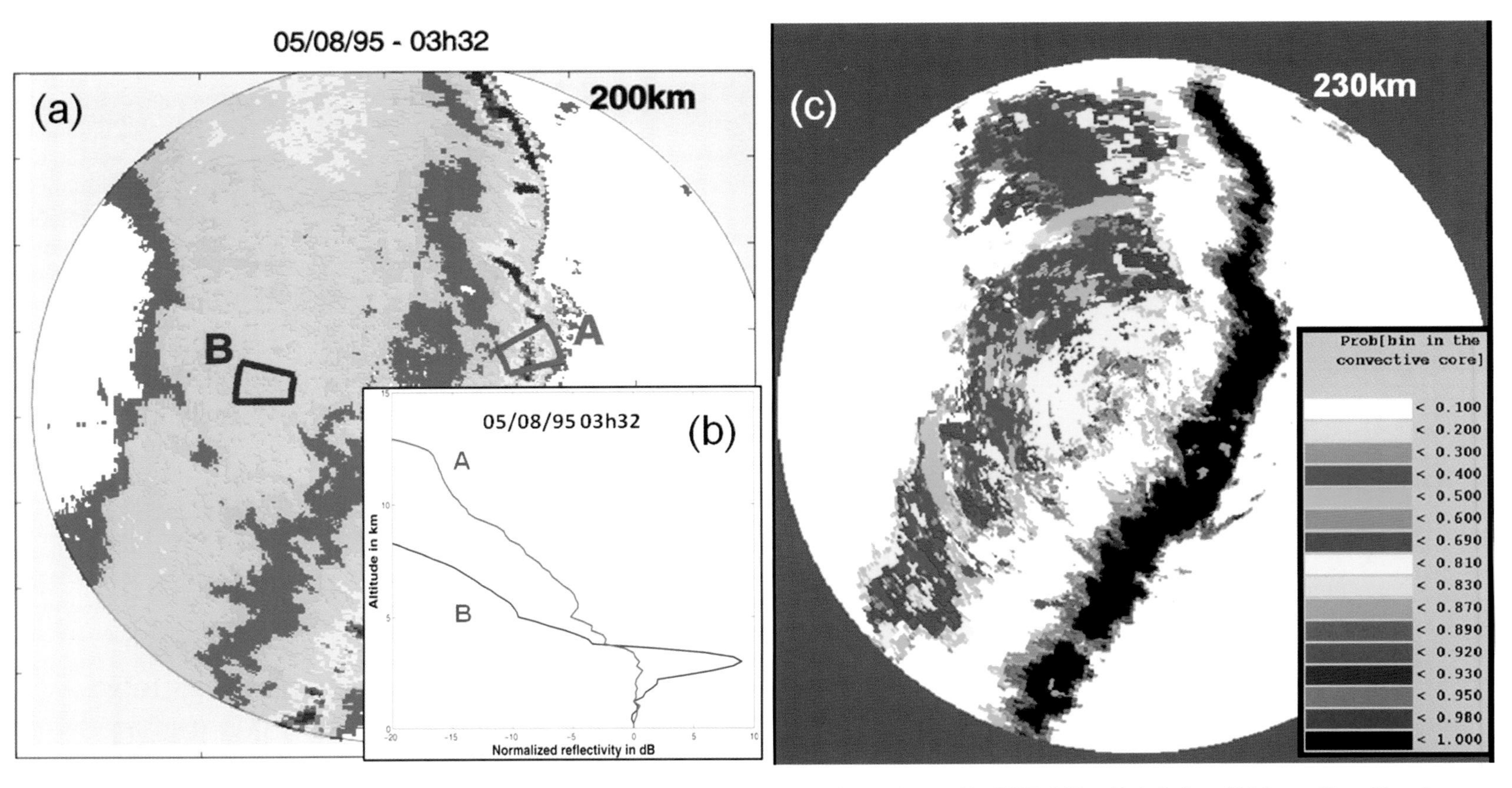

Plate 1. (a) Base reflectivity field for a very well developed squall line in the southern plains as observed by WSR-88D at Twin Lakes, Oklahoma [from *Vignal and Krajewski*, 2001]. Copyright American Meteorological Society. (b) Apparent mean vertical profiles of reflectivity (VPR) in the convective core (A) and trailing stratiform region (B) [from *Vignal and Krajewski*, 2001]. Copyright American Meteorological Society. (c) Convective-Stratiform Separation Algorithm (CSSA)-estimated probability of the azimuth-range bin belonging to convective core [from *Ding et al.*, 2003].

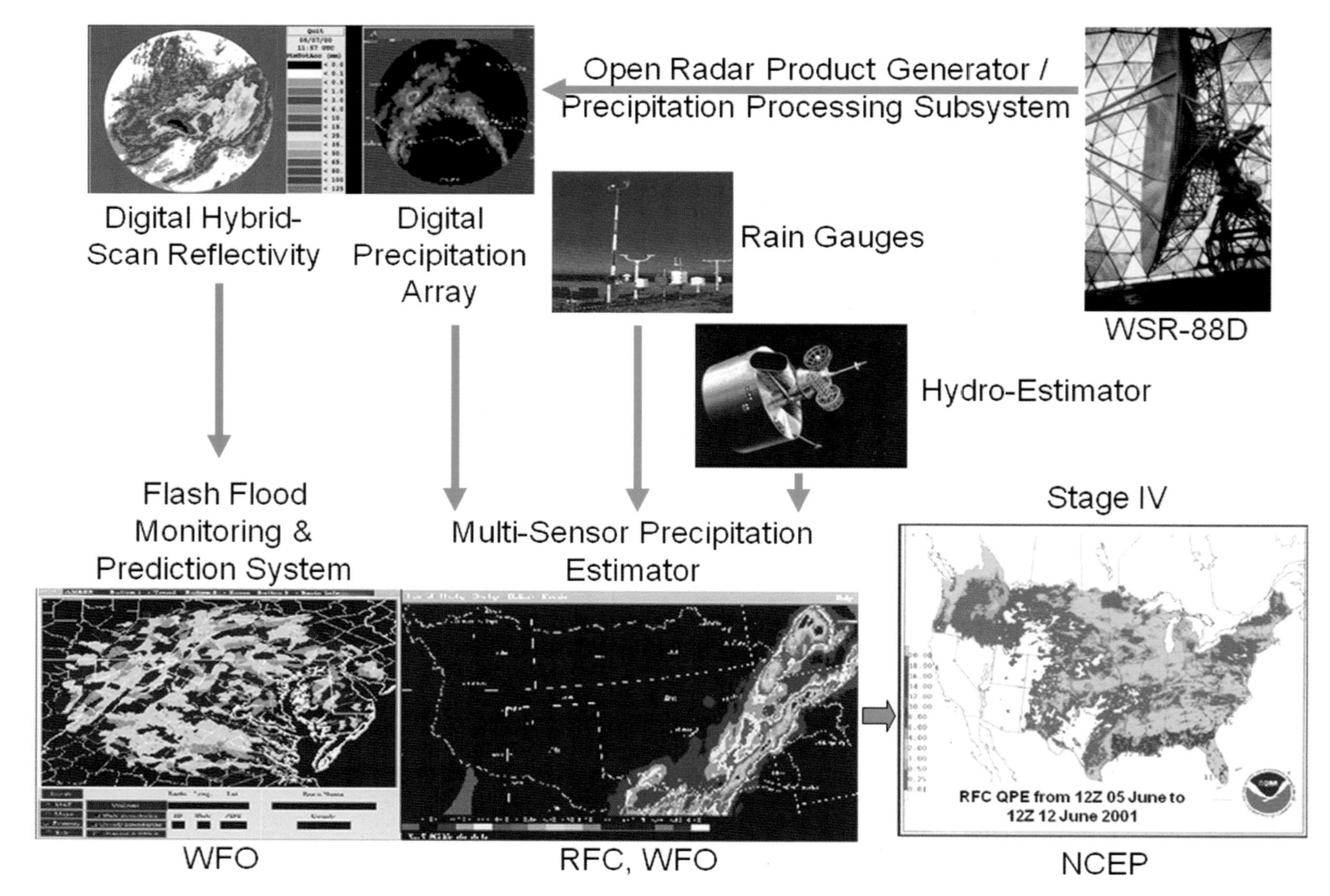

Plate 2. Use of radar rainfall data for quantitative hydrologic applications in NWS (see text for definition of the acronyms).

in the vertical extent of the precipitating cloud. By utilizing the full volume scan data, one may derive dynamic coverage maps based on echo-top height and beam blockage considerations. To illustrate, Plate 3b shows in solid line an example of such dynamically delineated coverage map for the WSR-88D at Seattle, Washington (KATX) [*Seo et al.*, 2000]. The azimuthal variations in the effective range reflect the radial-specific selection of an upper tilt due to severe blockages at lower elevations.

To reduce the VPR effects, numerous techniques have been developed over the years (see *Smith* [1986], *Fabry et al.* [1992b], *Kitchen et al.* [1994], *Andrieu and Creutin* [1995], *Seo et al.* [2000], *Vignal et al.* [1999], and *Germann and Joss* [2002], just to name several). The crux of VPR correction is to estimate the true VPR that would be observed by infinitely narrow and unobstructed radar beams at some representative space-time scale of sampling at which the predictive, or extrapolative, skill is maximized for estimating radar rainfall or reflectivity near the surface given that aloft. Such skill depends necessarily on the space-time variability of reflectivity morphology from intrastorm to climatological scales as captured by the particular scanning strategy of the radar under the local topography. Plate 3b (compare Plate 3a) shows how dynamic delineation of the effective coverage of the radar and VPR correction may be combined to produce radar rainfall maps in complex terrain. In Plate 3b, the close-range holes are due to rain shadowing by Mount Olympus. Owing to the fact that the height of the peak bright band enhancement is relatively high and the bright band profile is rather smooth, the VPR correction procedure works very well, and multiscan maximization [*Seo et al.*, 2000] produces a very realistic rainfall map almost everywhere within the effective coverage, including along the Cascade Mountains to the east where the uncorrected rainfall map (Plate 3a) shows highly irregular patterns of beam blockage and ground returns. Because VPR correction extrapolates the reflectivity profiles to the surface, VPR-corrected radar rainfall estimates are significantly more uncertain than those obtainable from the radar beam directly sampling the atmosphere near the ground. It may be expected that, for VPR correction to be operationally viable, reliable estimation of the uncertainty associated with VPR-corrected radar rainfall may be necessary in addition to characterization of the errors involved. Such uncertainty estimates may be obtained from the relationships of radar reflectivity or rainfall at different heights inferable from full-volume-scan reflectivity data [*Seo et al.*, 1996; *Berenguer and Zawadzki*, 2008, 2009].

Often, precipitating clouds consist of convective and stratiform areas for which the VPRs may differ greatly (see Plate 1). For VPR correction to be effective in mixed-type precipitation situations, it is therefore necessary first to identify areas of disparate reflectivity morphology before applying different VPR corrections and specific *Z-R* relationships to different areas. A number of techniques have been developed for convective-stratiform separation using radar data [*Steiner et al.*, 1995; *Sanchez-Diezma et al.*, 2000; *Mesnard et al.*, 2008; *Delrieu et al.*, 2009b]. Misidentification of the precipitation type and the resulting miscorrection of the VPR effects can carry a large penalty in real-time rainfall estimation for hydrologic applications because it may exacerbate errors in the raw radar rainfall estimates. As such, for convective-stratiform separation to be viable for operational hydrologic forecasting, the technique must be accurate at the space-time scale of the data (1° × 1 km, 5 to 10 min per volume scan for WSR-88D) and work well consistently for a variety of storms. The NWS Office of Hydrologic Development (OHD) has developed a technique for automatic identification of convective cores [*Ding et al.*, 2003, 2005] to be used with VPR correction. Referred to as the convective-stratiform separation algorithm (CSSA), it uses the maximum reflectivity in the vertical (dBZ), the vertically averaged local spatial correlation of reflectivity, the local spatial correlation of the top height (km) of apparent convective core (reflectivity exceeding 35 dBZ) and the vertically integrated liquid water (VIL) (kg m^{-2}) at each azimuthal-range bin to estimate the probability of the bin belonging to convective core. To estimate the probability, the CSSA employs a variant of indicator cokriging [*Deutsch and Journel*, 1992; *Seo*, 1996; *Brown and Seo*, 2010]. Plate 1c shows an example of the resulting probability map for the squall line shown in Plate 1a. The technique correctly identifies almost all areas of the convective front, but misidentifies a very small area in the stratiform region near the radar. In general, the technique works well for a variety of storms with limited training. For further details, the reader is referred to *Ding et al.* [2003].

Plate 3d shows example rainfall accumulation maps for a cool season event at Twin Lakes, Oklahoma (KTLX). It is obtained via convective-stratiform separation [*Ding et al.*, 2003] and VPR correction [*Seo et al.*, 2000] where the latter is applied only to the stratiform region. Plate 3d is to be compared with Plate 3c for which no VPR correction is applied. Without an extremely dense network of very high-quality rain gauges, it is not possible to quantitatively assess the goodness of the correction. Nevertheless, discerning qualitative assessment may be made by visual examination of the rainfall maps. Note in Plate 3d the reduced rainfall amounts in the bright band–enhanced areas, the increased amounts beyond the bright band, the easily identifiable convective cores following VPR correction, and the linear pattern of convective rainfall due to the advection of the storm

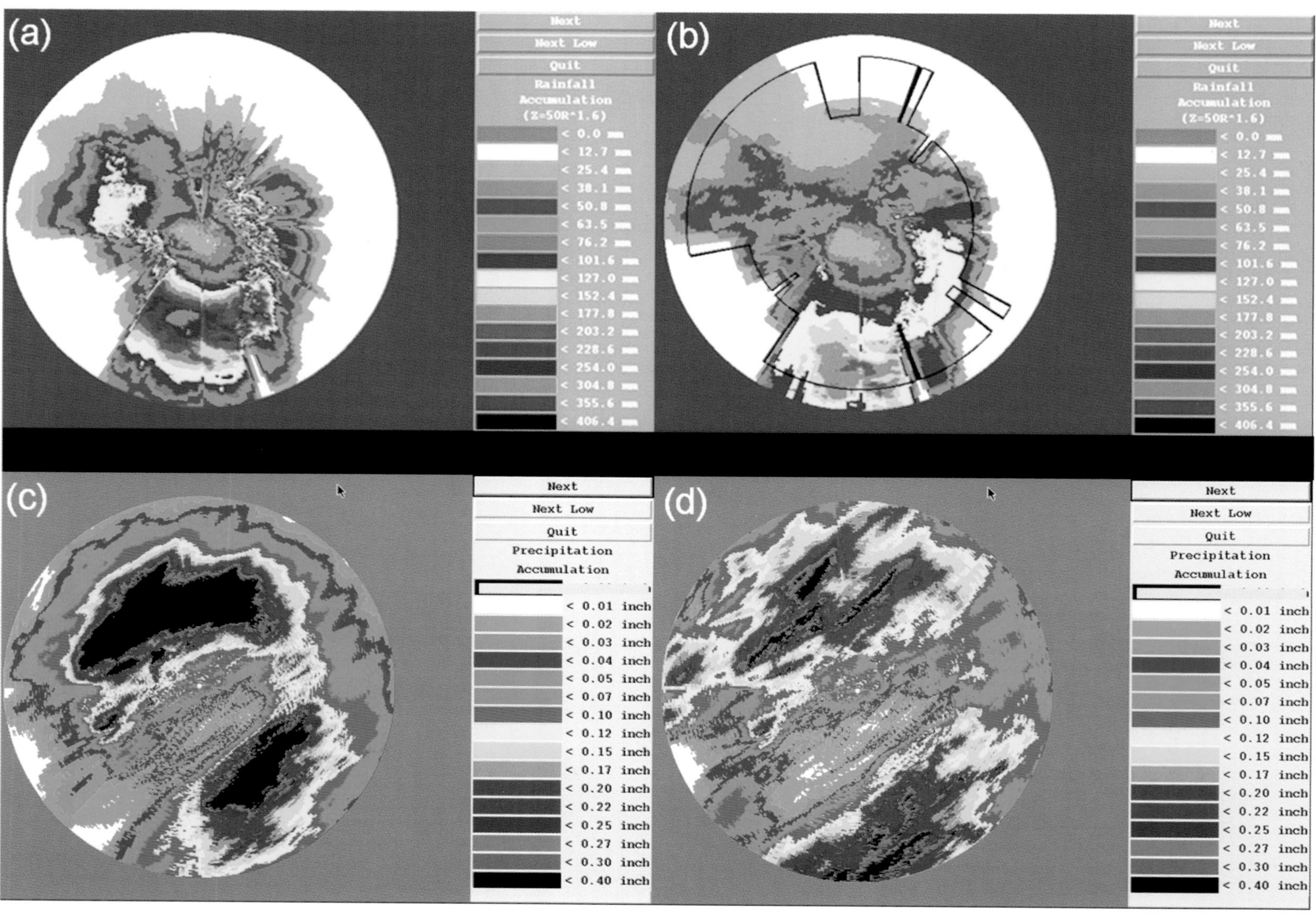

Plate 3. Example radar rainfall maps for a cool season event in complex terrain (Seattle, Washington) (a) without and (b) with VPR correction [from *Seo et al.*, 2000]. Example radar rainfall maps for the 5 January 2002, mixed-precipitation type event at Twin Lakes, Oklahoma, (c) without and (d) with convective-stratiform separation and VPR correction [from *Ding et al.*, 2004]. The maximum range is 230 km.

system. These observations suggest that the adjustment via a combination of convective-stratiform separation and VPR correction can significantly improve the quality of radar rainfall estimates, and serves to illustrate the importance of such correction, in particular, in the cool season. To ascertain the conditions under which such a two-step procedure may provide clear improvement, however, further research is necessary. For example, it should be recognized that the efficiency of convective-stratiform separation is inherently range-dependent due to the radar sampling characteristics. *Delrieu et al.* [2009b], in a mountainous setting in southern France, indicate a maximum range of about 80 km for the validity of their rain type separation; their trials to use the VPRs identified at close ranges for various rain types to improve rain type separation at greater ranges were only partially successful. Precipitation type-specific VPR correction as described above will also be necessary for rainfall estimation using dual-polarimetric radar, for which much additional research is needed as well.

4.2. Microphysics

Different types of precipitation are characterized not only by different reflectivity morphology but also by different DSDs (see section 2). In WSR-88D operations in the United States, five different *Z-R* relationships are used for different types of storms in different seasons and regions (see Table 1). Currently, the selection of the *Z-R* relationship is made by human forecasters based on guidance from the NWP models as well as real-time rain gauge observations. For large-scale events such as hurricanes and tropical storms, changes in the *Z-R* relationship are coordinated among all impacted WFOs and RFCs. In Australia, climatological *Z-R* relationships for convective and stratiform rainfall are derived for each radar in the network and a convective-stratiform classification algorithm is used to classify each pixel in the radar reflectivity map prior to converting radar reflectivity into rain rate [*Chumchean et al.*, 2008]. To assess relative merits among different approaches, rigorous objective assessment is needed.

Convective storms usually contain hail aloft. Because the *Z-R* relationships used in single-polarized radars assume that they are very large raindrops, converting reflectivity associated with hail to rainfall results in gross overestimation of rainfall. For this reason, an upper bound, or hail cap, is usually placed on the maximum reflectivity that can be converted to rainfall [*Fulton et al.*, 1998]. While such strategy works well in precipitation systems that are clearly highly convective, it may seriously underestimate rainfall if the capped reflectivity values are, in reality, associated with raindrops. Accurate radar rainfall estimation in the presence of hail is very important not only for real-time QPE of extreme events [*Baeck and Smith*, 1998] but also for climatological and probabilistic studies and applications using radar rainfall data [*Durrans et al.*, 2002]. While dual-polarimetric radar is expected to reduce the uncertainties in radar rainfall estimation of microphysical origin significantly, additional research is needed to provide objective guidance for estimation and specification of microphysical parameters in single-polarized radar rainfall estimation.

One of the biggest challenges in operational hydrologic forecasting in the United States is estimation and prediction of rainfall from relatively small-scale, low-centroid (i.e., bearing large reflectivity in the lower to near-surface part of the storm system), often tropical moisture-fed and orographically enhanced, warm rain process storms that produce extremely large amounts of rainfall [*Smith et al.*, 1996a; *Baeck and Smith*, 1998]. Such storms test the limits of radar rainfall estimation in that they demand accurate accounting of both the VPR effects due to orographic enhancement and any partial obstructions of radar beam due to the terrain as well as accurate estimation of spatiotemporally varying microphysical parameters. Much additional research is needed to address these issues for both single- and dual-polarized radar rainfall estimation.

4.3. Hardware and Beam Propagation

Radar calibration errors most often manifest as systematic differences in rainfall accumulations from multiple radars over overlapping areas as well as in radar–rain gauge comparisons [*Smith et al.*, 1996b]. By comparing systematic biases in radar rainfall estimates between each pair of radars with overlapping coverage, one may infer relative calibration differences among multiple radars. The radar sampling characteristics and the VPRs, however, may complicate the interpretation of such differences. Real-time radar–rain gauge comparisons of rainfall accumulations can also help detect major calibration problems. It is important to keep in mind, however, that the errors in radar rainfall may come from other sources such as the *Z-R* relationship, VPR, etc. *Delrieu et al.* [1995] suggested using ground clutter for checking long-term stability of radar calibration. They also found mountain clutter valuable in detecting antenna positioning errors. *Anagnostou et al.* [2001] used the Tropical Rainfall Measuring Mission (TRMM) Precipitation Radar (PR) observations to determine calibration biases of ground-based radars within the PR's coverage. While ideal, absolute global calibration, which involves checking the radar system against a known reference target such as a metal ball at different ranges, is too expensive to be practical. Internal calibration involves comparing the reading of the

Table 1. Recommended *Z-R* Relationships for WSR-88D Rainfall Estimation[a]

Relationship	Optimum for	Also Recommended for
Marshall-Palmer ($Z = 200R^{1.6}$)	general stratiform precipitation	
East: cool stratiform ($Z = 130R^{2.0}$)	winter stratiform precipitation: east of continental divide	orographic rain: east
West: cool stratiform ($Z = 75R^{2.0}$)	winter stratiform precipitation: west of continental divide	orographic rain: west
WSR-88D convective ($Z = 300R^{1.4}$)	summer deep convection	other nontropical convection
Rosenfeld tropical ($Z = 250R^{1.2}$)	tropical convective systems	

[a]From the *National Weather Service* [2006].

receiver against the known power source injected into the circulator. For WSR-88Ds in the United States, internal calibration is performed quarterly, or whenever transmitter/receiver components are replaced, and antenna gain and pointing accuracy are regularly verified via Suncheck [*Crum and Urell*, 2001]. An excellent summary of the issues related to absolute calibration can be found in the proceedings of the American Meteorological Society Radcal 2001 workshop (Radar Calibration and Validation Specialty Meeting, Albuquerque, New Mexico, 13–14 January, available at http://cdserver.ametsoc.org/cd/010430_1/radcal_main.html). Comparison of rainfall estimates among overlapping radar coverage is useful for inferring relative calibration differences only if the radars involved operate under similar conditions. Such comparison may not be possible if different sites are using different adaptable parameters in the radar rainfall algorithm, or severe VPR effects and/or topographic variations exist. Additional research is needed to develop techniques for real-time as well as off-line estimation of relative bias based on reflectivity measurements [*Michelson*, 2001].

Rainfall estimation using X band and, to a lesser extent, C band radar is subject to attenuation. The heavier the rainfall, the more significant attenuation is. *Bouilloud et al.* [2009] report maximum path-integrated attenuation between 15 and 20 dB at C band for path-averaged rain rates between 10 and 15 mm h^{-1} over a 120 km path during a flash flood–producing storm in Slovenia. *Hitschfeld and Bordan* [1954] derived the analytical solution of the attenuation equation (referred to here as the HB algorithm). They found, however, that the correction procedure is "almost useless" at X band "unless the calibration error may be held within extremely narrow limits" because of the high instability of the solution. The HB algorithm is also sensitive to parameterization of the DSD and its variations along the propagation path [*Nicol and Austin*, 2003; *Borowska et al.*, 2009]. Referred to as the mountain reference technique, algorithms that use mountain returns for estimating path-integrated attenuations to constrain forward or backward correction schemes have been developed [*Delrieu et al.*, 1999; *Serrar et al.*, 2000; *Bouilloud et al.*, 2009]. *Berenguer et al.* [2002] have also explored the use of rain gauge measurements for such a purpose. Attenuation through a wet radome can also be significant at both X and C bands. *Kurri and Huuskonen* [2006] reported a 3 dB two-way signal attenuation for a C band radar with a dirty radome under 15 mm h^{-1} rainfall intensity. With the use of dual polarization and Doppler capabilities, researchers reported a breakthrough for attenuation correction [e.g., *Testud et al.*, 2000; *Matrosov et al.*, 2002].

Various approaches have been reported to mitigate the adverse effects of ground returns from anomalous propagation (AP), such as the use of the Doppler velocity information during signal processing [*Goss and Chrisman*, 1995] and the use of the pulse-to-pulse variability of reflectivity [*Delrieu et al.*, 2009b]. In situations where ground clutter is embedded with returns from precipitation, however, complete and clean removal of the former is difficult. Also, excessive suppression of ground clutter may lead to low bias in reflectivity data, for which correction techniques have been developed for WSR-88D [*Ellis*, 2001]. At the postsignal processing level, ground clutter may be removed or reduced by statistical techniques with or without the aid of the Doppler velocity information [*Grecu and Krajewski*, 2000; *Krajewski and Vignal*, 2001; *Steiner and Smith*, 2002; *Kessinger et al.*, 2003]. Often, such techniques are designed as a part of a larger data quality-control procedure for identification and removal of not only ground clutter but also other nonprecipitation targets such as insects, birds or windborne particles, and test and interference patterns [*Steiner and Smith*, 2002; *Zhang et al.*, 2004; *Fritz et al.*, 2006].

While not dealt with in this chapter, other sources of error in radar rainfall estimation include vertical air motion, temporal sampling errors, evaporation in the subcloud layer, and horizontal advection below the radar sampling volume. For a comprehensive list, the reader is referred to *Krajewski and Smith* [2002] and *Villarini and Krajewski* [2010] and references therein.

To understand and reduce errors in radar rainfall estimates effectively, it is necessary to quantify the uncertainties associated with radar rainfall estimation. For most users, scale-specific total uncertainty in the radar rainfall products, as assessed against available ground truth (or estimated ground truth in the case of areal rainfall), is of primary importance [*Ciach et al.*, 2007; *Villarini et al.*, 2008, 2009; *Kirstetter et al.*, 2010]. For sophisticated users, estimates of scale- and source-specific uncertainties are necessary [*Ciach and Krajewski*, 1999a, 1999b; *Krajewski et al.*, 2006a]. Quantification of many of these uncertainties depends on the spatiotemporal scale of aggregation, and hence on the spatiotemporal correlation structures of the errors involved [*Berenguer and Zawadzki*, 2008; *Mandapaka et al.*, 2008; *Villarini and Krajewski*, 2009]. Such uncertainty analysis and modeling are also necessary to generate reliable (i.e., probabilistically unbiased) ensembles of radar rainfall [*Germann et al.*, 2006b] for hydrologic ensemble forecasting [*Seo et al.*, 2006]. Much additional research is needed in these areas as well as in verification, both single-valued and probabilistic, of radar QPE.

5. MULTISENSOR RAINFALL ESTIMATION

Weather radars provide spatially continuous estimates of rainfall that are subject to various sources of error. Rain gauges measure rainfall in situ at the surface but only at the points where the gauges are located. Since weather radars were first viewed as a tool for quantitative rainfall estimation for hydrologic applications in the 1970s, a number of ideas were advanced to capitalize on the complementarity between the sampling and error characteristics of the two sensors (see *Brandes* [1975], *Wilson and Brandes* [1979], *Krajewski* [1987], *Creutin et al.* [1988], *Azimi-Zonooz et al.* [1989], and *Seo et al.* [1990a, 1990b], just to name several). When applied to multisensor precipitation estimation, estimation theory [*Jazwinski*, 1970; *Schweppe*, 1973] states that, if the individual sensors possess skill in estimating precipitation, observations from multiple sensors may be combined to produce estimates that are more accurate than those obtainable from the individual sensors alone. Ultimately, optimal multisensor precipitation estimation depends on accurate knowledge of the spatiotemporal distribution of the observation and estimation errors associated with each sensor for the current and recent rainfall fields. This is a difficult task as the error varies concurrently with the atmospheric conditions. For example, the radar rainfall error as a function of range depends on the elevation of the bright band and the echo-top height, and the degree of partial occultation due to terrain depends on the humidity and temperature profiles which vary in space and time. Further complications arise when the data from one sensor is contaminated, e.g., with ground clutter that has not been removed, and these errors may degrade otherwise clean observations from the other sensors. The purpose of this section is to present a real-time application of multisensor estimation in the United States, the multisensor precipitation estimator (MPE), through which we describe the fundamentals of radar-based multisensor precipitation estimation and identify areas of further research toward realizing its promise as a source of the highest-quality precipitation analysis for all seasons, for all terrains, and over a wide range of spatiotemporal scales of aggregation.

5.1. *Multisensor Precipitation Estimator*

The ideas and prototype techniques for multisensor precipitation estimation were advanced further in the late 1980s and developed into automatic algorithms for operational implementation in the Next Generation Weather Radar (NEXRAD) [*Ahnert et al.*, 1983] and Advanced Weather Interactive Processing Systems (AWIPS) [*Hudlow et al.*, 1983] in NWS. In the early 1990s, the first generation of algorithms for multisensor QPE, known as Stage II/III [*Hudlow*, 1988; *Breidenbach et al.*, 1998], became operational at the RFCs. Later, Stage II went into operation at the WFOs in support of flash flood monitoring and prediction [*Smith et al.*, 2000]. In 1995–1996, the NCEP implemented Stage II over the CONUS to provide precipitation data for assimilation into the NWP models [*Lin and Mitchell*, 2005], for input into the continental-scale land surface models [*Mitchell et al.*, 2004], and for verification of quantitative precipitation forecasts (QPF) [*Baldwin and Mitchell*, 1998].

As experience with WSR-88D rainfall estimation [*Hunter*, 1996; *Fulton et al.*, 1998] and Stage II/III [*Breidenbach et al.*, 1998] grew, it became increasingly clear that a new suite of algorithms was necessary which, while building on the strengths of Stage II/III, takes full advantage of the improved understanding of the errors in the WSR-88D rainfall estimates [*Smith et al.*, 1996b; *Seo et al.*, 1999; *Young et al.*, 1999, 2000; *Seo and Breidenbach*, 2002; *Westcott et al.*, 2008; *Habib et al.*, 2009], the coverage overlap in the WSR-88D network, the improved computational and communications capabilities, the availability of multiyear radar rainfall data archive, and the improved availability of real-time rain gauge data. The result of the ensuing development is the MPE [*Breidenbach and Bradberry*, 2001; *Breidenbach et al.*, 2001b, 2002; *Seo and Breidenbach*, 2002], which replaced Stage III and II in the early 2000s at the RFCs and shortly thereafter at the WFOs, respectively.

In the following sections, we present the key components of MPE as a way of describing the state of operational multisensor rainfall estimation and identifying the areas of further research. Because the focus of this chapter is radar-based rainfall estimation, we limit ourselves to multisensor QPE using radar and rain gauge data only. For details regarding the use of satellite data in MPE, the reader is referred to *Kondragunta and Seo* [2004] and *Kondragunta et al.* [2005]. The MPE produces hourly precipitation estimates on the Hydrologic Rainfall Analysis Project (HRAP) [*Greene et al.*, 1979] grid which is about 4 km a side in CONUS. Recently, the high-resolution MPE, or HPE, has been implemented for precipitation analysis and nowcasting on a 1 km grid with a 5 min update cycle in support of flash flood monitoring and prediction. The further interested reader is referred to *Kitzmiller et al.* [2008].

5.2. Delineation of Effective Coverage of Radar

Owing to the sampling geometry of radar beams (see sections 2 and 4), the areal extent over which the radar can "see" precipitation may vary greatly depending on the vertical extent and structure of the storm. Because bias in precipitation detection directly results in bias in precipitation amount being estimated, it is very important that the areal extent over which the radar can consistently detect precipitation be delineated as accurately as possible. Plate 4 shows examples of the multiyear climatology of $E[R(u)]$, where $R(u)$ denotes the radar precipitation estimate at location u and $E[\ ]$ denotes the expectation operation (i.e., time averaging). The examples are based on the WSR-88D Digital Precipitation Array (DPA) product [*Klazura and Imy*, 1993] at Pittsburgh, Pennsylvania (KPBZ), for warm (left column) and cool (right column) seasons. The DPA product provides hourly radar rainfall estimates on the HRAP grid. Note that the unconditional mean $E[R(u)]$ may be written as $E[R(u)|R(u)>0]\bullet\Pr[R(u)>0]$ where $\Pr[\ ]$ denotes the probability of occurrence of the event bracketed. Assuming for the moment that radar can detect precipitation at close ranges as well as a rain gauge can with a small minimum detectable threshold (i.e., $\Pr[R(u)>0] \approx \Pr[G(u)>0]$ where $G(u)$ denotes the gauge rainfall observation at location u), one may apply appropriate thresholds to the probability or, alternatively, (unconditional) mean rainfall maps shown in the top row of Plate 4 to identify the areas of sufficiently large $\Pr[R(u)>0]$. The resulting binary maps, shown in the bottom row of Plate 4, may then be considered as the effective coverage of the radar for rainfall estimation. Note that in Plate 4, for the warm/cool seasons, the effective coverage is larger/smaller because the storms tend to be taller/shallower.

Because the true map of the probability of occurrence of precipitation is not known in reality, delineating the effective coverage is necessarily a somewhat subjective process that requires visual examination and human interactions. To assist interactive delineation, a software tool called RadClim (Radar Climatology) has been developed [*Breidenbach et al.*, 1999, 2001a]. RadClim is used to display various statistics of DPA, including the probability of detection at various thresholds and conditional and unconditional mean fields, apply user-selected thresholds, display the resulting effective coverage, display the PRISM data [*Daley et al.*, 1994] which is used as a proxy for true precipitation climatology and, if necessary, interactively modify the effective coverage maps [*Breidenbach et al.*, 1999]. The Precipitation-elevation Regressions on Independent Slopes Model, or PRISM, is an analysis technique for orographic precipitation at a monthly scale. Static delineation of the effective coverage can be overly conservative or too liberal if the storm on hand is not representative of the climatology. As such, it is necessary also to delineate the effective coverage dynamically based on real-time volumetric reflectivity data under beam overshooting and blockage considerations (see Plate 3b). Estimation and the combined use of such statically and dynamically delineated effective coverage maps are areas of further research for improving QPE particularly in the cool season in complex terrain.

5.3. Mean Field Bias Correction

There are two main sources of radar umbrella-wide bias, inaccurate radar calibration and uncertain *Z-R* relationship. While multiple *Z-R* relationships (see Table 1) offer more choices for the forecasters, selecting the appropriate relationship and the timing of the changes are often not very clear-cut and hence present a difficult task in real-time environment. Operational experience also indicates that early detection and correction of radar calibration errors remain a difficulty. Until a significant event occurs and comparisons may be made with observations from neighboring sites with overlapping coverage or from real-time rain gauges, it is often difficult to independently ascertain the goodness of radar calibration. Mean field bias (MFB) correction, which applies a multiplicative adjustment factor to the entire effective coverage area within the radar umbrella, is mathematically equivalent to adjusting the multiplicative constant in the *Z-R* relationship in real time [*Smith and Krajewski*, 1991], and thus mitigates also the uncertainty in the *Z-R* relationship. Note that MFB implicitly assumes that the detection domain has been previously determined and range-dependent biases (essentially related to VPR and terrain blockages for the S band WSR-88D systems) have been corrected. Because MFB has a very large impact on radar rainfall estimates, it has been a topic of active research for many years (see *Ahnert et al.* [1986], *Collier* [1986], *Smith and Krajewski* [1991], *Anagnostou et al.* [1998], and *Seo et al.*, 1999, just to name several). The algorithm used in MPE [*Seo et al.*, 1999] is an extension of *Seo et al.* [1997] and represents a significant departure from many of the previous approaches (see *Seo et al.* [1997] for a critical review), in that it targets unbiasedness rather than minimum error variance (i.e., seeks a first-order, rather than a second-order, solution). The

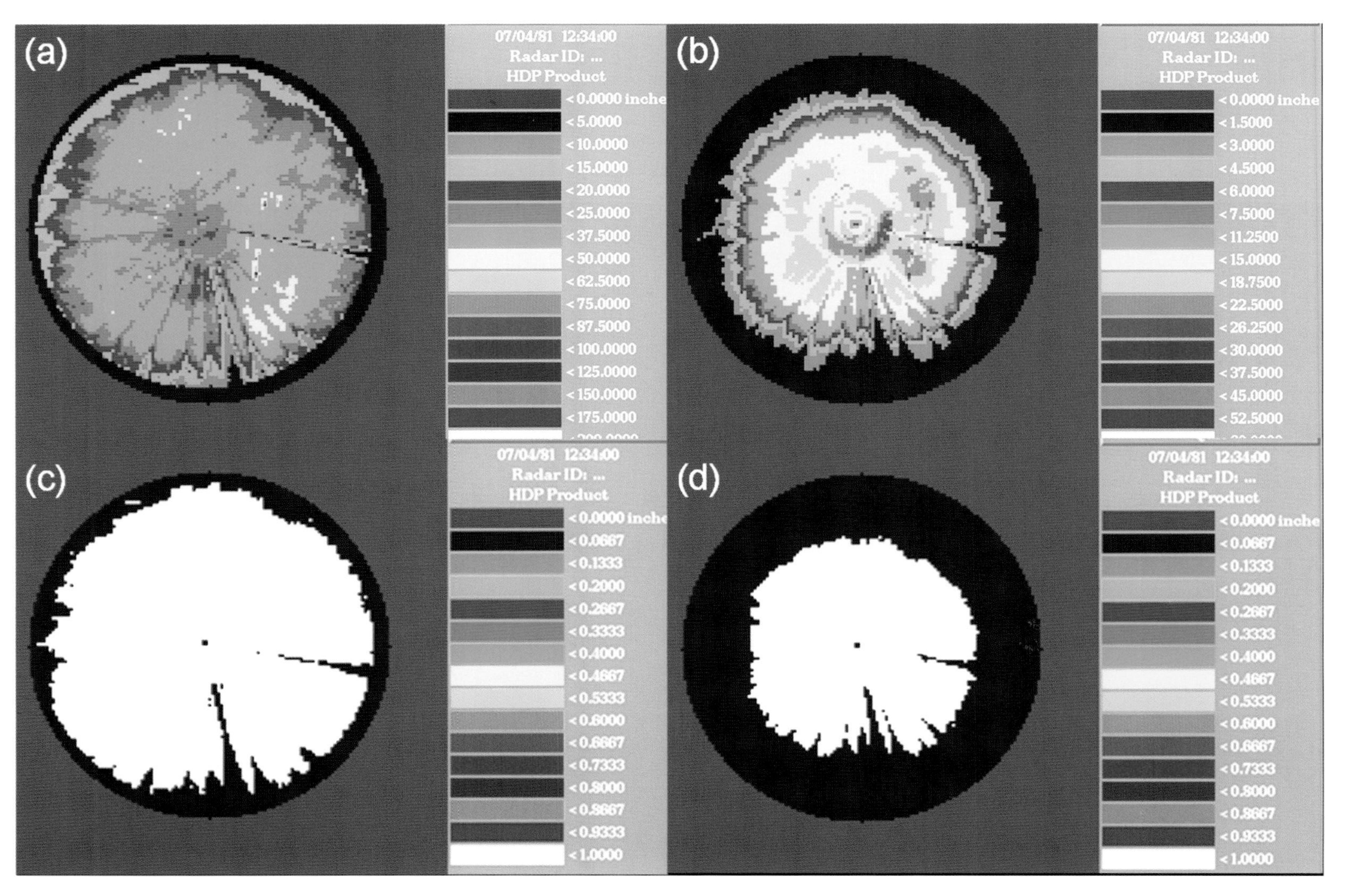

Plate 4. Multiyear climatology of (unconditional) mean radar rainfall at Pittsburgh, Pennsylvania, for (a) warm and (b) cool seasons. The effective coverage for radar rainfall estimation for (c) warm and (d) cool seasons as delineated via RadClim based on mean radar rainfall.

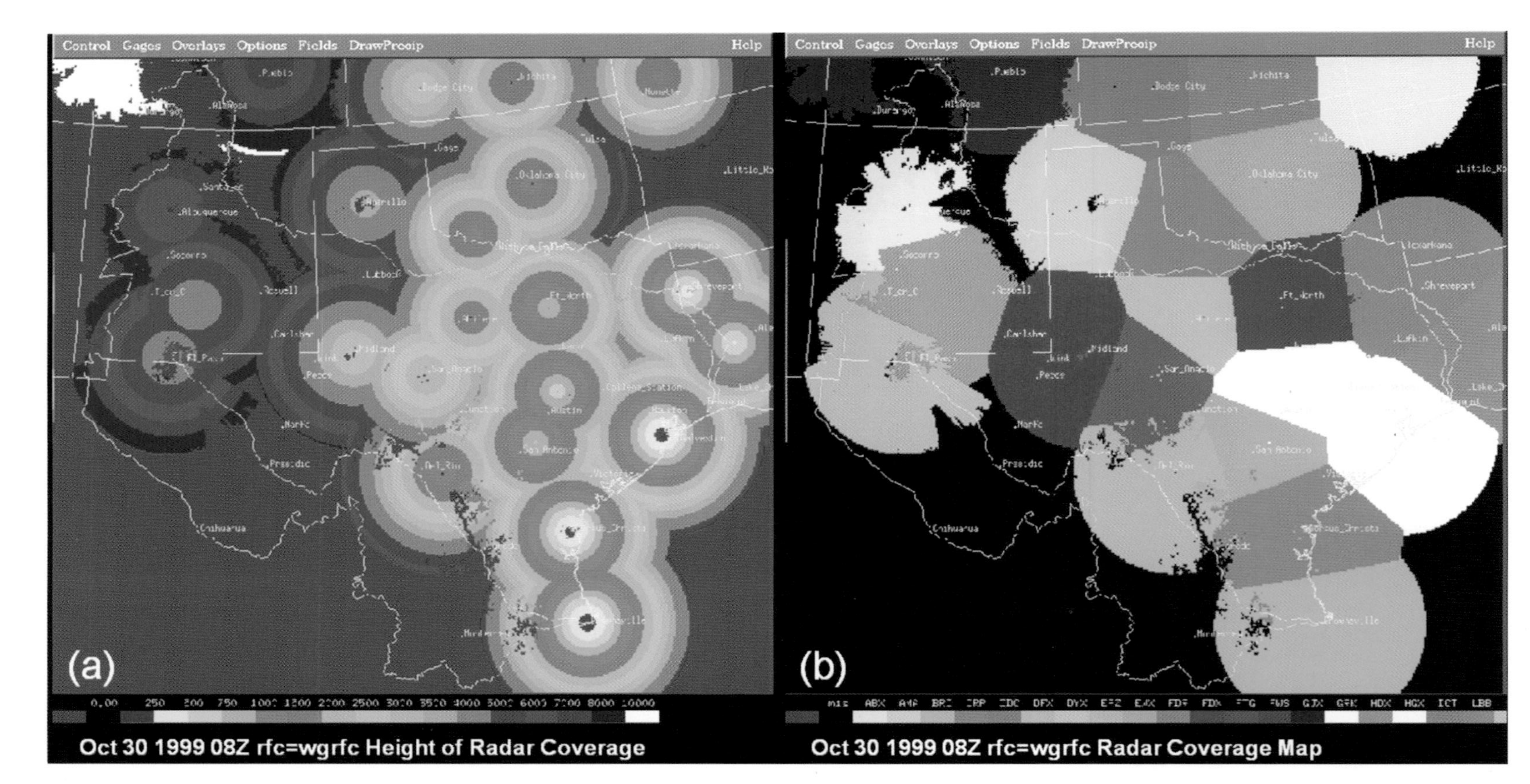

Plate 5. Examples of the height and index fields (see text for explanation) in MPE. The maps encompass the WGRFC's service area of Texas and vicinity. The area shown is bounded by 24.0° and 38.4° in latitude and by −110.3° and −90.4° in longitude.

algorithm estimates the mean field bias at hour k, β_k, defined as $\beta_k = \int_{A_c} G_k(u)du \Big/ \int_{A_c} R_k(u)du$, by recursively estimating the two terms in equation (5) at various scales of temporal averaging or aggregation:

$$\beta_k^* = E\left[\int_{A_c} G_k(u)du\right] \Big/ E\left[\int_{A_c} R_k(u)du\right]. \quad (5)$$

In the above, β_k^* denotes the estimate of β_k, A_c denotes the area commonly identified as raining by both the radar and the hypothetical infinitely dense network of rain gauges within the effective coverage of the radar, and $G_k(u)$ and $R_k(u)$ denote the hourly gauge and radar rainfall at location u at hour k, respectively. The algorithm produces a table of mean field bias estimates at various time scales of aggregation every hour for each WSR-88D site in the RFC's service area. From this table, the "best" bias estimate is selected by applying a threshold on the sample size. The result is that, in gauge-rich areas, the bias estimate is given by the sample bias for the current hour (i.e., the ratio of the sum of gauge rainfall to that of radar rainfall of all positive radar-gauge pairs from the current hour), whereas in gauge-poor areas it is given by the sample bias based on all positive radar-gauge pairs from as many preceding hours as necessary to reach the acceptably large sample size. A second-order solution approach to MFB is presented by *Chumchean et al.* [2006] where a Kalman filter was used to update the current estimate of the MFB using an observation error variance calculated at each 30 min time step based on the number of rain gauges reporting rainfall and their distance from the radar, and a simple first-order autoregressive model, or AR(1), that describes the temporal evolution of the MFB.

Necessarily, the effectiveness of MFB depends greatly on the availability and quality of rain gauge data. For MPE, the primary source of real-time hourly rain gauge data is the various gauge networks owned and operated by various government agencies and nongovernment organizations in the United States. They are collected and distributed to the forecast offices centrally via the Hydrometeorological Automated Data System (HADS) (http://www.nws.noaa.gov/oh/hads/) [*Kim et al.*, 2009]. The density of the collective gauge network varies greatly from region to region, and the majority of the data are delayed by a few or more hours because of limited bandwidth in the satellite communications channels and other factors. The design of the MFB and other algorithms used in MPE reflects these operational constraints so as to maximize the utility of all available rain gauge data.

5.4. Mosaicking Rainfall Estimates From Multiple Radars

Radar rainfall estimates are subject to increasingly large errors at far ranges. As such, mosaicking of radar data from multiple sites is a very important step in mitigating far-range degradation. Because MPE uses the two-dimensional DPA product rather than the three-dimensional full-volume-scan reflectivity data [*Zhang et al.*, 2005], mosaicking in MPE amounts to estimating radar rainfall in areas of overlapping coverage. For this, a number of approaches are possible. A more sophisticated approach may be to weight average all collocating DPA estimates from multiple radars based on VPR and beam blockage considerations. A simpler approach may be to choose a single estimate among the collocating DPA estimates based on beam blockage considerations alone. Currently, MPE employs the latter in which the rainfall estimate from the lowest unobstructed sampling volume is chosen as the "best" in areas of overlapping coverage. The rationale is that, barring beam obstructions due to natural or man-made objects [*O'Bannon*, 1997], the sampling volume closest to the surface should provide the most accurate estimate of surface rainfall.

Plate 5 shows examples of the map of the height of the sampling volume (above the mean sea level), referred to as the Height field (left plot), and the map of the contributing areas of coverage by individual radars, referred to as the Index field (right plot). In Plate 5a, the red-to-yellow transition at the low end of the legend corresponds to about 76 m (250 feet) above mean sea level (AMSL) and the purple-to-deep purple transition at the high end corresponds to about 2438 m (8000 feet) AMSL. The legend in Plate 5b shows the radar identifiers, which are not important in this paper. The examples in Plate 5 are for the West Gulf (WG) RFC, whose service area covers Texas and the neighboring states. Note in the Height field the widely varying sampling elevations in the radar beam due to topographic variations. The lowest are along the Gulf Coast to the bottom right corner of the map, and the highest are to the west of the continental divide in the western end of the map in New Mexico. Note in both the Height and Index fields the very significant coverage gaps in the western and southern parts of the WGRFC's service area. The western gap is due to the sparse WSR-88D network in sparsely populated areas and in high terrain. The southern gap is over Mexico for which MPE uses satellite precipitation estimates [*Kondragunta and Seo*, 2004; *Kondragunta et al.*, 2005].

The Index field is created in real time to reflect the availability of DPA products from all WSR-88D sites in the service area. The DPA products are then mosaicked according to the Index field to produce the Rmosaic field. The Index field is applied also to the mean field bias-corrected DPA data to produce the Bmosaic field. If significant calibration differences exist between the adjacent radars and/or different Z-R relationships are used, the Rmosaic field may exhibit discontinuities along the straight-line boundaries seen in the

Index field. The Bmosaic field may also exhibit similar artifacts if the gauge network is too sparse for MFB to be effective. These and other data processing artifacts may be best seen in long-term accumulations of precipitation estimates [see, e.g., *Nelson et al.*, 2010, Plate 5]. One might surmise that the use of multiple radar rainfall estimates over overlapping areas of coverage via some form of weighted averaging may readily provide significant improvement over the above procedure. Effectiveness of such an approach, however, depends greatly on how successful the correction of systematic, including range-dependent, biases may be in rainfall estimates from individual radars particularly in the cool season, for which much research is needed.

5.5. Local Bias Correction

If the spatial extent of systematic bias is localized because of space-time variability in the microphysical processes or the VPR effects, MFB may be of limited effectiveness (or even counterproductive [e.g., *Seo et al.*, 1999]). Local bias correction (LB) is an attempt to correct such locally varying systematic biases in place of MFB. In the United States, LB was first implemented operationally in the mid-1990s at the Arkansas-Red Basin River Forecast Center (ABRFC) at Tulsa, Oklahoma, using the P1 algorithm (Process 1 (B. Lawrence, personal communication, 2005)). Originally developed at the Tulsa District Office of the Army Corp of Engineers for analysis of daily precipitation (B. McCormick, personal communication, 2005), P1 performs bin-by-bin correction of DPA estimates using rain gauge data over the analysis domain (see *Seo and Breidenbach* [2002] for brief description). Operational experience indicates that P1 is particularly effective in gauge-rich areas for relatively uniform widespread precipitation in the cool season [*Young et al.*, 2000]. The LB algorithm used in MPE [*Seo and Breidenbach*, 2002] is a generalized local space-time smoother, intended for use in both gauge-rich and relatively gauge-poor areas. The LB algorithm estimates:

$$\beta_k^*(u_0) = \frac{E[G_k(u_0)|G_k(u_0) > 0, G_j(u_i), i = 1, \ldots, n_j, j = 1, \ldots, k, G_j(u_i) > 0]}{E[R_k(u_0)|R_k(u_0) > 0, R_j(u_i), i = 1, \ldots, n_j, j = 1, \ldots, k, R_j(u_i) > 0]}. \qquad (6)$$

In the above, $\beta_k^*(u_0)$ denotes the local bias estimate at the bin centered at u_0, n_j denotes the positive radar-gauge pairs at hour j, and $G_j(u_i)$ and $R_j(u_i)$ denote the gauge observation and radar rainfall estimate at location u_i at the jth hour, respectively. Note that equation (6) for LB differs from equation (5) for MFB in that the former yields spatially varying bias estimates. If, for example, the source of systematic bias is the uncertainty in the *Z-R* relationship, equation (6) is equivalent to adjusting the multiplicative constant in space and time in real time. As in MFB, equation (6) is estimated recursively at various scales of temporal averaging, and the "best" bias is selected based on the sample size consideration.

Two versions of the LB algorithm are available in MPE. The full version [*Seo and Breidenbach*, 2002] accounts for microscale variability of gauge precipitation (i.e., gauge measures "point" precipitation, whereas radar estimates are volume averaged) and different spatial correlation structures between gauge and radar rainfall, but is computationally expensive. The simple version is identical to the MFB algorithm as applied in a bin-by-bin manner, the difference being that the radar-gauge pairs are collected within a circle centered at each bin. The algorithm produces the local bias-corrected rainfall map, or the Lmosaic field, the local bias map, or the Locbias field, and the map of the time scale associated with the local bias selected, or the Locspan field. Depending on the adaptable parameter settings, the LB algorithm can be made to work similarly to P1 or the Brandes method [*Brandes*, 1975]. To account for regional and seasonal variations in the parameter settings, systematic parameter optimization for the algorithm is necessary, for which much additional work is necessary. Alternatively, real-time parameter optimization based on cross validation (see *Nelson et al.* [2010] for an example in a reanalysis mode) may be considered if sufficient computing power is available. Rather than estimating a simple local bias, *Bouilloud et al.* [2010] proposed, in the context of postevent analysis of localized flash flood–producing storms, to optimize an effective *Z-R* relationship from radar-rain gauge comparisons at the event time scale after correcting for range-dependent errors.

5.6. Gauge-Only Analysis

There are situations when radar rainfall data may not be available or contain extensive errors because of widespread ground clutter or bright band contamination. In such cases, rainfall analysis may have to be carried out using rain gauge data only. Gauge-only analysis has been a topic of research for almost a century [*Thiessen*, 1911] and numerous papers exist in the literature [see, e.g., *Creutin and Obled*, 1982; *Tabios and Salas*, 1985, and references therein]. Perhaps the most distinguishing aspect of gauge-only analysis in MPE is that the analysis domain is usually very large (O(10^5) km^2 or larger) whereas the temporal scale of analysis is only an hour. As such, precipitation almost always occurs only over parts of the analysis domain, and hence the analysis procedure needs to account for intermittency of precipitation [*Barancourt et al.*, 1992]. The algorithm used in MPE is a variant of the single optimal estimation (SOE) [*Seo*, 1998a] and accounts for intermittency in statistical modeling of the spatial variability of precipitation. The estimator used is analogous to ordinary kriging [*Journel and Huijbregts*, 1978] and has the following form:

$$G_k^*(u_0) = \sum_{i=1}^{n_g} \lambda_{gi} \frac{m_g(u_0)}{m_g(u_i)} G_k(u_i) \quad (7)$$

subject to

$$\sum_{i=1}^{n_g} \lambda_{gi} = 1. \quad (8)$$

In the above, $G_k(u_i)$ denotes the gauge rainfall at location u_i at hour k, n_g denotes the number of neighboring gauge data used in the spatial interpolation, λ_{gi} denotes the weight given to the ith gauge rainfall observation, and $G_k^*(u_0)$ denotes the estimated gauge rainfall at the ungauged bin centered at u_0 at hour k, and $m_g(u)$ denotes the climatological mean gauge rainfall at location u. The nearest gauge observations used in equation (7) are searched for within the user-specified radius of influence corresponding to the spatial scale of intermittency of precipitation [*Barancourt et al.*, 1992; *Seo and Smith*, 1996]. If no gauges exist within the radius, no precipitation is assumed at the bin centered in the radius of influence. The weights λ_{gi} are obtained by minimizing the conditional error variance of $G_k^*(u_0)$. Currently, the correlation structure is modeled as isotropic and invariant in space and time. Much research is needed to develop reliable, robust and computationally efficient procedures for improved modeling of the correlation structure which, most significantly, should reduce conditional bias. The climatological mean gauge precipitation, $m_g(u)$, is specified in MPE by the PRISM monthly precipitation climatology [*Daley et al.*, 1994]. Note that the constraint in equation (8) renders the estimate $G_k^*(u_0)$ climatologically unbiased in the mean, which may be easily verified by taking expectations on both sides of equation (7) and applying equation (8). While the specifics of selecting neighboring gauges and calculating the weights may differ, the estimator of the form of equations (7) and (8) represents a long-standing and time-tested procedure for gauge-only analysis in operational hydrology [*National Weather Service*, 2005]. The algorithm produced the gauge-only analysis, referred to as the Gmosaic field.

5.7. Radar-Gauge Analysis

The primary purpose of bias correction in MPE is to reduce systematic biases (e.g., the mean error) in radar rainfall estimates, rather than to reduce scatter between the bias-corrected estimates and ground truth (e.g., the error variance). As such, the bias-corrected estimates may not necessarily be an improvement over the raw estimates in the minimum error variance sense. Also, multiplicative bias correction such as MFB and LB is applicable only in areas where the radar can detect precipitation. Radar-gauge analysis is an attempt to address these issues by additively combining rain gauge and bias-corrected radar data. Multisensor merging or blending based particularly on geostatistical and related techniques is an active area of research, and many variations exist [see, e.g., *Sinclair and Pegram*, 2004; *Velasco-Forero et al.*, 2009]. The algorithm used in MPE is a variant of the bivariate extension [*Seo*, 1998b] of the single optimal estimation (SOE) [*Seo*, 1998a], and shares many of the same time-tested properties with the gauge-only analysis procedure:

$$G_k^*(u_0) = \sum_{i=1}^{n_g} \lambda_{gi} \frac{m_g(u_0)}{m_g(u_i)} G_k(u_i) + \sum_{j=1}^{n_r} \lambda_{rj} \frac{m_g(u_0)}{m_g(u_j)} \beta_k^* R_k(u_j) \quad (9)$$

subject to

$$\sum_{i=1}^{n_g} \lambda_{gi} + \sum_{j=1}^{n_r} \lambda_{rj} = 1. \quad (10)$$

In the above, the variables are as defined in equation (6) and the subscript "r" signifies that the variables are associated with radar data. In equation (9), instead of the MFB, β_k^*, the LB, $\beta_k^*(u_0)$, may also be used. Implicit in equations (9) and (10) is the assumption that, where the radar can detect rainfall, $\beta_k^* R_k(u_j)$ is climatologically unbiased with respect to gauge rainfall at that location. As in gauge-only analysis, the nearest gauge and radar rainfall data, n_g and n_r, respectively, are searched for only within the radius of influence. If no gauge or radar data exist within the radius, no precipitation is assumed at the bin centered in the radius of influence. The constraint equation (10) renders the estimate $G_k^*(u_0)$ in equation (9) unbiased in the mean sense. The weights given to the nearest gauge and radar data, λ_{gi} and λ_{rj}, in equation (9) are obtained by minimizing the conditional error variance of $G_k^*(u_0)$ (see *Seo* [1998b] for details). The resulting radar-gauge analysis map, referred to as the Mmosaic field, is generally considered the "best" analysis field because it is a product of both bias correction and error variance minimization.

Recently, *Velasco-Forero et al.* [2009] proposed a nonparametric technique based on Fast Fourier Transform (FFT) to obtain 2-D correlograms of the rainfall and those of the residuals to the so-called drift estimated from radar data. Coupled with kriging with external drift (KED), their approach was shown to outperform radar-only, rain gauge–only and other merging techniques such as collocated cokriging.

6. CHALLENGES IN MULTISENSOR RAINFALL ESTIMATION

The ultimate goal of multisensor QPE is to produce the most accurate precipitation estimates for all seasons and for all terrains at all space-time scales of aggregation using data from all available sources. Such multisensor QPE should be able to serve a wide range of hydrologic, hydroclimatological, water resources and other applications. Because the accuracy of multisensor estimates is tied to that of the ingredient data, the need to improve the accuracy of radar rainfall estimates [*Krajewski and Smith*, 2002; *Vasiloff et al.*, 2007; *Villarini and Krajewski*, 2010] and the availability and quality of rain gauge data [*Kim et al.*, 2009; *Nelson et al.*, 2010] is obvious. In this regard, rainfall estimates from dual-polarized radar should bring significant improvement to multisensor estimates within effective coverage [*Ryzhkov and Zrnic*, 1996; *Zrnic and Ryzhkov*, 1999; *Lang et al.*, 2009]. There remain, however, a number of fundamental issues that need to be addressed to realize the promise of multisensor precipitation estimation. Below, we describe some of them.

6.1. Detection Bias in Radar Rainfall Estimates

For quantitative hydrologic applications, by far the biggest issue with radar QPE is the systematic biases from various sources. From the perspective of multisensor estimation, one may group them into two: systematic bias in estimating rainfall amount given successful detection, and systematic bias in detecting rainfall. The estimation bias can be dealt with, to the extent that the space-time variability of rainfall and the availability of rain gauge data may allow, by corrective measures such as range-dependent bias correction, mean field bias correction and local bias correction. Correction of detection bias, on the other hand, requires having to create or remove areas of precipitation, and hence poses an estimation problem of a different nature. In theory, the detection bias may be accounted for by decomposing variability of precipitation into intermittency and inner variability, and carrying out estimation of probability of occurrence of precipitation and of conditional mean of precipitation separately [*Barancourt et al.*, 1992; *Seo*, 1998a]. Correcting for detection bias is particularly important in cool season, in complex terrain and for hydroclimatological applications. Accounting for detection bias may figure even more prominently when additional sources of data such as satellite, lightning and NWP model output are considered. As such, there exists a critical need for research and development of estimation procedures that explicitly account for detection bias in multisensor QPE.

6.2. Scale of Analysis

While the estimation theory–based statistical approach of gauge-only and radar-gauge analysis in MPE has proven effective in producing precipitation estimates at the scale of analysis of choice (e.g., hourly at HRAP), questions arise as to the goodness of the estimates when the hourly fields are aggregated to a larger temporal scale. For example, while the hourly MPE analysis may be reasonable anywhere in the analysis domain at a 1 h scale, the sum of the hourly analysis over a day, month, or year may not be, when compared with the analysis based on daily, monthly, or yearly observations, respectively. The above question arises because at the spatial scale of intermittency of hourly precipitation (i.e., the scale represented by the radius of influence used in gauge-only or radar-gauge analysis) the hourly rain gauge network may not be dense enough for the union of all radii of influence to encompass the entire analysis domain, whereas at the spatial scale of intermittency of daily, monthly, or yearly precipitation it is likely to be. Fundamentally, the above situation above is a consequence of the fact that the current operational multisensor estimation procedures are spatial only, rather than spatiotemporal. Much research is needed to develop spatiotemporal multisensor estimation procedures [*Gupta et al.*, 2006] in the context, e.g., of data assimilation, which would also accelerate the use of NWP model output in multisensor estimation.

6.3. Parameter Estimation

The algorithms used in MPE are designed to be parsimonious, and their adaptable parameters are chosen to be physically and statistically meaningful to produce reasonable results even with little or minimal parameter estimation. Note that the same may not be said for certain approaches such as the artificial neural network (ANN) [*Bellerby et al.*, 2000; *Grimes et al.*, 2003; *Hsu et al.*, 1997; *Chiang et al.*, 2007] for which training is a must. While the climatological sample statistics of the parameters make a good first guess, the optimal parameter values generally differ from the climatological sample statistics because of various nonlinear effects that are not accounted for in the estimation process [*Seo and Breidenbach*, 2002]. To realize the theoretically expected performance in terms of unbiasedness and minimum error variance, the adaptable parameters in MPE need to be optimized in a region- and seasonality-specific manner. A natural setup for such an effort may be an off-line version of MPE that allows routine long-term retrospective analysis (see below). Such a tool can also be used to quantify the uncertainty associated with the MPE products.

6.4. Reanalysis

In real-time rainfall estimation, a number of factors compromise the quality of the analysis: delays in data transmission, insufficient time for comprehensive quality control of the ingredient data, insufficient computational power to use better algorithms, etc. While timely analysis is critical to operational hydrologic forecasting, for many applications accuracy is by far the most important consideration. The purpose of reanalysis is to produce the highest-quality multisensor precipitation estimates retrospectively, taking full advantage of the additional rain gauge data that may not have been available in (near) real time, comprehensive quality control of the input data, post factum correction of systematic biases in the radar rainfall data, parameter optimization for improved accuracy, multiscale analysis to ensure consistency across a range of spatiotemporal scales of aggregation, etc.

Following reengineering of MPE for multisensor precipitation reanalysis (MPR), *Nelson et al.* [2010] have recently demonstrated over the Carolinas in the United States that such reanalysis can produce more accurate multisensor QPE than that obtainable operationally in real time. While their results are not representative of the entire United States, they drive home a number of the outstanding issues in radar and multisensor rainfall estimation described above, and serves to illustrate the potential of MPR for producing the highest-quality multisensor precipitation estimates. Below, we share some of these points through Figures 1 and 2. It is reminded that, in Figures 1 and 2, Stage IV refers to the CONUS mosaic of the RFC-produced QPE products (see Plate 2). Over the Carolinas, the Stage IV product is the same as the MPE product operationally produced at the Southeast (SE) RFC. It is very important to note here that, while the MPR and MPE products share the same names, both the quantity and the quality of the DPA products and, more importantly, the rain gauge data are not the same (e.g., only hourly gauge data are used for Stage IV whereas both hourly and daily gauge data are used for MPR), and the processing algorithms differ by varying degrees depending on the algorithm. For details, the reader is referred to *Nelson et al.* [2010].

Figure 1 shows the long-term bias in the MPR and Stage IV products in warm (April–September) and cool (October–March) seasons from 2002 to 2007 over the Carolinas. The bias is defined as the ratio of the sum of the estimates to that of the (withheld) daily gauge observations: bias $= \sum_{k=1}^{K} G_k / \sum_{k=1}^{K} R_k$, where G_k denotes the daily gauge observation for day

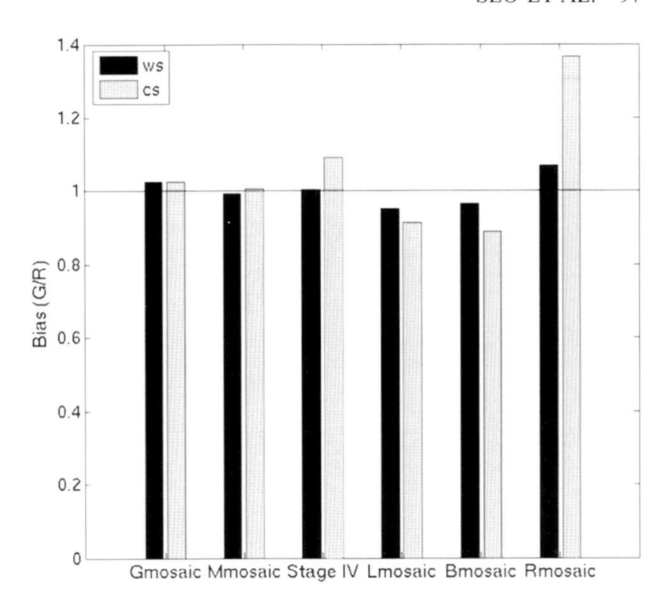

Figure 1. Long-term bias of various MPR products and stage IV relative to rain gauge observations in the Carolinas. From *Nelson et al.* [2010].

k, R_k denotes the precipitation estimate at the same location for the same 24 h period, and K denotes the total number of days in the cross-validation period. Note that the Stage IV product has little bias in the warm season, but is significantly biased in the cool season. The cool season bias is due to the underestimation in Rmosaic from the VPR effects and beam overshooting, and reflects the difficulty of addressing them in real time using a limited number of rain gauge observations. The high bias (i.e., overcorrection) in Lmosaic and Bmosaic also reflects the difficulty of gauge-based bias correction in the cool season in this area. Figures 2a and 2b show the root-mean-square error (RMSE) and correlation with the (withheld) daily gauge observation for each MPR product for warm and cool seasons, respectively. The smaller the RMSE and the larger the correlation is, the better the estimate is. Note that, in the warm season, Stage IV is consistently superior to Gmosaic, and Mmosaic consistently improves over Stage IV. For daily precipitation amounts greater than 25.4 mm, the margin of improvement by Mmosaic over Gmosaic is much greater, reflecting the great value that the radar offers for estimation of convective rainfall. In the cool season, however, Gmosaic is consistently superior to both Mmosaic and Stage IV while Mmosaic is consistently superior to Stage IV. That Mmosaic is inferior to Gmosaic is an indication that the quality of cool season radar QPE (i.e., the DPA products) needs to be improved significantly to add value to radar-gauge estimation. That Mmosaic very significantly improves over Stage IV reflects the value of utilizing all available rain gauge data [*Kim et al.*, 2009] and

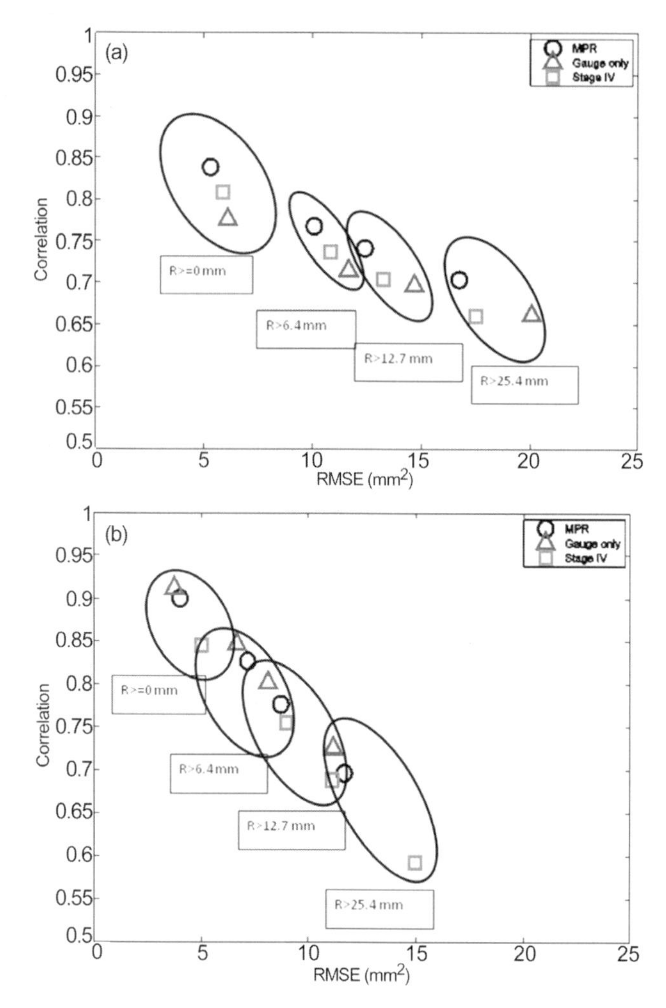

Figure 2. Root-mean-square error (RMSE) and correlation (with verifying gauge observation) of Mmosaic and Gmosaic from MPR (denoted in the legend as MPR and Gauge only, respectively), and stage IV over the Carolinas. The statistics are conditioned on the verifying gauge observation ≥0, >6.4, >12.7, and >25.4 (mm) for (a) warm and (b) cool seasons [from *Nelson et al.*, 2010].

the benefits of post factum bias correction and parameter optimization [*Nelson et al.*, 2010] afforded in reanalysis.

Development of such reanalysis capabilities and the associated error models have been identified as a critical area of research at the Weather Radar and Hydrology Workshop in 2008 (see the 2009 special issue, Weather Radar and Hydrology, in *Advances in Water Resources*, volume 32) for routine and widespread use of radar and radar-based rainfall products in hydrologic applications [*Delrieu et al.*, 2009a]. Finally, while not described in this section because of space limitations, much research is also needed in areas of ensemble and probabilistic approaches, and diagnostic evaluation and verification. For research directions and pertinent references on these and related topics, the reader is referred to *Krajewski and Smith* [2002] and *Vasiloff et al.* [2007].

7. CONCLUSIONS

While tremendous advances have been made in the science and technology of radar and multisensor rainfall estimation and the use of radar-based rainfall data and products, large challenges remain for the scientific, operational and user communities to capitalize on and realize fully the potential of radar-based rainfall estimation for a wide range of applications. In this chapter, we describe the foundations of radar and multisensor rainfall estimation, recent advances and real-world applications, and outstanding issues and emerging areas of research toward realizing their promises.

Experience with the WSR-88D network in the United States indicates that the most serious sources of error in radar rainfall estimation for operational hydrologic forecasting are the systematic biases arising from the sampling geometry of the radar beams and the reflectivity morphology of the precipitating clouds, and the uncertainties in the microphysical parameters and in discriminating hydrometeor type. The systematic biases are particularly important in the cool season and in complex terrain. The uncertainties are particularly important for small-scale low-centroid tropical moisture-fed events that produce extreme amounts of rainfall.

Estimation theory states that, if the individual sensors possess skill in estimating precipitation, observations from multiple sensors may be combined to produce estimates that are more accurate than those obtainable from the individual sensors alone. In this chapter, we present an operational application of the above principle, the multisensor precipitation estimator (MPE), through which the fundamentals of radar-based multisensor precipitation estimation are described and outstanding issues are identified toward realizing its promise as a source for the highest-quality precipitation analysis for all seasons, for all terrains, and over a wide range of spatiotemporal scales of aggregation. Challenges for multisensor precipitation estimation include accounting for the detection bias in radar rainfall estimates, ensuring statistical consistency across a wide range of scales of aggregation, improving accuracy through parameter optimization to address nonlinear effects, synergistically advancing real-time estimation and reanalysis, probabilistic and ensemble estimation, and verification and diagnostic evaluation.

Acknowledgments. The work carried out by the first author was supported by the NEXRAD Product Improvement (NPI), NEXRAD Technology Transfer, Advanced Weather Interactive Processing System (AWIPS), and Advanced Hydrologic Prediction

Service (AHPS) Programs of the NOAA National Weather Service (NWS) and the Climate Prediction Program for the Americas (CPPA) of the NOAA Climate Program Office (CPO). This support is gratefully acknowledged. The first author would like to thank David Kitzmiller and Yu Zhang of the NWS Office of Hydrologic Development (OHD) and the two anonymous reviewers for very helpful comments and suggestions, which significantly improved the paper.

REFERENCES

Ahnert, P. R., M. D. Hudlow, D. R. Greene, and E. R. Johnson (1983), Proposed on-site precipitation processing system for NEXRAD, paper presented at 21st Conference on Radar Meteorology, Am. Meteorol. Soc., Edmonton, Alta., Canada, 19–23 Sep.

Ahnert, P., W. Krajewski, and E. Johnson (1986), Kalman filter estimation of radar-rainfall field bias, paper presented at 23rd Conference on Radar Meteorology, Am. Meteorol. Soc., Snowmass, Colo.

Anagnostou, E. N., W. F. Krajewski, D.-J. Seo, and E. R. Johnson (1998), Mean-field rainfall bias studies for WSR-88D, *J. Hydrol. Eng.*, *3*, 149–159.

Anagnostou, E. N., C. A. Morales, and T. Dinku (2001), The use of TRMM precipitation radar observations in determining ground radar calibration biases, *J. Atmos. Oceanic Technol.*, *18*, 616–628.

Andrieu, H., and J. D. Creutin (1995), Identification of vertical profile of radar reflectivity for hydrological applications using an inverse method. Part I: Formulation, *J. Appl. Meteorol.*, *34*, 225–239.

Atlas, D., and C. W. Ulbrich (1977), Path- and area-integrated rainfall measurement by microwave attenuation in the 1–3 cm band, *J. Appl. Meteorol.*, *16*, 1322–1331.

Austin, P. M. (1987), Relation between measured radar reflectivity and surface rainfall, *Mon. Weather Rev.*, *115*, 1053–1070.

Azimi-Zonooz, A., W. F. Krajewski, D. S. Bowles, and D. J. Seo (1989), Spatial rainfall estimation by linear and non-linear co-kriging of radar-rainfall and raingage data, *Stochastic Hydrol. Hydraul.*, *3*, 51–67.

Baeck, M. L., and J. A. Smith (1998), Estimation of heavy rainfall by the WSR-88D, *Weather Forecasting*, *13*, 416–436.

Baldwin, M. E., and K. E. Mitchell (1998), Progress on the NCEP hourly multi-sensor U. S. precipitation analysis for operations and GCIP research, paper presented at 2nd Symposium on Integrated Observing Systems, 78th Annual Meeting, Am. Meteorol. Soc., Phoenix, Ariz., 11–16 Jan.

Barancourt, C., J. D. Creutin, and J. Rivoirard (1992), A method for delineating and estimating rainfall fields, *Water Resour. Res.*, *28*, 1133–1144.

Battan, L. J. (1973), *Radar Observation of the Atmosphere*, 324 pp., Univ. of Chicago Press, Chicago, Ill.

Bean, B., and E. Dutton (1968), *Radio Meteorology*, 435 pp., Dover, Mineola, N. Y.

Bellerby, T., M. Todd, D. Kniveton, and C. Kidd (2000), Rainfall estimation from a combination of TRMM precipitation radar and GOES multispectral satellite imagery through the use of an artificial neural network, *J. Appl. Meteorol.*, *39*, 2115–2128.

Berenguer, M., and I. Zawadzki (2008), A study of the error covariance matrix of radar rainfall estimates in stratiform rain, *Weather Forecasting*, *23*, 1085–1101.

Berenguer, M., and I. Zawadzki (2009), A study of the error covariance matrix of radar rainfall estimates in stratiform rain. Part II: Scale dependence, *Weather Forecasting*, *24*, 800–811.

Berenguer, M., G. W. Lee, D. Sempere-Torres, and I. Zawadzki (2002), A variational method for attenuation correction of radar signal, in *European Conference on Radar Meteorology (ERAD)*, *ERAD Publ. Ser.* 2, pp. 11–16, Copernicus, Göttingen, Germany.

Borowska, L., et al. (2009), Attenuation of radar signal in melting hail at C band, paper presented at 34th Conference on Radar Meteorology, Am. Meteorol. Soc., Williamsburg, Va., 5–9 Oct.

Bouilloud, L., G. Delrieu, B. Boudevillain, M. Borga, and F. Zanon (2009), Radar rainfall estimation for the post-event analysis of a Slovenian flash-flood case: Application of the Mountain Reference Technique at C-band frequency, *Hydrol. Earth Syst. Sci.*, *13*, 1349–1360.

Bouilloud, L., G. Delrieu, B. Boudevillain, and P. E. Kirstetter (2010), Radar rainfall estimation in the context of post-event analysis of flash floods, *J. Hydrol.*, in press.

Brandes, E. A. (1975), Optimizing rainfall estimates with the aid of radar, *J. Appl. Meteorol.*, *14*, 1339–1345.

Breidenbach, J. P., and J. S. Bradberry (2001), Multisensor precipitation estimates produced by National Weather Service River Forecast Centers for hydrologic applications, *Proceedings of the 2001 Georgia Water Resources Conference, March 26–27*, pp. 179–182, Inst. of Ecol., Univ. of Georgia, Athens, Ga.

Breidenbach, J. P., D.-J. Seo, and R. A. Fulton (1998), Stage II and III post processing of the NEXRAD precipitation estimates in the modernized National Weather Service, paper presented at 14th Conference on IIPS, Am. Meteorol. Soc., Phoenix, Ariz.

Breidenbach, J. P., D.-J. Seo, P. Tilles, and K. Roy (1999), Accounting for radar beam blockage patterns in radar-derived precipitation mosaics for river forecast centers, paper presented at 15th Conference on IIPS, Am. Meteorol. Soc., Dallas, Tex.

Breidenbach, J. P., D. J. Seo, P. Tilles, and C. Pham (2001a), Seasonal variation in multi-radar coverage for WSR-88D precipitation estimation in a mountainous region, paper presented at 81st Annual Meeting, Precipitation Extremes: Prediction, Impacts, and Responses, Am. Meteorol. Soc., Albuquerque, N. M.

Breidenbach, J. P., M. A. Fortune, D. J. Seo, and P. Tilles (2001b), Multi-sensor precipitation estimation for use by river forecast centers during heavy rainfall events, paper presented at 81st Annual Meeting, Precipitation Extremes: Prediction, Impacts, and Responses, Am. Meteorol. Soc., Albuquerque, N. M.

Breidenbach, J., D.-J. Seo, P. Tilles, and M. Fortune (2002), Multisensor precipitation estimation for use by the National Weather Service river forecast centers, paper presented at 16th Conference on Hydrology, Am. Meteorol. Soc., Orlando, Fla.

Brown, J. D., and D.-J. Seo (2010), A nonparametric post-processor for bias correcting ensemble forecasts of hydrometeorological and hydrologic variables, *J. Hydrometeorol.*, *11*, 642–665.

Chiang, Y.-M., F.-J. Chang, B. J.-D. Jou, and P.-F. Lin (2007), Dynamic ANN for precipitation estimation and forecasting from radar observations, *J. Hydrol.*, *334*, 250–261.

Chow, V. T., D. R. Maidment, and L. W. Mays (1988), *Applied Hydrology*, 570 pp., McGraw-Hill, New York.

Chumchean, S., A. Sharma, and A. Seed (2006), An integrated approach to error correction for real-time radar-rainfall estimation, *J. Atmos. Oceanic Technol.*, *23*, 67–79.

Chumchean, S., A. Seed, and A. Sharma (2008), An operational approach for classifying storms in real-time radar rainfall estimation, *J. Hydrol.*, *363*, 1–17, doi:10.1016/j.jhydrol. 2008.09.005.

Ciach, G. J., and W. F. Krajewski (1999a), On the estimation of radar rainfall error variance, *Adv. Water Resour.*, *22*, 585–595.

Ciach, G. J., and W. F. Krajewski (1999b), Radar-rain gauge comparisons under observational uncertainties, *J. Appl. Meteorol.*, *38*, 1519–1525.

Ciach, G. J., W. F. Krajewski, and G. Villarini (2007), Product-error-driven uncertainty model for probabilistic quantitative precipitation estimation with NEXRAD data, *J. Hydrometeorol.*, *8*, 1325–1347.

Cifelli, R., and V. Chandrasekar (2010), Dual-polarization radar rainfall estimation, in *Rainfall: State of the Science*, *Geophys. Monogr. Ser.*, doi: 10.1029/2010GM000930, this volume.

Cole, S. J., and R. J. Moore (2009), Distributed hydrological modelling using weather radar in gauged and ungauged basins, *Adv. Water Resour.*, *32*, 1107–1120.

Collier, C. G. (1986), Accuracy of rainfall estimates by radar, Part I: Calibration by telemetering raingauges, *J. Hydrol.*, *83*, 207–223.

Creutin, J. D., and C. Obled (1982), Objective analyses and mapping techniques for rainfall fields: An objective comparison, *Water Resour. Res.*, *18*, 413–431.

Creutin, J. D., G. Delrieu, and T. Lebel (1988), Rain measurement by raingage-radar combination: A geostatistical approach, *J. Atmos. Oceanic Technol.*, *5*, 102–115.

Crum, T., and B. Urell (2001), WSR-88D calibration changes and new approaches, paper presented at Radar Calibration Workshop, Radar Oper. Cent., Albuquerque, N. M., 13 Jan.

Daley, C., R. P. Neilson, and D. L. Phillips (1994), A statistical-topographic model for mapping climatological precipitation over mountainous terrain, *J.Appl. Meteorol.*, *33*, 140–158.

Delrieu, G., J. D. Creutin, and H. Andrieu (1995), Simulation of radar mountain returns using a digitized terrain model, *J. Atmos. Oceanic Technol.*, *12*, 1038–1049.

Delrieu, G., S. Serrar, E. Guardo, and J. D. Creutin (1999), Rain measurement in hilly terrain with X-band radar systems: Accuracy of mountain derived path-integrated attenuation, *J. Atmos. Oceanic Technol.*, *16*, 405–416.

Delrieu, G., et al. (2009a), Weather radar and hydrology, *Adv. Water Resour.*, *32*, 969–974.

Delrieu, G., B. Boudevillain, J. Nicol, B. Chapon, P.-E. Kirstetter, H. Andrieu, and D. Faure (2009b), Bollène 2002 experiment: Radar rainfall estimation in the Cevennes-Vivarais region, *J. Appl. Meteorol. Clim.*, *48*, 1422–1447.

Deutsch, C. V., and A. G. Journel (1992), *GSLIB Geostatistical Software Library andUser's Guide*, 340 pp., Oxford Univ. Press, Oxford, U. K.

Ding, F., D.-J. Seo, R. Fulton, and D. Kitzmiller (2003), Annual report of Office of Hydrologic Development to the Radar Operations Center, report, Natl. Weather Serv., Silver Spring, Md. (Available at www.nws.noaa.gov/oh/hrl/papers/papers.htm#wsr88d)

Ding, F., D.-J. Seo, D. Riley, R. Fulton, and D. Kitzmiller (2004), Annual report of Office of Hydrologic Development to the Radar Operations Center, report, Natl. Weather Serv., Silver Spring, Md. (Available at www.nws.noaa.gov/oh/hrl/papers/papers.htm#wsr88d)

Ding, F., D. Kitzmiller, D.-J. Seo, D. Riley, C. Dietz, C. Pham, and D. Miller (2005), A multi-site evaluation of the Range Correction and Convective-Stratiform Separation Algorithms for improving WSR-88D rainfall estimates, paper presented at 19th Conference on Hydrology, Am. Meteorol. Soc., San Diego, Calif., 9–13 Jan.

Doviak, R. J., and D. S. Zrnic (1993), *Doppler Radar and Weather Observations*, 458 pp., Academic, San Diego, Calif.

Dubé, I. (2003), From mm to cm: Study of snow/liquid ratio over Quebec, technical note, 127 pp., Meteorol. Serv. of Can. Que. Reg., Rimouski, Que., Canada. (Available at http://www.meted.ucar.edu/norlat/snowdensity/from_mm_to_cm.pdf)

Durrans, S. R., L. T. Julian, and M. Yekta (2002), Estimation of depth-area relationships using radar-rainfall data, *J. Hydrol. Eng.*, *7*, 356–367.

Ellis, S. M. (2001), Compensating reflectivity for clutter filter bias in the WSR-88D, paper presented at 17th International IIPS Conference, Am. Meteorol. Soc., Albuquerque, N. M., 14–18 Jan.

Fabry, F., and I. Zawadzki (1995), Long-term radar observations of the melting layer of precipitation and their interpretation, *J. Atmos. Sci.*, *52*(7), 838–851.

Fabry, F., G. L. Austin, and D. Tees (1992a), The accuracy of rainfall estimates by radar as a function of range, *Q. J. R. Meteorol. Soc.*, *118*, 435–453.

Fabry, F., G. L. Austin, and M. R. Duncan (1992b), Correction for the vertical profile of reflectivity using a vertically pointing radar, in *Hydrological Applications of Weather Radar*, edited by Fabry, F., and C. G. Collier, pp. 296–304, Ellis Horwood, Chichester, U. K.

Fritz, A., V. Lakshmanan, T. Smith, E. Forren, and B. Clarke (2006), A validation of radar reflectivity quality control methods, paper presented at 22nd International Conference on Information Processing Systems for Meteorology, Oceanography, and Hydrology, Am. Meteorol. Soc., Atlanta, Ga.

Fulton, R. A., J. P. Breidenbach, D.-J. Seo, and D. A. Miller (1998), The WSR-88D rainfall algorithm, *Weather Forecasting*, *13*, 377–395.

Germann, U., and J. Joss (2002), Mesobeta profiles to extrapolate radar precipitation measurements above the Alps to the ground level, *J. Appl. Meteorol.*, *41*, 542–557.

Germann, U., G. Galli, M. Boscacci, and M. Bolliger (2006a), Radar precipitation measurement in a mountainous region, *Q. J. R. Meteorol. Soc.*, *132*, 1669–1692.

Germann, U., M. Berenguer, D. Sempere-Torres, and G. Salvadè (2006b), Ensemble radar precipitation estimation—A new topic on the radar horizon, paper presented at Fourth ERAD Conference, Serv. Meteorol. de Catalunya, Barcelona, Spain.

Golding, B. W. (1998), Nimrod: A system for generating automated very short range forecasts, *Meteorol. Appl.*, *5*, 1–16.

Goss, S. M., and J. N. Chrisman (1995), An introduction to WSR-88D clutter suppression, and some tips for effective suppression utilization, operational report, Natl. Weather Serv., Silver Spring, Md.

Grecu, M., and W. Krajewski (2000), An efficient methodology for detection of anomalous propagation echoes in radar reflectivity data using neural networks, *J. Atmos. Oceanic Technol.*, *17*, 121–129.

Greene, D. R., M. D. Hudlow, and R. K. Farnsworth (1979), A multiple sensor rainfall analysis system, paper presented at Third Conference on Hydrometeorology, Am. Meteorol. Soc., Bogota.

Grimes, D. I. F., E. Coppola, M. Verdecchia, and G. Visconti (2003), A neural network approach to real-time rainfall estimation for Africa using satellite data, *J. Hydrometeorol.*, *4*, 1119–1133.

Gupta, R., V. Venugopal, and E. Foufoula-Georgiou (2006), A methodology for merging multisensor precipitation estimates based on expectation-maximization and scale-recursive estimation, *J. Geophys. Res.*, *111*, D02102, doi:10.1029/2004JD005568.

Habib, E., B. F. Larson, and J. Graschel (2009), Validation of NEXRAD multisensor precipitation estimates using an experimental dense rain gauge network in south Louisiana, *J. Hydrol.*, *373*, 463–478.

Harrison, D. L., S. J. Driscoll, and M. Kitchen (2000), Improving precipitation estimates from weather radar using quality control and correction techniques, *Meteorol. Appl.*, *6*, 135–144.

Hitschfeld, W., and J. Bordan (1954), Errors inherent in the radar measurement of rainfall at attenuating wavelengths, *J. Meteorol.*, *11*, 58–67.

Hsu, K. L., X. Gao, S. Sorooshian, and H. V. Gupta (1997), Precipitation estimation from remotely sensed information using artificial neural networks, *J. Appl. Meteorol.*, *36*, 1176–1190.

Hudlow, M. D. (1988), Technological developments in real-time operational hydrologic forecasting in the United States, *J. Hydrol.*, *102*, 69–92.

Hudlow, M. D., D. R. Greene, P. R. Ahnert, W. F. Krajewski, T. R. Sivaramakrishnan, E. R. Johnson, and M. R. Dias (1983), Proposed off-site precipitation processing system for NEXRAD, paper presented at 21st Conference on Radar Meteorology, Am. Meteorol. Soc., Edmonton, Alta., Canada, 19–23 Sep.

Hunter, S. M. (1996), WSR-88D radar rainfall estimation: Capabilities, limitations and potential improvements, *Natl. Weather Dig.*, *20*(4), 26–38.

Inter-Agency Committee on the Hydrological Use of Weather Radar in the United Kingdom (2007), Bibliography on hydrological applications of weather radar in the United Kingdom, report, Wallingford, U. K. (Available at http://www.iac.rl.ac.uk/Documents/Bibliography_IAC_2007.pdf)

Jazwinski, A. H. (1970), *Stochastic Processes and Filtering Theory*, Academic, San Diego, Calif.

Joss, J., and A. Waldvogel (1990), Precipitation measurement and hydrology, in *Radar inMeteorology: Battan Memorial and 40th AnniversaryRadar Meteorology Conference*, edited by Joss, J., pp. 577–606, Am. Meteorol. Soc., Boston, Mass.

Journel, A. G., and C. J. Huijbregts (1978), *Mining Geostatistics*, 600 pp., Academic, San Diego, Calif.

Judson, A., and N. Doesken (2000), Density of freshly fallen snow in the central Rocky Mountains, *Bull. Am. Meteorol. Soc.*, *81*, 1577–1587.

Kessinger, C., S. Ellis, and J. Van Andel (2003), The radar echo classifier: A fuzzy logic algorithm for WSR-88D, paper presented at 3rd Conference on Artificial Intelligence Applications to the Environmental Science, Am. Meteorol. Soc., Long Beach, Calif.

Kim, D., B. Nelson, and D.-J. Seo (2009), Characteristics of reprocessed Hydrometeorological Automated Data System (HADS), hourly precipitation data, *Weather Forecasting*, *24*, 1287–1296.

Kirstetter, P.-E., G. Delrieu, B. Boudevillain, and C. Obled (2010), Toward an error model for radar quantitative precipitation estimation in the Cévennes-Vivarais region, France, *J. Hydrol.*, in press.

Kitchen, M., and P. M. Jackson (1993), Weather radar performance at long range–Simulated and observed, *J. Appl. Meteorol.*, *32*, 975–985.

Kitchen, M., R. Brown, and A. G. Davies (1994), Real-time correction of weather radar data for the effects of bright band, range and orographic growth in widespread precipitation, *Q. J. R. Meteorol. Soc.*, *120*, 1231–1254.

Kitzmiller, D., F. Ding, S. Guan, D. Riley, M. Fresch, D. Miller, Y. Zhang, and G. Zhou (2008), Multisensor precipitation estimation in the NOAA National Weather Service: Recent advances, paper presented at World Environmental and Water Resources Congress, Am. Soc. of Civil Eng., Honolulu, Hawaii, 12–16 May.

Klazura, G. E., and D. A. Imy (1993), A description of the initial set of analysis products available from the NEXRAD WSR-88D system, *Bull. Am. Meteorol. Soc.*, *74*, 1293–1311.

Kondragunta, C., and D.-J. Seo (2004), Toward integration of satellite precipitation estimates into the multisensor precipitation estimator algorithm, paper presented at 18th Conference on Hydrology, Am. Meteorol. Soc., Seattle, Wash., 11–15 Jan.

Kondragunta, C., D. Kitzmiller, D.-J. Seo, and K. Shrestha (2005), Objective integration of satellite, rain gauge, and radar precipitation estimates in the multisensor precipitation estimator algorithm, paper presented at 19th Conference on Hydrology, Am. Meteorol. Soc., San Diego, Calif., 9–13 Jan.

Krajewski, W. F. (1987), Radar rainfall data quality control by the influence function method, *Water Resour. Res.*, *23*, 837–844.

Krajewski, W. F., and J. A. Smith (2002), Radar hydrology: Rainfall estimation, *Adv. Water Resour.*, *25*, 1387–1394.

Krajewski, W., and B. Vignal (2001), Evaluation of anomalous propagation echo detection in WSR-88D data: A large sample case study, *J. Atmos. Oceanic Technol.*, *18*, 807–814.

Krajewski, W. F., G. J. Ciach, and G. Villarini (2006a), Towards probabilistic quantitative precipitation WSR-88D algorithms: Data analysis and development of ensemble generator model: Phase 4, final report, 202 pp., Off. of Hydrol. Dev. of the Natl. Weather Serv., Silver Spring, Md.

Krajewski, W. F., A. A. Ntelekos, and R. Goska (2006b), A GIS-based methodology for the assessment of weather radar beam blockage in mountainous regions: Two examples from the US NEXRAD network, *Comput. Geosci.*, *32*, 283–302.

Krajewski, W. F., G. Villarini, and J. A. Smith (2010), Radar-rainfall uncertainties, where are we after thirty years of effort?, *Bull. Am. Meteorol. Soc.*, *91*, 87–94.

Kurri, M., and A. Huuskonen (2006), Measurement of the transmission loss of a radome at different rain intensities, paper presented at 4th European Conference on Radar in Meteorology and Hydrology (ERAD 2006), Cent. de Recerca Apl. en Hidrometeorol., Barcelona, Spain, 18–22 Sep.

Lang, T. J., S. W. Nesbitt, and L. D. Carey (2009), On the correction of partial beam blockage in polarimetric radar data, *J. Atmos. Oceanic Technol.*, *26*, 943–957.

Lee, G., and I. Zawadzki (2005), Variability of drop size distributions: Time scale dependence of the variability and its effects on rain estimation, *J. Appl. Meteorol.*, *44*, 241–255.

Lin, Y., and K. E. Mitchell (2005), The NCEP stage II/IV hourly precipitation analyses: Development and applications, paper presented at 19th Conference on Hydrology, Am. Meteorol. Soc., San Diego, Calif., 9–13 Jan.

Macpherson, B. (2001), Operational experience with assimilation of rainfall data in the Met Office Mesoscale Model, *Meteorol. Atmos. Phys.*, *76*, 3–8.

Maddox, R. A., J. Zhang, J. J. Gourley, and K. W. Howard (2002), Weather radar coverage over the contiguous United States, *Weather Forecasting*, *17*, 927–934.

Mandapaka, P. V., W. F. Krajewski, G. J. Ciach, G. Villarini, and J. A. Smith (2008), Estimation of radar-rainfall error spatial correlation, *Adv. Water Resour.*, *32*, 1020–1030.

Matrosov, S. Y., K. A. Clark, B. E. Martner, and A. Tokay (2002), X-band polarimetric radar measurements of rainfall, *J. Appl. Meteorol.*, *41*(9), 941–952.

Mesnard, F., O. Pujol, H. Sauvageot, N. Bon, C. Costes, and J.-P. Artis (2008), Discrimination between convective and stratiform precipitation in radar-observed rainfield using fuzzy logic, paper presented at 5th European Conference on Radar in Meteorology and Hydrology (ERAD2008), Helsinki, 30 June to 4 July.

Michelson, D. (2001), Normalizing a heterogeneous radar network for BALTEX, paper presented at Radar Calibration and Validation Specialty Meeting, Am. Meteorol. Soc., Albuquerque, N. M., 13–14 Jan.

Mitchell, K. E., et al. (2004), The multi-institution North American Land Data Assimilation System (NLDAS): Utilizing multiple GCIP products and partners in a continental distributed hydrological modeling system, *J. Geophys. Res.*, *109*, D07S90, doi:10.1029/2003JD003823.

National Weather Service (2005), The National Weather Service River Forecast System user manual documentation, report, Off. of Hydrol. Dev., Silver Spring, Md.

National Weather Service (2006), Guidance on adaptable parameters Doppler meteorological radar WSR-88D, in *WSR-88D Handbook*, vol. 4, Fed. Aviat. Admin. Tech. Issuance 6345.1, 186 pp., Radar Oper. Cent., Norman, Okla.

Nelson, B., D.-J. Seo, and D. Kim (2010), Multisensor precipitation reanalysis, *J. Hydrometeorol.*, *11*, 666–682.

Nicol, J. C., and G. L. Austin (2003), Attenuation correction constraint for single-polarisation weather radar, *Meteorol. Appl.*, *10*, 245–354.

O'Bannon, T. (1997), Using a 'terrain-based' hybrid scan to improve WSR-88D precipitation estimates, paper presented at 28th Conference on Radar Meteorology, Am. Meteorol. Soc., Austin, Tex., 7–12 Sep.

Pellarin, T., G. Delrieu, G. M. Saulnier, H. Andrieu, B. Vignal, and J. D. Creutin (2002), Hydrologic visibility of weather radar systems operating in mountainous regions: Case study for the Ardèche catchment (France), *J. Hydrometeorol.*, *3*, 539–555.

Pruppacher, H. R., and J. D. Klett (1997), *Microphysics of clouds and precipitation*, 2nd ed., 954 pp., Kluwer Acad., Dordrecht, Netherlands.

Rasmussen, R., M. Dixon, S. Vasiloff, F. Hage, S. Knight, J. Vivekanandan, and M. Xu (2003), Snow nowcasting using a real-time correlation of radar reflectivity with snow gauge accumulation, *J. Appl. Meteorol.*, *42*, 20–36.

Rosenfeld, D., and C. W. Ulbrich (2003), Cloud microphysical properties, processes, and rainfall estimation opportunities, in *Radar and Atmospheric Science: A Collection of Essays in Honor of David Atlas*, edited by R. M. Wakimoto and R. Srivastava, *Meteorol. Monogr.*, *30*, 237–258.

Ryzhkov, A. V., and D. S. Zrnic (1996), Assessment of rainfall measurement that uses specific differential phase, *J. Appl. Meteorol.*, *35*, 2080–2090.

Sánchez-Diezma, R., I. Zawadzki, and D. Sempere-Torres (2000), Identification of the bright band through the analysis of volumetric radar data, *J. Geophys. Res.*, *105*, 2225–2236.

Sauvageot, H. (1982), *Radarméteorologie*, 296 pp., Eyrolles, Paris.

Schweppe, F. C. (1973), *Uncertain Dynamic Systems*, 563 pp., Prentice-Hall, Englewood Cliffs, N. J.

Seo, D.-J. (1996), Nonlinear estimation of spatial distribution of rainfall—An indicator cokriging approach, *Stochastic Hydrol. Hydraul.*, *10*, 127–150.

Seo, D.-J. (1998a), Real-time estimation of rainfall fields using rain gage data under fractional coverage conditions, *J. Hydrol.*, *208*, 25–36.

Seo, D.-J. (1998b), Real-time estimation of rainfall fields using radar rainfall and rain gage data, *J. Hydrol.*, *208*, 37–52.

Seo, D.-J., and J. P. Breidenbach (2002), Real-time correction of spatially nonuniform bias in radar rainfall data using rain gauge measurements, *J. Hydrometeorol.*, *3*, 93–111.

Seo, D.-J., and J. A. Smith (1996), Characterization of the climatological variability of mean areal rainfall through fractional coverage, *Water. Resour. Res.*, *32*, 2087–2095.

Seo, D.-J., W. F. Krajewski, and D. S. Bowles (1990a), Stochastic interpolation of rainfall data from rain gages and radar using cokriging: 1. Design of experiments, *Water Resour. Res.*, *26*, 469–477.

Seo, D.-J., W. F. Krajewski, A. Azimi-Zonooz, and D. S. Bowles (1990b), Stochastic interpolation of rainfall data from rain gages and radar using Cokriging: 2. Results, *Water Resour. Res.*, *26*, 915–924.

Seo, D.-J., R. A. Fulton, J. P. Breidenbach, M. Taylor, and D. A. Miller (1996), Interagency memorandum of understanding among the NEXRAD Program, WSR-88D Operational Support Facility, and National Weather Service Office of Hydrologic Development, final report, 64 pp., Off. of Hydrol. Dev., Natl. Weather Serv., Silver Spring, Md.

Seo, D.-J., R. A. Fulton, and J. P. Breidenbach (1997), Interagency memorandum of understanding among the NEXRAD Program, WSR-88D Operational Support Facility, and National Weather Service Office of Hydrologic Development Hydrology Laboratory, final report, Off. of Hydrol. Dev., Natl. Weather Serv., Silver Spring, Md.

Seo, D.-J., J. P. Breidenbach, and E. R. Johnson (1999), Real-time estimation of mean field bias in radar rainfall data, *J. Hydrol.*, *223*, 131–147.

Seo, D.-J., J. P. Breidenbach, R. A. Fulton, D. A. Miller, and T. O'Bannon (2000), Real-time adjustment of range-dependent bias in WSR-88D rainfall data due to nonuniform vertical profile of reflectivity, *J. Hydrometeorol.*, *1*, 222–240.

Seo, D.-J., H. D. Herr, and J. C. Schaake (2006), A statistical post processor for accounting of hydrologic uncertainty in short range ensemble streamflow prediction, *Hydrol. Earth Syst. Sci. Discuss.*, *3*, 1987–2035.

Serrar, S., G. Delrieu, J. D. Creutin, and R. Uijlenhoet (2000), Mountain reference technique: Use of mountain returns to calibrate weather radars operating at attenuating wavelengths, *J. Geophys. Res.*, *105*, 2281–2290.

Sinclair, S., and G. Pegram (2004), Combining radar and rain gauge rainfall estimates using conditional merging, *Atmos. Sci. Lett.*, *6* (1), 19–22.

Smith, C. J. (1986), The reduction of errors caused by bright bands in quantitative rainfall measurements made using radar, *J. Atmos. Oceanic Technol.*, *3*, 129–141.

Smith, J. A., and W. F. Krajewski (1991), Estimation of the mean field bias of radar rainfall estimates, *J. Appl. Meteorol.*, *30*, 397–412.

Smith, J. A., M. L. Baeck, M. Steiner, and A. J. Miller (1996a), Catastrophic rainfall from an upslope thunderstorm in the central Appalachians: The Rapidan Storm of June 27, 1995, *Water Resour. Res.*, *32*, 3099–3113.

Smith, J. A., D. J. Seo, M. L. Baeck, and M. D. Hudlow (1996b), An intercomparison study of NEXRAD precipitation estimates, *Water Resour. Res.*, *32*, 2035–2045.

Smith, J. A., M. L. Baeck, M. Steiner, B. Bauer-Messmer, W. Zhao, and A. Tapia (1996c), Hydrometeorological assessments of the NEXRAD rainfall algorithms, final report, 59 pp., Off. of Hydrol., Hydrol. Res. Lab., Natl. Weather Serv., Silver Spring, Md.

Smith, M. B., K. P. Georgakakos, and X. Liang (2004), The distributed model intercomparison project (DMIP), *J. Hydrol.*, *298*(1–4), 1–334.

Smith, S. B., M. T. Filiaggi, M. Churma, J. Roe, M. Glaudemans, R. Erb, and L. Xin (2000), Flash flood monitoring and prediction in AWIPS 5 and beyond, paper presented at 5th Conference on Hydrology, Am. Meteorol. Soc., Long Beach, Calif., 9–14 Jan.

Steiner, M., and J. Smith (2002), Use of three-dimensional reflectivity structure for automated detection and removal of non-precipitating echoes in radar data, *J. Atmos. Oceanic Technol.*, *19*, 673–686.

Steiner, M., R. A. Houze, Jr., and S. E. Yuter (1995), Climatological characterization of three-dimensional storm structure from operational radar and rain gauge data, *J. Appl. Meteorol.*, *34*, 1978–2007.

Super, A. B., and E. W. Holroyd (1998), Snow accumulation algorithm for the WSR-88D radar: Final Report, *Rep. R-98-05*, 75 pp., Bur. of Reclam., Denver, Colo.

Tabary, P. (2007), The new French operational radar rainfall product: Part I: Methodology, *Weather Forecasting*, *22*, 393–408.

Tabary, P., J. Desplats, K. Dokhac, F. Eideliman, C. Guéguen, and J.-C. Heinrich (2007), The new French operational radar rainfall product. Part 2: Validation, *Weather Forecasting*, *22*, 409–427.

Tabios, G. Q., and J. D. Salas (1985), A comparative analysis of techniques for spatial interpolation of precipitation, *Water Resour. Bull.*, *21*, 365–380.

Testud, J., E. Le Bouar, E. Obligis, and A. Ali-Mehenni (2000), The rain profiling algorithm applied to polarimetric weather radar, *J. Atmos. Oceanic Technol.*, *17*, 332–356.

Thiessen, A. H. (1911), Precipitation averages for large areas, *Mon. Weather Rev.*, *39*, 1082–1089.

Uijlenhoet, R., M. Steiner, and J. A. Smith (2003), Variability of raindrop size distributions in a squall line and implications for radar rainfall estimation, *J. Hydrometeorol.*, *4*, 43–61.

Vasiloff, S. V., et al. (2007), Improving QPE and very short term QPF: An Initiative For A Community-Wide Integrated Approach, *Bull. Am. Meteorol. Soc.*, *88*, 1899–1911.

Velasco-Forero, C. A., D. Sempere-Torres, E. F. Cassiraga, and J. J. Gomez-Hernandez (2009), A non-parametric automatic blending methodology to estimate rainfall fields from rain gauge and radar data, *Adv. Water Resour.*, *32*, 986–1002.

Vignal, B., and W. F. Krajewski (2001), Large-sample evaluation of two methods to correct range-dependent error for WSR-88D rainfall estimates, *J. Hydrometeorol.*, *2*, 490–504.

Vignal, B., H. Andrieu, and J. D. Creutin (1999), Identification of vertical profiles of reflectivity from volume scan radar data, *J. Appl. Meteorol.*, *38*, 1214–1228.

Villarini, G., and W. F. Krajewski (2009), Empirically based modeling of uncertainties in radar rainfall estimates for a C-band weather radar at different time scales, *Q. J. R. Meteorol. Soc.*, *135*, 1424–1438.

Villarini, G., and W. F. Krajewski (2010), Review of the different sources of uncertainty in single polarization radar-based estimates of rainfall, *Surv. Geophys.*, *31*, 107–129, doi:10.1007/s10712-009-9079-x.

Villarini, G., F. Serinaldi, and W. F. Krajewski (2008), Modeling radar-rainfall estimation uncertainties using parametric and non-parametric approaches, *Adv. Water Resour.*, *31*, 1674–1686.

Villarini, G., W. F. Krajewski, G. J. Ciach, and D. L. Zimmerman (2009), Product-error-driven generator of probable rainfall conditioned on WSR-88D precipitation estimates, *Water Resour. Res.*, *45*, W01404, doi:10.1029/2008WR006946.

Ware, E., D. Schultz, and H. Brooks (2006), Improving snowfall forecasting by accounting for the climatological variability of snow density, *Weather Forecasting*, *21*, 94–103.

Westcott, N. E., H. V. Knapp, and S. D. Hilberg (2008), Comparison of gage and multi-sensor precipitation estimates over a range of spatial and temporal scales in the midwestern United States, *J. Hydrol.*, *351*, 1–12.

Wilson, J. W., and E. Brandes (1979), Radar measurement of rainfall —A summary, *Bull. Am. Meteorol. Soc.*, *60*, 1048–1058.

Wood, V. T., R. A. Brown, and S. V. Vasiloff (2003), Improved detection using negative elevation angles for mountaintop WSR-88Ds. Part II: Simulations of the three radars covering Utah, *Weather Forecasting*, *18*, 393–403.

Young, C. B., B. R. Nelson, A. A. Bradley, J. A. Smith, C. D. Peters-Lidard, A. Kruger, and M. L. Baeck (1999), An evaluation of NEXRAD precipitation estimates in complex terrain, *J. Geophys. Res.*, *104*, 19,691–19,703.

Young, C. B., A. A. Bradley, W. F. Krajewski, and A. Kruger (2000), Evaluation NEXRAD multisensor precipitation estimates for operational hydrologic forecasting, *J. Hydrometeorol.*, *1*, 241–254.

Zhang, J., S. Wang, and B. Clarke (2004), WSR-88D reflectivity quality control using horizontal and vertical reflectivity structure, paper presented at 11th Conference on Aviation, Range and Aerospace Meteorology, Am. Meteorol. Soc., Hyannis, Mass.

Zhang, J., K. Howard, and J. J. Gourley (2005), Constructing three-dimensional multiple radar reflectivity mosaics: Examples of convective storms and stratiform rain echoes, *J. Atmos. Oceanic Technol.*, *22*, 30–42.

Zrnic, D. S., and A. V. Ryzhkov (1999), Polarimetry for weather surveillance radar, *Bull. Am. Meteorol. Soc.*, *80*, 389–406.

G. Delrieu, Laboratoire d'étude des Transferts en Hydrologie en Environnement, Grenoble F-38041, France.

A. Seed, Australian Bureau of Meteorology, GPO Box 1289K, Melbourne, Vic 3001, Australia.

D.-J. Seo, Department of Civil Engineering, University of Texas at Arlington, Arlington, TX 76019, USA. (djseo@uta.edu)

Dual-Polarization Radar Rainfall Estimation

Robert Cifelli

Cooperative Institute for Research in the Atmosphere, Colorado State University, Fort Collins, Colorado, USA
Earth System Research Laboratory, NOAA, Boulder, Colorado, USA

V. Chandrasekar

Department of Electrical and Computer Engineering, Colorado State University, Fort Collins, Colorado, USA

Dual-polarization radar is a critical tool for weather research applications, including rainfall estimation, and is at the verge of being a key instrument for operational meteorologists. This new radar system is being integrated into radar networks around the world, including the planned upgrade of the U.S. National Weather Service Weather Surveillance Radar, 1988 Doppler radars. Dual polarization offers several advantages compared to single-polarization radar systems, including additional information about the size, shape, and orientation of hydrometeors. This information can be used to more accurately retrieve characteristics of the drop size distribution, identify types of hydrometeors, correct for signal loss (attenuation) in heavy precipitation, and more easily identify spurious echo scatterers. In addition to traditional backscatter measurements, differential propagation phase characteristics allow for rainfall estimation that is immune to absolute calibration of the radar system, attenuation effects, as well as partial beam blocking. By combining different radar measurements, rainfall retrieval algorithms have developed that minimize the error characteristics of the different rainfall estimators, while at the same time taking advantage of the data quality enhancements. Although dual-polarization techniques have been applied to S band and C band radar systems for several decades, polarization diversity at higher frequencies including X band are now widely available to the radar community. This chapter provides an overview of dual-polarization rainfall estimation applications that are typically utilized at X, C, and S bands. The concept of distinguishing basic and applied science issues and their impact on rainfall estimation is introduced. Various dual-polarization radar rainfall techniques are discussed, emphasizing the strengths and weaknesses of various estimators at different frequencies.

1. INTRODUCTION

1.1. Overview

In this chapter, we discuss dual-polarization techniques with regard to rainfall estimation. Historically, radar rainfall

Rainfall: State of the Science
Geophysical Monograph Series 191

10.1029/2010GM000930

algorithms have been delineated by frequency band, which indirectly implies an emphasis on the basic science aspect of the problem. In contrast, this chapter emphasizes the delineation of both the applied and basic science aspects of rainfall estimation. The chapter is organized as follows. In section 1, an overview of basic and applied science issues involved in rainfall estimation as well as the relationship of basic polarization measurements to rainfall microphysics is provided. Subsequently, a brief discussion of dual-polarization applications to radar data quality control (QC) and hydrometeor classification is addressed in section 2. With this foundation, we discuss both statistical and physically based rainfall retrievals from polarization measurements. Finally, in section 3, we present topics of active research and future directions in dual-polarization radar rainfall estimation.

1.2. Weather Radar Rainfall Estimation

The discussion of weather radar rainfall estimation is provided by *Seo et al.* [this volume]. Additional background is included in this chapter to provide context for the description of rainfall estimation with dual-polarization techniques.

Radars have been used to detect precipitation echos since the beginning of World War II. Quantitative estimates of rain rate from radar can be traced back to *Wexler and Swingle* [1947] and *Marshall et al.* [1947], using relations between radar reflectivity factor (Z) and rain rate (R). Rainfall relations between Z and R are referred to as "Z-R" relations and are the most common technique to estimate rainfall with radar. An excellent summary of the beginning of radar applications to precipitation is provided in the work of *Atlas and Ulbrich* [1990]. An example of a commonly used National Weather Service (NWS) Weather Surveillance Radar, 1988 Doppler (WSR-88D) Z-R relation [*Fulton et al.*, 1998] is

$$R(Z_h) = 0.017(Z_h)^{0.714} \quad (\text{mm h}^{-1}), \qquad (1)$$

where R is rain rate and Z_h is the radar reflectivity factor at horizontal polarization in $\text{mm}^6\ \text{m}^{-3}$.

The most important reason for using radar to estimate precipitation is the fact that, compared to a network of rain gauges, the radar (or combinations of radars) can sample a large area (>30,000 km^2 for a weather radar sampling out to 100 km) over a short period of time (<5 min) as well as provide information on the movement and evolution of precipitating systems. However, radar is a remote measurement tool, using microwaves to detect echo targets (e.g., precipitation particles). Therefore, assumptions are necessary to convert the radar measurements into an estimate of rainfall intensity or accumulation.

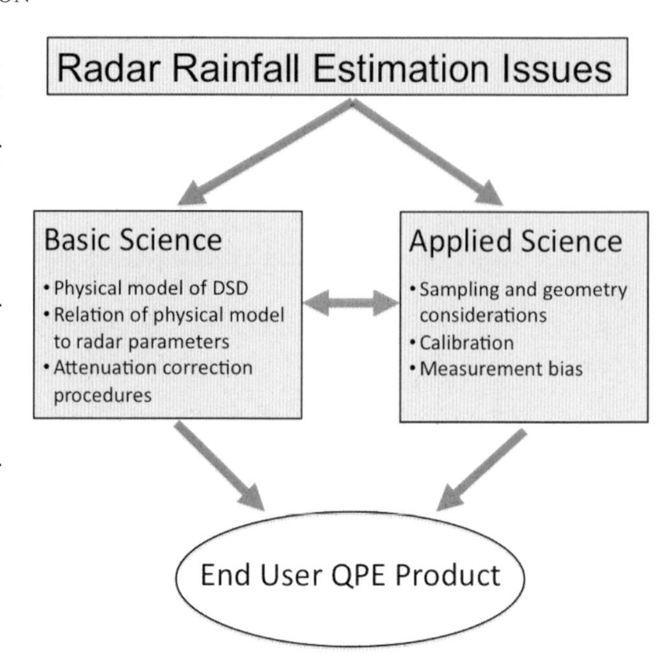

Figure 1. Conceptual diagram of the issues involved in radar rainfall estimation.

Factors that influence radar rainfall estimation can be routed into two broad categories: "basic science" and "applied science" (Figure 1). It is useful to distinguish these categories in order to understand how practical considerations imposed by applied science issues limit the improvement in rainfall estimation anticipated from theoretical expectations of dual-polarization radar retrievals (basic science). To be successful at radar rainfall estimation, both the basic and applied science issues must be addressed. Applied science issues include topics such as geometric and sampling considerations of the radar compared to ground-based measurements (rain gauges). The radar measures a sample volume that varies in range but is nevertheless many orders of magnitude larger than the single point measurement of a rain gauge (Figure 2). Also, differences between gauges and radar estimates could arise due to modification of the drop size distribution (DSD) between the location where the radar sampled the precipitation and the ground. These modifications include microphysical processes such as evaporation, drop coalescence and breakup. In addition, issues such as calibration and measurement bias are all included in applied science. For a more detailed discussion on the challenges of estimating precipitation with radar, the reader is referred to the work of *Zawadzki* [1984].

The basic science issues in radar rainfall estimation refer to the physical models that are used to represent the DSD and their relationship to the observed radar variables. The issue of

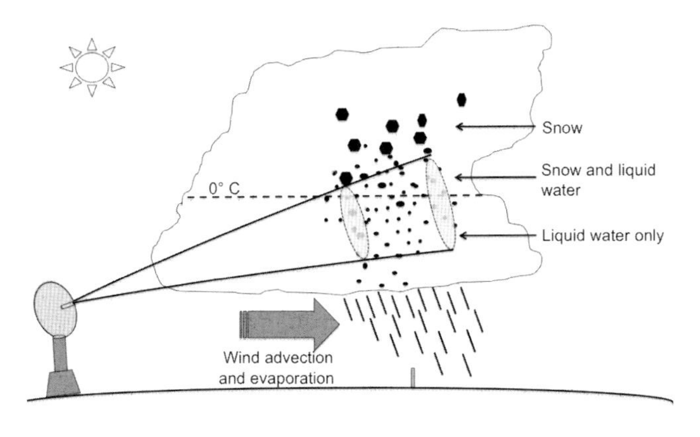

Figure 2. Illustration showing the impact of applied science issues on radar rainfall estimation.

the DSD can be illustrated in $R(Z_h)$ relations. The reflectivity factor Z is expressed as

$$Z_{h,v} = \frac{\lambda^4}{\pi^5|K|^2}\int_0^{D_{max}} \sigma_{h,v}(D)N(D)dD \quad (mm^6\ m^{-3}), \quad (2)$$

where K is the dielectric factor of water, $\sigma_{h,v}$ is the backscattering cross-sections at horizontal (h) and vertical (v) polarization, λ is the wavelength, $N(D)$ is the DSD, and D is the particle equivalent diameter. If the Rayleigh approximation is valid, the reflectivity factor can be expressed as [*Bringi and Chandrasekar*, 2001]

$$Z_{h,v} = \int_D N(D)D^6 dD \quad (mm^6\ m^{-3}). \quad (3)$$

The reflectivity factor is often expressed in "dBZ" ($10 \times \log10(Z_{h,v})$). Equation (3) shows that the reflectivity factor is related to the sixth moment of the DSD and is therefore strongly dependent on particle size. Rain rate is the volume flux of water per unit area and is given by

$$R = 0.6\pi \times 10^{-3}\int N(D)D^3 v(D)dD \quad (mm\ h^{-1}), \quad (4)$$

where $v(D)$ is the particle fall speed. Experimental data show that $v(D)$ can be described by an exponential function with exponent ~0.67, indicating that R is approximately proportional to the 3.67 moment of the DSD [*Atlas and Ulbrich*, 1977]. Thus, Z_h and R represent different moments of the DSD. The DSD can vary considerably within storms (e.g., convective versus stratiform rain), among different storms, and from region to region. *Battan* [1973] lists over 50 different *Z-R* relations, reflecting variability in the DSD. If a single $R(Z_h)$ relation is used, DSD variability limit the radar measurement accuracy to ~30–40% [*Chandrasekar et al.*, 2003]. The accuracy can be improved if the $R(Z_h)$ is modified to account for changes in the precipitation type.

$R(Z_h)$ estimates are also sensitive to applied science issues such as absolute radar calibration, beam blockage, and contamination from ice in the sampling volume. The applied science issues described above for traditional radar rainfall estimation also apply to dual-polarization radar. However, as described below, dual-polarization radar systems provide more information about the DSD compared to the traditional single-polarization systems, while at the same time, the differential phase measurements are immune to calibration biases and partial beam blocking. Thus, it can be seen that dual-polarization radar systems can help with both basic science and applied science issues of radar rainfall estimation.

1.3. Dual-Polarization Radar Measurements of Rainfall

Dual-polarization radar systems have, until recently, been exclusively within the realm of the research community. However, in the last decade, dual-polarization radar has moved into operational environments. The National Oceanic and Atmospheric Administration has already deployed a prototype of the Next Generation Weather Radar (NEXRAD) radars upgraded to dual-polarization capabilities. Herein, we briefly describe dual-polarization radar systems. For a historical perspective on dual-polarization radar development, the reader is referred to the work of *Seliga and Humphries* [1990].

Radar antennas transmit electromagnetic waves, and the plane containing the electric field vector determines the polarization state. Although several polarization states such as linear (in any direction) or circular polarizations are possibilities, horizontal polarization is probably the most common for single polarized weather radar applications. Conventional weather radars (e.g., NEXRAD) transmit and receive pulses in the horizontal plane and are therefore horizontally polarized. Because they transmit and receive in the same plane, they are referred to as single polarized systems, capable of receiving signal in the same polarization state as the transmitted state (i.e., measuring the co-polar component of the received electromagnetic signal).

In contrast to single-polarization systems, dual-polarization radars allow for variation in the polarization of the transmitted and/or received wave or allow for dual-channel reception of orthogonally polarized waves [*Bringi and Chandrasekar*, 2001]. These systems, therefore, provide additional information about the scatterer characteristics. For precipitation, these characteristics are related to the size, shape, phase state (liquid or ice), and orientation of hydrometeors in the radar resolution volume, as well as the polarization of the transmitted wave. As described below, dual polarization also provides important information that allows for the discrimination between clutter and precipitation echo targets as well as methods to correct for attenuation of radar signals due to heavy rain and partial beam blocking.

Dual-polarization systems can be of any arbitrary polarization state; however, most are either circular or linear, depending on the intended mode of operation. Although circular polarization radar systems have been used for hydrometeor identification and rainfall estimation, more recently, the linear polarization systems at (h) and (v) are more common because they are eigen polarization states of the precipitation medium for electromagnetic wave propagation [*Bringi and Chandrasekar*, 2001]. That is, the polarization states do not change as a function of propagation. In this chapter, we will concentrate on linear polarization systems. For more discussion on circular polarization techniques, the reader is referred to the works of *McCormick and Hendry* [1975], *Hendry and Antar* [1984], and references therein. For the remainder of this chapter, it is assumed that dual polarization refers to radar systems that transmit and/or receive electromagnetic energy in the h and v polarization states.

1.4. Description of Dual-Polarization Radar Parameters

In this section, we review the important radar parameters used in dual-polarization rainfall applications. These parameters are inherently related to assumptions about the DSD and the relationship between drop shape and size (basic science properties). Therefore, we begin our discussion with a brief description of the DSD and size-shape relationships.

A number of parametric forms of the DSD have been used, including exponential, lognormal, and gamma. The gamma form is representative of a wide range of naturally occurring DSDs [*Ulbrich*, 1983] and is commonly used in the literature. This form of the DSD will also be adopted herein. It should be noted that the discussions in this chapter are not limited to any specific form. The gamma distribution can be represented with three parameters as

$$N(D) = N_0 D^{\mu} exp^{(-\Lambda D)}, \tag{5}$$

where N_0 (mm$^{[-1-\mu]}$ m^{-3}) is a scaling parameter, μ is a distribution shape parameter, Λ (mm^{-1}) is a slope term and D (mm) is the spherical equivalent volume diameter.

As noted above, the form used to describe the shape of raindrops is an important basic science consideration in the interpretation of rainfall polarization parameters. Raindrop shape is determined by a balance of aerodynamic, hydrostatic, and surface tension forces [*Green*, 1975]. Raindrops larger than ~1 mm flatten as they fall, and their shape is therefore described as an oblate spheroid [see *Jones et al.*, this volume, Figure 3] with semimajor and semiminor axes a and b, respectively (oblateness is usually expressed by the ratio b:a). Determining the exact shape is complicated by the fact that observations of raindrops under natural conditions are difficult to obtain; therefore, a number of shape-size relationships are found in the literature. These relations were developed using both observations and modeling of raindrop shapes (see *Jones et al.* [this volume] for a comprehensive overview). These relations vary according to the ranges of drops sizes used, and the most appropriate description of the size-shape relation is a topic of active research. The simplest form is a linear relation:

$$\frac{b}{a} = 1.03 - \beta D \quad 1 \le D \le 9 \text{ mm}, \tag{6}$$

where β is the slope of the shape-size relation. *Pruppacher and Beard* [1970] found that β = 0.062. *Gorgucci et al.* [2000] developed an algorithm to estimate β from simultaneous measurements of Z_h, differential reflectivity (Z_{dr}), and specific differential phase (K_{dp}) and found that β decreased with increasing reflectivity, possibly due to the effect of drop oscillations. *Matrosov et al.* [2006] expanded the above relation to include the nonlinear relation between drop diameter and axis ratio suggested by *Andsager et al.* [1999] for $D < 4.4$ mm and a linear relation for large drops (>4.4 mm):

$$\frac{b}{a} = 1.012 - 0.144D - 1.03D^2 (D < 4.4 \text{ mm}) \tag{7a}$$

$$\frac{b}{a} = 1.02 - 0.62D (D \ge 4.4 \text{ mm}). \tag{7b}$$

More elaborate polynomial fits to the observations of raindrops [e.g., *Thurai and Bringi*, 2005] have also been developed. For example, *Brandes et al.* [2004] suggest a size-shape relation of the form

$$\frac{b}{a} = 0.9951 + 0.02510D - 0.03644D^2 + 0.005303D^3 - 0.0002492D^4. \tag{8}$$

Gorgucci et al. [2009] obtained estimates of the polynomial coefficients of axis ratio versus size directly from radar observations.

Given the shape-size model, it is possible to calculate the resulting polarimetric variables. In addition to the reflectivity factor described above, the most common polarization measurements for rainfall estimation are Z_{dr} and K_{dp}.

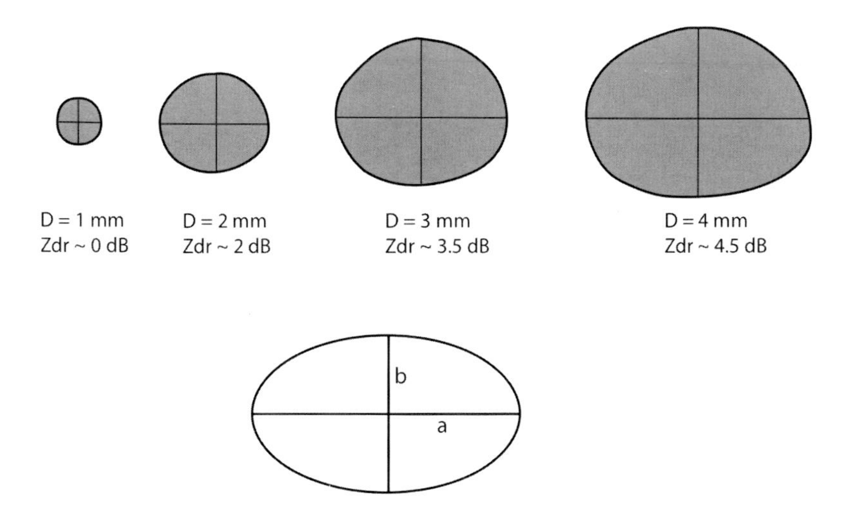

Figure 3. Cartoon illustrating differential reflectivity (Z_{dr}) as a function of drop diameter (D). Here "b" and "a" refer to the semiminor and semimajor axes, respectively.

1.4.1. Differential reflectivity. The differential reflectivity (Z_{dr}) is defined as

$$Z_{dr} = 10\log\left(\frac{Z_h}{Z_v}\right) \quad dB, \tag{9}$$

where Z_h and Z_v are defined in equation (2). Z_{dr} is positive (negative) for particles whose major axes are close to horizontal (vertical). Z_{dr} is zero for particles that are spherical or for nonspherical particles with a random distribution of orientations. As described above, the equilibrium shape of a falling raindrop is oblate. Raindrops tend to flatten and orient themselves with their major axes close to horizontal. For raindrops bigger than ~0.5 mm, Z_{dr} ranges from ~0.3 to 3 dB. Figure 3 illustrates how Z_{dr} can be used to discriminate different particle types.

Using observations of Z_{dr}, the inverse problem of identifying properties of the DSD can be pursued. *Seliga and Bringi* [1976] showed that Z_{dr} could be related to the median particle volume diameter for an assumed exponential particle size distribution. *Jameson* [1983] showed that Z_{dr} is related to the mean axis ratio (b/a) of raindrops:

$$10^{0.1(Z_{dr})} = (\overline{r})^{\frac{-3}{7}}, \tag{10}$$

where $\overline{r}$ is the reflectivity weighted mean rain drop axis ratio and Z_{dr} is in dB. Using the fit recommended by *Andsager et al.* [1999] for the Beard-Chuang equilibrium shape for S band observations, power law fits between particle mean mass diameter (D_m) and median volume diameter (D_0) can be derived as [*Bringi and Chandrasekar*, 2001; *Gorgucci et al.*, 2002]

$$D_m = 1.619 \times Z_{dr}^{0.485} \quad (\text{mm}) \tag{11a}$$

$$D_0 = 1.529 \times Z_{dr}^{0.467} \quad (\text{mm}), \tag{11b}$$

where Z_{dr} is in dB, and the relations are valid at S band frequency. For X band, *Matrosov et al.* [2005] used DSD data collected along the CA coast and simulations to develop a similar set of relations:

$$D_0 = 1.46(Z_{dr})^{0.49} \quad (\text{mm}) \tag{12a}$$

$$D_m = 1.63(Z_{dr})^{0.48} \quad (\text{mm}). \tag{12b}$$

Z_{dr} can also be used for discrimination of ice particles, though the contrast between Z_h and Z_v is usually weak due to the smaller dielectric constant and density of ice compared to liquid water.

1.4.2. Specific differential phase. Differential phase shift (Φ_{dp}) is a measure of the propagation phase difference between h and v polarized radar waves. For raindrops with an oblate spheroid shape, h polarized waves propagate more slowly than do v polarized waves. Electromagnetic waves returned to the receiver for the two polarizations exhibit different accumulative phase (time) shifts depending on hydrometeor size, shape, orientation, quantity, and distance from the radar. The radar estimated differential phase shift between two polarization states actually has two components, differential propagation phase (Φ_{dp}) and a component related to a backscatter differential phase (δ_{co}) [*Wang and Chandrasekar*, 2009]. Φ_{dp} is the important component for rainfall estimation. The range derivative of Φ_{dp} is the specific differential propagation phase K_{dp}:

$$K_{dp} = \frac{\Phi_{dp}(r_2) - \Phi_{dp}(r_1)}{2(r_2 - r_1)} \quad (^\circ\ \text{km}^{-1}). \tag{13}$$

K_{dp} is determined by the real part of the difference between the h and v polarization forward scattering amplitudes f_h and f_v. Because of the relationship to propagation, it is important that the backscattering component of the differential phase, as well as noise, be removed prior to calculating K_{dp} [*Wang and Chandrasekar*, 2009]. *Sachidananda and Zrnić* [1986] show that K_{dp} is related to ~4.24 power of the DSD. However, *Ryzhkov and Zrnić* [1996] show that the power law relation varies with drop size. Smaller drops are related to the 5.6 power of the DSD so that, at low rain rates with small drops, K_{dp} has a greater dependence on the DSD compared to higher rain rates (bigger drops). Nevertheless, because K_{dp} is closer to the DSD moment for rainfall compared to radar reflectivity factor, it is expected that $R(K_{dp})$ relations will be less susceptible to DSD variability compared to $R(Z_h)$.

It can be shown that K_{dp} is closely related to the DSD through the rainwater content and is relatively insensitive to ice in the radar volume. For Rayleigh conditions [*Jameson*, 1985]:

$$K_{dp} = \left(\frac{180}{\lambda}\right) 10^{-3} CW(1-\bar{r}_m) \quad (^\circ\ \mathrm{km}^{-1}), \qquad (14)$$

where C is a constant with magnitude ~3.75, λ is the radar wavelength in meters, W is the water content in g m^{-3}, and r_m is the mass weighted mean axis ratio. If the *Pruppacher and Beard* [1970] equilibrium axis ratio relation (equation (6)) is used, then K_{dp} can be expressed as [*Bringi and Chandrasekar*, 2001]

$$K_{dp} = \left(\frac{180}{\lambda}\right) 10^{-3} CW(0.062) D_m \quad (^\circ\ \mathrm{km}^{-1}), \qquad (15)$$

where D_m is the mass weighted mean diameter (mm).

1.4.3. Linear depolarization ratio (LDR) and copolar correlation coefficient (ρ_{hvco}). Though not directly related to rainfall estimation, we discuss two other polarimetric variables that are useful for hydrometeor identification (HID): LDR and copolar correlation coefficient (ρ_{hvco}). Hydrometeors whose principal axes are not aligned with the electrical field of the transmitted wave cause a small amount of the signal to be scattered back in the orthogonal polarization state. The cross-polar signal stems primarily from nonspheroidal particles that wobble or tumble as they fall, creating a distribution of orientations or canting angles. Signals are enhanced for wetted and melting particles. LDR (dB) is defined as the logarithm of the ratio of cross-polar and copolar signal returns, that is,

$$\mathrm{LDR} = 10\log_{10}\left(\frac{Z_{ex}}{Z_h}\right) \quad (\mathrm{dB}), \qquad (16)$$

where Z_{ex} is the signal received at the vertical polarization (cross-polar return) for a transmitted horizontally polarized wave. LDR is small (on the order of −35 to −15 dB, depending on the type of particles). Wet and melting snow can have an LDR from −15 to −20 dB.

The copolar correlation coefficient ρ_{hvco} is computed from backscattered signals at h and v polarization. This parameter is sensitive to the distribution of particle sizes, axis ratios, and shapes. Theoretical values are approximately 0.99 for raindrops and ice crystals. For melting aggregates, ρ_{hvco} can be less than 0.90. Because ρ_{hvco} and LDR are both sensitive to the presence of large wet particles, their melting-layer responses are similar.

2. RAINFALL ESTIMATION USING DUAL-POLARIZATION RADAR MEASUREMENTS

Dual-polarization techniques can be applied at multiple stages of radar rainfall estimation. We refer to these stages as preprocessing, classification, and quantification (Figure 4). Preprocessing includes QC procedures such as clutter removal and attenuation correction. Classification refers to the identification of hydrometeor types. Quantification is the rainfall estimation stage. Although this chapter emphasizes quantification, we provide brief descriptions of preprocessing and classification applications in order to provide context for the discussion on rainfall estimation. Without the prior two stages, the rainfall estimates are likely to be erroneous.

2.1. Preprocessing

In the analysis of radar data, QC (removal of nonprecipitation echos) can be critically important. This is especially true with traditional Doppler radars, where only radar reflectivity, radial velocity, and spectral width parameters may be available. One of the beneficial aspects of dual-polarization radar is that identification of anomalous propagation, ground clutter, and suppression of range-folded echos can be accomplished, with significantly better skill than single-polarization radar systems. There is extensive research in the literature for QC algorithms, and we cannot do it justice in this chapter (see *Hubbert et al.* [2009] and *Wang and Chandrasekar* [2010] for a review). For dual-polarization radar, *Ryzhkov and Zrnić* [1998a] discussed a technique based on thresholds of the cross-correlation coefficient and differential phase at S band. Fuzzy logic algorithms have proved especially useful for discriminating clutter from precipitation echo and for discriminating different hydrometeor types [*Straka and Zrnić*, 1993; *Höller et al.*, 1994; *Straka*, 1996; *Zrnić and Ryzkhov*, 1999; *Vivekanandan et al.*, 1999; *Liu and Chandrasekar*, 1998, 1999, 2000; *Kessinger et al.*, 2001;

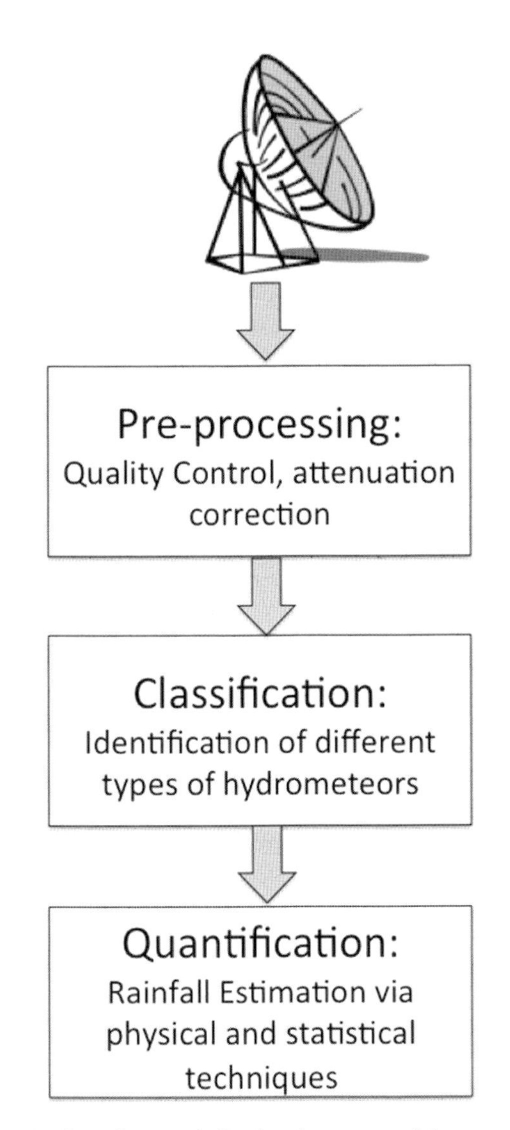

Figure 4. Flow diagram indicating the stages of data processing where dual-polarization radar can make important contributions.

Zrnić et al., 2001; *May and Keenan*, 2005; *Lim et al.*, 2005; *Gourley et al.*, 2007; *Hubbert et al.*, 2009]; (see also *Bringi and Chandrasekar* [2001] for an overview of fuzzy logic techniques).

Correction for loss due to attenuation of the radar signal is an important step before further quantitative analysis can take place. Attenuation becomes more severe at higher frequencies, so that X band systems must be corrected more than S band systems. However, even at S band, some signal loss in heavy rain can occur over long paths. There is an enormous body of literature covering this topic, and a full review is beyond the scope of this chapter. Nevertheless, some key articles that have addressed this issue at X, C, and S bands are the works of *Scarchilli et al.* [1993], *Ryzhkov and Zrnić* [1995], *Carey et al.* [2000a], *Testud et al.* [2000], *Keenan et al.* [2001], *Matrosov et al.* [2002], *Anagnostou et al.* [2004], *Park et al.* [2005], *Gorgucci and Chandrasekar* [2005], and *Vulpiani et al.* [2008]. Calibration monitoring is another important topic that is empowered by dual-polarization methods. For example, vertical pointing observations in light rain are used to calibrate Z_{dr} measurements [*Bringi and Chandrasekar*, 2001], and self-consistency or other methods have been used to check radar calibration [*Scarchilli et al.*, 1993].

2.2. Classification

After the completion of preprocessing procedures, classification of the precipitation can be accomplished. Classification is also referred to as HID. The most important classification for the purposes of rainfall estimation is the identification of ice in the radar volume. Ice can be especially problematic for rainfall estimation using $R(Z_h)$ methods. For precipitation ice such as hail and graupel, *Aydin et al.* [1986] describes a procedure, devised at S band, using a combination of Z_h and Z_{dr} measurements. Using both measured disdrometer and modeled DSDs, *Aydin et al.* [1986] showed that there is an upper bound of Z_h for given Z_{dr} in rain. Above this boundary, hail is likely. The boundary is described using the following equation

$$H_{dr} = Z_h - f(Z_{dr}) \quad (17a)$$

$$f(Z_{dr}) = \begin{cases} 27 & Z_{dr} \leq 0 \text{ (dB)} \\ 19Z_{dr} + 27 & 0 < Z_{dr} \leq 1.74 \text{ (dB)} \\ 60 & Z_{dr} > 1.74 \text{ (dB)}. \end{cases} \quad (17b)$$

Z_h is the measured reflectivity in dBZ and $f(Z_{dr})$ is a threshold line segment (at S band): This formulation is referred to as H_{dr}, the Z_{dr}-derived hail signal. It should be noted that, at C band, there is mounting evidence that Mie regime resonance effects of melting hail can prevent hail detection via equation (17) [*Höller et al.*, 1994].

Another polarimetric discriminator for rain and precipitation ice (hail or graupel) is the difference reflectivity (Z_{dp}) [*Golestani et al.*, 1989]. The assumption in this method is that $Z_h \sim Z_v$ for ice but $Z_h > Z_v$ for rain drops larger than ~1 mm.

$$Z_{dp} = 10\log_{10}(Z_h^{rain} - Z_v^{rain}) \quad \text{(dB)}. \quad (18)$$

The idea is to develop a relation between Z_h and Z_{dp} in rain-only regions

$$Z_{dp} = a(10\log_{10}Z_h^{rain}) + b. \quad (19)$$

This relation is then extrapolated to regions where ice can occur so that the fraction of the radar volume contributed by ice can be determined from

$$Z^{\text{ice}} = Z_{\text{h}}^{\text{obs}} - Z_{\text{h}}^{\text{rain}} \quad (20a)$$

$$\text{ice fraction} = 1 - 10^{(-0.1\Delta Z)}. \quad (20b)$$

This formulation is used in the CSU-ICE rainfall algorithm at Colorado State University (CSU) [*Cifelli et al.*, 2010] (described below in section 2.3.2.4). The technique works well for discrimination in convective rain where $Z_h > Z_v$; however, in light stratiform rain, Z_{dp} is noisy and can produce spurious ice signatures.

Discrimination of snow from rain can also be accomplished using a combination of Z_h, K_{dp}, Z_{dr}, and ρ_{hvco} [*Ryzhkov and Zrnić*, 1998b]. They show that a radar bright band with a minimum in ρ_{hvco} and maximum in Z_{dr} can be used to differentiate rain from snow. Alternatively, in the absence of a bright band (lack of aggregates), the rain snow line can be determined from a sharp change in Z_{dr}. *Lim et al.* [2005] show that the varying melting layer can be identified using vertical variations of Z_{dr}. *Brandes and Ikeda* [2004] outline a procedure to determine the freezing level using a combination of Z_h, ρ_{hvco}, and LDR. *Matrosov et al.* [2007] show that, at X band, the ρ_{hvco} provides a robust discriminator between rain, melting hydrometeors, and snow.

2.3. Quantification

With a basic introduction into the parameters used to estimate rainfall with dual-polarization radar and an overview of preprocessing and classification techniques, the topic of rainfall algorithms can now be discussed. Rainfall estimation for both single- and dual-polarized radar can be broadly classified into physically and statistically based approaches (Figure 5). Physical methodologies utilize parametric relations between radar measurements and assumed characteristics of the DSD and do not rely on ground measurements to calibrate the radar observations. Statistical approaches, on the other hand, implicitly assume that precipitation on the ground depends on the 4D variability of precipitation aloft and relies on ground (rain gauge or disdrometer) data to tune the radar measurements [*Bringi and Chandrasekar*, 2001]. The strength of dual-polarization radar is to bring more insight into microphysical processes; therefore, the physically based approach is much more common for rainfall estimation. However, statistical techniques show great promise for operational applications and are necessarily included in this chapter's rainfall discussion.

2.3.1. Statistical approach. Statistical radar rainfall estimation attempts to provide an optimal rainfall estimate at ground level, recognizing that the radar measurements are obtained aloft [*Chandrasekar et al.*, 2003]. As discussed in the Introduction, rainfall measured at the ground is likely to be different from rainfall measured aloft for a number of reasons that are related to applied and basic science issues. The philosophy for the statistical approach is that the best radar rainfall estimate is derived using feedback from ground data to "tune" the radar algorithm.

Statistical approaches can be broadly divided into probability distribution matching and neural network approaches. In the traditional probability matching method for single-polarization radar [*Calheiros and Zawadzki*, 1987; *Rosenfeld et al.*, 1993], cumulative probability distributions of radar reflectivity and rainfall are matched in order to determine the best coefficient and exponent pair in the $R(Z_h)$ relation. *Gorgucci et al.* [1995] extended this methodology to include Z_{dr} in order to estimate rainfall for one event in Florida. Using a network of rain gauges, *Gorgucci et al.* [1995] showed that the multiparameter approach performed better (in terms of mean square error) compared to the Z_h-only method. Moreover, the optimal $R(Z_h, Z_{dr})$ that was derived in this study was very close to the theoretical relationship for a wide range of gamma DSDs (discussed below).

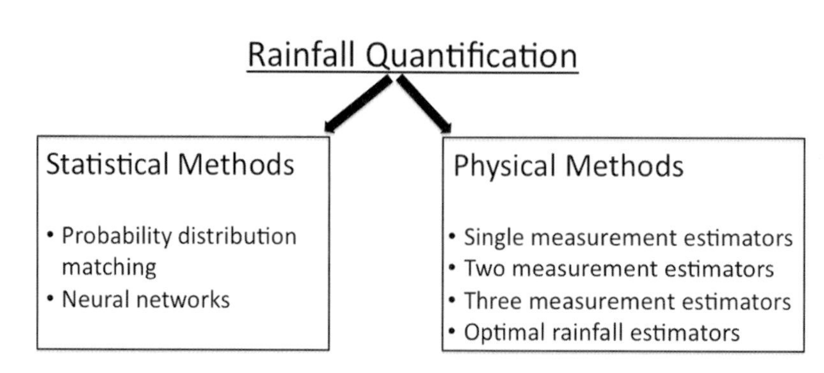

Figure 5. Classification of techniques utilized in radar rainfall estimation.

The other statistical estimation technique is the neural network approach [*Xiao and Chandrasekar*, 1995, 1997; *Liu et al.*, 2001; *Xu and Chandrasekar*, 2005; *Vulpiani et al.*, 2006, 2009]. In the neural network approach, rain gauge measurements are compared to radar rainfall estimates, and the difference is used to adjust the radar relation parameters until the network converges on an optimal solution. Once this "training" process is completed, a relationship between rainfall rate and the radar measurements is completed, and the network is ready for operational applications [*Liu et al.*, 2001].

The first application of the neural network technique to multiparameter observations was by *Xiao and Chandrasekar* [1995]. They used Z_h and Z_{dr} measurements from a dual-polarization radar operating in central Florida and a network of rain gauges to show that the neural network outperformed both $R(Z_h)$ and $R(Z_h,Z_{dr})$ parametric relations as well as a neural network using Z_h alone. *Vulpiani et al.* [2009] extended the multiparameter approach to include K_{dp} and tested the methodology on an Oklahoma data set. They found that the neural network had comparable results to a parametric approach involving Z_h, Z_{dr}, and K_{dp}. These results indicate that neural networks involving multiparameter estimators have great potential for rainfall estimation. Nevertheless, neural network techniques have not seen wide application so far.

2.3.2. Physical approach. Physical-based techniques are built on models of how the radar observations relate to the DSD and are the most commonly used application of dual-polarization radar rainfall estimation. The three variables that are used most often in rainfall applications are Z_h, Z_{dr}, and K_{dp}, and this chapter will emphasize rainfall applications based on these parameters. Each parameter has certain advantages and disadvantages for rainfall estimation, based on the basic and applied science issues described in section 1.2. Z_h is an easy quantity to observe but is sensitive to radar calibration (applied science). Z_{dr} is a relative measurement that is closely related to drop size and not dependent on absolute calibration. However, its application for rainfall is closely tied to relative calibration of Z_v and Z_h. In high-frequency radar systems (e.g., C and X bands), attenuation can produce biases in Z_h and Z_v, and correction schemes must be applied (applied science). K_{dp} is useful because of the immunity to absolute calibration and precipitation ice as well as the fact that it is more closely related to R (in terms of DSD moments) compared to Z_h (basic science). However, K_{dp} is not a point estimate but rather is measured along a specified range (applied science). As such, it is sensitive to filtering and the specified path length and can be a noisy measurement, especially at lower-frequency bands like S band (basic science).

Table 1. Dual-Polarization Rainfall Estimators Classified According to the Number of Radar Measurements Utilized

Single Measurement Estimators	Two Measurement Estimators	Three Measurement Estimators
$R(Z_h)$	$R(Z_h,Z_{dr})$	$R(Z_h,Z_{dr},K_{dp})$
$R(K_{dp})$	$R(K_{dp},Z_{dr})$	
$R(Z_{dr})$[a]	$R(Z_h,K_{dp})$	

[a]This estimator is not used since Z_{dr} is a relative measurement.

Rainfall estimation with dual-polarization radar can be accomplished by using the above parameters individually or in combination; therefore, it is helpful to divide the estimators into different classes as shown below in Table 1.

2.3.2.1. Single measurement estimators. There are two rainfall algorithms that fall into the single measurement class: $R(Z_h)$ and $R(K_{dp})$. We will focus our discussion on $R(K_{dp})$, since $R(Z_h)$ is treated in depth by *Seo et al.* [this volume]. As discussed above, K_{dp} has certain advantages and disadvantages for rainfall estimation. Because it is derived over a path as opposed to being a point estimator, K_{dp} can be noisy, especially in low-intensity rainfall. *Chandrasekar et al.* [1990] evaluated the threshold of usefulness of K_{dp} for rainfall estimation, suggesting that K_{dp} must be at least 0.4° km^{-1} (at S band) and must be range averaged over 3–4 km to reduce noise. However, at higher frequencies like X band, K_{dp} is especially useful because of its immunity to partial attenuation and the increased sensitivity of propagation phase (see equations (14)–(15)). As stated in section 1.4.2, K_{dp} is less sensitive to variations in the DSD compared to Z_h [*Sachidananda and Zrnić*, 1986].

The relation between R and K_{dp} can be expressed in terms of radar frequency

$$R(K_{dp}) = 129\left(\frac{K_{dp}}{f}\right)^{b} \quad (\mathrm{mm\ h^{-1}}), \tag{21}$$

with frequency f in GHz. *Bringi and Chandrasekar* [2001] show that at 3 GHz, equation (21) reduces to

$$R(K_{dp}) = 40.5(K_{dp})^{0.85} \quad (\mathrm{mm\ h^{-1}}) \tag{22}$$

for Beard and Chuang and Pruppacher and Beard equilibrium shapes, respectively. At X band, *Matrosov et al.* [2005, 2006] derive an $R(K_{dp})$ relation of the form

$$R(K_{dp}) = 14\text{–}15(K_{dp})^{0.76\text{–}0.80} \quad (\mathrm{mm\ h^{-1}}), \tag{23}$$

where the dash represents a range of values. Equation (23) is based on observed DSD data collected in California and

Colorado and using a linear drop shape model (see equation (6)). *Wang and Chandrasekar* [2010] used a relation at X band translated from an S band algorithm (recall equation (21)) for Oklahoma precipitation events. Their relationship

$$R(K_{dp}) = 18.15(K_{dp})^{0.791} \quad (\mathrm{mm\ h^{-1}}) \qquad (24)$$

showed very good agreement with rain gauge estimates for selected storms. *Matrosov et al.* [2002] showed, using calculations from DSDs collected in Virginia, that the coefficient in $R(K_{dp})$ relation is sensitive to the drop shape factor β (see equation (6)). The relation between rainfall rate and specific differential phase for varying shape factor (β) can be expressed as

$$R(\mathrm{mm\ h^{-1}}) \approx 8.2(\beta)^{-0.82} K_{dp}^{0.81}(°\ \mathrm{km^{-1}}) \quad (\beta : \mathrm{cm^{-1}}). \qquad (25)$$

The method to retrieve the shape factor is described below in section 3.

A number of studies have shown the efficacy of K_{dp} in rainfall estimation. *Chandrasekar et al.* [1990] used simulations to model random errors in S band radar measurements as well as DSD variability for assumed gamma distributions and found that $R(K_{dp})$ had the best performance of any rainfall estimator when the rainfall rate exceeded 70 mm h^{-1}. *Ryzhkov and Zrnić* [1996] utilized S band data and found that $R(K_{dp})$ performed significantly better than $R(Z_h)$ in Oklahoma storms. However, $R(K_{dp})$ results have not been uniformly positive at S band. Results from *Brandes et al.* [2001], using S band radar data and a network of rain gauges, showed that $R(K_{dp})$ performed similarly to a well-calibrated $R(Z_h)$ for a selection of Colorado and Kansas storms. In this study, both $R(K_{dp})$ and $R(Z_h)$ were significantly biased for storms with small drops, indicating that $R(K_{dp})$ is not immune to DSD variability at S band.

In contrast to S band, some of the best performance of $R(K_{dp})$ has been demonstrated at X band. In a study of convective and stratiform events in Virginia, *Matrosov et al.* [2002] showed that $R(K_{dp})$ outperformed mean $R(Z_h)$ relations in terms of bias and fractional error. *Matrosov et al.* [2005] showed similar results for precipitation events in California. *Matrosov et al.* [2006] found that the difference in coefficient for the $R(K_{dp})$ relation is slightly more than that expected by frequency scaling (equation (21)) due to non-Rayleigh effects. By comparing relationships derived from both S and X band and DSD data, this study showed that $R(K_{dp})$ at X band could be useful down to 2.5–3 mm h^{-1} (corresponding to 27–28 dBZ) compared to 8–10 mm h^{-1} (37–38 dBZ) at S band. X band K_{dp} was found to be useful (above the noise level) down to 0.09° km^{-1} compared to 0.3° km^{-1} at S band. *Wang and Chandrasekar* [2010] presented a comprehensive study based on 3 years of Collaborative Adaptive Sensing of the Atmosphere (CASA) radar observations. They found that $R(K_{dp})$ demonstrated a factor of three reduction in standard error (SE) compared to the S band results published for the KOUN dual-polarization radar in the same region of Oklahoma [*Ryzhkov et al.*, 2005a] and that the X band results had negligible bias. The results from these studies suggest that $R(K_{dp})$ has become the algorithm of choice for X band radar rainfall estimation.

Based on the above studies, it is expected that the performance of $R(K_{dp})$ at C band will be in between the results for X and S band. *Scarchilli et al.* [1993] presented a comprehensive analysis of dual-polarization radar rainfall estimation at C band. In addition to deriving the various radar rainfall algorithms, including $R(K_{dp})$, they developed error structure statistics of various estimators and studied the impact of backscatter differential phase for rainfall estimation. *May et al.* [1999] showed, based on accumulations from four heavy rain events, that the RMS error using $R(Z_h)$ was approximately 50% larger than similar statistics using $R(K_{dp})$. In this case, the $R(K_{dp})$ relation was derived from disdrometer data collected near Darwin Australia [*Keenan et al.*, 1997].

2.3.2.2. Two measurement estimators. Increasing the number of radar measurements from one to two means that more information about the DSD is acquired and that fewer assumptions are necessary for rainfall estimation purposes. However, this also means there are more measurement errors that must be accounted for. For a two measurement estimator to outperform a single measurement estimator, the increased information provided about the physical model of raindrop behavior (basic science) must overcome the increased error in the measurements (applied science).

The two measurement rainfall estimator with the longest history in the literature is $R(Z_h,Z_{dr})$. The observational studies of *Aydin et al.* [1987, 1990] and *Gorgucci et al.* [1995] compared $R(Z_h,Z_{dr})$ and $R(Z_h)$ using disdrometers and networks of rain gages in midlatitude and subtropical environments. These studies showed that the two measurement estimator outperformed the single measurement estimator in terms of bias and SE. *Chandrasekar and Bringi* [1988] used disdrometer and radar data to simulate the performance of the $R(Z_h,Z_{dr})$ and $R(Z_h)$ estimators. They found that $R(Z_h,Z_{dr})$ is susceptible to random errors in Z_{dr}, and at rain rates <20 mm h^{-1}, $R(Z_h,Z_{dr})$ did not outperform $R(Z_h)$. Also, the $R(Z_h,Z_{dr})$ estimator is sensitive to contamination from precipitation ice [*Ryzhkov and Zrnić*, 1995], so caution must be applied before using it in situations where graupel, hail, or the radar bright band might be present.

The relationship between Z_h and Z_{dr} and rainfall can be seen as follows. From equation (11a) above, D_0 can be related to Z_{dr}

through a power law relationship. Rain rate (R) can then be expressed as a function of Z_h and Z_{dr}

$$R = c_1 Z_h^{a1} 10^{0.1 b1 Z_{dr}} \quad (\text{mm h}^{-1}) \qquad (26)$$

where Z_h is in mm^6 m^{-3}, Z_{dr} is in dB, and the constants a1, b1, and c1 are a function of radar frequency [*Chandrasekar et al.*, 1990; *Scarchilli et al.*, 1993; *Gorgucci et al.*, 1995]. For a gamma DSD, simulations have been performed to determine the coefficients at different frequencies (see *Bringi and Chandrasekar* [2001] for a description on how the simulations were performed). These coefficients are shown in Table 2.

Another two measurement rainfall estimator can be derived from a combination of Z_{dr} and K_{dp} measurements. The advantage of an $R(K_{dp},Z_{dr})$ algorithm is that it is not sensitive to absolute calibration of the radar, and it is less sensitive to variations in the effective rain drop axis ratio. The variation in the axis ratio (due to oscillations or canting or presence of ice cores) leads to changes in the same direction in K_{dp} and Z_{dr} that partially offset each other [*Ryzhkov and Zrnić*, 1995; *Gorgucci et al.*, 2001].

The $R(Z_{dr},K_{dp})$ estimator has the lowest error structure compared to any of the other single or two measurement estimators because of its close correspondence to the DSD [*Chandrasekar et al.*, 1993; *Ryzhkov and Zrnić*, 1995]. Theoretically, this estimator should have the best performance for rainfall estimation. However, as noted previously, K_{dp} is noisy at low rainfall rates, and Z_{dr} can be biased. These tradeoffs are frequency dependent: K_{dp} noise is more pronounced at S band due to decreased sensitivity, whereas Z_{dr} error is more pronounced at higher frequencies like X band due to the heavier reliance on attenuation correction procedures. Thus, the application of this algorithm should depend on the amount of error anticipated in Z_{dr} and the rainfall environment. The $R(Z_{dr},K_{dp})$ algorithm is usually expressed in the form

$$R = a_1 K_{dp}^{b1} Z_{dr}^{c1} \quad (\text{mm h}^{-1}), \qquad (27)$$

where a1, b1, and c1 are 52.0, 0.960, and −0.447, respectively [*Ryzhkov and Zrnić*, 1995], Z_{dr} is in dB, and K_{dp} in ° km^{-1}. These coefficients are representative for S band. $R(Z_{dr},K_{dp})$ has not found widespread application at X or C band.

Table 2. Coefficients in the $R(Z_h,Z_{dr})$ Relationship (equation (26)) as a Function of Radar Frequency[a]

Frequency (GHz)	a_1	b_1	c_1
3.0	0.93	−3.43	6.7×10^{-3}
5.45	0.91	−2.09	5.8×10^{-3}
10.0	1.07	−5.97	3.9×10^{-3}

[a]From *Bringi and Chandrasekar* [2001]. Copyright 2001 Cambridge University Press. Reprinted with permission.

There are several observational studies at S band that have evaluated the performance of $R(Z_{dr},K_{dp})$ with other rainfall estimators. *Ryzhkov and Zrnić* [1995] compared $R(Z_{dr},K_{dp})$ with $R(K_{dp})$, $R(Z_h,Z_{dr})$, and $R(Z_h)$ for an Oklahoma squall line event. They found that $R(Z_{dr},K_{dp})$ had the lowest bias and RMS error of all the rainfall estimators. In a flash flood event in Colorado, *Petersen et al.* [1999] showed that the combined estimator $R(K_{dp},Z_{dr})$ did much better at estimating storm total precipitation compared to the WSR-88D $R(Z_h)$. The $R(K_{dp},Z_{dr})$ totals were 85% of the gauge total compared to 50–65% of the gauge total using the WSR-88D $R(Z_h)$, depending on which set of radar data was utilized. *Chandresekar et al.* [1993] and *Ryzhkov and Zrnić* [1995] showed, using simulations with radar and disdrometer data, that the $R(K_{dp},Z_{dr})$ estimator has the lowest error structure in the rain rate range of 10–90 mm h^{-1}. As noted above, higher-frequency applications of $R(Z_{dr},K_{dp})$ are limited due to the increased sensitivity of Z_{dr} to attenuation correction error. Similarly, at low frequencies such as S band $R(Z_{dr},K_{dp})$ may be affected by error in K_{dp}.

Although theoretically possible, the two measurement estimator $R(Z_h,K_{dp})$ has not undergone widespread application. *Gorgucci et al.* [2006] showed that by normalizing K_{dp} with Z_h, the actual drop shape variability could be ascertained from the combination of Z_h, Z_{dr}, and K_{dp} measurements. In this procedure, the concentration parameter N_w is removed from the normalized form of the gamma DSD equation, and no drop shape model assumptions are required to recover the parameters of the DSD. The extension of this methodology for rainfall applications has yet to be demonstrated.

2.3.2.3. Three measurement estimators. It is not common to utilize all three parameters (Z_h, Z_{dr}, and K_{dp}) directly for rainfall estimation. This is due to the increased measurement uncertainties that result from combining all of the parameters together. Rainfall is more often determined indirectly by using the three parameters to retrieve parameters of the DSD and estimate rainfall from equation (4). *Gorgucci et al.* [2000, 2001, 2002] have shown that parameters of the gamma DSD form can be estimated by treating the drop axis ratio as a variable and then computing drop shape and the DSD parameters using combined measurements of Z_h, Z_{dr}, and K_{dp}. The procedure developed from simulations with a variable β in equation (6), and random distributions of the gamma DSD parameters are used to establish relationships with Z_h, Z_{dr}, and K_{dp}. First, β is estimated according to

$$\beta = 2.08 Z_h^{-0.365} K_{dp}^{0.380} \xi_{dr}^{0.965} \quad (\text{mm}^{-1}) \qquad (28)$$

where $\zeta_{dr} = 10^{0.1Zdr}$ is the differential reflectivity in linear scale, and Z_h is the horizontal reflectivity in mm^6 m^{-3}. Then, for equilibrium axis ratios and S band frequency,

$$D_0 = 0.56 Z_h^{0.064} \zeta_{dr}^{1.245} \quad \text{(mm)}. \tag{29}$$

N_w is estimated from

$$\log_{10} N_w = 5.99 K_{dp}^{0.065} \xi_{dr}^{-1.057}. \tag{30}$$

The shape parameter μ is then estimated from

$$\mu = \frac{a_5 D_0^{b5}}{(\xi_{dr} - 1)} - c_5 (\xi_{dr})^{d5}, \tag{31}$$

where

$$a_5 = 200\beta^{1.89} \tag{32a}$$

$$b_5 = 2.23\beta^{0.039} \tag{32b}$$

$$c_5 = 3.16\beta^{-0.046} \tag{32c}$$

$$d_5 = 0.374\beta^{-0.355}. \tag{32d}$$

At X band, the shape factor relation is

$$\beta(\text{cm}^{-1}) \approx 12 Z_h^{-0.36} (\text{mm}^6\ \text{m}^{-3}) K_{dp}^{0.40} (^\circ\ \text{km}^{-1}) Z_{dr}^{1.02}, \tag{33}$$

where Z_{dr} is in linear units [*Matrosov et al.*, 2002, 2005]. *Matrosov et al.* [2002] then derived a three measurement rainfall estimator by inserting equation (33) into equation (25) to obtain

$$R \approx c(h) 1.06 (Z_h)^{0.3} (K_{dp})^{0.5} (Z_{dr})^{-0.84}, \tag{34}$$

where $c(h)$ is an altitude correction coefficient based on air density (ρ)

$$c(h) \approx 1.1 \rho(h)^{-0.45}. \tag{35}$$

Rainfall accumulations using equation (34) were shown to have the closet agreement with rain gauge data compared to $R(Z_h)$ and $R(K_{dp})$ for convective and stratiform rain events in Virginia. A similar three estimator approach was adopted by *Anagnostou et al.* [2004] and also found to produce superior performance compared to $R(Z_h)$.

The other common method to retrieve DSD parameters from dual-polarization measurements is the constrained gamma technique. The constrained gamma technique is based on the notion that the three parameters of the DSD are not independent [*Ulbrich*, 1983; *Chandrasekar and Bringi*, 1987; *Kozu and Nakamura*, 1991; *Haddad et al.*, 1997]. In particular, *Zhang et al.* [2001] analyzed disdrometer DSD data and showed a high correlation between the shape parameter μ and slope Λ of the DSD that led to the development of an empirical μ-Λ relationship of the form

$$\Lambda = 0.0365\mu^2 + 0.735\mu + 1.935. \tag{36}$$

Equation (36) essentially reduces the three-parameter gamma DSD to a two-parameter model [*Brandes et al.*, 2003]. However, the validity of such a constraint is currently a topic of debate [*Moisseev and Chandrasekar*, 2007]. The μ-Λ relationship, Z_{dr}, and Z_h are then used to retrieve the three-gamma DSD parameters. The method assumes that the raindrop axis ratio is constant and is expressed as in equation (8). Expressions for median drop diameter (D_0) and liquid water content (W) are

$$D_0 = 0.171(Z_{dr}^3) - 0.725(Z_{dr}^2) + 1.479(Z_{dr}) + 0.717 \quad \text{(mm)} \tag{37}$$

$$W = 5.589 \times 10^{-4} (Z_h) 10^{(0.223 Z_{dr}^2 - 1.124 Z_{dr})} \quad (\text{g m m}^{-3}), \tag{38}$$

where Z_h is in linear units, and Z_{dr} is in dB. More study is needed to verify the validity of the μ-Λ relation [*Brandes et al.*, 2004].

Based on the above description of single and multiple measurement estimators, it is apparent that some rainfall algorithms are more effective at certain frequencies compared to others. The relative performance differences can be understood in terms of a combination of the applied and basic science issues discussed in section 1.2. Any rainfall algorithm that uses Z_h will be subject to bias errors, independent of frequency. Similarly, algorithms that use Z_{dr} will be impacted by differential bias errors. At frequencies such as S band, the bias error is predominantly a calibration error (applied science). However, at C and X bands, bias can be caused by calibration or performance of attenuation correction procedures (basic science). Therefore, it is anticipated that these higher frequencies will be subject to both basic and applied science issues. The K_{dp} rainfall estimator comes with a unique set of advantages and disadvantages for rainfall estimation. As stated previously, K_{dp} is immune to all bias errors but needs to be estimated over a specified path (applied science), and the sensitivity is frequency dependent (basic science). While $R(K_{dp})$ is less sensitive at S band, it is widely used at X band (due to the basic science property), whereas $R(Z_h)$ and $R(Z_h, Z_{dr})$ perform poorly at X band due to applied science issues. For C band, the impact of both basic and applied science is in between S and X band. Although the above discussion is simplified to bias and parameterization errors, additional applied science issues such as ice/bright band contamination and clutter rejection need to be considered.

2.3.2.4. Optimal rainfall estimators. From the above discussion, it is obvious that all measurement estimator algorithms are useful and have strengths and weaknesses depending on the domain of operation. The best methodology for radar rainfall estimation is a synthesis approach that combines the strengths of all the measurement estimators with reference to both basic and applied science considerations. Several investigators have given different names for this combination approach (e.g., "synthetic," "optimal," or "blended"). In this chapter, we refer to any algorithm that utilizes one or more rainfall estimators and selects an individual rainfall relation depending on the radar-observed characteristics (Z_h, Z_{dr}, K_{dp}) as an optimization rainfall algorithm. This approach is most advanced at S band, and we now provide an overview of the development of two selected S band optimization algorithms.

The theoretical basis for this approach was discussed by *Chandrasekar et al.* [1993], and an optimization algorithm was developed at CSU to estimate rainfall for a flash flood event in Fort Collins, Colorado, using S band radar data [*Petersen et al.*, 1999]. This study introduced the concept of a combination rainfall algorithm that utilized $R(K_{dp},Z_{dr})$ of *Ryzhkov and Zrnić* [1995] in moderate to heavy rain when $Z_h > 38$ dBZ, a linearly weighted $(R(K_{dp},Z_{dr})/R(Z_h)$ in light rain when 35 dBZ $\leq Z_h \leq$ 38 dBZ and a $R(Z_h)$-only method when $Z_h <$ 35 dBZ. The performance of this algorithm in the flash flood event was similar to the $R(K_{dp}, Z_{dr})$ alone.

The CSU algorithm was extended to tropical rainfall during Tropical Rainfall Measuring Mission-Large-Scale Biosphere-Atmosphere Experiment (TRMM-LBA) [*Cifelli et al.*, 2002] and included an $R(Z_h,Z_{dr})$ relation where excellent agreement between gauge totals and S band radar was observed [*Carey et al.*, 2000b]. Subsequent to the TRMM-LBA work, the CSU optimization rainfall algorithm was further modified to compute precipitation that is typically observed in the high plains of eastern Colorado. Unlike the flash flood event described above, rainfall in this region is frequently mixed with hail/graupel during the summer months [*Dye et al.*, 1974]. Misidentification of rain and precipitation ice often leads to poor rainfall estimation and has important implications for flood forecasting. In these situations, additional characteristics of the hydrometeors, as revealed by dual-polarization observations, become critical for proper discrimination of rain and hail. The challenges of the high plains meteorological environment resulted in the development of an algorithm guided by the precipitation ice fraction (see equations (20a) and (20b)) in the radar volume. This algorithm is referred to as CSU-ICE [*Cifelli et al.*, 2010] and is shown in Figure 6.

Preliminary results using a volunteer rain gauge network indicated that the CSU-ICE algorithm outperformed the NEXRAD $R(Z_h)$ in terms of bias and RMS difference [*Cifelli et al.*, 2003]. Not surprisingly, the results also showed that performance of CSU-ICE improved relative to the NEXRAD $R(Z_h)$ as the fraction of ice over the rain gauge network

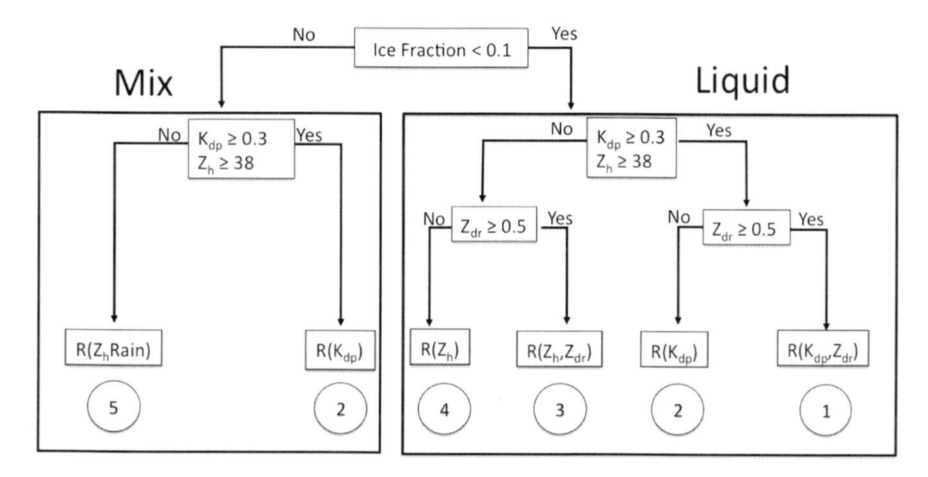

Figure 6. Flow chart describing the CSU-ICE optimization algorithm logic. The rainfall estimators corresponding to the circled numbers are identified in the text as equations (43)–(47), respectively.

increased. Subsequent analyses have shown, however, that ice fraction discrimination (described above in section 2.2.) can produce spurious results in situations where the reflectivity is moderate (38 < dBZ < 45).

As shown in Figure 6, the method of rainfall estimation in the CSU-ICE algorithm is based on thresholds of Z_h, Z_{dr}, and K_{dp} (the rainfall equations are discussed in section 3). The thresholds are derived for S band from *Petersen et al.* [1999] by visual inspection of colocated grid points to discriminate signal from noise and from *Bringi et al.* [1996]. The latter study quantified the error characteristics of selected polarization variables, including Z_h, Z_{dr}, and K_{dp}. The philosophy of rainfall estimation selection in CSU-ICE is to identify situations where a particular rainfall estimator's performance is maximized. For example, $R(K_{dp},Z_{dr})$ has the lowest error characteristics in liquid precipitation and is the preferred estimator if both K_{dp} and Z_{dr} are above their respective noise thresholds. $R(K_{dp})$ is the preferred estimator if ice is present, since Z_{dr} is usually near zero. $R(Z_h)$ is the default estimator in situations where K_{dp} and Z_{dr} are noisy or missing. The ability of the CSU-ICE algorithm to change estimators depending on the type of precipitation encountered is illustrated in Figure 7. The CSU-ICE methodology has recently been applied to C band [*Silvestro et al.*, 2009]. This study utilized the CSU-ICE rainfall estimators (described below in section 3) scaled to C band and demonstrated excellent performance in a variety of rainfall regimes, indicating its potential for operational applications.

A separate S band optimization algorithm scheme has been developed at the National Severe Storms Laboratory (NSSL) as a prototype for the upgrade of the NEXRAD WSR-88D radars [*Ryzhkov et al.*, 2005a, 2005b]. Using data collected in Oklahoma during the Joint Polarization Experiment (JPOLE), the authors developed an optimization rainfall estimation procedure combining measurements of Z_h, Z_{dr}, and K_{dp}. However, instead of assuming an explicit physical model relationship between axis ratio and drop shape, (as in CSU-ICE) the NSSL algorithm uses a semiempirical approach, where the impact of the drop shape model is implicit, to obtain the best $R(Z_h,Z_{dr})$ and $R(K_{dp},Z_{dr})$ relations for the Oklahoma data set. By comparing net areal rain rates ratios from radar $R(Z_h)$ and $R(K_{dp})$ and gauges as a function of an average Z_{dr}, functional relations between Z_{dr} and Z_h and K_{dp} can be established for the relations. In this method, the magnitude of the rain rate given by the standard WSR-88D $R(Z_h)$ formulation controls the selection of one of three rain rate expressions. The sequence of equations is relatively simple to implement and is described as follows:

1. The basic $R(Z_h)$ rate for rate equation selection is

$$R(Z_h) = .0170(Z_h^{0.714}) \quad (\text{mm h}^{-1}), \tag{39}$$

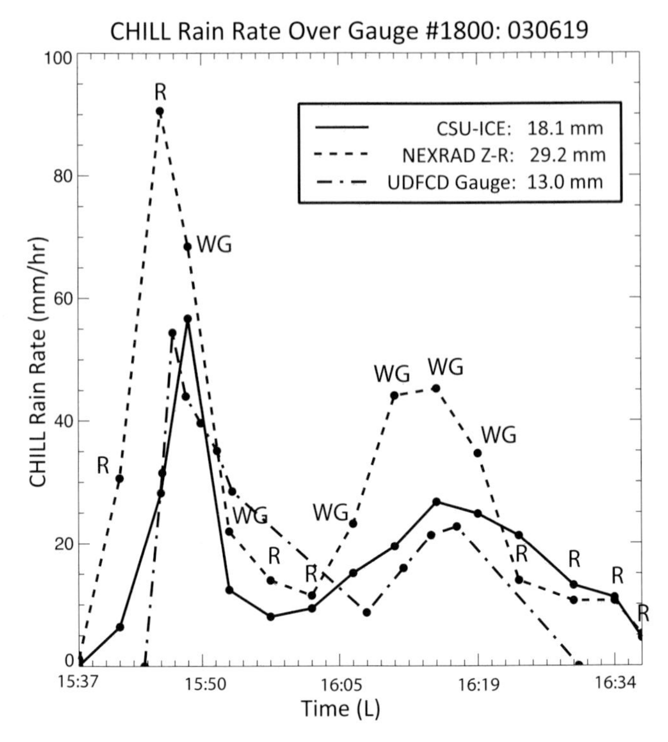

Figure 7. Time series of rain rate over a rain gauge in the ALERT network near Denver, Colorado, on 19 June 2003. "WG" and "R" refer to wet graupel and rain, respectively, as identified from a CSU-CHILL radar hydrometeor identification (HID) algorithm. The solid and dashed lines represent radar rain rate using the CSU-ICE and WSR-88D algorithms, respectively.

where Z_h is in mm^6 m^{-3}. As per NWS procedures, Z_h is limited to the linear scale equivalent of 53 dBZ to limit hail contamination.

2. Three dual-polarization rate equations based on $R(Z_h)$ are the following: If $R(Z_h) < 6$ mm h^{-1}, then

$$R = R(Z_h)/(0.4 + 5.0|Z_{dr}-1|^{1.3}) \quad (\text{mm h}^{-1}); \tag{40}$$

if $6 < R(Z_h) < 50$ mm h^{-1}, then

$$R = R(K_{dp})/(0.4 + 3.5|Z_{dr}-1|^{1.7}) \quad (\text{mm h}^{-1}); \tag{41}$$

if $R(Z_h) > 50$ mm h^{-1}, then $R = R(K_{dp})$, where

$$R(K_{dp}) = 44.0|K_{dp}|^{0.822}\text{sign}(K_{dp}) \quad (\text{mm h}^{-1}). \tag{42}$$

In equations (41) and (42), Z_{dr} is a linear scale value, and K_{dp} is in ° km^{-1}. A fuzzy logic classifier was used to remove spurious echoes and for HID during JPOLE, although the HID did not guide the rainfall estimation selection. As described by *Ryzhkov et al.* [2005a], the NSSL optimization technique exhibited superior performance compared to a number of single and two measurement rainfall estimators. Both the CSU and NSSL algorithms need to be tested in different meteorological regimes in order to evaluate the robustness of the methodologies.

3. GAPS IN CURRENT SCIENCE AND RECENT DEVELOPMENTS

Dual-polarization radar has become a common tool for researchers and will soon become widespread in many operational settings. Although dual polarization has proven advantages in preprocessing, classification, and quantification compared to single-polarization systems, many challenges remain, providing ample development opportunities. Future research for dual-polarization rainfall estimation can be roughly divided into high- and low-frequency issues. At low frequency (S band), the preprocessing, classification, and quantification algorithms are relatively mature. Future efforts will move toward the integration of classification algorithms to guide the selection of appropriate rainfall estimators and to extend the applications into cold season precipitation environments. Encouraging results in this area have already been demonstrated. *Giangrande and Ryzhkov* [2008] developed a polarization echo classification scheme (including HID as well as clutter and biological echos) and used this classification to determine which single or double measurement estimator to use in both cold and warm season Oklahoma precipitation events. The classifications scheme is based on Z_{h}, Z_{dr}, ρ_{hvco}, and a texture algorithm that describes the small-scale fluctuations of Z_{h} along the radial direction. The following precipitation estimators are used, based on the echo classification

$$R = 0;$$

if nonmeteorological echo is classified,

$$R = R(Z_{\mathrm{h}}, Z_{\mathrm{dr}});$$

if light or moderate rain is identified,

$$R = R(Z_{\mathrm{h}}, Z_{\mathrm{dr}});$$

if heavy rain or big drops are classified,

$$R = R(K_{\mathrm{dp}});$$

if rain-hail is classified and range $<R_t$,

$$R = 0.62R(Z_{\mathrm{h}});$$

if wet snow classified,

$$R = 0.8R(Z_{\mathrm{h}});$$

if graupel or rain-hail is classified at range $\geq R_t$,

$$R = R(Z_{\mathrm{h}});$$

if dry snow is classified at range $< R_t$,

$$R = 2.8R(Z_{\mathrm{h}});$$

if dry snow or ice crystals are classified at range $\geq R_t$, where R_t is the minimal slant range at which the entire radar resolution volume is above the freezing level [*Giangande and Ryzhkov*, 2008]. For a selection of Oklahoma precipitation events, significant improvement over the conventional WSR-88D $R(Z_{\mathrm{h}})$ algorithm was found at ranges <100 km from the radar and in regions of bright band contamination. Less improvement occurred above the melting level in snow. The degree of improvement of all relations gradually decreased with range and became insignificant beyond 200 km.

A similar effort to incorporate HID has been implemented at CSU. A new CSU optimization algorithm (CSU-HIDRO) has been developed in order to avoid problems with Z_{dp} described above in section 2.2 and to guide the rainfall estimation selection procedure based on HID as opposed to ice fraction [*Cifelli et al.*, 2010]. A three-class HID was developed specifically for this purpose (Figure 8). Although more classes can be recognized [e.g., *Liu and Chandrasekar*, 2000], for the purposes of rainfall estimation, it is only necessary to distinguish the presence of precipitation ice from pure liquid rain. Once the hydrometeor classification is performed, rainfall estimation can be computed.

The rainfall relations used in the CSU-HIDRO and CSU-ICE algorithms are shown here. Equations (43)–(46) are used in CSU-HIDRO, and equations (43)–(47) are used in CSU-ICE.

$$R(K_{\mathrm{dp}}, Z_{\mathrm{dr}}) = 90.8(K_{\mathrm{dp}})^{0.93} 10^{(0.169 \times Z_{\mathrm{dr}})} \tag{43}$$

$$R(K_{\mathrm{dp}}) = 40.5(K_{\mathrm{dp}})^{0.85} \tag{44}$$

$$R(Z_{\mathrm{h}}, Z_{\mathrm{dr}}) = 6.7 \times 10^{-3}(Z_{\mathrm{h}})^{0.927} 10^{(0.1 \times -3.433 \times Z_{\mathrm{dr}})} \tag{45}$$

$$R(Z_{\mathrm{h}}) = 0.0170(Z_{\mathrm{h}})^{0.7143} \tag{46}$$

$$R(Z_{\mathrm{h}}) = 0.0170(Z_{\mathrm{h}} - \text{rain only})^{0.7143}, \tag{47}$$

where R is in mm h^{-1}, Z_{h} is in mm^6 m^{-3}, Z_{dr} is in dB, and K_{dp} is ° km^{-1}.

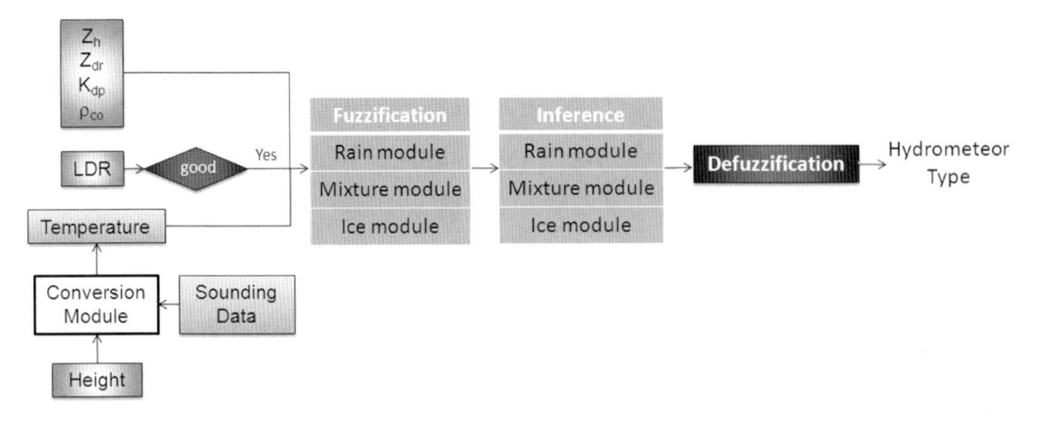

Figure 8. Flow chart of the three-class HID logic used in the CSU-HIDRO optimization algorithm, showing the inputs and steps to identify hydrometeors.

Equations (43)–(45) are physically based. The relationships were derived using simulations assuming a range of gamma DSD parameters that are typically found in observations, as described in section 1.4. Equations (43) and (45) assume that drop shape as a function of size follow the *Beard and Chuang* [1987] equilibrium model, which includes changes due to drop oscillations, a condition which commonly occurs in the high plains of Colorado. The coefficient in equation (44) assumes the *Pruppacher and Beard* [1971] equilibrium model, which assumes no oscillations. The latter was chosen due to the likely presence of rain drops >1 mm with ice cores that would act to stabilize the drops and reduce oscillations.

Equation (46) is the WSR-88D relationship described earlier. Equation (47) is identical to equation (46), except that the rain-only portion of Z_h (as determined from the ice fraction) is used.

The logic of the CSU-HIDRO algorithm is shown in Figure 9. As shown in Figure 10, the CSU-HIDRO algorithm produced encouraging results for a limited number of heavy rain events sampled in eastern Colorado. However, more testing is required to evaluate the efficacy of the algorithm in different meteorological environments. HID could also be applied to neural network systems to potentially improve rainfall estimation.

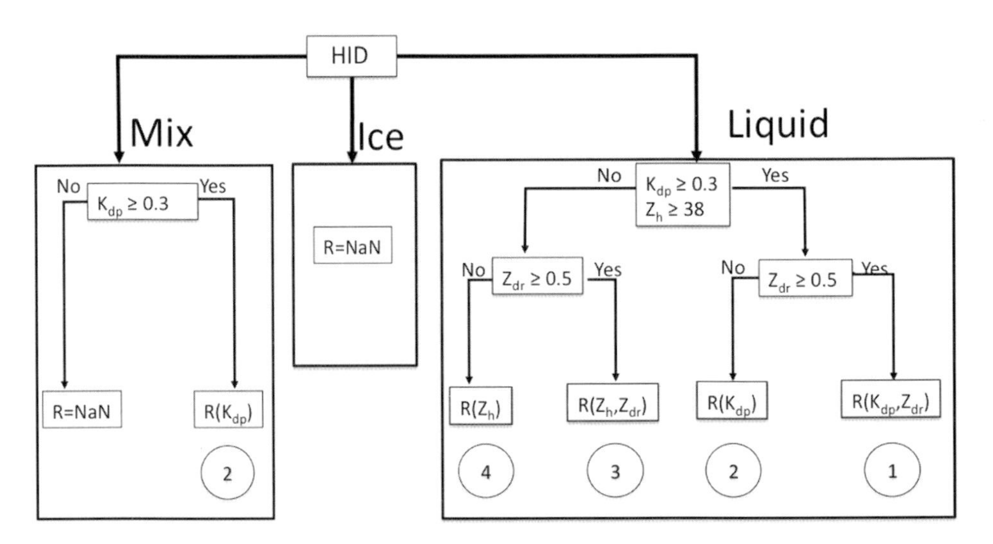

Figure 9. Flow chart describing the CSU-HIDRO optimization algorithm logic. The rainfall estimators corresponding to the circled numbers are the same as in Figure 6 and are described in the text.

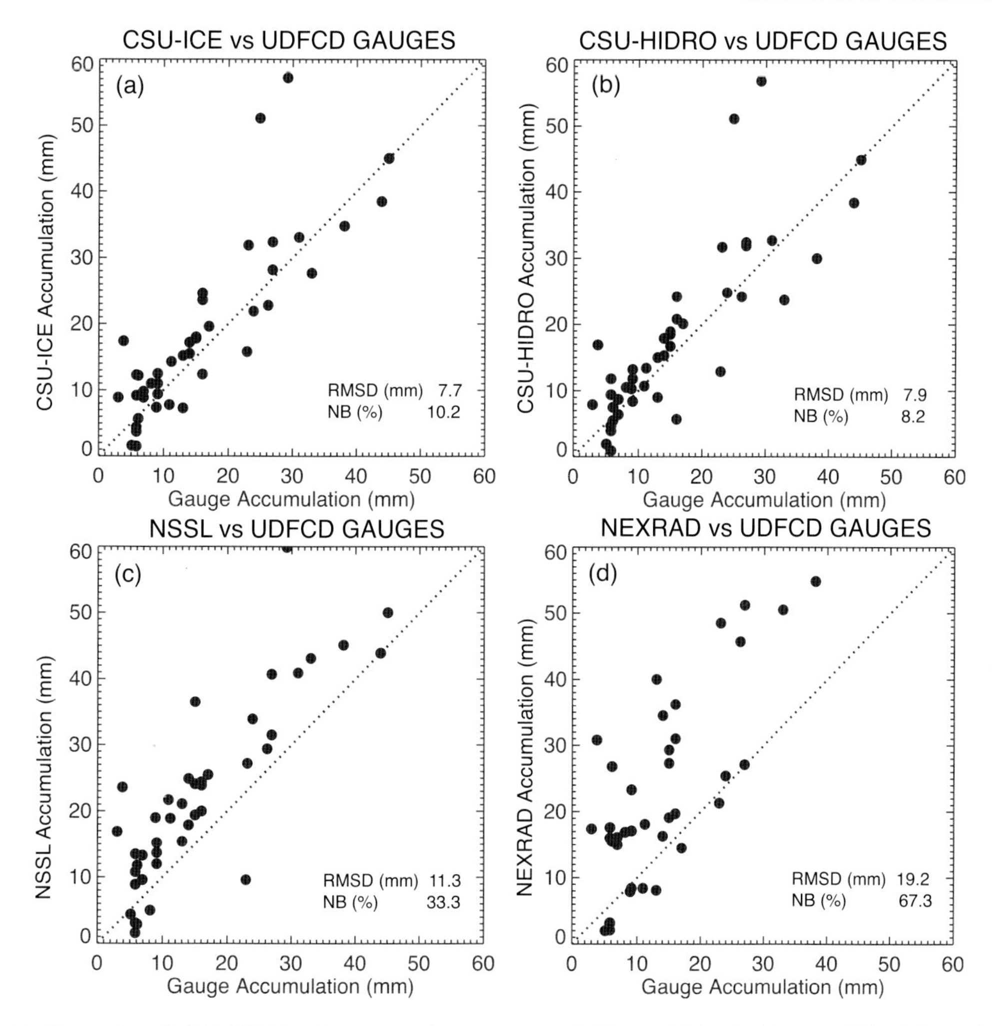

Figure 10. Scatterplot of CSU-CHILL radar versus rain gauge network (Denver Urban Drainage and Flood Control District (UDFCD)) rainfall accumulation for three events. Algorithms are (a) CSU-ICE, (b) CSU-HIDRO, (c) NSSL, and (d) WSR-88D. Root-mean-square difference (RMSD) and normalized bias (NB) results are indicated in each plot.

In addition to improving identification of ice contamination, for operational environments like NEXRAD, the challenge will be to extend improved precipitation estimation at long ranges from the radar. This is an especially challenging issue due to the fact that operational radars are typically separated by distances exceeding 100 km. *Giangrande and Rhyzhkov* [2008] have shown that the efficacy of radar measurements for rainfall estimation decreases significantly at ranges beyond 100–120 km due to the poor resolution and the fact that the radar beam is often well above the freezing level at long range. One solution to the sampling dilemma will be to supplement existing S and C bands networks with higher-frequency systems (e.g., X band) to "gap fill" regions that either have no coverage or are covered only with poor resolution measurements. The smaller antenna requirements of X band systems provide more portability, and networks of these systems may be a cost-effective option for improved rainfall estimation for radar networks with large separation distances [*McLaughlin et al.*, 2009]. X band networks could be especially useful for gap-filling purposes in areas of complex terrain. In situations where heavy rainfall is occurring over a small watershed that is blocked from a distant operational radar, a network of small radars may be the only way to adequately cover the watershed and identify potential flash flood situations.

An example of combined S and X band networks is the CASA Integrated Project 1 (IP1) network in Oklahoma [*Chandrasekar and Lim*, 2008]. In addition to providing increased temporal resolution and better coverage at low levels, the embedded networks can also provide important microphysical information on subpixel scales of S band systems so that the spatial variability of rainfall can be assessed. The high spatial and temporal resolution is important for many hydrologic applications, including runoff models that require accurate precipitation input. However, for X band networks to provide operational support, it will be necessary to improve the robustness of algorithms in all stages of application: preprocessing, classification, and quantification. Attenuation correction procedures must be developed that can be applied in different geographical regions. At the same time, future efforts at X band will be focused on improving HID and optimization rainfall algorithms. *Dolan and Rutledge* [2009] used a combination of CASA IP1 data and data from an S band dual-polarization radar to test an X band HID algorithm. The results were encouraging in the separation of ice and liquid. Future efforts will focus on extending the algorithm to new environment situations of mixed phase precipitation and situations where hail contamination is present.

4. SUMMARY

Dual-polarization radar is becoming an important tool for quantitative precipitation estimation (QPE) in operational networks. These radars offer a number of advantages over traditional single-polarization systems, including (1) more accurate physical models to represent the DSD and their relationship to the observed radar variables (basic science issues) and (2) measurements that are immune to absolute radar calibration and partial beam blocking (applied science issues) as well as help in data quality enhancement. Dual-polarization observations can be integrated into all three steps of the QPE process to improve rainfall estimation: preprocessing (data enhancement), classification (identification of different hydrometeor types), and quantification (rainfall estimation). By combining the different radar measurements, dual-polarization algorithms have been developed that take advantage of the strengths of different rainfall estimators in different precipitation environments with consideration to both basic and applied science aspects. Higher-frequency systems, such as X band, offer exciting opportunities for improved rainfall estimation, since these radars provide more portability and can serve to fill in coverage gaps within the longer-range and lower-frequency (S and C band) networks.

Acknowledgments. This chapter contains research results that have come from current and past funding from both the National Science Foundation and the National Aeronautical and Space Administration. The authors acknowledge the CSU CHILL facility for the various data sets presented in this paper. CSU-CHILL is supported by the National Science Foundation.

REFERENCES

Anagnostou, E. N., M. N. Anagnostou, W. F. Krajewski, A. Kruger, and B. J. Miriovsky (2004), High-resolution rainfall estimation from X-band polarimetric radar measurements, *J. Hydrometeorol.*, *5*, 110–128.

Andsager, K., K. V. Beard, and N. F. Laird (1999), Laboratory measurements of axis ratios for large drops, *J. Atmos. Sci*, *56*, 2673–2683.

Atlas, D., and C. W. Ulbrich (1977), Path- and area-integrated rainfall measurement by microwave attenuation in the 1–3 cm band, *J. Appl. Meteorol.*, *16*, 1322–1331.

Atlas, D., and C. W. Ulbrich (1990), Early foundations of the measurement of rainfall by radar, in *Radar in Meteorology*, edited by D. Atlas, pp. 86–97, Am. Meteorol. Soc., Boston, Mass.

Aydin, K., T. Seliga, and V. Balaji (1986), Remote sensing of hail with a dual linear polarization radar, *J. Appl. Meteorol.*, *25*, 1475–1484.

Aydin, K., H. Direskeneli, and T. A. Seliga (1987), Dual-polarization radar estimation of rainfall parameters compared with ground-based disdrometer measurements: October 29, 1982 Central Illinois Experiment, *IEEE Trans. Geosci. Remote Sens.*, *GE-25*, 834–844.

Aydin, K., Y. Zhao, and T. Seliga (1990), A differential reflectivity radar hail measurement technique: Observations during the Denver hailstorm of 13 June 1984, *J. Atmos. Oceanic Technol.*, *7*, 104–113.

Battan, L. J. (1973), *Radar Observations of the Atmosphere*, 324 pp., Univ. of Chicago Press, Chicago, Ill.

Beard, K. V., and C. Chuang (1987), A new model for the equilibrium shape of raindrops, *J. Atmos. Sci.*, *44*, 1509–1524.

Brandes, E. A., and K. Ikeda (2004), Freezing-level estimation with polarimetric radar, *J. Appl. Meteorol.*, *43*, 1541–1553.

Brandes, E. A., A. V. Ryzhkov, and D. S. Zrnić (2001), An evaluation of radar rainfall estimates from specific differential phase, *J. Atmos. Oceanic Technol.*, *18*, 363–375.

Brandes, E. A., G. Zhang, and J. Vivekanandan (2003), An evaluation of a drop distribution-based rainfall estimator, *J. Appl. Meteorol.*, *42*, 652–660.

Brandes, E. A., G. Zhang, and J. Vivekanandan (2004), Comparison of polarimetric radar drop size distribution retrieval algorithms, *J. Atmos. Oceanic Technol.*, *21*, 584–598.

Bringi, V. N., and V. Chandrasekar (2001), *Polarimetric Doppler Weather Radar: Principles and Applications*, 635 pp., Cambridge Univ. Press, New York

Bringi, V. N., L. Liu, P. C. Kennedy, V. Chandrasekar, and S. A. Rutledge (1996), Dual multiparameter radar observations of intense convective storms: The 24 June 1992 case study, *Meteorol. Atmos. Phys.*, *59*, 3–31.

Calheiros, R., and I. Zawadzki (1987), Reflectivity-rain rate relationships for radar hydrology in Brazil, *J. Appl. Meteorol.*, *26*, 118–132.

Carey, L. D., S. A. Rutledge, D. A. Ahijevych, and T. D. Keenan (2000a), Correcting propagation effects in C-band polarimetric radar observations of tropical convection using differential propagation phase, *J. Appl. Meteorol.*, *39*, 1405–1433.

Carey, L. D., R. Cifelli, W. A. Petersen, and S. A. Rutledge (2000b), Preliminary report on TRMM-LBA rainfall estimation using the S-POL radar, *Tech. Rep. 697*, 19 pp., Dept. Atmos. Sci., Colo. State Univ.

Chandrasekar, V., and V. N. Bringi (1987), Simulation of radar reflectivity and surface measurements of rainfall, *J. Atmos. Oceanic Technol.*, *4*, 464–478.

Chandrasekar, V., and V. N. Bringi (1988), Error structure of multiparameter radar and surface measurements of rainfall Part I: Differential reflectivity, *J. Atmos. Oceanic Technol.*, *5*, 783–795.

Chandrasekar, V., and S. Lim (2008), Retrieval of reflectivity in a networked radar environment, *J. Atmos. Oceanic Technol.*, *25*, 1755–1767.

Chandrasekar, V., V. Bringi, N. Balakrishnan, and D. Zrnić (1990), Error structure of multiparameter radar and surface measurements of rainfall. Part III: Specific differential phase, *J. Atmos. Oceanic Technol.*, *7*, 621–629.

Chandrasekar, V., E. Gorgucci, and G. Scarchilli (1993), Optimization of multiparameter radar estimates of rainfall, *J. Appl. Meteorol.*, *32*, 1288–1293.

Chandrasekar, V., R. Meneghini, and I. Zawadzki (2003), Global and local precipitation measurements by radar, *Meteorol. Monogr.*, *30*, 215–236.

Cifelli, R., W. A. Petersen, L. D. Carey, S. A. Rutledge, and M. A. F. da Silva Dias (2002), Radar observations of the kinematic, microphysical, and precipitation characteristics of two MCSs in TRMM LBA, *J. Geophys. Res.*, *107*(D20), 8077, doi:10.1029/2000JD000264.

Cifelli, R., D. Barjenbruch, D. Brunkow, L. Carey, C. Davey, N. Doesken, C. Gimmestad, T. Huse., P. Kennedy, and S. A. Rutledge (2003), Evaluation of an operational polarimetric rainfall algorithm, paper presented at 31st International Conference on Radar Meteorology, Am. Meteorol. Soc., Seattle, Wash., 6–11 Aug.

Cifelli, R., V. Chandrasekar, S. Lim, P. Kennedy, Y. Wang, and S. A. Rutledge (2010), A new dual-polarization radar rainfall algorithm: Application in Colorado precipitation events, *J. Atmos. Oceanic Technol.*, in press.

Dolan, B., and S. A. Rutledge (2009), A theory-based hydrometeor identification algorithm for X-band polarimetric radars, *J. Atmos. Oceanic Technol.*, *26*, 2071–2088.

Dye, J. E., C. A. Knight, V. Toutenhoofd, and T. W. Cannon (1974), The mechanism of precipitation formation in northeastern Colorado Cumulus III. Coordinated microphysical and radar observations and summary, *J. Atmos. Sci.*, *31*, 2152–2159.

Fulton, R. A., J. P. Breidenbach, D. J. Seo, D. A. Miller, and T. O'Bannon (1998), The WSR-88D rainfall algorithm, *Weather Forecast.*, *13*, 377–395.

Giangrande, S. E., and A. V. Ryzhkov (2008), Estimation of rainfall based on the results of polarimetric echo classification, *J. Appl. Meteorol. Climatol.*, *47*, 2445–2462.

Golestani, Y., V. Chandrasekar, and V. N. Bringi (1989), Intercomparison of multiparameter radar measurements, paper presented at 24th Conference on Radar Meteorology, Am. Meteorol. Soc., Boston, Mass.

Gorgucci, E., and V. Chandrasekar (2005), Evaluation of attenuation correction methodology for dual-polarization radars: Application to X-band systems, *J. Atmos. Oceanic Technol.*, *22*, 1195–1206.

Gorgucci, E., V. Chandrasekar, and G. Scarchilli (1995), Radar and surface measurement of rainfall during CaPE: 26 July 1991 Case Study, *J. Appl. Meteorol.*, *34*, 1570–1577.

Gorgucci, E., G. Scarchilli, V. Chandrasekar, and V. N. Bringi (2000), Measurement of mean raindrop shape from polarimetric radar observations, *J. Atmos. Sci.*, *57*, 3406–3413.

Gorgucci, E., G. Scarchilli, V. Chandrasekar, and V. N. Bringi (2001), Rainfall estimation from polarimetric radar measurements: Composite algorithms immune to variability in raindrop shape–size relation, *J. Atmos. Oceanic Technol.*, *18*, 1773–1786.

Gorgucci, E., V. Chandrasekar, V. N. Bringi, and G. Scarchilli (2002), Estimation of raindrop size distribution parameters from polarimetric radar measurements, *J. Atmos. Sci.*, *59*, 2373–2384.

Gorgucci, E., L. Baldini, and V. Chandrasekar (2006), What is the shape of a raindrop? An answer from radar measurements, *J. Atmos. Sci.*, *63*, 3033–3044.

Gorgucci, E., V. Chandrasekar, and L. Baldini (2009), Can a unique model describe the raindrop shape–size relation? A clue from polarimetric radar measurements, *J. Atmos. Oceanic Technol.*, *26*, 1829–1842.

Gourley, J. J., P. Tabary, and J. Parent du Chatelet (2007), A fuzzy logic algorithm for the separation of precipitating from nonprecipitating echoes using polarimetric radar observations, *J. Atmos. Oceanic Technol.*, *24*, 1439–1451.

Green, A. W. (1975), An approximation for the shapes of large raindrops, *J. Appl. Meteorol.*, *14*, 1578–1583.

Haddad, Z. S., D. A. Short, S. L. Durden, E. Im, S. Hensley, M. B. Grable, and R. A. Black (1997), A new parameterizing of raindrop size distribution, *IEEE Trans. Geosci. Remote Sens.*, *35*, 532–539.

Hendry, A., and Y. M. M. Antar (1984), Precipitation particle identification with centimeter wavelength dual-polarization radars, *Radio Sci.*, *19*, 115–122.

Höller, H., V. N. Bringi, J. Hubbert, M. Hagen, and P. F. Meischner (1994), Life cycle and precipitation formation in a hybrid-type hailstorm revealed by polarimetric and Doppler radar measurements, *J. Atmos. Sci.*, *51*, 2500–2522.

Hubbert, J. C., M. Dixon, and S. M. Ellis (2009), Weather radar ground clutter. Part II: Real-time identification and filtering, *J. Atmos. Oceanic Technol.*, *26*, 1181–1197.

Jameson, A. (1983), Microphysical interpretation of multi-parameter radar measurements in rain. Part I: Interpretation of polarization measurements and estimation of raindrop shapes, *J. Atmos. Sci.*, *40*, 1792–1802.

Jameson, A. (1985), Microphysical interpretation of multiparameter radar measurements in rain. Part III: Interpretation and measurement of propagation differential phase shift between Oorthogonal linear polarizations, *J. Atmos. Sci.*, *42*, 607–614.

Jones, B. K., J. R. Saylor, and F. Y. Testik (2010), Raindrop morphodynamics, in *Rainfall: State of the Science*, *Geophys. Monogr. Ser.*, doi: 10.1029/2009GM000928, this volume.

Keenan, T. D., D. Zrnić, L. Carey, P. May, and S. Rutledge (1997), Sensitivity of C-band polarimetric variables to propagation and backscatter effects in rain, paper presented at 28th Conference on Radar Meteorology, Am. Meteorol. Soc., Austin, Tex.

Keenan, T. D., L. D. Carey, D. S. Zrnić, and P. T. May (2001), Sensitivity of 5-cm wavelength polarimetric radar variables to raindrop axial ratio and drop size distribution, *J. Appl. Meteorol.*, *40*, 526–545.

Kessinger, C., S. Ellis, and J. Van Andel (2001), NEXRAD Data Quality Enhancements: The AP Clutter Mitigation Scheme. 30th International Conference on Radar Meteorology, AMS, Munich, Germany, 19–25 July 2001.

Kozu, T., and K. Nakamura (1991), Rain parameter estimation from dual-radar measurements combining reflectivity profile and path-integrated attenuation, *J. Atmos. Oceanic Technol*, *8*, 259–270.

Lim, S., V. Chandrasekar, and V. N. Bringi (2005), Hydrometeor classification system using dual-polarization radar measurements: Model improvements and in situ verification, *IEEE Trans. Geosci. Remote Sens.*, *43*, 792–801.

Liu, H., and V. Chandrasekar (1998), Classification of hydrometeor type based on multiparameter radar measurements, *Proceedings, Intl. Conf. on Cloud Phys.*, pp. 253– 256.

Liu, H., and V. Chandrasekar (1999), Classification of hydrometeor type based on multiparameter radar measurements, *29th International Radar Conference*, pp. 172–175, Am. Meteorol. Soc., Boston, Mass.

Liu, H., and V. Chandrasekar (2000), Classification of hydrometeor type based on multiparameter radar measurements: Development of a fuzzy logic and neuro fuzzy systems and in situ verification, *J. Atmos. Oceanic Technol.*, *17*, 140–164.

Liu, H., V. Chandrasekar, and G. Xu (2001), An adaptive neural network scheme for radar rainfall estimation from WSR-88D observations, *J. Appl. Meteorol.*, *40*, 2038–2050.

Marshall, J. S., R. C. Langille, and W. M. Palmer (1947), Measurement of rainfall by radar, *J. Atmos. Sci.*, *4*, 186–192.

Matrosov, S. Y., K. A. Clark, B. E. Martner, and A. Tokay (2002), X-band polarimetric radar measurements of rainfall, *J. Appl. Meteorol.*, *41*, 941–952.

Matrosov, S. Y., D. E. Kingsmill, B. E. Martner, and F. M. Ralph (2005), The utility of X-band polarimetric radar for quantitative estimates of rainfall parameters, *J. Hydrometeorol.*, *6*, 248–262.

Matrosov, S. Y., R. Cifelli, P. C. Kennedy, S. W. Nesbitt, S. A. Rutledge, V. N. Bringi, and B. E. Martner (2006), A comparative study of rainfall retrievals based on specific differential phase shifts at X- and S-band radar frequencies, *J. Atmos. Oceanic. Technol.*, *23*, 952–963.

Matrosov, S. Y., K. A. Clark, and D. E. Kingsmill (2007), A polarimetric radar approach to identify rain, melting-layer, and snow regions for applying corrections to vertical profiles of reflectivity, *J. Appl. Meteorol. Climatol.*, *46*, 154–166.

May, P. T., and T. D. Keenan (2005), Evaluation of microphysical retrievals from polarimetric radar with wind profiler data, *J. Appl. Meteorol.*, *44*, 827–838.

May, P. T., T. D. Keenan, D. S. Zrnić, L. D. Carey, and S. A. Rutledge (1999), Polarimetric radar measurements of tropical rain at a 5-cm wavelength, *J. Appl. Meteorol.*, *38*, 750–765.

McCormick, G. C., and A. Hendry (1975), Principles for radar determination of the polarization properties of precipitation, *Radio Sci.*, *10*, 421–434.

McLaughlin, D., et al. (2009), Short-wavelength technology and the potential for distributed networks of small radar systems, *Bull. Am. Meteorol. Soc.*, *90*, 1797–1817.

Moisseev, D. N., and V. Chandrasekar (2007), Examination of the μ–Λ relation suggested for drop size distribution parameters, *J. Atmos. Oceanic Technol.*, *24*, 847–855.

Park, S.-G., V. N. Bringi, V. Chandrasekar, M. Maki, and K. Iwanami (2005), Correction of radar reflectivity and differential reflectivity for rain attenuation at X band. Part I: Theoretical and empirical basis, *J. Atmos. Oceanic Technol.*, *22*, 1621–1632.

Petersen, W. A., L. D. Carey, S. A. Rutledge, J. C. Knievel, R. H. Johnson, N. J. Doesken, T. B. McKee, T. Vonder Haar, and J. F. Weaver (1999), Mesoscale and radar observations of the Fort Collins flash flood of 28 July 1997, *Bull. Am. Meteorol. Soc.*, *80*, 191–216.

Pruppacher, H. R., and K. V. Beard (1970), A wind tunnel investigation of the internal circulation and shape of water drops falling at terminal velocity in air, *Q. J. R. Meteorol. Soc.*, *96*, 247–256.

Pruppacher, H. R., and K. V. Beard (1971), A wind tunnel investigation of the internal circulation and shape of water drops falling at terminal velocity in air, *Q. J. R. Meteorol. Soc.*, *97*, 133–134.

Rosenfeld, D., D. B. Wolff, and D. Atlas (1993), General probability-matched relations between radar reflectivity and rain rate, *J. Appl. Meteorol.*, *32*, 50–72.

Ryzhkov, A. V., and D. S. Zrnić (1995), Comparison of dual-polarization radar estimators of rain, *J. Atmos. Oceanic Technol.*, *12*, 249–256.

Ryzhkov, A. V., and D. S. Zrnić (1996), Assessment of rainfall measurement that uses specific differential phase, *J. Appl. Meteorol.*, *35*, 2080–2090.

Ryzhkov, A. V., and D. S. Zrnić (1998a), Polarimetric rainfall estimation in the presence of anomalous propagation, *J. Atmos. Oceanic Technol.*, *15*, 1320–1330.

Ryzhkov, A. V., and D. S. Zrnić (1998b), Discrimination between rain and snow with a polarimetric radar, *J. Appl. Meteorol.*, *37*, 1228–1240.

Ryzhkov, A. V., S. E. Giangrande, and T. J. Schuur (2005a), Rainfall estimation with a polarimetric prototype of WSR-88D, *J. Appl. Meteorol.*, *44*, 502–515.

Ryzhkov, A. V., T. J. Schuur, D. W. Burgess, P. L. Heinselman, S. E. Giangrande, and D. S. Zrnić (2005b), The Joint Polarization Experiment: Polarimetric rainfall measurements and hydrometeor classification, *Bull. Am. Meteorol. Soc.*, *86*, 809–824.

Sachidananda, M., and D. S. Zrnić (1986), Differential propagation phase shift and rainfall rate estimation, *Radio Sci.*, *21*, 235–247.

Scarchilli, G., E. Gorgucci, V. Chandrasekar, and T. A. Seliga (1993), Rainfall estimation using polarimetric techniques at C-band frequencies, *J. Appl. Meteorol.*, *32*, 1149–1160.

Seliga, T. A., and V. N. Bringi (1976), Potential use of radar differential reflectivity measurements at orthogonal polarizations for measuring precipitation, *J. Appl. Meteorol.*, *15*, 69–76.

Seliga, T. A., and R. G. Humphries (1990), Polarization diversity in radar meteorology: Early developments, in *Radar in Meteorology*, edited by D. Atlas, pp. 109–121, Am. Meteorol. Soc., Boston, Mass.

Seo, D.-J., A. Seed, and G. Delrieu (2010), Radar and multisensor rainfall estimation for hydrologic applications, in *Rainfall: State of the Science, Geophys. Monogr. Ser.*, doi: 10.1029/2010GM000952, this volume.

Silvestro, F., N. Rebora, and L. Ferraris (2009), An algorithm for real-time rainfall rate estimation by using polarimetric radar: RIME, *J. Hydrometeorol.*, *10*, 227–240.

Straka, J. M. (1996), Hydrometeor fields in a supercell storm as deduced from dual-polarization radar, paper presented at 18th Conference on Severe Local Storms, Am. Meteorol. Soc., San Francisco, Calif.

Straka, J. M., and D. S. Zrnić (1993), An algorithm to deduce hydrometeor types and contents from multiparameter radar data, paper presented at 26th International Conference on Radar Meteorology, Am. Meteorol. Soc., Norman, Okla.

Testud, J., E. Le Bouar, E. Obligis, and M. Ali-Mehenni (2000), The rain profiling algorithm applied to polarimetric weather radar, *J. Atmos. Oceanic Technol.*, *17*, 332–356.

Thurai, M., and V. N. Bringi (2005), Drop axis ratios from a 2D video disdrometer, *J. Atmos. Oceanic Technol.*, *22*, 966–978.

Ulbrich, C. W. (1983), Natural variations in the analytical form of the raindrop size distribution, *J. Appl. Meteorol.*, *22*, 1764–1775.

Vivekanandan, J., S. M. Ellis, R. Oye, D. S. Zrnić, A. V. Ryzhkov, and J. Straka (1999), Cloud microphysics retrieval using S-band dual-polarization radar measurements, *Bull. Am. Meteorol. Soc.*, *80*, 381–388.

Vulpiani, G., F. S. Marzano, V. Chandrasekar, A. Berne, and R. Uijlenhoet (2006), Polarimetric weather radar retrieval of raindrop size distribution by means of a regularized artificial neural network, *IEEE Trans. Geosci. Remote Sens.*, *44*, 3262–3275.

Vulpiani, G., P. Tabary, J. Parent-Du-Chatelet, and F. S. Marzano (2008), Comparison of advanced radar polarimetric techniques for operational attenuation correction at C-band, *J. Atmos. Ocean. Technol.*, *25*, 1118–1135.

Vulpiani, G., S. Giangrande, and F. S. Marzano (2009), Rainfall estimation from polarimetric S-band radar measurements: Validation of a neural network approach, *J. Appl. Meteorol. Climatol.*, *48*, 2022–2036.

Wang, Y., and V. Chandrasekar (2009), Algorithm for estimation of the specific differential phase, *J. Atmos. Oceanic Technol.*, *26*, 2565–2578.

Wang, Y., and V. Chandrasekar (2010), Quantitative precipitation estimation in CASA X-band dual-polarization radar network, *J. Atmos. Oceanic Technol.*

Wexler, R., and D. Swingle (1947), Radar storm detection, *Bull. Am. Meteorol. Soc.*, *28*, 159–167.

Xiao, R., and V. Chandrasekar (1995), Multiparameter radar rainfall estimation using neural network techniques, *Proceedings, Intl. Conf. Radar Meteor., Vail, CO*, pp. 199–201, Am. Meteorol. Soc., Boston, Mass.

Xiao, R., and V. Chandrasekar (1997), Development of neural network based algorithm for rainfall estimation based on radar measurements, *IEEE Trans. Geosci. Remote Sens.*, *35*, 160–171.

Xu, G., and V. Chandrasekar (2005), Operational feasibility of neural-network-based radar rainfall estimation, *IEEE Geosci. Remote Sens. Lett.*, *2*, 13–17.

Zawadzki, I. (1984), Factors affecting the precision of radar measurements of rain, paper presented at 22nd Conference on Radar Meteorology, Am. Meteorol. Soc., Boston, Mass.

Zhang, G., J. Vivekanandan, and E. Brandes (2001), A method for estimating rain rate and drop size distribution from polarimetric radar measurements, *IEEE Trans. Geosci. Remote Sens.*, *39*, 830–841.

Zrnić, D. S., and A. V. Ryzhkov (1999), Polarimetry for weather surveillance radars, *Bull. Am. Meteorol. Soc.*, *80*, 389–406.

Zrnić, D. S., A. Ryzhkov, J. Straka, Y. Liu, and J. Vivekanandan (2001), Testing a procedure for automatic classification of hydrometeor types, *J. Atmos. Oceanic Technol.*, *18*, 892–913.

V. Chandrasekar, Department of Electrical Engineering, Colorado State University, Fort Collins, CO 80523-1373, USA.

R. Cifelli, Earth System Research Laboratory, NOAA, Boulder, CO 80305, USA. (rob.cifelli@noaa.gov)

Quantitative Precipitation Estimation From Earth Observation Satellites

Chris Kidd

School of Geography, Earth and Environmental Sciences, University of Birmingham, Birmingham, UK

Vincenzo Levizzani and Sante Laviola

Istituto di Scienze dell'Atmosfera e del Clima, ISAC-CNR, Bologna, Italy

The observation of the atmosphere by satellite instrumentation was one of the first uses of remotely sensed data nearly 50 years ago. Satellites offer an unrivalled vantage point to observe and measure Earth system processes and parameters. Observations of meteorological phenomena permit a more holistic view of the weather and climate that is not possible through conventional surface observations. Precipitation (rain and snow), in particular, benefit from such observations, since precipitation is spatially and temporally highly variable and overcome some of the deficiencies of conventional gauge and radar measurements. This paper provides an overall review of quantitative precipitation estimation, covering the basis of the satellite systems used in the observation of precipitation and the dissemination of this data, the processing of these measurements to generate the rainfall estimates, and the availability, verification, and validation of these precipitation estimates.

1. BACKGROUND

The measurement of precipitation (both rainfall and snowfall) is of great value to both science and society. Precipitation is a main component of the global water and energy cycle, helping to regulate the climate system. In addition, the availability of fresh water is vital to life on Earth. Measurement of global precipitation through conventional instrumentation uses networks of rain (or snow) gauges and, where available, weather radar systems. However, the distribution of these globally is uneven and often related to population density: over the land masses, the distribution and density of gauges is highly variable with some regions having "adequate" coverage, while others have few or no gauges. Over the oceans, a few gauges exist, and those that are placed on an island might be subject to local influences and not representative of the surrounding ocean. The availability of historical precipitation data sets can also be problematic, varying in availability, completeness and consistency as well availability for near real-time analysis.

The potential for observing our weather and climate from satellite systems was realized with the launch of the TIROS 1 in April 1960 and have proved to be an invaluable tool in providing measurements of various atmospheric parameters at regular intervals on a global scale. In 1963, the World Meteorological Organization (WMO) (see Table 1 for a detailed list of abbreviations) established the World Weather Watch program to coordinate a network of operational geostationary (GEO) and polar-orbiting meteorological satellites called the Global Observing System (GOS) [*World Meteorological Organization*, 2003, 2005]. This system was charged with providing long-term stable data sets required by international organizations and the user community; consequently, many of the early observations are compatible with current systems. Meteorological satellite sensors are

Rainfall: State of the Science
Geophysical Monograph Series 191

10.1029/2009GM000920

Table 1. Relevant Abbreviations and Definitions

Abbreviation	Definition
AE	Auto Estimator
AIP	Algorithm Intercomparison Programme
AIRS	Atmospheric InfraRed Sounder (NASA)
AMSR-E	Advanced Microwave Scanning Radiometer-EOS (JAXA)
AMSU	Advanced Microwave Sounding Unit (NOAA)
ANN	Artificial Neural Network
AOPC	Atmospheric Observation Panel for Climate (WMO)
AVHRR	Advanced Very High Resolution Radiometer (NOAA)
CALIOP	Cloud-Aerosol Lidar with Orthogonal Polarization (NASA-CNES)
CALIPSO	Cloud-Aerosol Lidar and Infrared Pathfinder Satellite Observation (NASA-CNES)
CAPSAT	Clouds-Aerosols-Precipitation Satellite Analysis Tool
CCD	Cold-Cloud Duration
CEOS	Committee on Earth Observations
CHOMPS	CICS High-Resolution Optimally Interpolated Microwave Precipitation from Satellites
CICS	Cooperative Institute for Climate Studies
CLW	Cloud Liquid Water
CMAP	CPC Merged Analysis of Precipitation
CMORPH	CPC MORPHing algorithm
CNES	Centre Nationale d'Etudes Spatiales
COMS	Korean Meteorological Satellite
CPC	Climate Prediction Center (NOAA)
CPR	Cloud Profiling Radar (NASA)
CRM	Cloud Resolving Model
CSH	Convective-Stratiform Heating
CWP	Cloud Water Path
DMSP	Defense Meteorological Satellite Program (U.S. Navy)
DPR	Dual-wavelength Precipitation Radar (JAXA)
Earth-CARE	ESA Clouds, Aerosol and Radiation Explorer
EGPM	European contribution to Global Precipitation Measurement (ESA)
EOS	Earth Observing System (NASA)
ESA	European Space Agency
EUMETSAT	European Organisation for the Exploitation of Meteorological Satellites
FAR	False Alarm Ratio
FLORAD	FLOwer constellation for mm-wave scanning RADiometers
GEO	Geostationary satellite
GEOSS	Global Earth Observation System of Systems
GEWEX	Global Energy and Water cycle EXperiment
GISS	Goddard Institute for Space Studies (NASA)
GMI	GPM Microwave Imager (NASA)
GMSRA	GOES Multi-Spectral Rainfall Algorithm
GOES	Geostationary Operational Environmental Satellite (NOAA)
GOS	Global Observing System
GPCP	Global Precipitation Climatology Project
GPI	Global Precipitation Index
GPM	Global Precipitation Measurement mission
GPROF	Goddard Profiling algorithm
GSFC	Goddard Space Flight Center (NASA)
GSMaP	Global Satellite Mapping of Precipitation
HE	Hydro Estimator
HH	Hydrometeor Heating
H-SAF	Satellite Applications Facility on Support to Operational Hydrology and Water Management (EUMETSAT)
HSB	Humidity Sounder for Brazil

Table 1. (continued)

Abbreviation	Definition
IPWG	International Precipitation Working Group
IR	Infrared
IWP	Ice Water Path
JAXA	Japan Aerospace eXploration Agency
JMA	Japan Meteorological Administration
LEO	Low Earth Orbit satellite
LMODEL	Lagrangian MODEL
LIS	Lightning Imaging Sensor (TRMM, NASA)
MIRA	Microwave-adjusted IR Algorithm
MIRRA	Microwave/Infrared Rain Rate Algorithm
MM5	Penn State/NCAR Mesoscale Model
MODIS	Moderate Resolution Imaging Spectroradiometer (NASA)
MSG	Meteosat Second Generation (EUMETSAT)
MSPPS	Microwave Surface and Precipitation Products System (NOAA)
MTG	Meteosat Third Generation (EUMETSAT)
MTSAT	Multifunctional Transport Satellite (JMA)
MW	Microwave
NASA	National Aeronautics and Space Administration
NCAR	National Center for Atmospheric Research
NCEP	National Centers for Environmental Prediction (NOAA)
NOAA	National Oceanic and Atmospheric Administration
NRC	National Research Council
NRL	Naval Research Laboratory
NWP	Numerical Weather Prediction
OLR	Outgoing Longwave Radiation
OPI	OLR Precipitation Index
OTD	Optical Transient Detector
PacNet	Pacific Lightning Detection Network
PCA	Principal Component Analysis
PCT	Polarization-Corrected Temperature
PEHRPP	Program for the Evaluation of High-Resolution Precipitation Products
PERSIANN	Precipitation Estimation from Remotely Sensed Information using Artificial Neural Network
PERSIANN-MSA	PERSIANN Multi-Spectral Analysis
PIP	Precipitation Intercomparison Project
PMIR	Passive Microwave InfraRed technique
PMW	Passive microwave
POD	Probability of Detection
PP	Precipitation Property algorithm
PR	Precipitation Radar
PWC	Precipitation Water Content
RGB	Red-Green-Blue
SCaMPR	Self-Calibrating Multivariate Precipitation Retrieval
SEVIRI	Spinning Enhanced Visible and InfraRed Imager (EUMETSAT)
SLH	Spectral Latent Heating algorithm
SOFM	Self-Organizing Feature Map
SSM/I	Special Sensor Microwave Imager (U.S. Navy)
SSMIS	Special Sensor Microwave Imager/Sounder (U.S. Navy)
Tb	Brightness Temperature
TIROS	Television Infrared Observation Satellite (NOAA)
TMI	TRMM Microwave Imager (NASA)
TMPA	TRMM Multi-Satellite Precipitation Algorithm

Table 1. (continued)

Abbreviation	Definition
TOGA/COARE	Tropical Ocean–Global Atmosphere/Coupled Ocean-Atmosphere Response Experiment
TPW	Total Precipitable Water
TRMM	Tropical Rainfall Measuring Mission (NASA)
VIS	Visible
WGNE	Working Group for Numerical Experimentation (WMO)
WMO	World Meteorological Organization
183-WSL	Water Vapor Strong Lines at 183 GHz algorithm
ΔTb	Brightness Temperature difference

characterized by the need to provide frequent large-scale measurements resulting in large swath widths, resolutions greater than ~1 km, and systems capable of regular/frequent observations. Table 2 shows the WMO requirements of operational programs.

Operational meteorological satellites can be divided into two main categories: GEO satellites and low Earth orbiting (LEO) satellites (which include polar-orbiting satellites) (see Figure 1). GEO satellites are stationed at about 35,800 km above the equator in an orbit such that they rotate at the same speed as the Earth and therefore appear stationary relative to a location on the Earth's surface. This enables each GEO satellite to view about one third of the Earth's surface on a frequent and regular basis. A total five operational GEO satellites are required to provide full west-east coverage: Figure 2 show images from each of the individual GEO satellites on 10 November 2002 together with a derived remapped composite product. Current GEO satellites include the Meteosat Second Generation (MSG) satellites [*Schmetz et al.*, 2002] from the European Organisation for the Exploitation of Meteorological Satellites (EUMETSAT), two U.S. GOES [*Menzel and Purdom*, 1994], three Feng-Yun-2 satellites from China (http://www.fas.org/spp/guide/china/earth/fy-2.htm#ref625), and the Japanese Multifunctional Transport Satellites series [*Yoshiro*, 2002]. Plans are currently underway for the Korean Meteorological Satellite (COMS; http://web.kma.go.kr/eng/about/abo_03_04.jsp), EUMETSAT's Meteosat Third Generation (MTG) satellites [*Stuhlmann et al.*, 2005], and the U.S. GOES-R mission [*Schmit et al.*, 2005]. (See Table 3 for a detailed list of Web sites.) Although the specification of the sensors on each GEO system varies, they share a number of common attributes: they typically carry visible (VIS) and IR sensors with resolutions of about 1 km × 1 km and 4 km × 4 km, respectively, acquiring images nominally every 30 min, with some GEO satellites such as MSG, providing imagery every 15 min. Furthermore, rapid scanning permits limited-area subminute sampling which, while useful for monitoring cloud evolution and motion, is rarely used for rainfall estimation. Although GEO orbits provide an unrivalled platform for continual observation over

Table 2. Observational Requirements for World Meteorological Organization Programs[a]

	Horizontal Resolution (km)		Observation Cycle (h)		Delay (hours)		Accuracy (mm h^{-1})			
Application	*Min*	**Ideal**	*Min*	**Ideal**	*Min*	**Ideal**	*Min*	**Ideal**	Confidence	Source
Agricultural meteorology	*50*	**10**	*72*	**24**	*48*	**24**	*10*[b]	**2**[b]	reasonable	WMO
AOPC	*500*	**100**	*6*	**3**	*12*	**3**	*2*	**0.6**	firm	GCOS
GEWEX	*250*	**50**	*12*	**1**	*1440*	**720**	*5*[b]	**0.5**[b]	reasonable	WCRP
Global NWP	*100*	**50**	*12*	**1**	*4*	**1**	*1*	**0.1**	tentative	WMO
Hydrology^{liquid}	*50*	**5**	*1*	**0.08**	*0.5*	**0.08**	*1*	**0.1**	firm	WMO
Hydrology^{solid}	*50*	**5**	*1*	**0.25**	*0.5*	**0.5**	*1*	**0.1**	firm	WMO
Regional NWP^{daily}	*250*	**10**	*12*	**0.5**	*720*	**24**	*5*	**0.5**	reasonable	WMO
Regional NWP^{liquid}	*50*	**10**	*6*	**0.5**	*2*	**0.5**	*1*	**0.1**	tentative	WMO
Regional NWP^{solid}	*100*	**10**	*12*	**0.5**	*2*	**0.5**	*1*	**0.1**	tentative	WMO
Synoptic meteorology	*100*	**20**	*6*	**3**	*6*	**0.25**	*1*	**0.1**	firm	WMO
Terrestrial climate	*10*	**1**	*6*	**3**	*120*	**24**	*0.1*	**0.05**	tentative	GTOS GCOS

[a]Data after *World Meteorological Organization* [2003]. Minimum (Min) values are italic; ideal values are boldface.
[b]Values are given in mm d^{-1}.

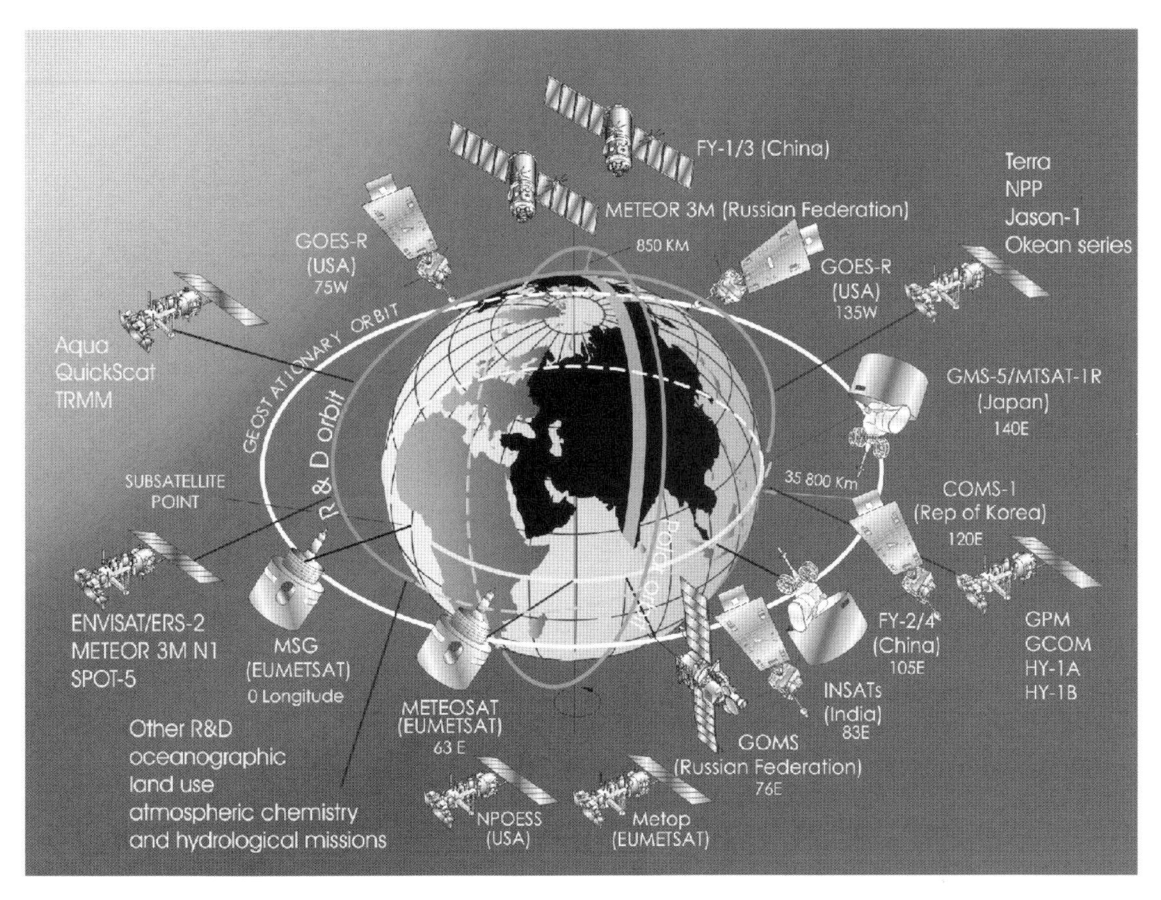

Figure 1. Current and projected configuration of the space-based WMO Global Observing System (GOS). Image courtesy of WMO (http://www.wmo.int/pages/prog/sat/index_en.html).

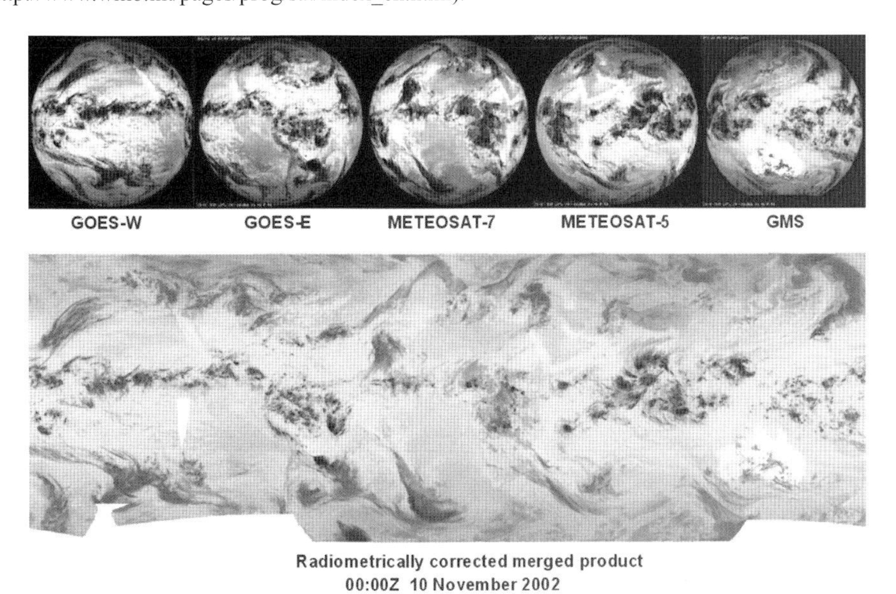

Figure 2. Geostationary satellite imagery for 10 November 2002. (top) Five individual satellite images. (bottom) A composite of these five individual images, reprojected and radiometrically corrected. The displayed images are radiometrically correct with the dark areas representing low levels of emitted radiation (low temperatures), with bright areas representing higher levels of radiation (higher temperatures).

Table 3. Pertinent Web Sites

	Location
International Precipitation Working Group IPWG	http://www.isac.cnr.it/~ipwg/
Program to Evaluate High Resolution Precipitation Products (PEHRPP)	http://essic.umd.edu/~msapiano/PEHRPP/
Validation all sites	http://cawcr.gov.au/bmrc/SatRainVal/validation-intercomparison.html
Validation Australia	http://www.bom.gov.au/bmrc/SatRainVal/sat_val_aus.html
Validation Europe	http://kermit.bham.ac.uk/~ipwgeu/
Validation Japan	http://www.radar.aero.osakafu-u.ac.jp/~gsmap/IPWG/sat_val_Japan.html
Validation South America	http://cics.umd.edu/~dvila/web/SatRainVal/dailyval.html
Validation USA	http://www.cpc.ncep.noaa.gov/products/janowiak/us_web.shtml
Coordination Group for Meteorological Satellites (CGMS)	http://www.wmo.ch/pages/prog/sat/CGMS/CGMS_home.html
WMO Space Programme	http://www.wmo.int/pages/prog/sat/index_en.html
Bureau of Meteorology	
Forecast Verification—Issues, Methods and FAQ	http://cawcr.gov.au/bmrc/wefor/staff/eee/verif/verif_web_page.html
RAINVAL QPF Verification	http://cawcr.gov.au/bmrc/mdev/expt/rainval/rainval_gui/rainval_gui.shtml
Colorado State University	
Climate Rainfall Data Center	http://rain.atmos.colostate.edu/CRDC/
Gridded Rainfall Maps	http://rain.atmos.colostate.edu/RAINMAP/index.html
Radiometer Level 1C Data	http://mrain.atmos.colostate.edu/LEVEL1C/
EUMETSAT	
H-SAF	http://www.eumetsat.int/Home/Main/What_We_Do/SAFs/Projects/SP_1124781691272
Multisensor Precipitation Estimate	http://www.eumetsat.int/Home/Main/Access_to_Data/Meteosat_Meteorological_Products/Product_List/SP_1119538666663
Global Precipitation Climatology Center (GPCC)	http://orias.dwd.de/GPCC/GPCC_Visualizer
Global Precipitation Climatology Project (GPCP)	
NASA-GSFC Global Precipitation Analysis	http://precip.gsfc.nasa.gov/
University of Maryland GPCP site	http://cics.umd.edu/~yin/GPCP/main.html
World Data Center	http://www.ncdc.noaa.gov/oa/wmo/wdcamet-ncdc.html
Japan	
GSMap	http://sharaku.eorc.jaxa.jp/GSMaP_crest/index.html
GSMaP Global Rainfall Map	http://sharaku.eorc.jaxa.jp/GSMaP/
Validation/intercomparison over Japan (GSMaP)	http://www.radar.aero.osakafu-u.ac.jp/~gsmap/IPWG/sat_val_Japan.html
NASA	
GES DISC Analysis Tool	http://disc2.nascom.nasa.gov/Giovanni/tovas/rain.ipwg.shtml
GES DISC DAAC	http://daac.gsfc.nasa.gov/
Global Precipitation Measurement (GPM) mission	http://gpm.gsfc.nasa.gov/
NSIDC AMSR-E Validation Data	http://nsidc.org/data/amsr_validation/
Tropical Rainfall Measuring Mission (TRMM)	http://trmm.gsfc.nasa.gov/
Naval Research Laboratory (NRL)	
NexSat	http://www.nrlmry.navy.mil/nexsat_pages/nexsat_home.html
Satellite Meteorology	http://www.nrlmry.navy.mil/sat_products.html
NOAA	
Microwave Surface and Precipitation Products Center	http://www.star.nesdis.noaa.gov/corp/scsb/mspps/main.html
National Climatic Data Center (NCDC)	http://www.ncdc.noaa.gov/oa/ncdc.html
NWS Global Precipitation Monitoring	http://www.cpc.ncep.noaa.gov/products/global_precip/html/web.shtml
SSM/I Global Gridded Products	http://lwf.ncdc.noaa.gov/oa/satellite/ssmi/ssmiproducts.html
University of Birmingham	
Infrared and Passive Microwave Matched Rainfall Results	http://kermit.bham.ac.uk/~kidd/matched/matched.html

Table 3. (continued)

	Location
University of California, Irvine	
Center for Hydrometeorology and Remote Sensing (CHRS)	http://chrs.web.uci.edu/
CHRS PERSIANN-GCCS Global 4 km product	http://fire.eng.uci.edu/GCCS/
HyDis G-Wadi Map Server	http://hydis.eng.uci.edu/gwadi/
PERSIANN CCS North America	http://fire.eng.uci.edu/CCS/
PERSIANN CCS Global	http://fire.eng.uci.edu/GCCS/
Precipitation Product Evaluation over CEOP Reference Sites	http://hydis0.eng.uci.edu/CEOP/eval/
University of Oklahoma	
Environmental Verification and Analysis Center (EVAC)	http://www.evac.ou.edu/
Surface Reference Data Center (SRDC)	http://srdc.evac.ou.edu/

much of the Earth, they are limited by the resolution of their imagery and the paucity of coverage in the polar regions.

Observations made by operational LEO meteorological satellites complement those made by GEO instrumentation. The orbits of these satellites are such that they cross the equator at the same local time on each orbit, providing about two overpasses each day. These satellites carry a range of instruments capable of precipitation retrievals. In addition to the multichannel VIS and IR sensors, passive microwave (PMW) sounders and imagers are included that are capable of more direct measurement of precipitation. Current operational polar-orbiting satellites include the NOAA series of satellites [*Goodrum et al.*, 2000] with NOAA 18 and 19 and EUMETSAT's MetOp series [*Klaes et al.*, 2007] providing current observations. Each of these Sun-synchronous satellites orbit the Earth about every 100 min at an altitude of about 850 km and carry a wide range of instrumentation including the third-generation of the VIS/IR advanced very high resolution radiometer and the advanced microwave sounding unit (AMSU) [*Goodrum et al.*, 2000].

In addition to the operational meteorological satellites, observations from a number of military and research satellite missions can be used for precipitation estimation. The Defense Meteorological Satellite Program (DMSP) satellites series of Special Sensor Microwave Imager (SSM/I) and Special Sensor Microwave Imager-Sounder (SSMIS) [*Kunkee et al.*, 2008] have provided measurements of the naturally emitted microwave (MW) radiation from the Earth and its atmosphere since 1987. The Aqua satellite sensors include the PMW Advanced Microwave Scanning Radiometer–EOS (AMSR-E) [*Kawanishi et al.*, 2003] and the VIS/IR Moderate Resolution Imaging Spectroradiometer (MODIS) [*NASA*, 1999], the latter also carried onboard the Terra and Aqua satellites. The MODIS instrument measures radiation across 36 spectral bands of the VIS and IR spectrum, enabling greater information on cloud properties (such as cloud drop radii and phase) to be retrieved at spatial resolutions up to 250 m × 250 m.

Of particular note is the Tropical Rainfall Measuring Mission (TRMM), launched in 1997, which is the first dedicated precipitation satellite [*Kummerow et al.*, 1998]. This satellite has been a key in the development and improvement of satellite rainfall estimation techniques due to the range of instruments that it carries, which allow direct comparisons to be made between VIS, IR, passive MW and active MW observations. The Precipitation Radar (PR) [*Iguchi et al.*, 2000] was the first spaceborne radar for precipitation measurement capable of sampling precipitation both vertically and horizontally to provide a three-dimensional (3-D) image of weather systems. In addition to the instrumentation, the non-Sun-synchronous nature of its orbit allows samples across the full diurnal cycle to be made. The other instruments are the TRMM Microwave Imager (TMI), the Visible and InfraRed Scanner (VIRS), and the lightning imaging sensor (LIS) (see Figure 3 for the swath details). Other current active instrumentation includes the cloud profiling radar (CPR) on CloudSat [*Stephens et al.*, 2008] and the Cloud-Aerosol Lidar with Orthogonal Polarization on the Cloud-Aerosol Lidar and Infrared Pathfinder Satellite Observation mission [*Winker et al.*, 2007; *Hunt et al.*, 2009]. These two latter missions, complemented by the Atmospheric InfraRed Sounder (AIRS) [*Gautier et al.*, 2003], are currently paving the way to a better global cloud characterization including what has been unknown to date, i.e., the vertical cloud height, amount and, most important for precipitation estimation, structure [e.g., *Kahn et al.*, 2008; *Mace et al.*, 2009; *Marchand et al.*, 2008; *Stubenrauch et al.*, 2008]. Note that most satellite rainfall estimation methods have been conceived with much uncertainty of

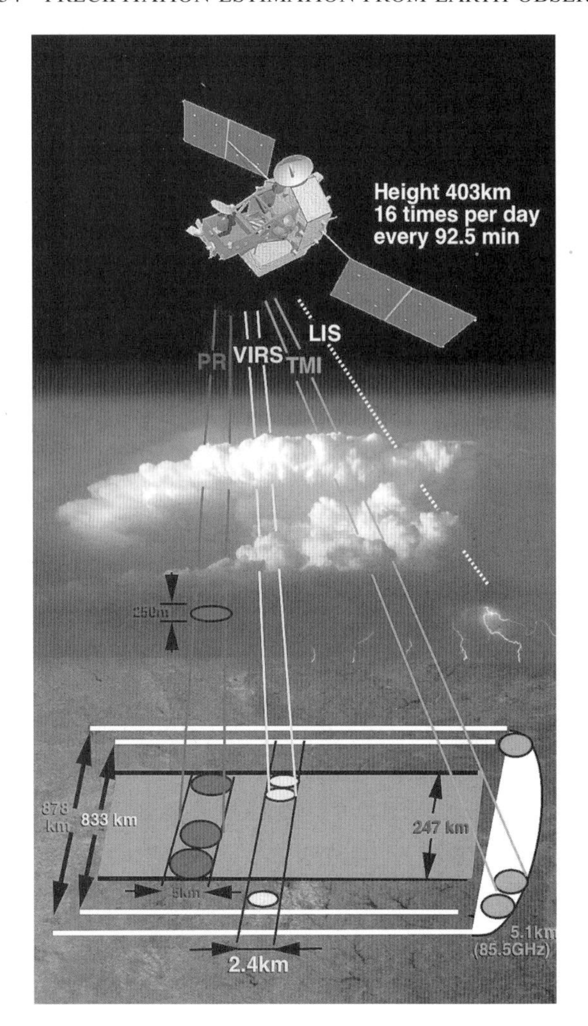

Figure 3. Tropical Rainfall Measuring Mission (TRMM) instrument payload and scan swaths. Image courtesy of NASA (http://trmm.gsfc.nasa.gov/).

cloud structure and precipitation formation mechanisms, even those that were based on cloud model output. This is especially true when dealing with cloud ice content.

The reader is also referred to other review papers on various aspects of precipitation retrievals: *Barrett and Martin* [1981], *Wilheit et al.* [1994], *Petty* [1995], *Kidd* [2001], *Levizzani et al.* [2007], *Michaelides* [2008], *Kidd et al.* [2009a, 2009b], and *Michaelides et al.* [2009].

2. DERIVATION OF QUANTITATIVE PRECIPITATION RETRIEVAL METHODOLOGIES

Satellite estimates of precipitation can be derived from a range of observations from many different sensors. The retrievals methodologies fall primarily into three main categories based upon type of observation, namely, VIS/IR techniques, PMW techniques, and multisensor techniques.

2.1. VIS/IR Methods

VIS and IR imagery remains the mainstay of operational meteorological Earth observations: Figure 4 shows six channels from the MODIS sensor covering the VIS through thermal IR channels. Observations in the VIS part of the spectrum provide the most directly comparable imagery to what we would observe with our eyes: from above clouds appear relatively bright against the surface of the Earth due to their high albedo. Rainfall can be inferred from VIS images (e.g., Figure 4a), since the brightness of a cloud is dependent upon its thickness, and thick clouds are more likely to be associated with surface rainfall. The main drawback of VIS imagery is that it is only available for, at most, during daylight and requires careful processing to account for changes in the Sun angles. Cloud top texture, however, derived from VIS imagery, can provide useful information on the cloud types being observed: stratus (either high-level cirro-stratus or low-level stratus) is typically smooth, while clouds associated with strong updraughts tend to have more texture. Properties of cloud top particles, such as the size and whether the particles are liquid or ice, can be obtained from multi-near-IR channels data. In particular, the use of reflected/emitted radiances around 1.6, 2.1, and 3.9 μm have proved very useful in studying the microphysical properties of clouds [e.g., *Rosenfeld and Gutman*, 1994; *Rosenfeld and Lensky*, 1998]. The main drawback of these channels when used alone is that they are limited by solar illumination, restricting their usage to full daylight operations. However, the Clouds-Aerosols-Precipitation Satellite Analysis Tool (CAPSAT) recently introduced by *Lensky and Rosenfeld* [2008] avails itself of all the VIS/IR channels to classify the imagery using red-green-blue lookup tables specifically tailored to "day microphysical," "day natural colors," and "night microphysical" conditions. The results are classified scenes that allow for a quick cloud scene classification, which can be potentially useful for rain area delineation based on cloud microphysics. Recently, *Chen et al.* [2008] have proposed a new method for the detection of drizzling clouds using MODIS imagery based on the cloud microphysics vertical structure.

Thermal IR imagery (e.g., Figure 4e) measures the emissions from objects and is therefore potentially more useful, being available both day and night. Heavier rainfall tends to be associated with larger, taller clouds, which have colder cloud tops: by observing cloud top temperatures, a simple rainfall estimate can be derived. However, the relationship between the cloud top temperature and rainfall is indirect, with significant variations in the relationship

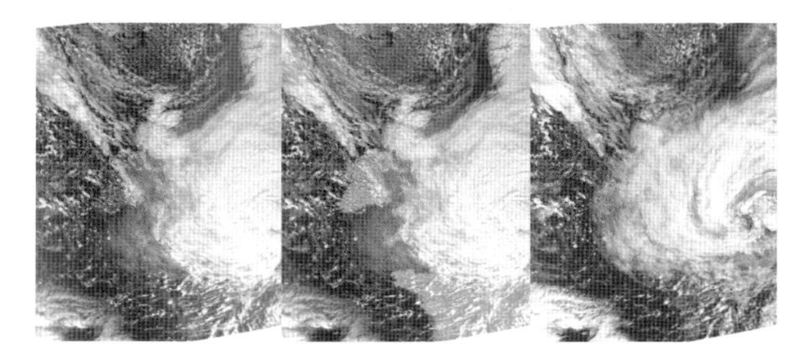

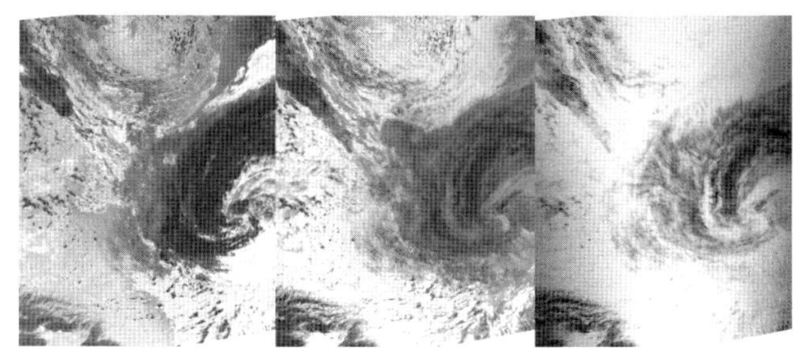

Figure 4. Visible (VIS) and infrared (IR) imagery over the U. K. for selected channels of the Moderate Resolution Imaging Spectroradiometer (MODIS) onboard the Aqua satellite for 13:15 on 28 February 2010. (top) Images in the reflective region of the spectrum (blue, near infrared and mid-infrared, from left to right). (bottom) Images in the emissive, mid- to far-infrared region of the spectrum. The western edge of Storm Xynthia can be seen on the right-hand side of the imagery. All images are shown as the correct radiative grey scales; that is, low radiation is dark, high radiation is light.

occurring during the lifetime of a rainfall event, between rain systems, and between climatological regimes. Despite this, IR-based techniques benefit from a degree of simplicity coupled with the availability of the data. Techniques have been developed to exploit one or more VIS/IR channels; *Barrett and Martin* [1981] give a comprehensive overview of the early days algorithms and methods. One of the simplest methods is the cold-cloud duration technique whose rainfall is related to the occurrence of cold clouds below a set threshold. An example of this is the global precipitation index (GPI) [*Arkin*, 1979; *Arkin and Meisner*, 1987], which derives rainfall fields from thermal IR observations by assigning a constant rain rate (3 mm h^{-1}) to the fraction of clouds below a set threshold (235 K or −38°C) over large time/spatial scales, usually 1 month and 2.5°. Another example of IR-based technique but for hydrological and real-time applications is the auto estimator (AE) [*Vicente et al.*, 1998]. The AE uses GOES 10.7 μm band to compute real-time precipitation amounts based on a power law regression algorithm. This regression is derived from a statistical analysis between surface radar-derived instantaneous rainfall estimates and satellite-derived IR cloud top temperatures collocated in time and space. Other examples of IR techniques are those by *Griffith et al.* [1978], *Negri et al.* [1984], *Scofield* [1987], *Tsonis* [1987], *Adler and Negri* [1988], *O'Sullivan et al.* [1990], *Bendix* [1997], *Anagnostou et al.* [1999], *Grimes et al.* [1999], *Todd et al.* [1999], *Reudenbach et al.* [2001]. Such simple techniques, although not without problems, provide a useful benchmark against which to compare other techniques.

A special method was adopted by *Xie and Arkin* [1998] to link outgoing longwave radiation (OLR) estimated from polar orbiting satellites and precipitation using monthly data, the so-called OLR precipitation index. The authors found that the mean annual cycle of OLR in the tropics is dominated by changes in cloudiness and exhibits a strong negative correlation with precipitation, while in the extratropics, the strongest influence on the annual cycle of OLR is surface temperature, and a positive correlation with precipitation is found. However, the anomaly of OLR exhibits a negative correlation with precipitation over most of the globe.

Multichannel techniques, such as the GOES Multi-Spectral Rainfall Algorithm [*Ba and Gruber*, 2001], utilize five channels of data from GOES (VIS, near IR, water vapor, and two thermal IR channels) to extract information on the cloud and rain extent, derive rainfall, and apply corrections depending upon the type of cloud/precipitation regime. *Lensky and Rosenfeld* [2003a, 2003b] have extended their daytime-only retrieval method to night time by using the brightness temperature difference (ΔTb) between the thermal IR 10.8 μm and the 3.7 μm channels, which reveals information on the microphysical structure and the precipitation potential of cloud at nighttime. Note that this kind of algorithm is more suited for the delineation of rain areas than for a quantitative precipitation estimate; in this sense, they can be very instrumental as ancillary cloud analysis methods when estimating precipitation with passive of active MW methods.

A method was proposed for the delineation of precipitation using cloud properties for the detection of mainly convective precipitation by means of the commonly used connection between IR cloud top temperature and rainfall probability, but also for the detection of stratiform precipitation [*Nauss and Kokhanovski*, 2006; *Thies et al.*, 2008a]. The scheme is based on the conceptual model that precipitating clouds must have both a sufficient vertical extent and large-enough droplets; the authors derived an autoadaptive threshold to be computed every time and linking optical thickness and cloud microphysics to precipitation potential at the ground. Based on this work, *Thies et al.* [2008b] have introduced a new method for rain area delineation and rainfall intensity retrieval based on MSG. In the case of advective-stratiform precipitation areas, the authors assume that areas with a higher cloud water path (CWP) and more ice particles in the upper parts are characterized by higher rainfall intensities. The rain area is separated into areas of convective and advective-stratiform precipitation processes. Next, both areas are divided into subareas of differing rainfall intensities. The classification of the convective area relies on information about the cloud top height gained from water vapor-IR differences and the IR cloud top temperature. The discrimination of the advective-stratiform area is based on information about the CWP and the particle phase in the upper cloud portions. At nighttime, suitable combinations of temperature differences ($\Delta T_{3.9-10.8}$, $\Delta T_{3.9-7.3}$, $\Delta T_{8.7-10.8}$, $\Delta T_{10.8-12.1}$) are incorporated to infer information about the CWP, while a VIS and a near-IR channel are considered during daytime. $\Delta T_{8.7-10.8}$ and $\Delta T_{10.8-12.1}$ are particularly included to supply information about the cloud phase. Rainfall intensity is attributed on the basis of a calibration with a ground radar network.

Another approach was followed by *Roebeling and Holleman* [2009] with the precipitation properties (PP) algorithm to detect precipitating clouds and estimate rain rates from cloud physical properties retrieved from MSG Spinning Enhanced Visible and InfraRed Imager (SEVIRI). The algorithm uses information on cloud water path (CWP), particle effective radius, and cloud thermodynamic phase to detect precipitating clouds, while CWP and cloud top height are used to estimate rain rates. An independent data set of weather radar data is used to determine the optimum settings of the PP algorithm and calibrated it. During the observation period, the spatial extents of precipitation over the study area from SEVIRI and weather radar are highly correlated (correlation $\approx$ 0.90), while weaker correlations (correlation $\approx$ 0.63) are found between the spatially mean rain rate retrievals from these instruments. The combined use of information on CWP, cloud thermodynamic phase, and particle size for the detection of precipitation results in an increase in explained variance (~10%) and decrease in false alarms (~15%), compared to detection methods that are solely based on a threshold CWP. At a pixel level, the SEVIRI retrievals have an acceptable accuracy (bias) of about 0.1 mm h^{-1} and a precision (standard error) of about 0.8 mm h^{-1}.

However, perhaps the most effective use of multispectral techniques is in rain area delineation. *Behrangi et al.* [2009a] have devised a neural network method for the improvement of rain/no rain detection based on the early work of *Hsu et al.* [1999]. The algorithm uses powerful classification features of the self-organizing feature map (SOFM), along with probability matching techniques to map single or multispectral input space into rain/no rain maps. The results show that during daytime, the VIS channel at 0.65 μm can yield significant improvements in rain/no rain detection when combined with any of the other four GOES 12 channels. At nighttime, the combination of two IR channels, particularly at 6.5 and 10.7 μm, results in significant performance gain over any single IR channel. In both cases, however, the authors show that using more than two channels leads only to marginal improvements over two-channel combinations. A surface rain rate classifier based on artificial neural network (ANN) for MSG SEVIRI has been proposed by *Capacci and Porcù* [2009] using five rain classes and calibrating with the Nimrod ground radar network over the British Isles; a comparison of the product with the rain product from AMSR-E shows that the method is lightly better than AMSR-E in detecting rain areas, while the opposite is true for rain rate classification.

2.2. *Passive Microwave Methods*

Although the observation sampling in the VIS/IR is good, the relationship between the cloud top temperature and the surface rainfall is indirect. This often results in cold, high cloud (e.g., cirrus) being classified as rain-bearing cloud, while warm, low-level rain cloud (e.g., stratus) is missed. In the MW part of the spectrum, the Earth naturally emits (hence "passive") low levels of MW radiation, which is attenuated in the presence of precipitation-sized particles. Two distinct processes are used to identify precipitation: emission from rain droplets, which leads to an increase in MW radiation and scattering, causing precipitating ice particles, which leads to a decrease in MW radiation; the magnitude of these effects is dependent upon size and concentration of the particles. However, the background surface effectively dictates which process can be exploited to retrieve surface precipitation. Over water, the background emissions, referred to as emissivity, ε, tend to be low ($\varepsilon \sim$ 0.4–0.5) and constant, thus additional emissions from raindrops can be used to identify and quantify the rainfall using low-frequency channels (<20 GHz, e.g., Figure 5a). Over land surface, additional emissions from rain droplets are negligible in comparison to the much higher background emissivity ($\varepsilon \sim$ 0.9): here the scattering caused by ice particles, which results in a decrease in received radiation at high frequencies (>35 GHz), must be utilized (e.g., Figure 5f). Very little information is available on land surface emissivity [e.g., *Weng et al.*, 2001], and thus, this remains one of the open problems for precipitation estimation over land [*Wang et al.*, 2009]. It should therefore be noted that care is needed in interpreting retrievals over land/ocean regions, since different techniques are often used for each region. In addition, while the emission-based techniques provide a measure of the rainfall through the total atmospheric column, the scattering-based techniques provide a measure of ice particles: neither observes the actual surface rainfall. However, this latter is a problem that characterizes all remote sensing rainfall retrieval methods, including those from ground-based radars Figure 6.

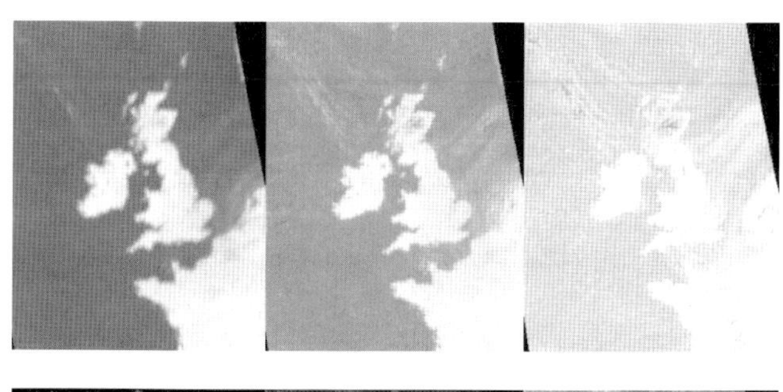

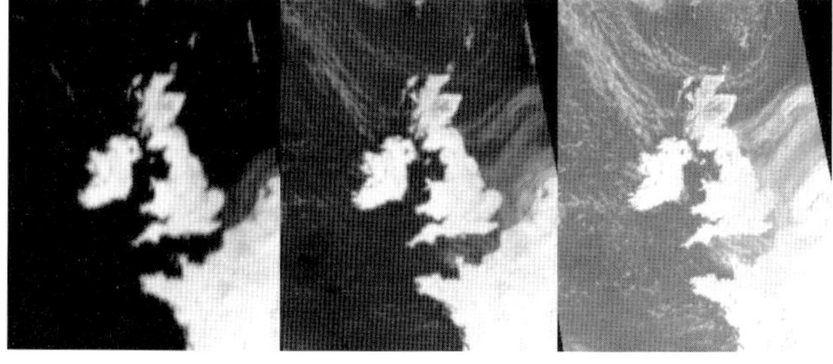

Figure 5. Selected passive microwave imagery derived from the Advanced Microwave Scanning Radiometer (AMSR) onboard the Aqua satellite for 13:15 on 28 February 2010: the western edge of Storm Xynthia can be seen on the right-hand side of the images. These images illustrate the increasing sensitivity to smaller and smaller hydrometeors with increasing frequency. Over Scotland, surface snow cover can be seen, having a similar signature at 89 GHz to the frozen hydrometeors as observed in showers over France. (H indicates horizontal channels; V indicates vertical channels.)

Another important problem to be solved before applying any kind of retrieval algorithm is the proper identification of rain areas and the elimination of those areas that produce a signal similar to that of precipitation. *Ferraro et al.* [1998] present a discussion on methods that identify areas of rain, snow cover, deserts, and semiarid conditions over land and rain, sea ice, strong surface winds, and clear, calm conditions over ocean. The main drawback of these PMW-based techniques is that observations are currently only available from LEO satellites, typically resulting in two observations per day per satellite. In addition, the resolution of the observations is not ideal: the spatial resolution of the low-frequency channels that are used over the ocean is of the order of 50 km × 50 km, while for the higher-frequency channels used over the land resolutions are typically no better than 10 km × 10 km. Early research on PMW radiative transfer in clouds and its application to rainfall retrieval are, among others, those by *Weinman and Guetter* [1977], *Wilheit et al.* [1977], *Ferraro et al.* [1986], *Kummerow and Weinman* [1988], *Olson* [1989], *Spencer et al.* [1989], *Alishouse et al.* [1990a, 1990b], *Mugnai et al.* [1990], *Weng and Grody* [1994], and *Weng et al.* [1994].

Algorithms to retrieve precipitation using PMW observations can be generally classified following a PMW sensor channel approach or a methodological approach. When following the channel-based classification, the algorithms can be thought as belonging to one of three classes: (1) "emission" type algorithms that make use of low-frequency channels to detect the increased radiances against a "cold" oceanic background [e.g., *Berg and Chase*, 1992; *Wilheit et al.*, 1991; *Chang et al.*, 1999], (2) "scattering" type algorithms that correlate rainfall to radiance depression caused by ice scattering [e.g., *Spencer et al.*, 1989; *Grody*, 1991; *Ferraro and Marks*, 1995], and (3) "multichannel inversion" algorithms [e.g., *Olson*, 1989; *Mugnai et al.*, 1993; *Kummerow and Giglio*, 1994; *Smith et al.*, 1994; *Petty*, 1994; *Bauer et al.*, 2001; *Kummerow et al.*, 2001]. If we follow the methodological classification, algorithms can be attributed to one of two main groups: empirical techniques that are calibrated against surface data sets and physical techniques that minimize the difference between modeled and observed radiation [see *Kidd et al.*, 1998; *Levizzani et al.*, 2007]. The empirical techniques are relatively simple to implement and incorporate many artifacts associated with PMW observations (such as beam-filling/inhomogeneous field-of-view, absolute calibration issues, and resolution differences). However, regional calibration may be necessary due to variations in the physical nature of precipitation systems. Physical techniques rely upon radiative transfer calculations to model the observed radiation. This can be done either through the inverse modeling of the observed radiation, to retrieve information about the precipitation particles, or through the use of a priori databases of model-generated atmospheric profiles which are compared with the satellite observations [e.g., *Bauer*, 2001; *Bauer et al.*, 2001]. The foundations for such statistical-physical algorithms were set by *Smith et al.* [1992] and *Mugnai et al.* [1993]. The Goddard profiling technique (GPROF) [*Kummerow et al.*, 2001] is one of the most successful of such techniques and specifically developed for the TMI. GPROF V6 incorporates a probability of convective rain, in conjunction with a convective and stratiform set of rain rate radiance vectors, which are then used to compute a final surface rain rate. *Ferraro and McCollum* [2003] show GPROF results with TMI and AMSR-E over the United States on the instantaneous scale with low bias errors and high correlations when compared with rain gauge-adjusted radar rainfall estimates. Further development of these physical methodologies has continued to ensure consistency between physical retrievals from PR and PMW brightness temperatures (Tbs) through the use of cloud resolving models by calculating the path integrated attenuation with drop-size distributions [*Masunaga and Kummerow*, 2005]. The main advantage of physical techniques is that

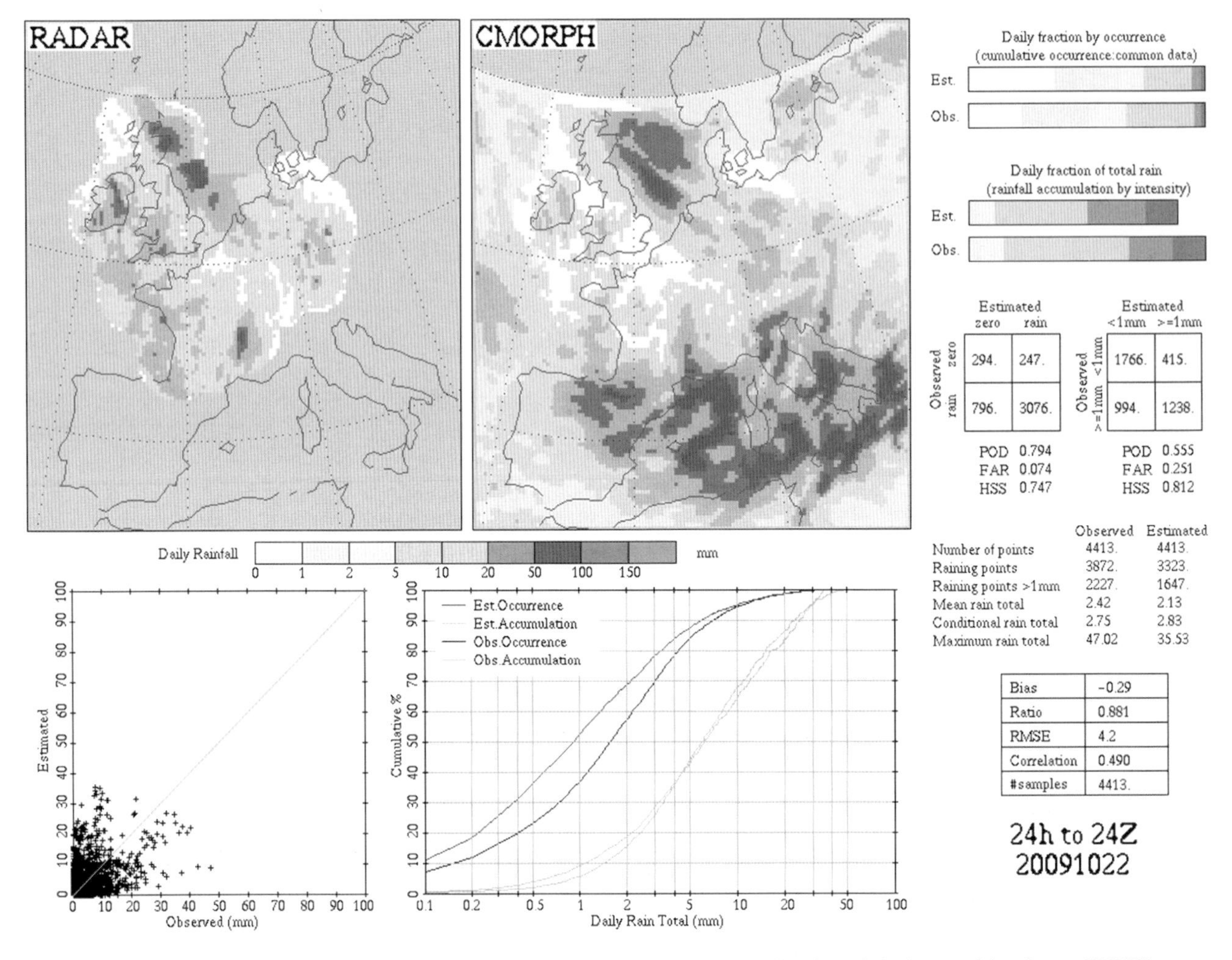

Figure 6. CMORPH verification on 22 October 2009 from the European International Precipitation Working Group (IPWG) validation site (http://kermit.bham.ac.uk/~ipwgeu/).

they provide more information about the precipitation system than just the surface rainfall.

Note, however, that the classification of PMW algorithms into empirical and physical is somehow artificial. An example of mixed type algorithm to ensure global applicability on a pixel-by-pixel basis is that of *Wilheit et al.* [2003] for AMSR-E: over ocean, the algorithm is physically based, while over land, it is generated empirically.

High biases exist for rain estimation algorithms over land areas for all the available sensors (TMI, SSM/I, and AMSR-E) [*McCollum et al.*, 2000, 2002]. *McCollum and Ferraro* [2003] have proposed a recalibration of the relationships between rain rates and PMW Tbs at 85 GHz using collocated TMI and PR data. They also propose a procedure for discriminating between convective and stratiform rainfall as a mean to reduce the ambiguity of possible rainfall rates for a given Tb. While such a methodology generally eliminates global high biases as previously identified, regional and seasonal biases still remain.

More recently, the availability of high-frequency PMW observations, such as those available from the AMSU-B instrument, has led to the development of new techniques. The AMSU data are important for rainfall retrievals, since it provides additional PMW frequencies that are not available on PMW conical scanning instruments: these frequencies are more sensitive to light precipitation and to cloud micro-physical properties. Moreover, since the AMSU is a sounding instrument whose channel weighting functions peak well above the surface of the Earth, it is relatively insensitive to the surface, thereby providing a unique capability for retrieving precipitation over problematic surface backgrounds, such as snow/ice and coastal regions. Note that these instruments were designed primarily for profiling atmospheric temperature and moisture [e.g., *Kakar*, 1983;

English, 1999]. However, the high-frequency PMW channels were soon used for the research retrieval algorithms for total precipitable water [*Wang and Wilheit*, 1989], for hydrological parameters in general (precipitation, cloud liquid water, snow cover, sea ice concentration) [*Grody et al.*, 2000], for total precipitable water (TPW) and cloud liquid water (CLW) [*Grody et al.*, 2001], for precipitation over all surfaces [*Bennartz et al.*, 2002], and for precipitating ice particles [*Bennartz and Bauer*, 2003]. *Staelin and Chen* [2000] found a promising agreement over land and sea between Next Generation Weather Radar (NEXRAD) radar observations of precipitation rate and retrievals based on simultaneous AMSU PMW observations at 50–191 GHz. Based on these studies, *Chen and Staelin* [2003] constructed one of the first global techniques to exploit these sounding channels to derive precipitation rates with 15 and 50 km resolution from AIRS, AMSU, and the Humidity Sounder for Brazil using a neural network approach trained by a U.S. weather radar network data set over a full year. A further approach to improve these rainfall retrieval algorithms was conducted by *Surussavadee and Staelin* [2007] by comparing Tb histograms from AMSU with predictions from the Pennsylvania State University/National Center for Atmospheric Research (Penn State/NCAR) Mesoscale Model (MM5). This initial study on sensitivity to model assumptions led the authors to conceive a global millimeter-wave precipitation retrieval using MM5 as a cloud-resolving numerical weather prediction (NWP) model [*Surussavadee and Staelin*, 2008a, 2008b] using neural networks trained to retrieve hydrometeor water paths, peak vertical wind, and 15 min average surface precipitation rates for rain and snow at 15 km resolution at all viewing angles. Different estimators were trained for land and sea, and surfaces classed as snow or ice were generally excluded.

At the same time at NOAA, *Weng et al.* [2003] proposed a method based on a radiative transfer model for a single-layered cloud using a two-stream approximation and utilized for the retrievals of CLW and TPW, cloud ice water path (IWP), and particle effective diameter. This latter study showed that the AMSU-derived ice water path is directly correlated to surface rain rate and, thus, instrumental for global precipitation estimations and paved the way to develop a suite of products that are generated through the Microwave Surface and Precipitation Products System. This includes precipitation rate, total precipitable water, land surface emissivity, and snow cover [*Ferraro et al.*, 2005].

Laviola and Levizzani [2008, 2009] have proposed a simple thresholding technique, the Water Vapor Strong Lines at 183 GHz (183-WSL) algorithm, which first discriminates between rain and no-rain areas and then retrieves precipitation divided in convective and stratiform. A peculiarity of the algorithm is the identification of water vapor areas and of the regions where cloud droplets are formed around the developing cloud and that were normally wrongly attributed to precipitation; these are the areas were nucleation happens and that act as reservoirs of cloud water for precipitation formation inside the cloud. *Di Tomaso et al.* [2009] have also published a similar algorithm that exploits the 89, 150, and 183 GHz channels whose ΔTbs are analyzed though a radiative transfer model for over land and over ocean rainfall retrieval. The probability of detection (POD) of precipitation is 75% and 90% for rain rates greater than 1 and 5 mm h^{-1}, respectively.

The inclusion of the AMSU-B data sets, along with the DMSP SSMIS observations, now provides important information in the generation of many precipitation products, in particular, those which combine multisensor and multifrequency information.

However, all the algorithm intercomparison exercises conducted, so far, have not identified the so-called "best" algorithm of all, since every algorithm has its strengths and weaknesses that are generally related to the specific application it is conceived for. The launch of future satellite constellations such as the Global Precipitation Measurement (GPM) mission requires the availability of a transparent, parametric algorithm that can ensure uniform precipitation intensity products across all the available sensors [*Kummerow et al.*, 2007]; error characterization is obviously of utmost importance, especially when using a sensor to calibrate the other belonging to the constellation. The algorithm should be applicable to any sensor, thus avoiding the need for a specific frequency dictated by any of the sensors.

Note that the physical diffraction law limits the ground resolution for a given satellite PMW antenna size, thus limiting the ground resolution of PMW observations and their repetition, since PMW sensors can only be hosted on board LEO satellites. The advent of higher frequency channels and/or synthetic aperture antenna concepts opens up the possibility to board the next generation of PMW sensors on GEO orbit spacecrafts such as the Geostationary Synthetic Thinned Aperture Radiometer/Precipitation and All-weather Temperature and Humidity mission [*Lambrigtsen et al.*, 2007]. It remains to be demonstrated whether these concepts will provide sufficient complementary information to LEO systems and IR sounding on board GEO satellites to warrant the substantial technical effort. In this sense, *Staelin and Surussavadee* [2007] have evaluated 11 alternative PMW sensors to retrieve surface precipitation rates and hydrometeor water paths. Five instruments observed selected frequencies from 116 to 429 GHz, with a filled-aperture antenna, and six instruments observed from 52 to 191 GHz with a U-shaped aperture synthesis array. The analysis is based on the *Surussavadee and Staelin* [2007] ANN retrieval method. Several instruments showed considerable promise in retrieving hydrometeor water paths and 15 min average precipitation rates ~1–100 mm h^{-1} with spatial resolutions that vary from ~15 km × 15 km to ~50 km × 50 km.

Precipitation estimation over coastal areas has always represented a problem, since they represent transition areas where the land algorithm stops working and is substituted by the ocean algorithm: the first PMW algorithms simply omitted estimates over coastal pixels. *Kidd* [1998] describes a technique for the delineation and retrieval of rainfall from PMW data using the polarization-corrected temperature (PCT) algorithm, which derives from the earlier concept of "polarization algorithm" by *Grody* [1984]. The main advantage of the PCT is its ability to reduce the effect of background surface emissivities making it possible to delineate areas of rainfall over varying surface types, especially over coastal areas with emissivity signatures from sea, land, and combinations of these two. The PCT was adopted also by *McCollum and Ferraro* [2005] for improving the GPROF V6 algorithm over coastal areas. Higher window (89 and 150 GHz), opaque oxygen (53.6 GHz), and water vapor absorption (183 *F* 1, *F*3 and *F*7 GHz) channels that are less impacted by surface emissivity variations were used with some success in the strategy for coastal areas by *Kongoli et al.* [2007]. Rain extent is determined using a number of scattering measures computed from linear combinations of Tbs at the selected high-frequency channels. Next, the IWP parameter is estimated from a quadratic relationship with the Tb at the 183 *F* 7 GHz water vapor absorption channel and the cosine of the local zenith angle. Rain rate is computed from the IWP.

2.3. *Active Microwave Methods*

The measurement of precipitation through active MW techniques is perhaps the most direct of all methods. Despite this, the use of radar technology for spaceborne precipitation estimation has been very limited, with only the TRMM PR being specifically designed for retrieving information on precipitation characteristics over a narrow swath of 215 km across. The PR, as all radar systems, relies upon the interpretation of the backscatter of radiation from, in this case, precipitation. The amount of backscatter is broadly proportional to the number of precipitation-sized particles and, therefore, the intensity. However, the relationship of precipitation intensity to backscatter is not constant, being dependent upon the size distribution of the particles, which, in the case of satellite measurements, is also variable along the vertical. Moreover, the attenuation effects are very significant at the 13.8 GHz frequency of the PR and thus must be taken into account, since the nonuniform rain distribution within the rain cell may become a large source of error when the attenuation is severe [*Iguchi et al.*, 2000]. Nonuniform beam filling effects also contribute to increase the errors of this kind of rainfall measurements.

The PR has been extensively used as a primary source of high-quality rainfall estimates for evaluating the differences of rainfall regimes over land and over the ocean. *Ikai and Nakamura* [2003] have shown that there are systematic differences between TMI and PR rainfall estimates, which are due to the freezing level assumption of TMI algorithm in winter at midlatitudes, inadequate Z-R relationships for convective and stratiform precipitation types in the PR algorithm, and the incorrect interpretation of the rain layer when the freezing level is low and convective rain is involved. *Furuzawa and Nakamura* [2005] found that the rain rate near the surface for the TMI is smaller than that for the PR in winter, and it is also smaller from 09:00 to 18:00 LT. These dependencies show some differences at various latitudes or local times. Moreover, rain rate estimates for the PR and the TMI depend on the storm height regardless of local time and latitude, and the effect is more pronounced in convective rather than in stratiform rain. An interesting result is that the PR rain for the cases with no TMI rain amounts to about 10–30% of the total but that the TMI rain for the cases with no PR rain accounts for only a few percent of the TMI rain. The authors explain this fact by the difficulty of detecting shallow rain with the TMI. This is also part of the reason why the PR has rapidly become a sort of "truth" against which satellite rainfall estimates are evaluated.

Combined radar-radiometer algorithms have been devised using a rain profiling approach. *Haddad et al.* [1997] produced the TRMM "day 1" combined rain profiling algorithm, which gives importance to the PR or the TMI as is warranted by their intrinsic ambiguities. The algorithm estimates the rain profile using the radar reflectivities, while at the same time constraining the inversion to be consistent with the estimate of the total attenuation provided by the radiometer. *Grecu et al.* [2004] have investigated a technique for estimating vertical profiles of precipitation from multifrequency, multiresolution active and passive MW observations. The technique is applicable to TRMM observations, and it is based on models that simulate high-resolution Tbs as functions of observed reflectivity profiles and a parameter related to the raindrop size distribution. The modeled high-resolution Tbs are used to determine normalized Tb polarizations at the TMI resolution. An optimal estimation procedure is employed to minimize the differences between the simulated and observed normalized polarizations by adjusting the drop size distribution parameter. *Viltard et al.* [2006] have proposed a rainfall retrieval technique for the

TMI using a database of about 35,000 Tb vectors collocated with the corresponding PR rain rate profiles.

A comparison of rainfall products from TMI and PR was conducted by *Masunaga et al.* [2002] to find the reason of systematic mean monthly rainfall from version 5 of the GPROF. The authors found that the inconsistencies arise from TMI underestimating near surface precipitation water content in midlatitudes winter and the PR underestimating the precipitation water path in the tropics. The disagreement in the rainfall products between TMI and PR essentially is a combined result of the intrinsic bias originating from the different physical principles between TMI and PR measurements and the purely algorithmic bias inherent in the conversion from precipitation water to rain rate. Recently, *Haddad and Park* [2009] have critically examined the reasons why the training of PMW retrievals with PR data fails on certain occasions over the ocean: they found that the reason is the unknown signature of the sea surface in the portion of the beam not filled with precipitation and proposed a filtering approach for the problematic cases.

The future GPM mission [*Hou et al.*, 2008] will host on board a dual-wavelength precipitation radar (DPR) at 13.6 and 35.5 GHz, which will have a high sensitivity to detect weak rain and snow, the capability to discriminate solid precipitation from liquid one, and a better accuracy of rain retrieval than the TRMM PR. The swath width at 13.6 GHz will be the same as that of the PR, while at 35.5 GHz, it will be 100 km; the spatial beam matching of the two channels is essential. The rain retrieval algorithms are under development and will require a substantial validation effort using ground-based and airborne radars [*Grecu and Anagnostou*, 2004; *Nakamura and Iguchi*, 2007].

An interesting comparison between cloud and rainfall information from the TRMM VIRS and PR observations was presented by *Inoue and Aonashi* [2000] using the ratio of 0.6 and 1.6 μm channels, the ΔTb between 11 and 12 μm, the ΔTb between 3.8 and 11 μm, and the 11 μm Tb. Several thresholds were found for the delineation of rain areas confirming the potential of VIS/IR channels for this specific task prior to the application of any PMW or radar rainfall estimation method. *Torricella et al.* [2008] and *Cattani et al.* [2009] have also found similar relationships for convective clouds over the Mediterranean using MODIS-derived cloud properties and rain products from AMSR-E and AMSU-B, respectively.

Recently, the CloudSat CPR [*Stephens et al.*, 2008] has provided some insightful, although limited precipitation retrieval capability. Unlike the TRMM PR, which operates at 13.8 GHz, the CPR operates at 94 GHz and therefore is much more sensitive to cloud hydrometeors and tends to saturate in regions of dense cloud or rainfall. However, through the use of attenuation-correction algorithms and surface reflectivity modeling, it has proved to be very useful in the identification of light rainfall and snowfall. *Haynes et al.* [2009] have demonstrated a notable potential at midlatitudes over the ocean. In particular, the greater sensitivity of the CPR over the TRMM PR has shown that the occurrence of precipitation over the tropics is greater. For rain rates less than 0.8 mm h^{-1} the CPR produces nearly three times the rainfall occurrence than the TRMM PR. The improved sensitivity of the CPR to light rainfall means that a more representative distribution function of precipitation can be produced. Moreover, the increased cloud structure information content of CloudSat observations has led *Todini et al.* [2009] to refine the detection of precipitating clouds over snow-covered areas at high latitudes decreasing the false alarm ratio of the AMSU-B PMW precipitation retrieval method of *Surussavadee and Staelin* [2008a, 2008b] by as much as 30% while at the same time considerably increasing the POD.

Other techniques have used neural networks to derived precipitation estimates through combining information from multichannel and multisensor observations like the Precipitation Estimation from Remotely Sensed Information using Artificial Neural Network (PERSIANN) [*Hsu et al.*, 1997; *Hong et al.*, 2004]. Recently, *Behrangi et al.* [2009b] have used the rain/no rain multispectral detection method of *Behrangi et al.* [2009a] to introduce a new method called PERSIANN-Multi-Spectral Analysis (PERSIANN-MSA). The proposed approach uses SOFM [*Behrangi et al.*, 2009a] to classify multidimensional input information, extracted from each grid box and corresponding textural features of multispectral bands. The authors use principal component analysis to reduce the dimensionality to a few independent features. The method shows that multispectral data have a potential to improve rain detection and estimation skills with an average of more than 20–50% gain in Equitable Threat Score for rain/no-rain detection, and more than 20% gain in the correlation coefficient associated with rain rate estimation. A recent study of *Tadesse and Anagnostou* [2009] has used storm tracking information to evaluate error statistics of the Multisatellite Precipitation Analysis-Real Time (TRMM 3B42-RT) product of the NASA and that of PERSIANN at different maturity stages of storm life cycles; rain gauge calibrated radar rainfall estimates over the United States. Results show that PERSIANN exhibits consistently improved correlations relative to the 3B42-RT for all storm life durations and maturity stages, thus supporting the hypothesis that incorporating cloud type information into the retrieval (done by PERSIANN) can help improve the satellite retrieval accuracy.

2.4. Multisensor Techniques

To overcome the deficiencies of individual satellite systems, a number of techniques have been developed to exploit the synergy between different satellite observations. Although the LEO PMW retrievals provide more direct observations of the hydrometeors, their temporal sampling is relatively infrequent and unevenly distributed across the diurnal cycle. Conversely, GEO VIS/IR observations have much better temporal sampling, but are less directly associated with precipitation. Techniques developed to exploit VIS/IR and PMW observations essentially fall into those that use the PMW to calibrate the IR observations and those that use the IR to derive cloud motion to move PMW precipitation estimates.

Techniques that adjust the IR or generate calibration curves to map IR radiances using the other data sets (such as the PMW) are termed "blended" algorithms [e.g., *Kidd et al.*, 2003; *Turk et al.*, 2000]. These techniques involve the correction or adjustment of IR cloud top temperature information by some other data set, such as radar, gauge, or another satellite data set. In many ways, all IR-derived estimates can be attributed to this category: even the relatively simple GPI technique was originally determined through the comparison of thresholded cold cloud duration and surface data. More complex techniques include the TRMM Multi-Satellite Precipitation Algorithm (TMPA) [*Huffman et al.*, 2007], which generates a precipitation product at 3-hourly, 0.25° resolution. This technique ingests data from PMW imaging and sounding sensors and GEO IR data with adjustments made for the different satellite retrievals before combination into the single precipitation product. Other techniques include the Passive Microwave InfraRed technique [*Kidd et al.*, 2003], the Microwave-adjusted IR Algorithm (MIRA) [*Todd et al.*, 2001], the Naval Research Laboratory technique [*Turk et al.*, 2000], the Microwave-Infrared Combined Rainfall Algorithm [*Marzano et al.*, 2004], the Self-Calibrating Multivariate Precipitation retrieval [*Kuligowski*, 2002], and the Microwave/Infrared Rain Rate Algorithm [*Miller et al.*, 2001].

Tapiador et al. [2004] describe a neural network technique to generate high temporal and spatial resolution rainfall estimates from both IR and PMW data, while *Grimes et al.* [2003] used ANN to improve a standard IR-based cold cloud duration technique with information from a NWP model over Zambia. *Hong et al.* [2005] have routinely adjusted the model parameters of PERSIANN using coincident rainfall derived from the TMI for a better representation of the diurnal cycle.

Although the techniques that calibrate the IR observations show great promise, they are ultimately limited by the indirectness of the IR to sense rainfall itself. However, the IR data can provide a reasonable measure of cloud movement, which can then be used to advect, or morph, the more direct PMW data between the successive satellite overpasses. These techniques have developed more recently due to increased computing power, which has enabled the evolution of practical motion-based (or advection) techniques. Examples of these current state-of-the-art methodologies are the Climate Prediction Center Morphing (CMORPH) technique [*Joyce et al.*, 2004] and the Global Satellite Mapping of Precipitation (GSMaP) [*Kubota et al.*, 2007]. The main drawback of this methodology is that the motion of the clouds might not represent the true motion of the precipitation at the surface, particularly if changes in the surface precipitation occur between the infrequent PMW observations.

A multisensor satellite rainfall monitoring methodology called Lagrangian Model is proposed by *Bellerby et al.* [2009a, 2009b] and is based on a conceptual cloud development model driven by GEO satellite imagery and is locally updated using MW-based rainfall measurements from LEO platforms. The model uses single-band thermal IR GEO satellite imagery to characterize cloud motion, growth, and dispersal at high spatial resolution (4 km) to drive a simple, linear, semi-Lagrangian, conceptual cloud mass balance model, incorporating separate representations of convective and stratiform processes. The model is locally updated against MW satellite data using a two-stage process that scales precipitable water fluxes into the model and then updates model states using a Kalman filter.

For many applications, the combination of all available data sets is ideal: products derived from the various satellite observations, gauge data sets and, where available, surface radar data. Various combination schemes can be used, most taking account of the expected errors inherent in each of the products. There is some experimentation with the use of Kalman filters to blend products derived from observations taken from different sources at different times to better account for the errors in the component estimates [*Joyce et al.*, 2009; *Ushio and Okamoto*, 2009; *Ushio et al.*, 2009]. One satellite rainfall product that is widely used is that of the Global Precipitation and Climatology Project (GPCP) [*Huffman et al.*, 1997], similar to the TMPA product, but incorporates surface data sets, where available, to produce a largely homogenous global precipitation product. Another data set widely used for NWP model verification, climate studies, and hydrological applications is the Climate Prediction Center Merged Analysis of Precipitation (CMAP) [*Xie and Arkin*, 1997], which incorporates satellite IR along with gauge and reanalyses data from the National Centers for Environmental Prediction (NCEP) and NCAR.

Among the ancillary data that can be used to improve the algorithm performances, attention should be given to the lightning data either from ground-based detection networks or from space sensors such as the TRMM LIS or the Optical Transient Detector (OTD). Unfortunately, these satellite-based sensors were active only for a limited amount of time; a new sensor is under investigation for the MTG payload. *Grecu et al.* [2000] were among the first to investigate the combined use of cloud-to-ground lightning and satellite IR data for rainfall estimation. Based on the analysis of the correlation between satellite PMW and IR rainfall estimates and on the number of strikes in "contiguous" areas with lightning, an empirical algorithm was developed for convective rainfall estimation. The authors show results with a reduction of about 15% in the root-mean-square error of the estimates of rain volumes defined by convective areas associated with lightning. *Morales and Anagnostou* [2003] put the emphasis on enhancing the capabilities in continuous rainfall monitoring using long-range lightning detection networks over large regions at high spatio/temporal resolutions and in separating precipitation type into convective and stratiform components. *Cecil et al.* [2005] have used TRMM PR, TMI, and LIS data over the tropics and the subtropics to examine the precipitation features observed by the satellite during the first 3 years in orbit. Significant variability exists between locations, seasons, and meteorological regimes, but the first important result concerns the known differences in bulk lightning flash rates over land and ocean, which result from the frequency of occurrence of intense storms and the magnitude of those intense storms that do occur. Even when restricted to storms with the same Tb, same size, or same radar reflectivity aloft, the storms over water are considerably less likely to produce lightning than are comparable storms over land. The wealth of information contained in lightning data was recently examined by *Pessi and Businger* [2009] who compared lightning data from the Pacific Lightning Detection Network and TRMM LIS with PR products, latent heating, and hydrometeor data. Their results show a consistent logarithmic increase in convective rainfall rate with increasing lightning rates; other storm characteristics such as radar reflectivity, storm height, ice water path, and latent heat show a similar logarithmic increase. In addition, the height of the echo tops showed a strong logarithmic correlation with lightning rate. These results have applications over data-sparse ocean regions by allowing lightning rate data to be used as a proxy for related storm properties, which can be used for driving satellite rainfall estimation methods.

Finally, the development of all these techniques has greatly benefited from the increasing availability of data sets from online resources. In addition, the longevity of some of the data sets is starting to allow long-term data sets to be established. Archives of precipitation data sets, such as that of the GPCP [*Adler et al.*, 2003] start in 1979 due to the availability of well-calibrated satellite IR data sets [*Janowiak et al.*, 2001], while from 1987, PMW data are available: this represents the start date of many satellite-derived precipitation estimates. A new high-resolution, consistent precipitation data set, based only on PMW satellite estimates, has been developed by the Cooperative Institute for Climate Studies (CICS) and called the CICS High-Resolution Optimally Interpolated Microwave Precipitation from Satellites [*Joseph et al.*, 2009]; when the data set is long enough, it will represent a natural tool for studying longer modes of climate variability. The availability of these data sets has also led to "operational" precipitation products to be routinely produced through concerted programs, such as the EUMETSAT Satellite Applications Facility on Support to Operational Hydrology and Water Management (H-SAF) [*Bizzarri et al.*, 2008] and Japan's GSMaP [*Kubota et al.*, 2007].

2.5. Snowfall Retrievals

An open, complex problem for precipitation estimation from space is the detection of snowfall, which has so far been out of reach of all the proposed algorithms. The possibility of measuring snow is of utmost importance for the closure of the water cycle, as ~5% of average global annual precipitation falls as snow [*ESA*, 2004]. However, the occurrence and accumulation of snow at mid-high latitudes is very significant, since frontal and stratiform systems are predominant over these areas. North of 60–70°, snow largely dominates: for example, in Canada, the average total annual precipitation is 535 mm, of which 36% falls as snow [*ESA*, 2004, see Figures 2 and 3].

Several studies demonstrated the feasibility of estimating snow from space using high-frequency channels [e.g., *Liu and Curry*, 1997; *Ferraro et al.*, 2000; *Katsumata et al.*, 2000; *Staelin and Chen*, 2000; *Weng and Grody*, 2000; *Bennartz and Petty*, 2001; *Wang et al.*, 2001; *Bennartz et al.*, 2002]. Precipitating snow above such surfaces is difficult to observe using window channels at low frequencies (<100 GHz). These problems are less important at high frequencies (>100 GHz), where water vapor screens the surface emission, and sensitivity to frozen hydrometeors is significant. However, the scattering effect of snowfall at higher frequencies is also impacted by water vapor in the upper atmosphere. Moreover, *Burns et al.* [1997] simulated the extinction of ice and rain at AMSU frequencies used for water vapor profile retrievals. *Greenwald and Christopher* [2002] examined the effect of cold clouds on measurements at 183 GHz. *Zhao and Weng* [2002] proposed a retrieval method for IWP from AMSU data. *Liu* [2004] devised new methods for the approximation of the scattering properties of ice and snow particles at high MW frequencies.

The scheme of *Kongoli et al.* [2003] takes advantage of the unique combination of AMSU measurements in the atmospheric window (23, 31, 89, and 150 GHz), opaque water vapor (183 ±1, ±3, and ±7 GHz) and oxygen absorption (50–60 GHz) regions to discriminate the scattering features over land surfaces (especially over snow cover) and that of the atmosphere (from precipitation-sized ice particles).

The scattering by randomly oriented dry snow particles at high MW frequencies [*Skofronick-Jackson et al.*, 2004] appears to be better described by considering snow as a concatenation of "equivalent" ice spheres rather than as a sphere with the effective dielectric constant of an air-ice mixture. *Kim et al.* [2008] use the discrete dipole approximation (DDA) to calculate the single scattering parameters for snow particles using five different snow particle models. Snow particle size distributions are assumed to vary with air temperature, and Tbs at AMSU-B frequencies for a snow blizzard are calculated using the DDA calculated single scattering parameters and particle size distributions. The vertical profiles of pressure, temperature, relative humidity, and hydrometeors are provided by MM5 model simulations. These profiles are finally treated as the a priori database in the Bayesian retrieval algorithm.

A snowfall retrieval algorithm based on Bayes' theorem was developed by *Noh et al.* [2006] using high-frequency PMW data. Observational data from both airborne and surface-based radars were used to construct an a priori database of snowfall profiles, which were then used as input to a forward radiative transfer model to obtain Tbs at high frequencies. In the radiative transfer calculations, two size distributions for snowflakes and 10 observed atmospheric sounding profiles are used with snowfall profiles from observations. In addition, the scattering properties of the snowflakes were calculated on the basis of realistic nonspherical shapes using the DDA.

Combined cloud radar and PMW radiometer data are used for a retrieval algorithm by *Grecu and Olson* [2008] for snowfall detection over ocean. The algorithm is based on physical models to simulate cloud radar and millimeter-wave radiometer observations from basic atmospheric variables such as hydrometeor content, temperature, and relative humidity profiles and works with an optimal estimation technique to retrieve these variables from actual observations. Note that, however, the research is still ongoing, as the problem is very difficult to solve, and further studies are needed especially in light of the GPM launch and the necessity to extend further north the current global precipitation retrievals.

Liu [2008] has recently developed a retrieval methodology for snowfall from CloudSat consisting of two parts: (1) determining whether a radar echo corresponds to snowfall (instead of rainfall) and (2) converting radar reflectivity to snowfall rate. The first part is a snow-rain threshold based on multiyear land station and shipboard weather reports, and the second part is based on backscatter computations of nonspherical ice particles and in situ measured particle size distributions.

The International Precipitation Working Group (IPWG) has recognized the importance of snowfall retrieval research and has given an impulse to it over the past few years through the International Workshop on Space-Based Snowfall Measurements [*Bennartz and Ferraro*, 2005; *Bennartz et al.*, 2008]. Several high-priority recommendations were advanced on modeling, data assimilation, use of combined active and passive sensor strategies, increased sensitivity of sensors to detect snow down to 100–200 m from the ground, need for new instruments and combinations of channels, and use of dedicated and enhanced ground validation for snowfall [*Bennartz et al.*, 2008].

3. APPLICATIONS

The mapping and monitoring of rainfall on a global scale has been one of the main goals of meteorological satellite missions. Consequently, many applications have been able to directly use precipitation products generated from satellite observations. The availability of new satellite systems, such as the GEO data from MSG SEVIRI, has greatly enhanced frequency of observation and the multispectral potential of rainfall retrievals [*Levizzani et al.*, 2001]. In particular, many data sets are now readily available in near real time [*Turk et al.*, 2008; *Sapiano and Arkin*, 2009] and have raised the potential for improved and new applications. A few applications are now discussed such as hydrology and water cycle, process studies, snowfall estimation, latent heat evaluation, and climate studies. This list is not exhaustive; e.g., assimilation into NWP models at all scales is to be considered as one of the emerging applications; *Michaelides et al.* [2009] provide an up-to-date account of the topic.

3.1. Hydrology and the Water Cycle

The importance of using satellites operationally, particularly for disaster management and mitigation, has been a main application of satellite precipitation products, including crop forecasting and assessment, water resource management, flood forecasting, and geohazards. The usability of such products tends to rely upon the satellite data being available in real time (or very near real time). Precipitation products, such as the GPCP's monthly precipitation analysis [*Adler et al.*, 2006, 2008], have been used for a range of applications including flood and landslide prediction [*Huffman et al.*, 2007] and drought monitoring [*Kidd et al.*, 2009a]. The use of satellite precipitation estimates for hydrological and water resource applications [*Kidd et al.*, 2009b; *Sorooshian et al.*, 2009] is very much dependent upon the type of application and the accuracy, spatial resolution, temporal resolution, and latency of the estimates: different applications will have different data requirements. Hydrological requirements for precipitation estimates can be divided into two main categories: high-resolution/short-duration estimates and lower-resolution/longer-term estimates. Flash flood events, with rapid catchment response, necessitate fine spatial and temporal scales together with timely delivery of the estimates. The availability of observations at 4 km × 4 km every 15 min is available from GEO observations (with the potential for 1 km × 1 km, 1 min rapid scan imagery). However, due to the indirect nature of the cloud top to surface rainfall relationship, such estimates would be subject to a degree of error. Nevertheless, techniques, such as the Hydro-Estimator based upon the GOES IR observations [*Scofield and Kuligowski*, 2003], have proved useful for operational use over the United States. Fluvial flooding and water resources are characterized by relatively long lead times; therefore, satellite-derived precipitation products can be of great benefit. Static surface parameters (e.g., geology, soil type, and relief), dynamic surface parameters (e.g., soil moisture, vegetation, and groundwater) as well as the precipitation (satellite and surface) and meteorological/climatological conditions can be brought together on a global scale to provide a comprehensive hydrometeorological database.

3.2. Process Studies

The availability of finer resolution observations has also provided new insights into precipitation events and process. *Hawkins et al.* [2008] describe the use of the SSMIS to aid the observation and monitoring of hurricanes and cyclones, while *Posselt et al.* [2008] apply CloudSat to study the classical view of extratropical cyclones. The capability of the TRMM PR to provide long-term (available since 1997), high spatial resolution precipitation measurements has been demonstrated by *Kidd and McGregor* [2007]. Although some sampling issues remain, this 4 km/10 year product reveals small-scale precipitation features closely linked to orography as well as persistent larger-scale features. *Zipser et al.* [2006] looked at the characteristics of precipitation from TRMM observations, highlighting the distribution of intense precipitation regimes. *Geerts and Dejene* [2005] studied TRMM PR data over Africa to determine the regional and diurnal variability in the vertical structure of precipitation systems concluding that precipitation systems in Africa are deeper than the Amazon. Due to the non-Sun-synchronous nature of the TRMM satellite, the diurnal variation in precipitation can be studied: *Sanderson et al.* [2006] compared the performance of PMW techniques to detect the diurnal cycle. Studies by *Kidd and McGregor* [2007] over Hawaii, and *De Angelis et al.* [2004] over the Amazon, showed that there was great merit in looking at fine-scale regional mapping of rainfall from TRMM to reveal small-scale variations in rainfall distributions, showing significant links to surface morphology. Although the use of satellite estimates for flash flood events is challenging, observations such as those from TRMM, can provide information on the extremes of rainfall occurrence and characteristics at the local scale. In addition, such climatological products also provide useful information for geomorphological processes that are often rainfall-dependent.

Studies of anthropogenic influences of precipitation have also been studied: *Shepherd et al.* [2002] investigated the distribution of rainfall around U.S. cities to quantify the effects of urban areas upon precipitation distribution and characteristics. *Rosenfeld* [2000] and *Givati and Rosenfeld* [2004] investigated the suppression of rainfall and snow by air pollution through comparison of multispectral observations made by the TRMM precipitation sensors. These studies showed that aerosols can either suppress precipitation by creating a larger number of nonprecipitating droplets or suppress the onset of precipitation until significantly larger droplets precipitate creating intense rainfall events. The above-mentioned CAPSAT tool [*Lenksy and Rosenfeld*, 2008] exploits, for example, the capabilities of MSG SEVIRI, to enable insights into the precipitation processes through the examination of cloud top effective radii and temperatures.

3.3. Latent Heat Evaluation

When analyzing the water cycle, it is clear that fresh water provided by tropical rainfall and its variability can exert a large impact upon the structure of the upper ocean layer. Almost two thirds of global rainfall occurs in the tropics. Precipitation from convective cloud systems makes up a large portion of tropical heating and rainfall, and the vertical distribution of convective latent heat release modulates large-scale tropical circulations. In this perspective, the launch of TRMM in 1997 produced a leap forward in quantifying the tropical rainfall and the release of latent heat, two previously largely unknown variables.

A method for the remote sensing of three-dimensional latent heating distributions in precipitating tropical weather systems from satellite PMW observations was presented by *Olson et al.* [1999] using SSM/I data. Cloud model simulated hydrometeor/latent heating vertical profiles that have radiative characteristics consistent with a given set of multispectral microwave radiometric observations are composited to create a best estimate of the observed profile. A revised Bayesian method was then published by *Olson et al.* [2006].

A first estimation of the 4-D latent heating structure over the global tropics from TRMM data was attempted by *Tao et al.* [2001] using three different heating algorithms: the Goddard Space Flight Center convective-stratiform heating (CSH), the GPROF heating, and the hydrometeor heating. The horizontal distribution of patterns of latent heat release and the magnitude of the estimates are very similar among the three different methods, although the differences are found in the height of the maximum heating level.

The spectral latent heating algorithm by *Shige et al.* [2004] was introduced to estimate latent heat release from TRMM data using a cloud-resolving model (CRM). The method is based on heating profile lookup tables for three rain types, convective, shallow stratiform, and anvil rain (deep stratiform with a melting level). Further refinements followed and concerned the tropical oceans [*Shige et al.*, 2007], apparent moisture sinks over tropical oceans [*Shige et al.*, 2008], and the differences in the lookup tables between 2- and 3-D CRM runs [*Shige et al.*, 2009].

3.4. Climate Studies

With the longevity of satellite precipitation records becoming available, there are emerging studies into the use of long-term precipitation data sets for climates studies. Although some of the instrument records are now well established, data set continuity is an issue with known errors originating from individual sensors (such as scan-angle dependencies). In addition, the orbital drift of the satellite affects the local equatorial crossing times, and hence the sampling of the diurnal cycle which needs to be corrected. Methodologies are being developed to incorporate data not only from multiple sensors, but also to ensure that data from the same sensor is correctly combined to generate long-term data sets suitable for climate studies. Initial comparisons between model reanalysis and satellite precipitation data sets show significant long-term differences in the trends of global precipitation as well as year-on-year biases with the GPCP analysis producing an average of 2.6 mm d^{-1} globally compared with the NCEP reanalysis of ~3.2 mm d^{-1}. Importantly, most satellite-based long-term precipitation studies, incorporating corrections for multisatellite/sensor data sets, show little or no trend in global precipitation [*Gruber and Levizzani*, 2008]. Associations between global and regional precipitation and surface temperature anomalies on interannual and longer timescales were explored for the period of 1979–2006 using the GPCP precipitation product and the NASA-Goddard Institute for Space Studies (GISS) surface temperature data set by *Adler et al.* [2008b]. Positive (negative) correlations are generally confirmed between these two variables over tropical oceans (lands). El Niño–Southern Oscillation is the dominant factor in these interannual tropical relations. Away from the tropics, particularly in the Northern Hemisphere mid-high latitudes, this correlation relationship becomes much more complicated with positive and negative values of correlation tending to appear over both ocean and land, with a strong seasonal variation in the correlation patterns. Most intense long-term, linear changes in annual mean rainfall during the data record tend to be within the tropics. For surface temperature, however, the strongest linear changes are observed in the Northern Hemisphere mid-high latitudes, with much weaker temperature changes in the tropical region and Southern Hemisphere. Globally, the calculation results in a +2.3%/°C precipitation change, although the magnitude is sensitive to small errors in the precipitation data set and to the length of record used for the calculation [*Gruber and Levizzani*, 2008].

At the same time and availing themselves of these relatively long precipitation data sets, several studies were conducted on regional precipitation climatologies and annual, interannual, and seasonal variability. Studies on monsoon mechanisms, the onset and withdrawal for individual years, were conducted by a number of authors globally [*Janowiak and Xie*, 2003], over North America [*Gebremichael et al.*, 2007], over Taiwan and the Far East [*Wang and Chen*, 2008], around the Himalayas [*Bhatt and Nakamura*, 2005, 2006]. Other authors concentrated on specific areas such as Greece [*Hatzianastassiou et al.*, 2008], the tropics [*Takayabu*, 2002], tropical and subtropical South America [*De Angelis et al.*, 2004], China in summer [*Zhou et al.*, 2008], the tropical East Pacific [*Cifelli et al.*, 2008]. Several studies focused on the global scale concentrating on diurnal variations over land and sea [*Yang and Smith*, 2008; *Yang et al.*, 2008], the diurnal cycle of rainfall and convective intensity over the tropics [*Nesbitt and Zipser*, 2003], and on interannual and interdecadal variability [*Gu et al.*, 2007]. *Yang and Smith* [2006] concentrated on the diurnal variability of precipitation over the global tropics from TRMM data finding, in general, that most ocean areas show more rainfall at night, while during daytime, rainfall peaks over land areas with some important exceptions.

4. PRECIPITATION PRODUCT COMPARISONS AND VALIDATION

Numerous projects have taken place to intercompare and evaluate the performance of precipitation estimates derived from satellite observations and those from models. In the late 1980s, the WetNet project encouraged the interaction of scientists working on the then new SSM/I PMW data sets, including the validation of the rainfall retrievals. As a result, a series of Precipitation Intercomparison Projects (PIP) were organized: the first and third projects concentrated upon estimates derived at monthly, 0.25° × 0.25° resolutions [*Dodge and Goodman*, 1994; *Barrett et al.*, 1994; *Kniveton et al.*, 1994; *Adler et al.*, 2001], while the second investigated individual instantaneous case studies over a variety of different meteorological situations and geographically diverse regions [*Smith et al.*, 1998]. Parallel to these projects was the GEWEX/GPCP Algorithm Intercomparison Programme (AIP) [*Ebert et al.*, 1996; *Arkin and Xie*, 1994; *Ebert and Manton*, 1998]. These encompassed a range of precipitation estimates from various satellite observation sources as well as model output for specific coordinated observation periods.

The main conclusion from the PIP and AIP intercomparisons was that PMW techniques were clearly better than IR techniques for instantaneous estimates due to their more direct observation of the rainfall. However, this advantage deteriorated over longer timescales due to the poorer sampling of PMW observations compared to the IR-based techniques. The combined IR-PMW techniques, which were expected to do well, did not show a clear advantage at this stage over the IR-alone techniques. It was also noted, however, that all algorithms were very much dependent upon common underlying factors, such as surface background conditions, seasons/latitude and meteorological conditions, and therefore no single algorithm, technique or methodology could be deemed superior to any other.

The validation of precipitation products is seen as a critical part of the development and routine production of precipitation products from satellite observations. Validation data sets, or perhaps more correctly, surface reference data sets, generally rely upon gauge data or radar depending upon the temporal and spatial scales over which the estimates were being evaluated. In all cases, the surface data sets are not necessarily the "truth" since many factors affect the true value. However, through the careful selection of data, inconsistencies or shortcomings can be reduced. The AIP-3 project utilized data specifically collected for the project from the Tropical Ocean–Global Atmosphere/Coupled Ocean-Atmosphere Response Experiment region of the Pacific Ocean [*Ebert and Manton*, 1998] comprising of both radar and gauge data. This type of ground truth data collection (or ground validation) has become a major component of most validation/intercomparisons of satellite products. Specific ground validation campaigns [e.g., *Wolff et al.*, 2005] identify specific locations for the collection of (primarily) radar and gauge data sets: these campaigns are generally designed for improving the physical retrievals of the satellite estimates. However, more recent and ongoing validation rely upon mainly statistical intercomparisons with surface data sourced from routinely available data sets, such as the U.S. NEXRAD radar product or the European Nimrod radar product.

The IPWG provides a focus for operational and research satellite-based quantitative precipitation measurement issues and challenges, addressing a number of key objectives described by *Turk and Bauer* [2006]. It provides a forum for users of satellite precipitation measurements to exchange information on methods for measuring precipitation and the impact of spaceborne precipitation measurements for numerical weather prediction, hydrometeorology, and climate studies. The IPWG intends to build upon the expertise of international scientists currently involved in precipitation measurements to further the development of better measurements and utilization of precipitation products, improved scientific understanding, and the fostering of international partnerships. The IPWG currently provides ongoing near real-time validation of quasi-operational and operational satellite estimates, as well as NWP model outputs [see *Turk and Bauer*, 2006; *Ebert et al.*, 2007]. A series of intercomparison regions have been established to provide algorithm developers, and the wider user community, quantitative information on the performance of satellite rainfall products in near real time. The current IPWG intercomparison regional sites provide validation statistics for daily, 0.25° resolution estimates over the United States, Europe, Australia, Japan, and South America, using radar and/or gauge data as their surface ground truth. Figure 5 shows an example of CMORPH statistics over the European area (http://kermit.bham.ac.uk/~ipwgeu/) where the comparison is done against the Nimrod ground radar network (http://badc.nerc.ac.uk/data/nimrod/). Other validation efforts, which operate "off-line" include South Korea, China, South Africa, Ethiopia. These results [e.g., *Ebert et al.*, 2007] show that the NWP models and motion-based techniques outperform the standard satellite estimates of precipitation in cold-season environments (e.g., during midlatitude winters). However, warm-season performance studies tend to favor the satellite techniques, since these can capture the convective nature of the precipitation better than existing NWP models.

More recently, the Program for the Evaluation of High-Resolution Precipitation Products (PEHRPP) has been established to characterize the errors in high-resolution precipitation products (less than 0.25° × 0.25° resolution and 3 hourly or less) over different spatial, temporal, regional, and climate scales [see *Turk et al.*, 2008]. This project acts as an interface between the large-scale regional precipitation validation and the targeted physical validation campaigns, which concentrate on understanding the physical processes observed by satellite observation. This validation tends to rely upon single, instantaneous case studies: an overview of ground validation for TRMM is provided by *Wolff et al.* [2005], while *Lobl et al.* [2007] describe the AMSR-related Wakasa Bay field campaign. *Turk et al.* [2009] have discussed the need for adequate gauge networks with the proper instrument density and time sampling resolution.

In particular, investigations into the errors associated with rainfall effects have started to be addressed. *Bowman* [2005] studied the spatial and temporal averaging errors of the TRMM precipitation retrievals and ocean gauge data over the tropical Pacific Ocean, while *Gebremichael and Krajewski* [2005] investigated the sampling effects of rainfall from surface data sets. Other validation studies clustered around the TRMM products either using the TRMM ground validation sites or independent data sets are, among others, those by *Nicholson et al.* [2003a, 2003b], *Nesbitt et al.* [2004], *Wolff et al.* [2005], and *Marks et al.* [2009].

Relative contributions of error from different satellite sensors were studied by *Huffman et al.* [1997] as part of the GPCP combined precipitation product. Other validation exercises were conducted in the frame of GPCP by *Krajewski et al.* [2000] (quantification of the error variance at reference sites for the monthly product), *Gebremichael et al.* [2003] (evaluation of the error uncertainty of monthly product), *Gebremichael et al.* [2005] (evaluation of the 1° daily product over the Mississippi river basin), *McPhee and Margulis* [2005] (validation and error characterization of the 1° daily product over the United States), and *Bolvin et al.* [2009] (assessment of monthly and daily products against high-latitude rain gauges). *Yin et al.* [2004] compared the GPCP data set for the 1979–2001 period with the CMAP [*Xie and Arkin*, 1997] data set. For the long-term mean, major precipitation patterns are clearly demonstrated by both products, but there are differences in the pattern magnitudes. In the tropical ocean, CMAP precipitation is higher than GPCP's, but this is reversed in the high-latitude ocean. The GPCP-CMAP spatial correlation is generally higher over land than over the ocean.

Lin and Hou [2007] have used the TRMM PR together with surface radar and gauge measurements over the United States to quantitatively assess rainfall retrievals from cross-track humidity sounders relative to the conical scanning PMW imagers in view of the GPM planning. The results show that for future global missions, the cross-track scanners play an important role in filling in the time gaps left by conical instruments without a degradation in the quality of the precipitation estimates. A similar study was conducted by *Wolff and Fisher* [2009] using TRMM ground validation data.

What has always been considered a flaw in past validation exercises was the all-statistical character of the work. In other words, satellite rainfall estimation was compared with rainfall intensities measured by completely different instruments regardless of the physical content of the various data sources. This has been indicated as perhaps the most relevant problem of the TRMM ground validation sites. Regarding GPM, *Chandrasekar et al.* [2008] propose a new strategy based on the use of dual-polarization radars for addressing a number of open issues that arise in the validation process, especially those associated with cloud microphysics and algorithm development. In this sense, the GPM validation activity will focus on building validation sites specifically designed to (1) perform statistical validation of retrieved satellite surface precipitation products, (2) investigate precipitation processes, and (3) validate integrated hydrological applications.

The representations of the uncertainties in IR/PMW satellite precipitation products was addressed by *Bellerby and Sun* [2005], who developed a methodology deriving conditional probability distribution functions of rainfall on a pixel-by-pixel basis. The array of distribution functions is combined with a simple model of the spatio/temporal covariance structure of the uncertainty in the precipitation field to stochastically generate an ensemble precipitation product. Each element of the ensemble represents an equiprobable realization of the precipitation field that is consistent with the original satellite data, while containing a random element commensurate with the uncertainty in that field. A study along this line was also presented by *Hossain and Anagnostou* [2006a] trying to answer the question on how realistic is the ensemble generation of satellite rainfall products by a multidimensional satellite rainfall error model. A 2-D error model was then produced by *Hossain and Anagnostou* [2006b] with the aim of characterizing the multidimensional stochastic error structure of satellite rainfall estimates as a function of scale.

The assimilation of precipitation estimates into hydrological models is a natural application of satellite precipitation products: consequently, the error propagation into such models is clearly of capital importance. *Hossain and Anagnostou* [2004] examined TRMM PMW and IR-based satellite rainfall retrievals for flood prediction using a probabilistic error model, while *Hossain et al.* [2004] did a sensitivity analysis of satellite rainfall retrieval and sampling error on flood prediction uncertainty. *Hong et al.* [2006] have used a Monte Carlo assessment of such propagation by generating 100 ensemble members of precipitation data as forcing input to a conceptual rainfall-runoff hydrologic model and quantifying the resulting uncertainty in the streamflow prediction. *Hossain and Huffman* [2008] have examined the set of error metrics for satellite rainfall data that can advance the hydrologic application of new-generation, high-resolution rainfall products over land.

5. CONCLUSION

Precipitation products derived from satellite observations have reached a good level of maturity over the last decade.

Ongoing research and development continues to address the accuracy and the resolution (temporal and spatial) of these products, along with the generation of error estimates vital to hydrological modeling and water resource assessment. At present, there is a relatively data-rich environment for satellite rainfall estimates. VIS and IR data derived from GEO satellites is available nominally every 30 min, while a good number of PMW instruments, both imaging instruments and cross-track sounders are currently available, providing <3 h sampling.

The future direction of satellite observation of the atmosphere involves both the continuation of operational missions, critical to the routine observation of the weather, and provision of long-term data sets critical for climate monitoring and impact assessment [*Asrar et al.*, 2001]. In 2008, the National Research Council of the National Academies published their report on priorities for Earth observation missions over the next decade and beyond [*NRC*, 2008]. Three sections in this report relate to observation of the atmosphere: climate variability and change, weather science and applications, and water resources and the global hydrologic cycle. Within the climate section, they note that future systems must sustain the observational capabilities of key global climate parameters, together with management of these data sets. In addition, it was stressed that there is an increasing need for the observations and derived products to be related to societal needs, thus requiring both improvements in science, but also observational capabilities with the end user in mind. Observations relating to improvements in weather forecasting were noted in the weather science and applications section, in particular, observations that would help NWP model forecasts during severe weather events. In addition, the monitoring of pollution was also noted as a key requirement. Finally, the water resources and the global hydrologic cycle section highlighted the need for improved water cycle observations, including observations that would allow the better understanding of the interaction of water cycle parameters with other components of the Earth System.

A number of recommendations for future research and development of precipitation estimation from satellite observations can be identified:

1. The availability of high-quality satellite data sets made available in near real time is of critical importance, including those of GEO-based sensors. In particular, continuation of merged global composite, such as the Global-IR data set [*Janowiak et al.*, 2001] is deemed very important. Alongside the real-time data, existing data require permanent archives and reprocessing were necessary to ensure long-term usability of the data sets. New data sets should be exploited where possible, including not only new satellite data sets but model data and satellite data from nonprecipitation missions to provide additional information for retrieval methodologies and techniques. The protection of data sets into the future should also be assured, particularly those at PMW frequencies where radio interference is becoming an increasing issue.

2. The generation of precipitation products for the user community is necessary in order to mainstream the satellite precipitation products into operational use. In particular, the distribution of precipitation products through existing networks such as GEONETCast should be encouraged. In addition, data and software should be made freely available to enable new algorithm developers to become established. Further exploitation of satellite observations for data assimilation is needed, since relatively little data are exploited, while assimilation of precipitation products themselves is somewhat limited.

3. Future activities should build upon the existing strong intercomparison/validation record of the IPWG. In particular, a specific validation program aimed at evaluating precipitation products at subdaily timescales, such as that of the PEHRPP, is required. Additional precipitation products from models should be sought through the involvement of the Working Group for Numerical Experimentation. In terms of the validation itself, validation data should have a quality index associated with them, while the usefulness of different validation tools should be assessed. Validation should also be extended to the use of hydrological impact studies and should also be used as a means of validating the precipitation products, since these are often at similar spatial scales.

4. Of particular importance is the identification and retrieval of frozen precipitation. One route is through improved radiative transfer modeling of the interactions with snow particles and through the combined use of both active and passive observations (e.g., CloudSat, AMSR-E, and AMSU-B). However, the sensitivity to shallow precipitation remains an issue, particularly over highly variable surface backgrounds.

5. New satellites systems and sensors are planned with the potential to improve precipitation retrievals. The present radar systems (TRMM PR and CloudSat CPR) will be added to by the upcoming radars on the European Space Agency's Clouds, Aerosol and Radiation Explorer (EarthCare) and the GPM-core satellites. In addition, exploitation of new lightning sensors to be carried on new GEO satellites: the U.S. GOES-R, the European MTG and the Chinese FY-4. Moreover, new orbital concepts such as the flower constellation deploying minisatellites into orbit are now under study for water vapor and precipitation sounding (FLOwer constellation for millimeter-wave scanning RADiometers) [*Marzano et al.*, 2009] using millimeter-wave scanning radiometers: their development will considerably

help in ensuring an adequate space-time coverage of PMW observations over specific areas of interest.

Mapping of global precipitation will be centered upon the GPM mission led by NASA, and the Japan Aerospace Exploration Agency will start with the launch of a core satellite in 2013. In particular, the CEOS, the international coordinating body for earth observing satellite systems (available online at http://www.ceos.org/), declared precipitation to be an important measurement and identified GPM as a prototype of the Global Earth Observation System of Systems. This mission comprises a core satellite, including the GPM Microwave Imager (http://gpm.gsfc.nasa.gov/gmi.html), with a frequency range between 10 and 183 GHz and the DPR [*Nakamura et al.*, 2005] consisting of a Ku-band precipitation radar (KuPR, 13.6 GHz) and a Ka-band precipitation radar (KaPR, 35.5 GHz), together with a constellation of satellites from international partners. This will provide precipitation estimates over the globe with a temporal sampling period of 3 h or less. Coupled with the satellite effort, new techniques are being developed to combine satellite observations into standard and comparable measurements to ensure consistency across the different constellation sensors.

REFERENCES

Adler, R. F., and A. J. Negri (1988), A satellite infrared technique to estimate tropical convective and stratiform rainfall, *J. Appl. Meteorol.*, *27*, 30–51.

Adler, R. F., C. Kidd, G. Petty, M. Morrissey, and H. M. Goodman (2001), Intercomparison of Global Precipitation products: The third Precipitation Intercomparison Project (PIP-3), *Bull. Am. Meteorol. Soc.*, *82*, 1377–1396.

Adler, R. F., et al. (2003), The version-2 Global Precipitation Climatology Project (GPCP) monthly precipitation analysis (1979–present), *J. Hydrometeorol.*, *4*, 1147–1167.

Adler, R. F., Y. Hong, and G. J. Huffman (2006), Flood and landslide applications of high time resolution satellite rain products, paper presented at 3rd International Precipitation Working Group Workshop, Coord. Group for Meteorol. Satell., Melbourne, Vic., Australia, 23-27 Oct. (Available at http://www.isac.cnr.it/~ipwg/meetings/melbourne/melbourne2006-pres.html)

Adler, R. F., G. Gu, J.-J. Wang, G. J. Huffman, S. Curtis, and D. Bolvin (2008), Relationships between global precipitation and surface temperature on interannual and longer timescales (1979–2006), *J. Geophys. Res.*, *113*, D22104, doi:10.1029/2008JD010536.

Alishouse, J., S. A. Snyder, J. Vongsathorn, and R. R. Ferraro (1990a), Determination of oceanic total precipitable water from the SSM/I, *IEEE Trans. Geosci. Remote Sens.*, *28*, 811–816.

Alishouse, J. C., J. B. Snider, E. R. Westwater, C. T. Swift, C. S. Ruf, S. Snyder, J. Vongsathorn, and R. R. Ferraro (1990b), Determination of cloud liquid water content from the SSM/I, *IEEE Trans. Geosci. Remote Sens.*, *28*, 817–822.

Anagnostou, E. N., A. J. Negri, and R. F. Adler (1999), A satellite infrared technique for diurnal rainfall variability studies, *J. Geophys. Res.*, *104*, 31,477–31,488.

Arkin, P. (1979), Relationship between fractional coverage of high cloud and rainfall accumulations during GATE over the B-Scale array, *Mon. Weather Rev.*, *107*, 1382–1387.

Arkin, P. A., and B. N. Meisner (1987), The relationship between large-scale convective rainfall and cold cloud over the western hemisphere during 1982-84, *Mon. Weather Rev.*, *115*, 51–74.

Arkin, P. A., and P. Xie (1994), The Global Precipitation Climatology Project: First Algorithm Intercomparison Project, *Bull. Am. Meteorol. Soc.*, *75*, 401–419.

Asrar, G., J. A. Kaye, and P. Morel (2001), NASA Research strategy for Earth System Science: Climate component, *Bull. Am. Meteorol. Soc.*, *82*, 1309–1329.

Ba, M. B., and A. Gruber (2001), GOES multispectral rainfall algorithm (GMSRA), *J. Appl. Meteorol.*, *40*, 1500–1514.

Barrett, E. C., and D. W. Martin (1981), *The Use of Satellite Data in Rainfall Monitoring*, 340 pp., Academic, London.

Barrett, E. C., J. Dodge, H. M. Goodman, J. Janowiak, C. Kidd, and E. A. Smith (1994), The First WetNet Precipitation Intercomparison Project, *Remote Sens. Rev.*, *11*, 49–60.

Bauer, P. (2001), Over-ocean rainfall retrieval from multisensory data or the Tropical Rainfall Measuring Mission. Part I: Design and evaluation of inversion databases, *J. Atmos. Oceanic Technol.*, *18*, 1315–1330.

Bauer, P., P. Amayenc, C. D. Kummerow, and E. A. Smith (2001), Over-ocean rainfall retrieval from multisensory data or the Tropical Rainfall Measuring Mission. Part II: Algorithm implementation, *J. Atmos. Oceanic Technol.*, *18*, 1838–1855.

Behrangi, A., K.-L. Hsu, B. Imam, S. Sorooshian, and R. Kuligowski (2009a), Evaluating the utility of multispectral information in delineating the areal extent of precipitation, *J. Hydrometeorol.*, *10*, 684–700.

Behrangi, A., K.-L. Hsu, B. Imam, S. Sorooshian, G. J. Huffman, and R. J. Kuligowski (2009b), PERSIANN-MSA: A precipitation estimation method from satellite-based multi-spectral analysis, *J. Hydrometeorol.*, *10*(6), 1414–1429, doi:10.1175/2009JHM1139.1.

Bellerby, T. J., and J. Sun (2005), Probabilistic and ensemble representations of the uncertainties in an IR/microwave satellite precipitation product, *J. Hydrometeorol.*, *6*, 1032–1044.

Bellerby, T., K.-L. Hsu, and S. Sorooshian (2009a), LMODEL: A satellite precipitation methodology using cloud development modeling. Part I: Algorithm construction and calibration, *J. Hydrometeorol.*, *10*, 1081–1095.

Bellerby, T., K.-L. Hsu, and S. Sorooshian (2009b), LMODEL: A satellite precipitation methodology using cloud development modeling. Part II: Validation, *J. Hydrometeorol.*, *10*, 1096–1108.

Bendix, J. (1997), Adjustment of the Convective-Stratiform Technique (CST) to estimate 1991/93 El Niño rainfall distribution in Ecuador and Peru by means of Meteosat-3 data, *Int. J. Remote Sens.*, *18*, 1387–1394.

Bennartz, R., and P. Bauer (2003), Sensitivity of microwave radiances at 85–183 GHz to precipitating ice particles, *Radio Sci.*, *38*(4), 8075, doi:10.1029/2002RS002626.

Bennartz, R., and R. Ferraro (2005), Report on the IPWG/GPM/GRP Workshop on global microwave modeling and retrieval of snowfall, October 11–13, 2005, report, Univ. of Wis., Madison. (Available at http://www.isac.cnr.it/~ipwg/meetings/madison/ipwg_snowfall_workshop_report.pdf)

Bennartz, R., and G. W. Petty (2001), The sensitivity of microwave remote sensing observations of precipitation to ice particle size distributions, *J. Appl. Meteorol.*, *40*, 345–364.

Bennartz, R., A. Thoss, A. Dybbroe, and D. Michelson (2002), Precipitation analysis using the Advanced Microwave Sounding Unit in support of nowcasting applications, *Meteorol. Appl.*, *9*, 177–189.

Bennartz, R., G. J. Tripoli, and R. R. Ferraro (Eds.) (2008), Report on the Second International Workshop on Space-based Snowfall Measurement, Steamboat Ski Village, 31 March–4 April, report, 31 pp., Int. Precip. Working Group. (Available at http://www.isac.cnr.it/~ipwg/meetings/steamboat/iwssm_report-v8.pdf)

Berg, W., and R. Chase (1992), Determination of mean rainfall from the Special Sensor Microwave/Imager (SSM/I) using a mixed lognormal distribution, *J. Atmos. Oceanic Technol.*, *9*, 129–141.

Bhatt, B. C., and K. Nakamura (2005), Characteristics of monsoon rainfall around the Himalayas revealed by TRMM Precipitation Radar, *Mon. Weather Rev.*, *133*, 149–165.

Bhatt, B. C., and K. Nakamura (2006), A climatological-dynamical analysis associated with precipitation around the southern part of the Himalayas, *J. Geophys. Res.*, *111*, D02115, doi:10.1029/2005JD006197.

Bizzarri, B., and the H-SAF Consortium (2008), Update on the status of precipitation products in the EUMETSAT Satellite Application Facility on Support to Hydrology and Water Management, Proc. 4th IPWG Workshop on Precipitation Measurements, Beijing, 13-17 Oct. (Available at http://www.isac.cnr.it/~ipwg/meetings/beijing/beijing2008.html)

Bolvin, D. T., R. F. Adler, G. J. Huffman, E. J. Nelkin, and J. P. Poutiainen (2009), Comparison of GPCP monthly and daily precipitation estimates with high-latitude gauge observations, *J. Appl. Meteorol. Climatol.*, *48*, 1843–1857.

Bowman, K. P. (2005), Comparison of TRMM precipitation retrievals with rain gauge data from ocean buoys, *J. Clim.*, *18*, 178–190.

Burns, B. A., X. Wu, and G. R. Diak (1997), Effects of precipitation and cloud Ice on brightness temperatures in AMSU moisture channels, *IEEE Geosci. Remote Sens. Lett.*, *35*, 1429–1437.

Capacci, D., and F. Porcù (2009), Evaluation of a satellite multispectral VIS–IR daytime statistical rain-rate classifier and comparison with passive microwave rainfall estimates, *J. Appl. Meteorol. Climatol.*, *48*, 284–300.

Cattani, E., F. Torricella, S. Laviola, and V. Levizzani (2009), On the statistical relationship between optical and microphysical characteristics of storm clouds from AVHRR and rainfall intensity from a novel AMSU-B retrieval algorithm, *Nat. Hazards Earth Syst. Sci.*, *9*, 2135–2142.

Cecil, D. J., S. J. Goodman, D. J. Boccippio, E. J. Zipser, and S. W. Nesbitt (2005), Three years of TRMM precipitation features. Part I: Radar, radiometric, and lightning characteristics, *Mon. Weather Rev.*, *133*, 543–566.

Chandrasekar, V., A. Hou, E. A. Smith, V. N. Bringi, S. A. Rutledge, E. Gorgucci, W. A. Petersen, and G. Skofronick-Jackson (2008), Potential role of dual-polarization radar in the validation of satellite precipitation measurements, *Bull. Am. Meteorol. Soc.*, *89*, 1127–1145.

Chang, A. T. C., L. S. Chiu, C. D. Kummerow, J. Meng, and T. T. Wilheit (1999), First results of the TRMM Microwave Imager (TMI) monthly oceanic rain rate: Comparison with SSM/I, *Geophys. Res. Lett.*, *26*, 2379–2382.

Chen, F. W., and D. H. Staelin (2003), AIRS/AMSU/HSB precipitation estimates, *IEEE Trans. Geosci. Remote Sens.*, *41*, 410–417.

Chen, R., R. Wood, Z. Li, R. Ferraro, and F.-L. Chang (2008), Studying the vertical variation of cloud droplet effective radius using ship and space-borne remote sensing data, *J. Geophys. Res.*, *113*, D00A02, doi:10.1029/2007JD009596.

Cifelli, R., S. W. Nesbitt, S. A. Rutledge, W. A. Petersen, and S. Yuter (2008), Diurnal characteristics of precipitation features over the Tropical East Pacific: A comparison of the EPIC and TEPPS regions, *J. Clim.*, *21*, 4068–4086.

De Angelis, C. F., G. R. McGregor, and C. Kidd (2004), A 3 year climatology of rainfall characteristics over tropical and subtropical South America based on tropical rainfall measuring mission precipitation radar data, *Int. J. Climatol.*, *24*, 385–399.

Di Tomaso, E., F. Romano, and V. Cuomo (2009), Rainfall estimation from satellite passive microwave observations in the range 89 GHz to 190 GHz, *J. Geophys. Res.*, *114*, D18203, doi:10.1029/2009JD011746.

Dodge, J., and H. M. Goodman (1994), The WetNet Project, *Remote Sens. Rev.*, *11*, 5–21.

Ebert, E. E., and M. J. Manton (1998), Performance of satellite rainfall estimation algorithms during TOGA COARE, *J. Atmos. Sci.*, *55*, 1537–1557.

Ebert, E. E., M. J. Manton, P. A. Arkin, R. J. Allam, C. E. Holpin, and A. Gruber (1996), Results from the GPCP Algorithm Intercomparison Programme, *Bull. Am. Meteorol. Soc.*, *77*, 2875–2887.

Ebert, E. E., J. E. Janowiak, and C. Kidd (2007), Comparison of near-real-time precipitation estimates from satellite observations and numerical models, *Bull. Am. Meteorol. Soc.*, *88*, 47–64.

English, S. J. (1999), Estimation of temperature and humidity profile information from microwave radiances over different surface types, *J. Appl. Meteorol.*, *38*, 1526–1541.

ESA (2004), EGPM – European Contribution to Global Precipitation Measurement, *Eur. Space Agency Spec. Publ., ESA SP-1279(5)*, 60 pp. (Available at http://esamultimedia.esa.int/docs/SP_1279_5_EGPM.pdf)

Ferraro, R. R., and G. F. Marks (1995), The development of SSM/I rain-rate retrieval algorithms using ground-based radar measurements, *J. Atmos. Oceanic Technol.*, *12*, 755–770.

Ferraro, R. R., and J. R. McCollum (2003), Rainfall over land from the AMSR-E, *in Geoscience Remote Sensing Symposium IGARSS '03*, pp. 630–633, IEEE, New York.

Ferraro, R. R., N. C. Grody, and J. Kogut (1986), Classification of geophysical parameters using passive microwave satellite measurements, *IEEE Trans. Geosci. Remote Sens.*, *24*, 1008–1013.

Ferraro, R. R., E. A. Smith, W. Berg, and G. J. Huffman (1998), A screening methodology for passive microwave precipitation retrieval algorithms, *J. Atmos. Sci.*, *55*, 1583–1600.

Ferraro, R. R., F. Weng, N. C. Grody, and L. Zhao (2000), Precipitation characteristics over land from the NOAA-15 AMSU sensor, *Geophys. Res. Lett.*, *27*, 2669–2672.

Ferraro, R. R., F. Weng, N. C. Grody, L. Zhao, H. Meng, C. Kongoli, P. Pellegrino, S. Qiu, and C. Dean (2005), NOAA operational hydrological products derived from the Advanced Microwave Sounding Unit, *IEEE Trans. Geosci. Remote Sens.*, *43*, 1036–1049.

Furuzawa, F. A., and K. Nakamura (2005), Differences of rainfall estimates over land by Tropical Rainfall Measuring Mission (TRMM) Precipitation Radar (PR) and TRMM Microwave Imager (TMI)—Dependence on storm height, *J. Appl. Meteorol.*, *44*, 367–382.

Gautier, C., Y. Shiren, and M. D. Hofstadter (2003), AIRS/Vis Near IR instrument, *IEEE Trans. Geosci. Remote Sens.*, *41*, 330–342.

Gebremichael, M., and W. F. Krajewski (2005), Effect of temporal sampling on inferred rainfall spatial statistics, *J. Appl. Meteorol.*, *44*, 1626–1633.

Gebremichael, M., W. F. Krajewski, M. Morrissey, D. Langerud, G. J. Huffman, and R. Adler (2003), Error uncertainty analysis of GPCP monthly rainfall products: A data-based simulation study, *J. Appl. Meteorol.*, *42*, 1837–1848.

Gebremichael, M., W. F. Krajewski, M. Morrissey, G. J. Huffman, and R. F. Adler (2005), A detailed evaluation of GPCP 1° daily rainfall estimates over the Mississippi River basin, *J. Appl. Meteorol.*, *44*, 665–681.

Gebremichael, M., E. R. Vivoni, C. J. Watts, and J. C. Rodriguez (2007), Submesoscale spatiotemporal variability of North American monsoon rainfall over complex terrain, *J. Clim.*, *20*, 1751–1773.

Geerts, B., and T. Dejene (2005), Regional and diurnal variability of the vertical structure of precipitation systems in Africa based upon spaceborne radar data, *J. Clim.*, *18*, 893–916.

Givati, A., and D. Rosenfeld (2004), Quantifying precipitation suppression due to air pollution, *J. Appl. Meteorol.*, *43*, 1038–1056.

Goodrum, G., K. B. Kidwell, and W. Winston (Eds.) (2000), NOAA KLM user's guide with NOAA-N, N' supplement, report, NOAA, Silver Spring, Md. (Available at http://www2.ncdc.noaa.gov/docs/klm/index.htm).

Grecu, M., and E. N. Anagnostou (2004), A differential attenuation based algorithm for estimating precipitation from dual-wavelength spaceborne radar, *Can. J. Remote Sens.*, *30*, 697–705.

Grecu, M., and W. S. Olson (2008), Precipitating snow retrievals from combined airborne cloud radar and millimeter-wave radiometer observations, *J. Appl. Meteorol. Climatol.*, *47*, 1634–1650.

Grecu, M., E. N. Anagnostou, and R. F. Adler (2000), Assessment of the use of lightning information in satellite infrared rainfall estimation, *J. Hydrometeorol.*, *1*, 211–221.

Grecu, M., W. S. Olson, and E. N. Anagnostou (2004), Retrieval of precipitation profiles from multiresolution, multifrequency active and passive microwave observations, *J. Appl. Meteorol.*, *43*, 562–575.

Greenwald, T. J., and S. A. Christopher (2002), Effect of cold clouds on satellite measurements near 183 GHz, *J. Geophys. Res.*, *107* (D13), 4170, doi:10.1029/2000JD000258.

Griffith, C. G., W. L. Woodley, P. G. Grube, D. W. Martin, J. Stout, and D. N. Sikdar (1978), Rain estimation from geosynchronous satellite imagery—Visible and infrared studies, *Mon. Weather Rev.*, *106*, 1153–1171.

Grimes, D. I. F., E. Pardo-Iguzquiza, and R. Bonifacio (1999), Optimal areal rainfall estimation using raingauges and satellite data, *J. Hydrol.*, *222*, 93–108.

Grimes, D. I. F., E. Coppola, M. Verdecchia, and G. Visconti (2003), A neural network approach to real-time rainfall estimation for Africa using satellite data, *J. Hydrometeorol.*, *4*, 1119–1133.

Grody, N. C. (1984), Precipitation monitoring over land from satellites by microwave radiometry, paper presented at IGARSS'84 Symposium, IEEE Geosci. and Remote Sens. Soc., Strasbourg, France, 27–30 Aug.

Grody, N. C. (1991), Classification of snow cover and precipitation using the Special Sensor Microwave/Imager (SSM/I), *J. Geophys. Res.*, *96*, 7423–7435.

Grody, N. C., F. Weng, and R. R. Ferraro (2000), Application of AMSU for obtaining hydrological parameters, in *Microwave Radiometry and Remote Sensing of the Earth's Surface and Atmosphere*, edited by P. Pampaloni and S. Paloscia, pp. 339–352, VSP Int. Sci. Publ., Utrecht, Netherlands.

Grody, N. C., J. Zhao, and R. R. Ferraro (2001), Determination of precipitable water and cloud liquid water over oceans from the NOAA 15 advanced microwave sounding unit, *J. Geophys. Res.*, *106*, 2943–2953.

Gruber, A., and V. Levizzani (Eds.) (2008), Assessment of global precipitation products, *WCRP Ser. Rep. 128 and WMO TD-No. 1430*, 55 pp., World Clim. Res. Programme, Geneva, Switzerland. (Available at http://wcrp.wmo.int/documents/AssessmentGlobalPrecipitationReport.pdf)

Gu, G., R. F. Adler, G. J. Huffman, and S. Curtis (2007), Tropical rainfall variability on interannual-to-interdecadal and longer time scales derived from the GPCP monthly product, *J. Clim.*, *20*, 4033–4046.

Haddad, Z. S., and K.-W. Park (2009), Vertical profiling of precipitation using passive microwave observations: The main impediment and a proposed solution, *J. Geophys. Res.*, *114*, D06118, doi:10.1029/2008JD010744.

Haddad, Z. S., E. A. Smith, C. D. Kummerow, T. Iguchi, M. Farrar, S. Darden, M. Alves, and W. Olson (1997), The TRMM 'Day-1' radar/radiometer combined rain-profile algorithm, *J. Meteorol. Soc. Jpn.*, *75*, 799–808.

Hatzianastassiou, N., B. Katsoulis, J. Pnevmatikos, and V. Antakis (2008), Spatial and temporal variation of precipitation in Greece and surrounding regions based on Global Precipitation Climatology Project data, *J. Clim.*, *21*, 1349–1370.

Hawkins, J. D., F. J. Turk, T. F. Lee, and K. Richardson (2008), Observations of tropical cyclones with SSMIS, *IEEE Trans. Geosci. Remote Sens.*, *46*, 901–912.

Haynes, J. M., T. S. L'Ecuyer, G. L. Stephens, S. D. Miller, C. Mitrescu, N. B. Wood, and S. Tanelli (2009), Rainfall retrieval over the ocean with spaceborne W-band radar, *J. Geophys. Res.*, *114*, D00A22, doi:10.1029/2008JD009973.

Hong, Y., K.-L. Hsu, S. Sorooshian, and X. Gao (2004), Precipitation estimation from remotely sensed imagery using an artificial neural network cloud classification system, *J. Appl. Meteorol.*, *43*, 1834–1852.

Hong, Y., K.-L. Hsu, S. Sorooshian, and X. Gao (2005), Improved representation of diurnal variability of rainfall retrieved from the Tropical Rainfall Measurement Mission Microwave Imager adjusted Precipitation Estimation From Remotely Sensed Information Using Artificial Neural Networks (PERSIANN) system, *J. Geophys. Res.*, *110*, D06102, doi:10.1029/2004JD005301.

Hong, Y., K.-L. Hsu, H. Moradkhani, and S. Sorooshian (2006), Uncertainty quantification of satellite precipitation estimation and Monte Carlo assessment of the error propagation into hydrologic response, *Water Resour. Res.*, *42*, W08421, doi:10.1029/2005WR004398.

Hossain, F., and E. N. Anagnostou (2004), Assessment of current passive-microwave- and infrared-based satellite rainfall remote sensing for flood prediction, *J. Geophys. Res.*, *109*, D07102, doi:10.1029/2003JD003986.

Hossain, F., and E. N. Anagnostou (2006a), Assessment of a multidimensional satellite rainfall error model for ensemble generation of satellite rainfall data, *IEEE Geosci. Remote Sens. Lett.*, *3*, 419–423.

Hossain, F., and E. N. Anagnostou (2006b), A two-dimensional satellite rainfall error model, *IEEE Trans. Geosci. Remote Sens.*, *44*, 1511–1522.

Hossain, F., and G. Huffman (2008), Investigating error metrics for satellite rainfall data at hydrologically relevant scales, *J. Hydrometeorol.*, *9*, 563–575.

Hossain, F., E. N. Anagnostou, and T. Dinku (2004), Sensitivity analyses of satellite rainfall retrieval and sampling error on flood prediction uncertainty, *IEEE Trans. Geosci. Remote Sens.*, *42*, 130–139.

Hou, A. Y., G. Skofronick-Jackson, C. D. Kummerow, and J. M. Shepherd (2008), Global precipitation measurement, in *Precipitation: Advances in Measurement, Estimation and Prediction*, edited by S. Michaelides, pp. 131–169, Springer, New York.

Hsu, K., X. Gao, S. Sorooshian, and H. V. Gupta (1997), Precipitation estimation from remotely sensed information using artificial neural networks, *J. Appl. Meteorol.*, *36*, 1176–1190.

Hsu, K.-L., H. V. Gupta, X. Gao, and S. Sorooshian (1999), Estimation of physical variables from multichannel remotely sensed imagery using neural networks: Application to rainfall estimation, *Water Resour. Res.*, *35*, 1605–1618.

Huffman, G. J., R. F. Adler, P. Arkin, A. Chang, R. Ferraro, A. Gruber, J. Janowiak, A. McNab, B. Rudolf, and U. Schneider (1997), The Global Precipitation Climatology Project (GPCP) combined precipitation dataset, *Bull. Am. Meteorol. Soc.*, *78*, 5–20.

Huffman, G. J., R. F. Adler, D. T. Bolvin, G. Gu, E. J. Nelkin, K. P. Bowman, Y. Hong, E. F. Stocker, and D. B. Wolff (2007), The TRMM Multisatellite Precipitation Analysis (TMPA): Quasi-global, multiyear, combined-sensor precipitation estimates at fine scales, *J. Hydrometeorol.*, *8*, 38–55.

Hunt, W. H., D. M. Winker, M. A. Vaughan, K. A. Powell, P. L. Lucker, and C. Weimer (2009), CALIPSO lidar description and performance assessment, *J. Atmos. Oceanic Technol.*, *26*, 1214–1228.

Iguchi, T., T. Kozu, R. Meneghini, J. Awaka, and K. Okamoto (2000), Rain-profiling algorithm for the TRMM Precipitation Radar, *J. Appl. Meteorol.*, *39*, 2038–2052.

Ikai, J., and K. Nakamura (2003), Comparison of rain rates over the ocean derived from TRMM Microwave Imager and Precipitation Radar, *J. Atmos. Oceanic Technol.*, *20*, 1709–1726.

Inoue, T., and K. Aonashi (2000), A comparison of cloud and rainfall information from instantaneous Visible and InfraRed Scanner and Precipitation Radar observations over a frontal zone in East Asia during June 1998, *J. Appl. Meteorol.*, *39*, 2292–2301.

Janowiak, J. E., and P. Xie (2003), A global-scale examination of monsoon-related precipitation, *J. Clim.*, *16*, 4121–4133.

Janowiak, J. E., R. J. Joyce, and Y. Yarosh (2001), A real-time global half-hourly pixel-resolution infrared dataset and its applications, *Bull. Am. Meteorol. Soc.*, *82*, 205–217.

Joseph, R., T. M. Smith, M. R. P. Sapiano, and R. R. Ferraro (2009), A new high-resolution satellite-derived precipitation dataset for climate studies, *J. Hydrometeorol.*, *10*, 935–952.

Joyce, R. J., J. E. Janowiak, P. A. Arkin, and P. Xie (2004), CMORPH: A method that produces global precipitation estimates from passive microwave and infrared data at high spatial and temporal resolutions, *J. Hydrometeorol.*, *5*, 487–503.

Joyce, R. J., P. Xie, and J. Janowiak (2009), A Kalman filter approach to blend various satellite rainfall estimates in CMORPH, paper presented at Proceedings of 4th IPWG Workshop, Coord. Group for Meteorol. Satell., Beijing, 13–17 Oct. (Available at http://www.isac.cnr.it/~ipwg/meetings/beijing/4th-IPWG-Proceedings-web-March-2009.pdf)

Kahn, B. H., et al. (2008), Cloud type comparisons of AIRS, CloudSat, and CALIPSO cloud height and amount, *Atmos. Chem. Phys.*, *8*, 1231–1248.

Kakar, R. K. (1983), Retrieval of clear sky moisture profiles using the 183 GHz water vapor line, *J. Clim. Appl. Meteorol.*, *22*, 1282–1289.

Katsumata, M., H. Uyeda, K. Iwanami, and G. Liu (2000), The response of 36- and 89-GHz microwave channels to convective snow clouds over ocean: Observation and modeling, *J. Appl. Meteorol.*, *39*, 2322–2335.

Kawanishi, T., T. Sezai, Y. Ito, K. Imaoka, T. Takeshima, Y. Ishido, A. Shibata, M. Miura, H. Inahata, and R. W. Spencer (2003), The Advanced Microwave Scanning Radiometer for the Earth Observing System (AMSR-E), NASDA's contribution to the EOS for global energy and water cycle studies, *IEEE Trans. Geosci. Remote Sens.*, *41*, 184–194.

Kidd, C. (1998), On rainfall retrieval using polarization-corrected temperatures, *Int. J. Remote Sens.*, *19*, 981–996.

Kidd, C. (2001), Satellite rainfall climatology: A review, *Int. J. Climatol.*, *21*, 1041–1066.

Kidd, C., and G. R. McGregor (2007), Observation and characterisation of rainfall over Hawaii and surrounding region from the Tropical Rainfall Measuring Mission, *Int. J. Climatol.*, *27*, 541–553.

Kidd, C., D. Kniveton, and E. C. Barrett (1998), Advantages and disadvantages of statistical/empirical satellite estimation of rainfall, *J. Atmos. Sci.*, *55*, 1576–1582.

Kidd, C., D. R. Kniveton, M. C. Todd, and T. J. Bellerby (2003), Satellite rainfall estimation using a combined passive microwave and infrared algorithm, *J. Hydrometeorol.*, *4*, 1088–1104.

Kidd, C., V. Levizzani, and P. Bauer (2009a), A review of satellite meteorology and climatology at the start of the twenty-first century, *Prog. Phys. Geogr.*, *33*, 474–489.

Kidd, C., V. Levizzani, F. J. Turk, and R. R. Ferraro (2009b), Satellite precipitation measurements for water resource monitoring, *J. Am. Water Resour. Assoc.*, *45*, 567–579.

Kim, M.-J., J. A. Weinman, W. S. Olson, D.-E. Chang, G. Skofronick-Jackson, and J. R. Wang (2008), A physical model to estimate snowfall over land using AMSU-B observations, *J. Geophys. Res.*, *113*, D09201, doi:10.1029/2007JD008589.

Klaes, K. D., et al. (2007), An introduction to the EUMETSAT polar system, *Bull. Am. Meteorol. Soc.*, *88*, 1085–1096.

Kniveton, D. R., B. C. Motta, H. M. Goodman, M. Smith, and F. J. LaFontaine (1994), The First WetNet Precipitation Intercomparison Project: Generation of results, *Remote Sens. Rev.*, *11*, 243–302.

Kongoli, C., P. Pellegrino, R. R. Ferraro, N. C. Grody, and H. Meng (2003), A new snowfall detection algorithm over land using measurements from the Advanced Microwave Sounding Unit (AMSU), *Geophys. Res. Lett.*, *30*(14), 1756, doi:10.1029/2003GL017177.

Kongoli, C., R. R. Ferraro, P. Pellegrino, H. Meng, and C. Dean (2007), Utilization of the AMSU high frequency measurements for improved coastal rain retrievals, *Geophys. Res. Lett.*, *34*, L17809, doi:10.1029/2007GL029940.

Krajewski, W. F., G. J. Ciach, J. R. McCollum, and C. Bacotiu (2000), Initial validation of the Global Precipitation Climatology Project monthly rainfall over the United States, *J. Appl. Meteorol.*, *39*, 1071–1086.

Kubota, T., et al. (2007), Global precipitation map using satelliteborne microwave radiometers by the GSMaP Project: Production and validation, *IEEE Trans. Geosci. Remote Sens.*, *45*, 2259–2275.

Kuligowski, R. J. (2002), A self-calibrating real-time GOES rainfall algorithm for short-term rainfall estimates, *J. Hydrometeorol.*, *3*, 112–130.

Kummerow, C. D., and L. Giglio (1994), A passive microwave technique for estimating rainfall and vertical structure information from space. Part II: Applications to SSM/I data, *J. Appl. Meteorol.*, *33*, 19–34.

Kummerow, C. D., and J. A. Weinman (1988), Radiative properties of deformed hydrometeors for commonly used passive microwave frequencies, *IEEE Trans. Geosci. Remote Sens.*, *26*, 629–638.

Kummerow, C., W. Barnes, T. Kozu, J. Shiue, and J. Simpson (1998), The Tropical Rainfall Measuring Mission (TRMM) sensor package, *J. Atmos. Oceanic Technol.*, *15*, 809–817.

Kummerow, C. D., Y. Hong, W. S. Olson, S. Yang, R. F. Adler, J. McCollum, R. R. Ferraro, G. Petty, D.-B. Shin, and T. T. Wilheit (2001), The evolution of the Goddard Profiling Algorithm (GPROF) for rainfall estimation from passive microwave sensors, *J. Appl. Meteorol.*, *40*, 1801–1820.

Kummerow, C., H. Masunaga, and P. Bauer (2007), A next-generation microwave rainfall retrieval algorithm for use by TRMM and GPM, in *Measuring Precipitation From Space—EURAINSAT and the Future*, edited by V. Levizzani, P. Bauer, and F. J. Turk, pp. 235–252, Springer, Dordrecht, Netherlands.

Kunkee, D. B., G. A. Poe, D. J. Boucher, S. D. Swadley, Y. Hong, J. E. Wessel, and E. A. Uliana (2008), Design and evaluation of the first Special Sensor Microwave Imager/Sounder, *IEEE Trans. Geosci. Remote Sens.*, *46*, 863–883.

Lambrigtsen, B., A. Tanner, T. Gaier, P. Kangaslahti, and S. Brown (2007), Prototyping GeoSTAR for the PATH Mission, paper presented at NASA Science Technology Conference, NASA, Adelphi, Md., 19-21 June. (Available at http://esto.nasa.gov/conferences/nstc2007/papers/Lambrigtsen_Bjorn_B1P4_NSTC-07-0020.pdf).

Laviola, S., and V. Levizzani (2008), Rain retrieval using 183 GHz absorption lines, in *Microwave Radiometry and Remote Sensing of the Environment*, pp. 1–4, doi:101029/MICRAD.2008.4579505, Inst. of Atmos. Res. Sens., Natl. Res. Counc., Bologna, Italy.

Laviola, S., and V. Levizzani (2009), Observing precipitation by means of water vapor absorption lines; A first check of the retrieval capabilities of the 183-WSL rain retrieval method, *Italian J. Remote Sens.*, *41*(3), 39–49.

Lensky, I. M., and D. Rosenfeld (2003a), A night-rain delineation algorithm for infrared satellite data based on microphysical considerations, *J. Appl. Meteorol.*, *42*, 1218–1226.

Lensky, I. M., and D. Rosenfeld (2003b), Satellite-based insights into precipitation formation processes in continental and maritime convective clouds at nighttime, *J. Appl. Meteorol.*, *42*, 1227–1233.

Lenksy, I. M., and D. Rosenfeld (2008), Clouds-Aerosols-Precipitation Satellite Analysis Tool (CAPSAT), *Atmos. Chem. Phys.*, *8*, 6739–6753.

Levizzani, V., J. Schmetz, H. J. Lutz, J. Kerkmann, P. P. Alberoni, and M. Cervino (2001), Precipitation estimations from geostationary orbit and prospects METEOSAT Second Generation, *Meteorol. Appl.*, *8*, 23–41.

Levizzani, V., P. Bauer, and F. J. Turk (2007), *Measuring Precipitation from Space—EURAINSAT and the Future*, 722 pp., Springer, Dordrecht, Netherlands.

Lin, X., and A. Y. Hou (2007), Evaluation of coincident passive microwave rainfall estimates using TRMM PR and ground measurements as references, *J. Appl. Meteorol. Climatol.*, *47*, 3170–3187.

Liu, G. (2004), Approximation of single scattering properties of ice and snow particles for high microwave frequencies, *J. Atmos. Sci.*, *61*, 2441–2456.

Liu, G. (2008), Deriving snow cloud characteristics from CloudSat observations, *J. Geophys. Res.*, *113*, D00A09, doi:10.1029/2007JD009766.

Liu, G., and J. A. Curry (1997), Precipitation characteristics in Greenland-Iceland-Norwegian Seas determined by using satellite microwave data, *J. Geophys. Res.*, *102*, 13,987–13,997.

Lobl, E. S., K. Aonashi, B. Griffith, C. Kummerow, G. Liu, M. Murakami, and T. Wilheit (2007), Wakasa Bay—An AMSR precipitation validation campaign, *Bull. Am. Meteorol. Soc.*, *88*, 551–558.

Mace, G. G., Q. Zhang, M. Vaughan, R. Marchand, G. L. Stephens, C. Trepte, and D. Winker (2009), A description of hydrometeor layer occurrence statistics derived from the first year of merged Cloudsat and CALIPSO data, *J. Geophys. Res.*, *114*, D00A26, doi:10.1029/2007JD009755.

Marchand, R., G. G. Mace, T. Ackerman, and G. L. Stephens (2008), Hydrometeor detection using Cloudsat—An Earth-orbiting 94-GHz cloud radar, *J. Atmos. Oceanic Technol.*, *25*, 519–533.

Marks, D. A., D. B. Wolff, D. S. Silberstein, A. Tokay, J. L. Pippitt, and J. Wang (2009), Availability of high-quality TRMM ground validation data from Kwajalein, RMI: A practical application of the relative calibration adjustment technique, *J. Atmos. Oceanic Technol.*, *26*, 413–429.

Marzano, F. S., M. Palmacci, D. Cimini, G. Giuliani, and F. J. Turk (2004), Multivariate statistical integration of satellite infrared and microwave radiometric measurements for rainfall retrieval at the geostationary scales, *IEEE Trans. Geosci. Remote Sens.*, *42*, 1018–1032.

Marzano, F. S., D. Cimini, A. Memmo, M. Montopoli, T. Rossi, M. D. Sanctis, M. Lucente, D. Mortari, and S. D. Michele (2009), Flower constellation of millimeter-wave radiometers for tropospheric monitoring at pseudogeostationary scale, *IEEE Trans. Geosci. Remote Sens.*, *47*, 3107–3122.

Masunaga, H., and C. D. Kummerow (2005), Combined radar and radiometer analysis of precipitation profiles for a parametric retrieval algorithm, *J. Atmos. Oceanic Technol.*, *22*, 909–929.

Masunaga, H., T. Iguchi, R. Oki, and M. Kachi (2002), Comparison of rainfall products derived from TRMM Microwave Imager and Precipitation Radar, *J. Appl. Meteorol.*, *41*, 849–862.

McCollum, J. R., and R. R. Ferraro (2003), Next generation of NOAA/NESDIS TMI, SSM/I, and AMSR-E microwave land rainfall algorithms, *J. Geophys. Res.*, *108*(D8), 8382, doi:10.1029/2001JD001512.

McCollum, J. R., and R. R. Ferraro (2005), Microwave rainfall estimation over coasts, *J. Atmos. Oceanic Technol.*, *22*, 497–512.

McCollum, J. R., A. Gruber, and M. B. Ba (2000), Discrepancy between gauges and satellite estimates of rainfall in equatorial Africa, *J. Appl. Meteorol.*, *39*, 666–679.

McCollum, J. R., W. F. Krajewski, R. R. Ferraro, and M. B. Ba (2002), Evaluation of biases of satellite rainfall estimation algorithms over the continental United States, *J. Appl. Meteorol.*, *41*, 1065–1080.

McPhee, J., and S. A. Margulis (2005), Validation and error characterization of the GPCP-1DD precipitation product over the contiguous United States, *J. Hydrometeorol.*, *6*, 441–459.

Menzel, W. P., and J. F. W. Purdom (1994), Introducing GOES-I: The first of a new generation of geostationary operational environmental satellites, *Bull. Am. Meteorol. Soc.*, *75*, 757–781.

Michaelides, S. (2008), *Precipitation: Advances in Measurement, Estimation and Prediction*, 540 pp., Springer, Berlin.

Michaelides, S., V. Levizzani, E. N. Anagnostou, P. Bauer, T. Kasparis, and J. E. Lane (2009), Precipitation: Measurement, remote sensing, climatology and modelling, *Atmos. Res.*, *94*(4), 512–533, doi:10.1016/j.atmosres.2009.08.017.

Miller, S. W., P. A. Arkin, and R. J. Joyce (2001), A combined microwave/infrared rain rate algorithm, *Int. J. Remote Sens.*, *22*, 3285–3307.

Morales, C. A., and E. N. Anagnostou (2003), Extending the capabilities of high-frequency rainfall estimation from geostationary-based satellite infrared via a network of long-range lightning observations, *J. Hydrometeorol.*, *4*, 141–159.

Mugnai, A., H. J. Cooper, E. A. Smith, and G. J. Tripoli (1990), Simulation of microwave brightness temperatures of an evolving hailstorm at SSM/I frequencies, *Bull. Am. Meteorol. Soc.*, *71*, 2–13.

Mugnai, A., E. A. Smith, and G. J. Tripoli (1993), Foundations for statistical-physical precipitation retrieval from passive microwave satellite measurements. Part II: Emission-source and generalized weighting-function properties of a time-dependent cloud-radiation model, *J. Appl. Meteorol.*, *32*, 17–39.

Nakamura, K., and T. Iguchi (2007), Dual-wavelength radar algorithm, in *Measuring Precipitation from Space—EURAINSAT and the Future*, edited by V. Levizzani, P. Bauer, and F. J. Turk, pp. 225–234, Springer, Dordrecht, Netherlands.

Nakamura, K., T. Iguchi, M. Kojima, and E. A. Smith (2005), Global Precipitation Mission (GPM) and Dual-Wavelength Radar (DPR), paper presented at General Assembly, Union Radio Sci. Int., Ghent, Belgium. 23–29 Oct. (Available at http://www.ursi.org/Proceedings/ProcGA05/pdf/F10.1(0803).pdf)

NASA (1999), *EOS Reference Handbook: A Guide to NASA's Earth Science Enterprise and the Earth Observing System*, edited by M. King, and R. Greenstone, 355 pp., EOS Project Science Office, NASA/GSFC, Greenbelt, Md.

Nauss, T., and A. A. Kokhanovsky (2006), Discriminating raining from non-raining clouds at mid-latitudes using multispectral satellite data, *Atmos. Chem. Phys.*, *6*, 5031–5036.

Negri, A. J., R. F. Adler, and P. J. Wetzel (1984), Rain estimation from satellite: An examination of the Griffith-Woodley technique, *J. Clim. Appl. Meteorol.*, *23*, 102–116.

Nesbitt, S. W., and E. J. Zipser (2003), The diurnal cycle of rainfall and convective intensity according to three years of TRMM measurements, *J. Clim.*, *16*, 1456–1475.

Nesbitt, S. W., E. J. Zipser, and C. D. Kummerow (2004), An examination of Version-5 rainfall estimates from the TRMM Microwave Imager, Precipitation Radar, and rain gauges on global, regional, and storm scales, *J. Appl. Meteorol.*, *43*, 1016–1036.

Nicholson, S. E., et al. (2003a), Validation of TRMM and other rainfall estimates with a high-density gauge dataset for West Africa. Part I: Validation of GPCC rainfall product and pre-TRMM satellite and blended products, *J. Appl. Meteorol.*, *42*, 1337–1354.

Nicholson, S. E., et al. (2003b), Validation of TRMM and other rainfall estimates with a high-density gauge dataset for West Africa. Part II: Validation of TRMM rainfall products, *J. Appl. Meteorol.*, *42*, 1355–1368.

Noh, Y.-J., G. Liu, E.-K. Seo, J. R. Wang, and K. Aonashi (2006), Development of a snowfall retrieval algorithm at high microwave frequencies, *J. Geophys. Res.*, *111*, D22216, doi:10.1029/2005JD006826.

NRC (2008), *Earth Science and Application from Space: National Imperatives for the Next Decade and Beyond*, 456 pp., The Natl. Acad. Press, Washington, D. C.

Olson, W. S. (1989), Physical retrieval of rainfall rates over the ocean by multispectral microwave radiometry: Application to tropical cyclones, *J. Geophys. Res.*, *94*, 2267–2280.

Olson, W. S., C. D. Kummerow, Y. Hong, and W.-K. Tao (1999), Atmospheric latent heating distributions in the tropics derived from satellite passive microwave radiometer measurements, *J. Appl. Meteorol.*, *38*, 633–664.

Olson, W. S., et al. (2006), Precipitation and latent heating distributions from satellite passive microwave radiometry. Part I: Improved method and uncertainties, *J. Appl. Meteorol. Climatol.*, *45*, 702–720.

O'Sullivan, F., C. H. Wash, M. Stewart, and C. E. Motell (1990), Rain estimation from infrared and visible GOES satellite data, *J. Appl. Meteorol.*, *29*, 209–223.

Pessi, A. T., and S. Businger (2009), Relationships among lightning, precipitation, and hydrometeor characteristics over the North Pacific Ocean, *J. Appl. Meteorol. Climatol.*, *48*, 833–848.

Petty, G. W. (1994), Physical retrievals of over-ocean rain rate from multichannel microwave imaging. Part II: Algorithm implementation, *Meteorol. Atmos. Phys.*, *54*, 101–121.

Petty, G. W. (1995), The status of satellite-based rainfall estimation over land, *Remote Sens. Environ.*, *51*, 125–137.

Posselt, D. J., G. L. Stephens, and M. Miller (2008), CLOUDSAT adding a new dimension to a classical view of extratropical cyclones, *Bull. Am. Meteorol. Soc.*, *89*, 599–609.

Reudenbach, C., G. Heinemann, E. Heuel, J. Bendix, and M. Winiger (2001), Investigation of summertime convective rainfall in Western Europe based on a synergy of remote sensing data and numerical models, *Meteorol. Atmos. Phys.*, *76*, 23–41.

Roebeling, R. A., and I. Holleman (2009), SEVIRI rainfall retrieval and validation using weather radar observations, *J. Geophys. Res.*, *114*, D21202, doi:10.1029/2009JD012102.

Rosenfeld, D. (2000), Suppression of rain and snow by urban and industrial air pollution, *Science*, *287*, 1793–1796.

Rosenfeld, D., and G. Gutman (1994), Retrieving microphysical properties near the tops of potential rain clouds by multispectral analysis of AVHRR data, *Atmos. Res.*, *34*, 259–283.

Rosenfeld, D., and I. M. Lensky (1998), Satellite-based insights into precipitation formation processes in continental and maritime convective clouds, *Bull. Am. Meteorol. Soc.*, *79*, 2457–2476.

Sanderson, V. L., C. Kidd, and G. R. McGregor (2006), A comparison of TRMM microwave techniques for detecting the diurnal rainfall cycle, *J. Hydrometeorol.*, *7*, 687–704.

Sapiano, M. R. P., and P. A. Arkin (2009), An intercomparison and validation of high-resolution satellite precipitation estimates with 3-hourly gauge data, *J. Hydrometeorol.*, *10*, 149–166.

Schmetz, J., P. Pili, S. Tjemkes, D. Just, J. Kerkmann, S. Rota, and A. Ratier (2002), An introduction to Meteosat Second Generation (MSG), *Bull. Am. Meteorol. Soc.*, *83*, 977–992.

Schmit, T. J., J. Li, S. A. Ackerman, and J. J. Gurka (2005), High-spectral- and high-temporal-resolution infrared measurements from geostationary orbit, *J. Atmos. Oceanic Technol.*, *26*, 2273–2292.

Scofield, R. A. (1987), The NESDIS operational convective precipitation technique, *Mon. Weather Rev.*, *115*, 1773–1792.

Scofield, R. A., and R. J. Kuligowski (2003), Status and outlook of operational satellite precipitation algorithms for extreme precipitation events, *Weather Forecast.*, *18*, 1037–1051.

Shepherd, J. M., H. Pierce, and A. J. Negri (2002), Rainfall modification by major urban areas: Observations from spaceborne rain radar on the TRMM satellite, *J. Appl. Meteorol.*, *41*, 689–701.

Shige, S., Y. N. Takayabu, W.-K. Tao, and D. E. Johnson (2004), Spectral retrieval of latent heating profiles from TRMM PR data. Part I: Development of a model-based algorithm, *J. Appl. Meteorol.*, *43*, 1095–1113.

Shige, S., Y. N. Takayabu, W.-K. Tao, and C.-L. Shie (2007), Spectral retrieval of latent heating profiles from TRMM PR data. Part II: Algorithm improvement and heating estimates over tropical ocean regions, *J. Appl. Meteorol. Climatol.*, *46*, 1098–1124.

Shige, S., Y. N. Takayabu, and W.-K. Tao (2008), Spectral retrieval of latent heating profiles from TRMM PR data. Part III: Estimating apparent moisture sink profiles over tropical oceans, *J. Appl. Meteorol. Climatol.*, *47*, 620–640.

Shige, S., Y. N. Takayabu, S. Kida, W.-K. Tao, X. Zeng, C. Yokoyama, and T. L'Ecuyer (2009), Spectral retrieval of latent heating profiles from TRMM PR data. Part IV: Comparisons of lookup tables from two- and three-dimensional cloud-resolving model simulations, *J. Clim.*, *22*, 5577–5594.

Skofronick-Jackson, G. M., M.-J. Kim, J. A. Weinman, and D.-E. Chang (2004), A physical model to determine snowfall over land by microwave radiometry, *IEEE Trans. Geosci. Remote Sens.*, *42*, 1047–1058.

Smith, E. A., A. Mugnai, H. J. Cooper, G. J. Tripoli, and X. Xiang (1992), Foundations for statistical-physical precipitation retrieval from passive microwave satellite measurements. Part I: Brightness-temperature properties of a time-dependent cloud-radiation model, *J. Appl. Meteorol.*, *31*, 506–531.

Smith, E. A., X. Xiang, A. Mugnai, and G. J. Tripoli (1994), Design of an inversion-based precipitation profile retrieval algorithm using an explicit cloud model for initial guess microphysics, *Meteorol. Atmos. Phys.*, *54*, 53–78.

Smith, E. A., et al. (1998), Results of WetNet PIP-2 Project, *J. Atmos. Sci.*, *55*, 1483–1536.

Sorooshian, S., K.-L. Hsu, E. Coppola, B. Tomassetti, M. Verdecchia, and G. Visconti (2009), *Hydrological Modelling and the Water Cycle: Coupling the Atmospheric and Hydrological Models*, 291 pp., Springer, Berlin

Spencer, R. W., H. M. Goodman, and R. E. Hood (1989), Precipitation retrieval over land and ocean with SSM/I. Part I: Identification and characteristics of the scattering signal, *J. Atmos. Oceanic Technol.*, *6*, 254–273.

Staelin, D. H., and F. W. Chen (2000), Precipitation observations near 54 and 183 GHz using the NOAA-15 satellite, *IEEE Trans. Geosci. Remote Sens.*, *38*, 2322–2332.

Staelin, D. H., and C. Surussavadee (2007), Precipitation retrieval accuracies for geo-microwave sounders, *IEEE Trans. Geosci. Remote Sens.*, *45*, 3150–3159.

Stephens, G. L., et al. (2008), The CloudSat Mission: Performance and early science after the first year of operation, *J. Geophys. Res.*, *113*, D00A18, doi:10.1029/2008JD009982.

Stubenrauch, C. J., S. Cros, N. Lamquin, R. Armante, A. Chedin, C. Crevoisier, and N. A. Scott (2008), Cloud properties from Atmospheric Infrared Sounder and evaluation with Cloud--Aerosol Lidar and Infrared Pathfinder Satellite Observations, *J. Geophys. Res.*, *113*, D00A10, doi:10.1029/2008JD009928.

Stuhlmann, R., A. Rodriguez, S. Tjemkes, J. Grandell, A. Arriaga, J.-L. Bezy, D. Aminou, and P. Bensi (2005), Plans for EUMETSAT's Third Generation Meteosat geostationary satellite programme, *Adv. Space Res.*, *36*, 975–981.

Surussavadee, C., and D. H. Staelin (2007), Millimeter-wave precipitation retrievals and observed-versus-simulated radiance distributions: Sensitivity to assumptions, *J. Atmos. Sci.*, *64*, 3808–3826.

Surussavadee, C., and D. H. Staelin (2008a), Global millimeter-wave precipitation retrievals trained with a cloud-resolving numerical weather prediction model, Part I: Retrieval design, *IEEE Trans. Geosci. Remote Sens.*, *46*, 99–108.

Surussavadee, C., and D. H. Staelin (2008b), Global millimeter-wave precipitation retrievals trained with a cloud-resolving numerical weather prediction model, Part II: Performance evaluation, *IEEE Trans. Geosci. Remote Sens.*, *46*, 109–118.

Tadesse, A., and E. N. Anagnostou (2009), The effect of storm life cycle on satellite rainfall estimation error, *J. Atmos. Oceanic Technol.*, *26*, 769–777.

Takayabu, Y. N. (2002), Spectral representation of rain profiles and diurnal variations observed with TRMM PR over the equatorial area, *Geophys. Res. Lett.*, *29*(12), 1584, doi:10.1029/2001GL014113.

Tao, W.-K., S. Lang, W. S. Olson, R. Meneghini, S. Yang, J. Simpson, C. Kummerow, E. Smith, and J. Halverson (2001), Retrieved vertical profiles of latent heat release using TRMM rainfall products for February 1988, *J. Appl. Meteorol.*, *40*, 957–982.

Tapiador, F. J., C. Kidd, V. Levizzani, and F. S. Marzano (2004), A neural networks-based fusion technique to estimate half-hourly rainfall estimates at 0.1° resolution from satellite passive microwave and infrared data, *J. Appl. Meteorol.*, *43*, 576–594.

Thies, B., T. Nauss, and J. Bendix (2008a), Discriminating raining from non-raining clouds at mid-latitudes using Meteosat Second Generation daytime data, *Atmos. Chem. Phys.*, *8*, 2341–2349.

Thies, B., T. Nauss, and J. Bendix (2008b), Precipitation process and rainfall intensity differentiation using Meteosat Second Generation Spinning Enhanced Visible and Infrared Imager data, *J. Geophys. Res.*, *113*, D23206, doi:10.1029/2008JD010464.

Todd, M. C., E. C. Barrett, M. J. Beaumont, and T. J. Bellerby (1999), Estimation of daily rainfall over the upper Nile river basin using a continuously calibrated satellite infrared technique, *Meteorol. Appl.*, *6*, 201–210.

Todd, M. C., C. Kidd, and T. J. Bellerby (2001), A combined satellite infrared and passive microwave technique for estimation of small-scale rainfall, *J. Atmos. Oceanic Technol.*, *18*, 742–755.

Todini, G., R. Rizzi, and E. Todini (2009), Detecting precipitating clouds over snow and ice using a multiple sensors approach, *J. Appl. Meteorol. Climatol.*, *48*, 1858–1867.

Torricella, F., E. Cattani, and V. Levizzani (2008), Rain area delineation by means of multispectral cloud characterization from satellite, *Adv. Geosci.*, *17*, 43–47.

Tsonis, A. A. (1987), Determining rainfall intensity and type from GOES imagery in the midlatitudes, *Remote Sens. Environ.*, *21*, 29–36.

Turk, F. J., G. D. Rohaly, J. Hawkins, E. A. Smith, F. S. Marzano, A. Mugnai, and V. Levizzani (2000), Meteorological applications of precipitation estimation from combined SSM/I, TRMM and infrared geostationary satellite data, in *Microwave Radiometry and Remote Sensing of the Earth's Surface and Atmosphere*, edited by P. Pampaloni and S. Paloscia, pp. 353–363, VSP Int. Sci. Publ., Utrecht, Netherlands.

Turk, F. J., P. Arkin, E. E. Ebert, and M. Sapiano (2008), Evaluating high-resolution precipitation products, *Bull. Am. Meteorol. Soc.*, *89*, 1911–1916.

Turk, F. J., B.-J. Sohn, H.-J. Oh, E. E. Ebert, V. Levizzani, and E. A. Smith (2009), Validating a rapid-update satellite precipitation analysis across telescoping space and time scales, *Meteorol. Atmos. Phys.*, *105*, 99–108.

Turk, J., and P. Bauer (2006), The International Precipitation Working Group and its role in the improvement of quantitative precipitation measurements, *Bull. Am. Meteorol. Soc.*, *87*, 643–647.

Ushio, T., and K. Okamoto (2009), *GSMAP_MVK(+), IPWG Algorithm Inventory*, 4 pp. (Available at http://www.isac.cnr.it/~ipwg/algorithms/inventory/GSMaP.pdf)

Ushio, T., et al. (2009), A Kalman filter approach to the Global Satellite Mapping of Precipitation (GSMaP) from combined passive microwave and infrared radiometric data, *J. Meteorol. Soc. Jpn.*, *87A*, 137–151.

Vicente, G. A., R. A. Scofield, and W. P. Menzel (1998), The operational GOES infrared rainfall estimation technique, *Bull. Am. Meteorol. Soc.*, *79*, 1883–1898.

Viltard, N., C. Burlaud, and C. D. Kummerow (2006), Rain retrieval from TMI brightness temperature measurements using a TRMM PR-based database, *J. Appl. Meteorol. Climatol.*, *45*, 455–466.

Wang, J. R., and T. T. Wilheit (1989), Retrieval of total precipitable water using radiometric measurements near 92 and 183 GHz, *J. Appl. Meteorol.*, *28*, 146–154.

Wang, J. R., P. E. Racette, and M. E. Triesky (2001), Retrieval of precipitable water vapor by the millimeter-wave imaging radiometer in the Arctic region during FIRE-ACE, *IEEE Geosci. Remote Sens. Lett.*, *39*, 595–605.

Wang, N. Y., C. T. Liu, R. Ferraro, D. Wolff, E. Zipser, and C. Kummerow (2009), TRMM 2A12 Land Precipitation Product—Status and Future Plans, *J. Meteorol. Soc. Jpn.*, *87*, 237–253.

Wang, S.-Y., and T.-C. Chen (2008), Measuring East Asian summer monsoon rainfall contributions by different weather systems over Taiwan, *J. Appl. Meteorol. Climatol.*, *47*, 2068–2080.

Weinman, J. A., and P. J. Guetter (1977), Determination of rainfall distributions from microwave radiation measured by the Numbus-7 ESMR, *J. Appl. Meteorol.*, *16*, 437–442.

Weng, F., and N. C. Grody (1994), Retrieval of cloud liquid water using the Special Sensor Microwave Imager (SSM/I), *J. Geophys. Res.*, *99*, 25,535–25,551.

Weng, F., and N. C. Grody (2000), Retrieval of ice cloud parameters using a Microwave Imaging Radiometer, *J. Atmos. Sci.*, *57*, 1069–1081.

Weng, F., R. R. Ferraro, and N. C. Grody (1994), Global precipitation estimations using Defense Meteorological Satellite Program F10 and F11 Special Sensor Microwave Imager data, *J. Geophys. Res.*, *99*, 14,493–14,502.

Weng, F., B. Yan, and N. C. Grody (2001), A microwave land emissivity model, *J. Geophys. Res.*, *106*, 20,115–20,123.

Weng, F., L. Zhao, R. R. Ferraro, G. Poe, X. Li, and N. C. Grody (2003), Advanced microwave sounding unit cloud and precipitation algorithms, *Radio Sci.*, *38*(4), 8068, doi:10.1029/2002RS002679.

Wilheit, T. T., A. T. C. Chang, M. S. V. Rao, E. B. Rodgers, and J. S. Theon (1977), A satellite technique for quantitatively mapping rainfall rates over the oceans, *J. Appl. Meteorol.*, *16*, 551–560.

Wilheit, T. T., A. T. C. Chang, and L. S. Chiu (1991), Retrieval of monthly rainfall indices from microwave radiometric measurements using probability distribution functions, *J. Atmos. Oceanic Technol.*, *8*, 118–136.

Wilheit, T. T., et al. (1994), Algorithms for the retrieval of rainfall from passive microwave measurements, *Remote Sens. Rev.*, *11*, 163–194.

Wilheit, T. T., C. D. Kummerow, and R. R. Ferraro (2003), Rainfall algorithms for AMSR-E, *IEEE Trans. Geosci. Remote Sens.*, *41*, 204–214.

Winker, D. M., W. H. Hunt, and M. J. McGill (2007), Initial performance assessment of CALIOP, *Geophys. Res. Lett.*, *34*, L19803, doi:10.1029/2007GL030135.

Wolff, D. B., and B. L. Fisher (2009), Assessing the relative performance of microwave-based satellite rain-rate retrievals using TRMM ground validation data, *J. Appl. Meteorol. Climatol.*, *48*, 1069–1099.

Wolff, D. B., D. A. Marks, E. Amitai, D. S. Silberstein, B. L. Fisher, A. Tokay, J. Wang, and J. L. Pippitt (2005), Ground validation for the Tropical Rainfall Measuring Mission (TRMM), *J. Atmos. Oceanic Technol.*, *22*, 365–380.

World Meteorological Organization (2003), The role of satellites in WMO programmes in the 2010s, *WMO-TD1177*, WMO Space Programme, Geneva, Switzerland.

World Meteorological Organization (2005), World Weather Watch—Twenty-Second Status Report on Implementation, 60 pp., Secr. of the World Meteorol. Organ., Geneva, Switzerland.

Xie, P., and P. A. Arkin (1997), Global precipitation: A 17-year monthly analysis based on gauge observations, satellite estimates, and numerical model outputs, *Bull. Am. Meteorol. Soc.*, *78*, 2539–2578.

Xie, P., and P. A. Arkin (1998), Global monthly precipitation estimates from satellite-observed outgoing longwave radiation, *J. Clim.*, *11*, 137–164.

Yang, S., and E. A. Smith (2006), Mechanisms for diurnal variability of global tropical rainfall observed from TRMM, *J. Clim.*, *19*, 5190–5226.

Yang, S., and E. A. Smith (2008), Convective–stratiform precipitation variability at seasonal scale from 8 yr of TRMM observations: Implications for multiple modes of diurnal variability, *J. Clim.*, *21*, 4087–4114.

Yang, S., K.-S. Kuo, and E. A. Smith (2008), Persistent nature of secondary diurnal modes of precipitation over oceanic and continental regimes, *J. Clim.*, *21*, 4115–4131.

Yin, X., A. Gruber, and P. Arkin (2004), Comparison of the GPCP and CMAP merged gauge–satellite monthly precipitation products for the period 1979–2001, *J. Hydrometeorol.*, *5*, 1207–1222.

Yoshiro, K. (2002), Meteorological mission of MTSAT, *IEIC Tech. Rep.*, *102*, 55–58.

Zhao, L., and F. Weng (2002), Retrieval of ice cloud parameters using the Advanced Microwave Sounding Unit, *J. Appl. Meteorol.*, *41*, 384–395.

Zhou, T., R. Yu, H. Chen, A. Dai, and Y. Pan (2008), Summer precipitation frequency, intensity, and diurnal cycle over China: A comparison of satellite data with rain gauge observations, *J. Clim.*, *21*, 3997–4010.

Zipser, E. J., D. J. Cecil, C. Liu, S. W. Nesbitt, and D. P. Yorty (2006), Where are the most intense thunderstorms on Earth?, *Bull. Am. Meteorol. Soc.*, *87*, 1057–1071.

C. Kidd, School of Geography, Earth and Environmental Sciences, University of Birmingham, Birmingham B15 2TT, U.K. (C.Kidd@bham.ac.uk)

S. Laviola and V. Levizzani, Istituto di Scienze dell'Atmosfera e del Clima, ISAC-CNR, Bologna I-40129, Italy.

Intensity-Duration-Frequency Curves

S. Rocky Durrans

Department of Civil, Construction, and Environmental Engineering, University of Alabama, Tuscaloosa, Alabama, USA

Intensity-duration-frequency (IDF) curves, or tables, are perhaps the most commonly used method for presentation of the characteristics of extreme rainfall events. This chapter provides an historical overview of developments in modeling and representation of IDF curves. Direction is provided to sources of information for IDF curve estimation, and alternative methods of modeling and representation of IDF relationships are discussed. Finally, this chapter provides an overview of contemporary IDF studies that are currently underway, as well as brief discussions of some emerging technologies that are leading to improvements in IDF curve estimation.

1. INTRODUCTION

Intensity-duration-frequency (IDF) curves, or tables, are perhaps the most commonly used method for presentation of the characteristics of extreme rainfall events. (In the following the terminology IDF curves will be used repeatedly, though the text should be read to infer that this includes IDF tables or other methods of representation.) IDF curves, examples of which are illustrated in Figure 1, are usually constructed for a geographical site of interest with the storm duration (in units of time) on the horizontal axis and with the intensity (depth of precipitation per unit time) on the vertical axis. A family of curves is provided to depict the dependence of intensity on not only the duration of interest but also the frequency (return period or exceedance probability) at which that intensity has been observed at the site. The rainfall intensity indicated by a curve is the average intensity over the corresponding duration and does not reflect in any way the variations of intensity that may occur during that duration in an actual rainfall event. In any case, if any two of the intensity, duration, and frequency variables are known or given, the IDF curves may be applied to estimate the value of the third variable.

Rainfall: State of the Science
Geophysical Monograph Series 191

10.1029/2009GM000919

Intensity-duration-frequency curves are constructed to represent point rainfall characteristics and are strictly applicable only to the gauging site for which they have been assembled. They do not represent spatially averaged rainfall characteristics over sizable areas, nor are they applicable to geographical sites remote from the one for which they are constructed. The degree to which a set of IDF curves accurately represents spatial rainfall characteristics, or the rainfall characteristics at nearby sites, depends on local conditions.

This chapter provides an historical overview of the development and use of IDF curves. The presentation then turns to the pragmatic issues of how to construct a set of IDF curves for a particular locale using official publications and how they can be represented mathematically for use in computer software. Some methods and issues involved in development of IDF curves by subjecting raw rainfall data series to frequency analyses are also presented, but the reader is referred to *El Adlouni and Ouarda* [this volume] for more detailed information on this topic. This chapter closes with a brief description of some emerging methods and technologies for modeling and representing rainfall data.

2. HISTORICAL DEVELOPMENT

With the early efforts made in urban storm water management in the late 1800s and early 1900s, corresponding efforts began to be made in rainfall data collection and analysis. The primary motivation of these efforts was to study

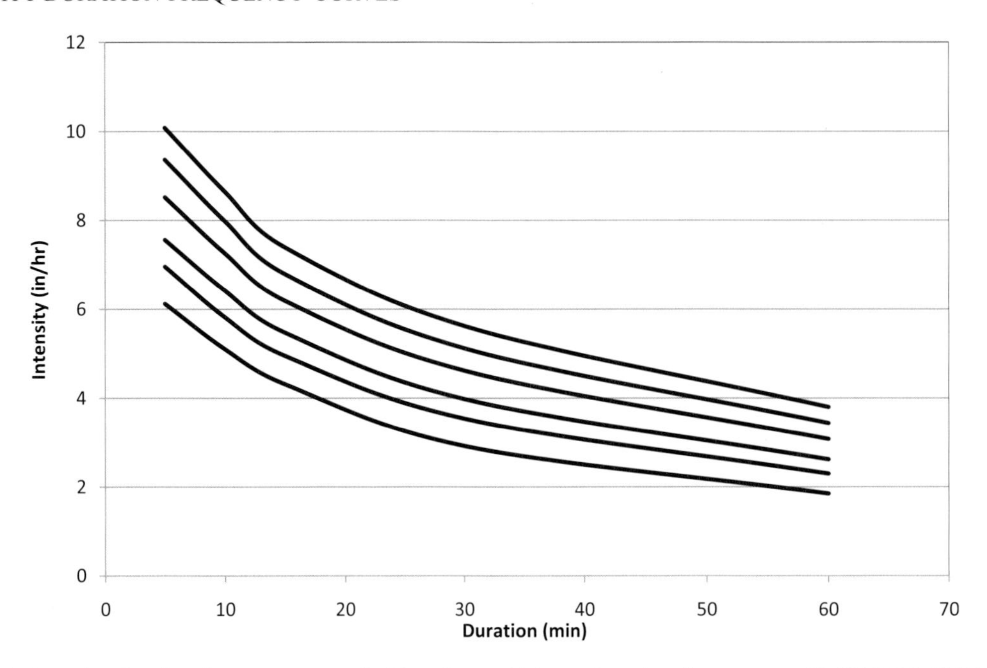

Figure 1. Intensity-duration-frequency curves for Tuscaloosa, Alabama, based on data presented in HYDRO 35 [*Frederick et al.*, 1977b].

the relationship between rainfall intensity and duration. *Burian et al.* [1999] describe how A. N. Talbot, in 1899, plotted storm intensities for various durations on cross-section paper. Talbot then sketched two curves, one depicting very rare rainfalls and the other representing more ordinary rainfalls. While these were not IDF curves as we know them today, Talbot's results were clearly a forerunner to present-day IDF curves.

2.1. Official National Weather Service Studies

In the United States, the earliest extensive study of rainfall extremes was undertaken by *Yarnell* [1935]. He constructed generalized rainfall maps for the coterminous United States for durations from 5 min to 24 h and for return periods from 2 to 100 years. His maps were based on data from about 200 first-order Weather Bureau stations and were considered the basic source of information on rainfall extremes until the 1950s.

In the early 1950s, the Weather Bureau published Technical Paper 24 (TP 24) in two parts [*Weather Bureau*, 1953, 1954] using data from a network of U.S. Army Corps of Engineers gauges as well as Weather Bureau gauges. TP 24 was significant in its demonstration of the importance of the large amounts of data required to define short-duration rainfall characteristics in the mountainous regions of the western states.

Technical Paper 25 (TP 25) was published in 1955 [*Weather Bureau*, 1955]. That document contained IDF curves for durations from 5 min to 24 h and for return periods from 2 to 100 years for each of 203 locations in the United States, Alaska, Hawaii, and Puerto Rico. Those curves were developed by fitting the Gumbel (extreme value type 1) distribution to annual maxima, with no smoothing or regionalization of the relationships.

To support the needs of the Soil Conservation Service, Technical Paper 28 (TP 28) was published in 1956 [*Weather Bureau*, 1956]. That document presented rainfall characteristics for local drainage designs in the western United States (locations west of the 105th meridian) for durations from 20 min to 24 h and for return periods from 1 to 100 years.

Technical Paper 29 (TP 29) [*Weather Bureau*, 1957, 1958a, 1958b, 1959, 1960] was published in five parts to provide coverage of various portions of the United States east of the 90th meridian. This series of publications provided limited information on seasonal variations in rainfall characteristics and, for the first time, included areal reduction curves to permit point rainfall values represented by IDF curves to be converted to spatially averaged values. Publication of TP 29 appears to be the first time that seasonality in rainfall characteristics was explicitly considered.

In May 1961, the Weather Bureau published Technical Paper 40 (TP 40) [*Hershfield*, 1961]. That document was "intended as a convenient summary of empirical relationships, working guides, and maps, useful in practical problems requiring rainfall frequency data" and covered the entire coterminous

United States. TP 40 presents isohyetal maps and seasonal variation diagrams for rainfall durations from 30 min to 24 h and for return periods from 1 to 100 years. Since its publication, TP 40 has been considered a standard source of rainfall information for use by practicing engineers and hydrologists. An extension of TP 40 for durations from 2 to 10 days and for return periods from 2 to 100 years was published as Technical Paper 49 (TP 49) [*Miller*, 1964]. Additional extensions for short and long durations for Alaska were published as Technical Papers 47 and 52 [*Miller*, 1963, 1965].

By the middle to late 1970s, the growing awareness of environmental issues had increased the demand for hydrologic planning and design for small drainage areas having very short times of concentration. It was also recognized that for storm durations of less than 1 h, ratios of subhourly to hourly rainfall values that had been published in TP 40 were in need of revision as they had a discernible geographic pattern. These issues led to the publication of the NOAA Technical Memorandum NWS HYDRO-35 in June 1977 [*Frederick et al.*, 1977b]. HYDRO-35, as it is commonly known, presents information on hourly and subhourly rainfall extremes for the eastern and central United States.

Because of the orographic effects caused by the high mountain ranges in the western United States, spatial variations in rainfall in that region are more complex than in other parts of the nation. Recognizing that TP 40 did not adequately address this issue, the National Weather Service (NWS) published the NOAA Atlas 2 [*Miller et al.*, 1973] in 11 volumes, with each volume being applicable to one of the western states. Extensions of that work for short durations (less than 1 h) were published for California [*Frederick and Miller*, 1979] and for the remaining western states [*Arkell and Richards*, 1986].

The National Weather Service is currently conducting rainfall frequency studies on a regional basis to update TP 40, HYDRO-35, and the NOAA Atlas 2 and is publishing the results of the new studies as the NOAA Atlas 14. The first four volumes of the NOAA Atlas 14 are complete as of this writing, and they cover the semiarid southwestern United States (including part of southern California) [*Bonnin et al.*, 2004a], the Ohio River region [*Bonnin et al.*, 2004b], Puerto Rico and the U.S. Virgin Islands [*Bonnin et al.*, 2006], and the Hawaiian Islands [*Perica et al.*, 2009]. Additional studies currently underway cover the balance of California, selected Pacific Islands, the southeastern and midwestern states, and Alaska.

All currently applicable studies that have been completed by the National Weather Service (or its predecessor, the Weather Bureau) can be obtained online using the Web site for the NWS' Hydrometeorological Design Studies Center (HDSC), where the studies are organized by state (or U.S. territory) and duration.

2.2. *Other Studies*

Around 1990, growing concerns over the ages of TP 40, HYDRO-35, and the NOAA Atlas 2 provided motivation for a number of states and regions of the United States to fund studies to update those documents. In the roughly 20 years that had passed since those publications were developed, the lengths of historical rainfall records had increased significantly, and statistical methods for treatment of the data had also advanced.

Schaefer [1990] completed a study to support dam safety analyses in the State of Washington. A nearly identical study was completed in Montana [*Parrett*, 1997]. Other rainfall frequency studies were completed in the Midwest [*Huff and Angel*, 1992], northeast [*Wilks and Cember*, 1993; *McKay and Wilks*, 1995], Florida [*Wanielista et al.*, 1996a, 1996b], Louisiana [*Naghavi and Yu*, 1995], Texas [*Asquith*, 1998], Oklahoma [*Tortorelli et al.*, 1999], and Alabama [*Durrans and Brown*, 2001; *Durrans and Kirby*, 2004]. Unfortunately, this fragmentation of efforts resulted in different methods of raw data treatment, data analysis, and presentations of results from one study to another. Indeed, the most consistent thing that can be said about these state and regional studies is that they are inconsistent with one another.

Despite these consistency issues, these newer studies did yield some innovations in the ways in which frequency studies were presented. The Texas [*Asquith*, 1998] study, for example, presented contour maps of the parameters of the probability distributions of extreme rainfall rather than contours of rainfall amounts. These parameters, along with the equations of the probability distributions used, could then be used to estimate rainfall depths for any desired return period. One of the products of the Oklahoma study [*Tortorelli et al.*, 1999] was a database for use with geographic information systems in which rainfall depths for selected durations and return periods were reported on a regular grid. Finally, in the case of the Alabama study [*Durrans and Brown*, 2001; *Durrans and Kirby*, 2004], a point-and-click Internet-based server was developed. Users could simply point at a geographical location of interest on a computer-displayed map, and IDF curves for that location would be generated on demand. This tool eventually became the model for the Precipitation Frequency Data Server (PFDS) currently being deployed by the National Weather Service (see section 7).

3. SPATIAL PRECIPITATION FREQUENCY

As pointed out in the introduction, IDF curves are developed using rain gauge data for a particular location and hence are point representations of rainfall characteristics. However, real drainage basins are not points and may cover considerable areas. For this reason, methods are needed whereby point rainfall intensities, for a particular duration and return period, can be converted to spatially averaged values for the same duration and return period.

The problem of point-to-area rainfall conversion historically has been addressed using areal reduction curves, of which two types are generally recognized. The two types are commonly known as storm-centered and geographically fixed relationships and perhaps are best distinguished from one another by descriptions in Technical Paper 29, Part 2 [*Weather Bureau*, 1958a], Technical Paper 49 [*Miller*, 1964], and NOAA Atlas 2 [*Miller et al.*, 1973]. A third approach is known as an annual maxima–centered approach, and this has been presented by *Asquith and Famiglietti* [2000]. However, this latter approach does not provide information on the return period associated with the spatially averaged rainfall amount [*Durrans et al.*, 2002], and hence its practical utility is limited.

Storm-centered relationships represent profiles of discrete storms and are generally used in probable maximum flood studies where there is a need to construct and/or transpose rainfall patterns; they are not discussed further here. On the other hand, geographically fixed areal reduction relationships are usually estimated using dense networks of gauges. However, the annual maxima at the individual gauges seldom, if ever, occur during the same storm events, even when they are closely spaced. The net effect is that geographically fixed curves represent aggregated or composite, as opposed to individual, storm behaviors. Geographically fixed areal reduction relationships are generally used with information from precipitation frequency studies.

Figure 2 is a representation of the set of areal reduction curves originally published in Part 1 of Technical Paper 29 [*Weather Bureau*, 1957]. Those curves present, for a fixed or given rainfall duration and averaging area, the ratio of the spatially averaged rainfall depth to the average point rainfall depth over the same area. The return period of the rainfall is not explicitly shown, as it was determined that the curves were essentially independent of the return period. The curves were also found to be independent of geographic location and time of year.

Several subsequent studies by the Weather Bureau and its successor, the National Weather Service, revisited the problem of estimation of areal reduction curves. However, no firm evidence has been found indicating that the curves originally published in 1957 (Figure 2) are in need of revision. *Myers and Zehr* [1980], in NOAA Technical Report NWS 24, did find some evidence that areal reduction ratios should decrease with increases in the return period, but the large uncertainties associated with the curves have precluded their revision.

A number of studies other than those completed by the NWS have considered areal reduction curve development. These include the studies of *Roche* [1963], *Rodriguez-Iturbe and Mejia* [1974], *Osborn et al.* [1980], *Omolayo* [1993], *Srikathan* [1995], *Siriwardena and Weinmann* [1996], *Sivapalan and Blöschl* [1998], and *DeMichele et al.* [2001].

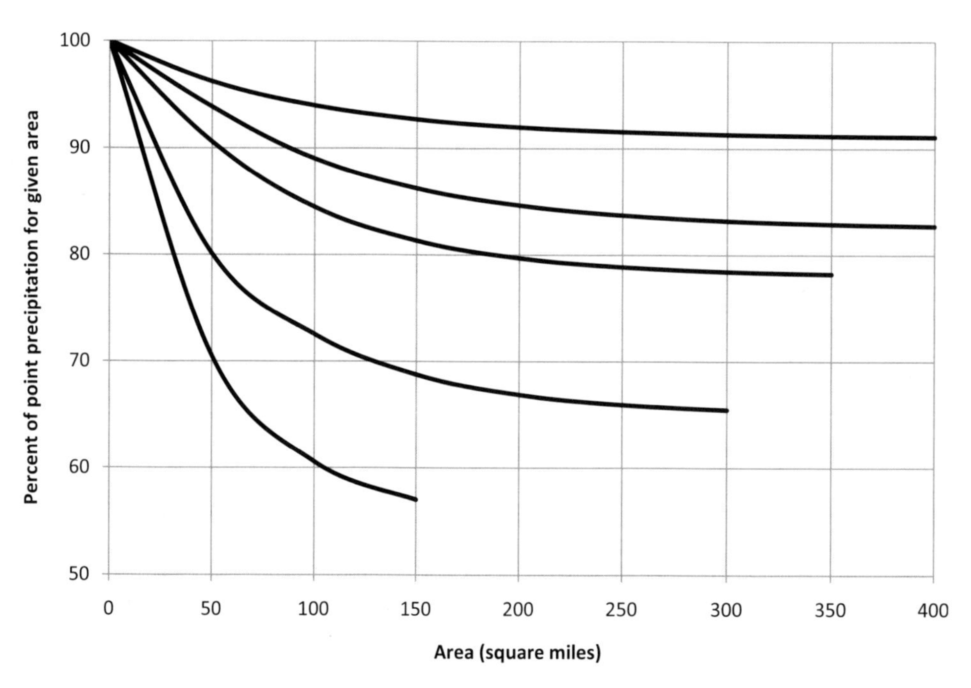

Figure 2. Areal reduction curves.

Frederick et al. [1977a] employed limited radar rainfall data, as opposed to gauge data, for estimation. *Durrans et al.* [2002], using a larger database, did the same but concluded that radar rainfall data remain problematical for studies of this nature. *Bacchi and Ranzi* [1996] and *Olivera et al.* [2007] also have used radar rainfall data.

This section is brought to a closure by noting that *Eagleson* [1972] suggested the following equation for representation of the areal reduction curves in Figure 2:

$$K = 1 - exp(-1.1t^{0.25}) + \exp(-1.1t^{0.25} - 0.01A). \quad (1)$$

Here K is the areal reduction factor (the vertical axis of Figure 2), t is the rainfall duration, in hours, and A is the averaging area, in square miles. A graph of this equation does not agree particularly well with the curves shown in Figure 2, especially for short durations. However, the large uncertainties associated with the curves may make use of the equation acceptable.

4. IDF CURVES FROM OFFICIAL NWS PUBLICATIONS

The currently applicable official NWS precipitation frequency publications, available from the HDSC Web site (see section 2.1), contain isohyetal maps from which rainfall depths can be obtained for selected durations and return periods. An exception to this is the new NOAA Atlas 14, where tables and graphs of precipitation depths or intensities, along with their confidence limits, are produced for various durations and return periods. On the basis of the information available from these sources, average rainfall intensities can be computed for the corresponding durations and then graphed as shown in Figure 1 to develop IDF curves. To aid the computations and graphing, most of the official publications contain charts or supplementary equations to enable interpolation of the mapped values for intermediate durations and/or return periods. The exact procedures for interpolation differ from one official publication to another, but the procedure outlined in HYDRO-35 will be presented here as an example.

HYDRO-35 contains six isohyetal maps, representing rainfall depths for durations of 5, 15, and 60 min, and for return periods of 2 and 100 years. For any chosen return period, rainfall depths corresponding to durations of 10 and 30 min may be estimated as

$$\text{10 min value} = 0.59(\text{15 min value}) + 0.41(\text{5 min value}) \quad (2a)$$

$$\text{30 min value} = 0.49(\text{60 min value}) + 0.51(\text{15 min value}). \quad (2b)$$

Finally, for any chosen duration, one may obtain rainfall depths for return periods between 2 and 100 years as

$$\text{5 year value} = 0.278(\text{100 year value}) + 0.674(\text{2 year value}), \quad (3a)$$

$$\text{10 year value} = 0.449(\text{100 year value}) + 0.496(\text{2 year value}), \quad (3b)$$

$$\text{25 year value} = 0.669(\text{100 year value}) + 0.293(\text{2 year value}), \quad (3c)$$

$$\text{50 year value} = 0.835(\text{100 year value}) + 0.146(\text{2 year value}). \quad (3d)$$

These equations are applicable to any geographical location within the spatial domain covered by HYDRO-35.

5. EQUATIONS FOR IDF CURVES

Modern computer software for hydrologic simulation and prediction often requires that IDF curves be represented using mathematical equations, though some such equations predate the computer era. Most of these equations are of the general form [*Chow et al.*, 1988]

$$i = aT^d/(t^c + b), \quad (4)$$

where i is the rainfall intensity, t is the duration, T is the return period, and a, b, c and d are coefficients that vary from one location to another. Alternative equations are [*Haestad Methods*, 2002]:

$$i = a/(t + b)^c, \quad (5)$$

$$i = aT^d/(t + b)^c, \quad (6)$$

$$i = a + b(ln\ t) + c(ln\ t)^2 + d(ln\ t)^3. \quad (7)$$

Equations (5) and (7) are not explicit functions of the return period, and hence their coefficients depend on both location and return period. An alternative to equations such as these for implementation in computer software is to tabulate intensities as a function of duration and return period and to use interpolating equations or splines to obtain values for nontabulated durations and/or return periods.

Numerical values of the coefficients in the above equations can be found in commonly used hydrology and water resources engineering texts for a number of cities in the United States. However, caution should be exercised in using those published values as significant variability in the coefficients can be observed over fairly short distances. Therefore, it is recommended that official NWS publications be used to estimate IDF curves for a locale of interest and that the coefficients for that locale then be estimated on the basis of the local IDF curves.

6. DEVELOPMENT OF IDF CURVES FROM RAW DATA

An alternative to estimation of IDF curves from official publications of the National Weather Service is to estimate them by subjecting raw rainfall data time series to frequency analyses. This is a laborious and time-consuming process, and it is advised that it should be undertaken only if there is reason to be concerned about the validity of a published value for a location of interest and if there is a suitably long and reliable record of rainfall available. This section is intentionally brief, and provides only an overview of the tasks involved in frequency analysis. More detailed information pertaining to frequency analysis is presented by *El Adlouni and Ouarda* [this volume], as well as by *Stedinger et al.* [1993] and *Hosking and Wallis* [1997].

6.1. Annual Series Versus Partial Duration Series

A rainfall frequency analysis for a chosen site and duration may be performed using an annual series or a partial duration series. An annual series consists of the largest observed rainfall depths in each year of a period of record. If the record is n years in length, the annual series consists of n data values, one from each year.

In assembly of an annual series, it is frequently observed that the second or even third largest event(s) in some years are larger than the maximum events in other years. Use of an annual series results in the neglect of these events. To overcome this, a partial duration series may be constructed by extraction of the largest events without regard for their occurrence times. The number of events included in the partial duration series may be equal to n, the number of years of record at a site, or it may be different from n. The only restriction placed on the individual events is that they be independent of one another.

An empirically observed relationship between the return periods of events based on annual and partial duration series is illustrated in Table 1. The differences between the two are negligible for return periods greater than about 10 years. However, where they are different (for return periods of about 10 years or less), the use of a partial duration series results in a larger estimated rainfall depth (or intensity) for a given duration and return period than does use of an annual maximum series. Additional information on annual versus partial duration series, and some guidance for selection of one type versus the other, is provided by *Kite* [1977], *Rosbjerg* [1985], *Rasmussen and Rosbjerg* [1989], and *Stedinger et al.* [1993].

Table 1. Empirically Observed Relationship Between Partial Duration and Annual Series With Respect to Return Period[a]

Partial Duration Series (years)	Annual Series (years)
0.5	1.2
1.0	1.6
1.5	2.0
2.0	2.5
5.0	5.5
10.0	10.5

[a]From *Kite* [1977], in turn, from *Dalrymple* [1960].

6.2. Data Adjustments and Corrections

Time series of raw rainfall data typically contain many errors and omissions, the presence of which can adversely affect estimates of depths for chosen durations and return periods. For example, in cases where many of the data at a site are recorded as missing, the possibility exists that an annual or partial duration series maximum during any period of time was among the missing values, particularly since gauge failures can often be associated with heavy storms and/or lightning strikes. In essence, the likelihood of not recording a true annual or partial duration series maximum increases with the number of missing values at a site. R. Durrans, in unpublished work in 1998, found that neglect of this effect could bias estimated rainfall depths by about 15%; however, much more research is needed to define objective criteria for inclusion/exclusion of data when many of them are missing.

Other adjustments and corrections to data may be necessary when gauging sites have been moved during the period of record or when gauges exist in close proximity to one another. In the case of a gauge being moved, one should perform statistical tests to ascertain whether the periods of record before and after the move are homogeneous. Similarly, one should test records for time trends that might be caused, for instance, by growth of a tree or other obstacle near the gauge. Where gauges exist in close proximity to one another, the records might be combined into a single, and possibly longer, record if it can be shown that they are homogeneous with one another.

A data correction that is often quite significant is caused by the fixed times at which rainfall depths are recorded. Consider, for example, the problem of determination of the maximum 24 h rainfall depth during a year using data recorded at daily time intervals and assume that the daily values are reported for a fixed time interval of midnight to midnight. If the true maximum 24 h rainfall amount in a year actually occurred from 5:00 P.M. on one day to 5:00 P.M. on the next, then that true 24 h amount would have been recorded as two daily amounts, both of which would have been lower than the true 24 h amount. This bias that is caused by fixed recording times has the effect of depressing (reducing) actual maxima and has been recognized for many years. The National Weather Service, in its official publications, has typically accounted for the bias using empirically estimated correction factors. *Weiss* [1964], in a theoretical study of this problem, developed correction factors that agree quite closely with empirically determined ones.

6.3. *Frequency Analysis*

Raw data time series of annual or partial duration maxima that have been adjusted and/or corrected as briefly described above may be subjected to frequency analyses to estimate depths of rainfall for chosen durations and return periods at gauging sites. If desired, confidence limits may also be placed on the estimated depths at this stage of the analysis. Accomplishment of a frequency analysis, as detailed more fully by *El Adlouni and Ouarda* [this volume], involves selection of a probability distribution for modeling of the data and estimation of the parameters of the chosen distribution in a suitable way. Methods of regionalization [see, e.g., *Hosking and Wallis*, 1997] typically are also applied to temper the variability that is typically observed in the results from one gauge to another.

When estimating the parameters of chosen distributions for various durations at a given gauging site, it sometimes occurs that the resulting distributions are inconsistent in the sense that they do not always yield increasing rainfall depths with increasing duration (or decreasing rainfall intensities with increasing duration). This problem occurs because the data for different durations are modeled independently of one another. Mitigation of this problem generally requires implementation of ad hoc adjustments or some sort of smoothing algorithm. As an alternative, and as discussed later in section 8.2, methods of mathematical modeling involving scaling also can be applied to combat this problem.

6.4. *Spatial Smoothing*

Results of frequency analyses of gauge data are uncertain and represent point estimates applicable only to the location of the gauge itself. In practice, hydrologists usually need rainfall depth or intensity estimates at ungauged locations. The isohyetal contour maps presented in most rainfall publications enable this estimation to be accomplished, but construction of those maps requires development and implementation of some sort of "surface" modeling to enable prediction of the rainfall depth variable as a function of two spatial variables (latitude and longitude).

There are many spatial modeling methods than have been and could be applied for this purpose, and different rainfall publications, including official ones of the NWS, have employed different methods. In any case, it is recommended here that any approach employed should be a spatial smoothing method rather than a spatial interpolation method because of the inherent uncertainties in the point estimates at individual gauges (see *Chapra and Canale* [2010] for a description of the difference between smoothing and interpolation). For example, the classical reciprocal-distance-squared method, an interpolator used widely in some types of rainfall analyses, is not a recommended choice for this task.

7. PRECIPITATION FREQUENCY DATA SERVER

Historically, publications dealing with estimates of extreme rainfall have been published in hard copy (i.e., paper) form. A more modern approach is to make use of the Internet. This has the benefits of making the results available to a broad audience, and it minimizes problems and issues associated with report publication, distribution, and updating.

The National Weather Service, along with their current efforts in completion of the new NOAA Atlas 14, is now making information on extreme rainfall available via their Precipitation Frequency Data Server (PFDS). The PFDS can be accessed via the Web site for NWS' Hydrometeorological Design Studies Center. The PFDS is not yet operational for all U.S. states and territories, but it will be in the future as additional studies are completed.

Users of the PFDS can interactively choose a state where information is needed and can then choose a location in that state (either by clicking on a map, by entering latitude/longitude coordinates, or by selection from a list of gauging sites). The user also chooses whether depth-duration-frequency or intensity-duration-frequency curves are desired, whether U.S. customary or SI units are desired, and whether results are desired for annual or partial duration series maxima. With these choices, the PFDS generates a depth-duration-frequency or intensity-duration-frequency table and graph, including confidence limits, for the chosen location. Supplementary location maps showing the location of the chosen site are also generated.

8. EMERGING ISSUES AND TECHNOLOGIES

There are several areas of active research where efforts are being made to improve methods for characterization of extreme precipitation. A few of these are described below, but the presentation here is intentionally very brief as these areas of research are more closely associated with the problem of rainfall frequency analysis (the topic of *El Adlouni and Ouarda* [this volume]) than with the problem of IDF curve estimation. The volume of literature on these topics is also growing at a rapid pace, making it nearly prohibitive to provide a complete accounting of the current state of the art.

8.1. Seasonality

The types and characteristics of rainfall in a region often exhibit distinct differences from one season to another. In Alabama, for example, convective precipitation is common during the hot summer months, tropical storms are frequent during the late summer and fall months, and frontal events are common during the winter months. In Alabama, the most intense storms tend to occur in the late summer months, while the largest storms in terms of depth of precipitation tend to occur in the winter months. The summer storms, because they are often convective in nature, tend to have short durations (often only a couple of hours), whereas the winter storms tend to have much longer durations (often a day or more). These differences in storm characteristics can have significance for hydrologic design, as the designs of some types of structures (such as sewers and culverts) are more sensitive to rainfall intensities than they are to depths, whereas designs of other types of facilities (such as detention basins) are more sensitive to total rainfall depths than to intensities.

Given these seasonal differences in storm characteristics, one would expect that information on rainfall extremes should be provided on a seasonal basis. Unfortunately, however, seasonal analyses have been difficult to achieve, with the result that only limited information on seasonality is generally available. Among official NWS studies, seasonal variation diagrams have been presented in TP 29 [*Weather Bureau*, 1957, 1958a, 1958b, 1959, 1960] and TP 40 [*Hershfield*, 1961]. However, those diagrams present only the likelihood of observing a *T*-year event in a given month (or season). No information is provided on an IDF relationship for a particular month or season.

Estimation of probability distributions of rainfall depths for each season of a year, as well as for the annual maximum for the year, is a problem that seems to need additional attention in the literature. However, estimation of these distributions should be accomplished jointly (i.e., simultaneously) because of inconsistencies that usually arise in their estimation. These consistency problems have been illustrated and have been solved only approximately for joint seasonal/annual flood frequency analyses by *Durrans et al.* [2003]. Additional recent studies pertaining to seasonality, but again for flood applications, have been completed by *McCuen and Beighley* [2003], *Cunderlik et al.* [2004], *Ouarda et al.* [2006], *Fang et al.* [2007], *Strupczewski and Kochanek* [2008], and *Strupczewski et al.* [2009]. The rainfall estimation problem is more difficult than the flood problem because of the need to address not only seasonal and annual consistencies but also because of the need to address multiple rainfall durations. Nevertheless, the flood studies cited above should provide valuable guidance to those seeking to perform similar rainfall frequency studies.

8.2. Scaling

Fractals, and the associated notions of simple and multi-scaling, have received a huge amount of attention in the hydrological literature in the last couple of decades. In fact, scaling ideas seem to be almost custom-tailored for one of the problems encountered in rainfall frequency analyses, namely, the problem of inconsistencies caused by separate modeling of rainfall data series for different durations (see *Xu and Tung* [2009] and section 6.3 of this chapter). There have been many applications of scaling theories to rainfall analysis, including recent studies by *Burlando and Rosso* [1996], *Veneziano and Furcolo* [2002], *Borga et al.* [2005], *Bougadis and Adamowski* [2006], and *Langousis and Veneziano* [2007], but it is often found that scaling principles may not apply over the full range of storm durations of interest in a particular application. This could be viewed as no more than a nuisance in practice, but this field appears to be a ripe one for further insights and contributions.

8.3. Multivariate Analyses

Probability distributions of rainfall depths (or intensities) for a given duration are conditional univariate distributions (conditional on the duration). IDF curves represent "contours" of those conditional distributions for selected exceedance probabilities. With the recognition that the duration of a storm, like its average intensity, is a random variable, efforts have been made to model the joint distribution of both depth and duration.

Historically, modeling of bivariate (and multivariate) distributions has been confounded by the usual requirement that all of the univariate marginal distributions be from the same parametric family (normal, lognormal, exponential, etc.). Data transformations (usually to normality) have been used to approximately satisfy this need, but they invariably introduce biases into the analysis. Further, even in the seemingly rare cases when the marginals are nonnormal but from the same parametric family, it has been seen that distributional models may not be capable of representing the correlation structure between the variables. For example, the bivariate Gumbel model is capable of representing correlation coefficients only in the range $0 < \rho < 2/3$ [*Oliveria*, 1975; *Yue*, 2000].

A significant development in multivariate modeling occurred in 1959 when A. Sklar introduced the notion of a copula and proved a theorem that now bears his name. Use of a copula eliminates the requirement for the marginals to be of the same family and permits one to model the dependence structure among the random variables separately from the marginal distributions. This flexibility is of great practical utility and has garnered interest from investigators across a broad spectrum of professional disciplines.

In hydrology, copulas have been applied only in about the last 10 years, but interest in them is growing rapidly as evidenced by the online bibliography published by the Statistics in Hydrology Working Group of the International Association of Hydrological Science. The *Journal of Hydrologic Engineering* recently devoted an entire special issue to the subject of copulas [*Singh and Strupczewski*, 2007]. There is obviously much more to be learned about copulas and their use in hydrologic applications. The reader is encouraged to become somewhat familiar with them as they appear to have the power to provide solutions to what previously have been some very thorny problems in multivariate analysis.

REFERENCES

Arkell, R. E., and F. Richards (1986), Short duration rainfall relations for the western United States, *Proceedings of the Conference on Climate and Water Management, Asheville, North Carolina, August 4–7*, pp. 136–141, Am. Meteorol. Soc., Boston, Mass.

Asquith, W. H. (1998), Depth-duration frequency of precipitation for Texas, , *U.S. Geol. Surv. Water Resour. Invest. Rep., 98-4044*.

Asquith, W. H., and J. S. Famiglietti (2000), Precipitation areal-reduction factor estimation using an annual-maxima centered approach, *J. Hydrol., 230*, 55–69.

Bacchi, B., and R. Ranzi (1996), On the derivation of the areal reduction factor of storms, *Atmos. Res., 42*(1–4), 123–135.

Bonnin, G. M., D. Martin, B. Lin, T. Parzybok, M. Yekta, and D. Riley (2004a), *Precipitation-Frequency Atlas of the United States*, vol. *1*, *NOAA Atlas*, vol. *14*, NOAA, Silver Spring, Md.

Bonnin, G. M., D. Martin, B. Lin, T. Parzybok, M. Yekta, and D. Riley (2004b), *Precipitation-Frequency Atlas of the United States*, vol. *2*, *NOAA Atlas*, vol. *14*, NOAA, Silver Spring, Md.

Bonnin, G. M., D. Martin, B. Lin, T. Parzybok, M. Yekta, and D. Riley (2006), *Precipitation-Frequency Atlas of the United States*, vol. *3*, *NOAA Atlas*, vol. *14*, NOAA, Silver Spring, Md.

Borga, M., C. Vezzani, and G. D. Fontana (2005), Regional rainfall depth-duration-frequency equations for an alpine region, *Nat. Hazards*, *36*, 221–235.

Bougadis, J., and K. Adamowski (2006), Scaling model of a rainfall intensity-duration-frequency relationship, *Hydrol. Processes*, *20* (17), 3747–3757.

Burian, S. J., S. J. Nix, S. R. Durrans, R. E. Pitt, C.-Y. Fan, and R. Field (1999), Historical development of wet-weather flow management, *J. Water Resour. Plann. Manage.*, *125*(1), 3–13.

Burlando, P., and R. Rosso (1996), Scaling and multiscaling models of depth-duration-frequency curves for storm precipitation, *J. Hydrol.*, *187*, 45–64.

Chapra, S. C., and R. P. Canale (2010), *Numerical Methods for Engineers*, 6th ed. McGraw-Hill, New York.

Chow, V. T., D. R. Maidment, and L. W. Mays (1988), *Applied Hydrology,* McGraw-Hill, New York.

Cunderlik, J. M., T. B. M. J. Ouarda, and B. Bobée (2004), Determination of flood seasonality from hydrological records, *Hydrol. Sci. J.*, *49*(3), 511–526.

Dalrymple, T. (1960), Flood frequency analysis, *U.S. Geol. Surv. Water Supply Pap.*, *1543-A*, 47 pp.

De Michele, C., N. T. Kottegoda, and R. Rosso (2001), The derivation of areal reduction factor of extreme storm rainfall from its scaling properties, *Water Resour. Res.*, *37*(12), 3247–3252.

Durrans, S. R., and P. A. Brown (2001), Estimation and internet-based dissemination of extreme rainfall information, *Transp. Res. Rec.*, *1743*, 41–48.

Durrans, S. R., and J. T. Kirby (2004), Regionalization of extreme precipitation estimates for the Alabama rainfall atlas, *J. Hydrol.*, *295*, 101–107.

Durrans, S. R., L. T. Julian, and M. Yekta (2002), Estimation of depth-area relationships using radar-rainfall data, *J. Hydrol. Eng.*, *7*(5), 356–367.

Durrans, S. R., M. A. Eiffe, W. O. Thomas, Jr., and H. M. Goranflo (2003), Joint seasonal/annual flood frequency analysis, *J. Hydrol. Eng.*, *8*(4), 181–189.

Eagleson, P. S. (1972), Dynamics of flood frequency, *Water Resour. Res.*, *8*(4), 878–898.

El Adlouni, S., and T. B. M. J. Ouarda (2010), Frequency analysis of extreme rainfall events, in *Rainfall: State of the Science*, *Geophys. Monogr. Ser.*, doi: 10.1029/2010GM000976, this volume.

Fang, B., S. Guo, S. Wang, P. Liu, and Y. Xiao (2007), Non-identical models for seasonal flood frequency analysis, *Hydrol. Sci. J.*, *52* (5), 974–991.

Frederick, R. H., and J. F. Miller (1979), Short duration rainfall frequency relations for California, *Proceedings of the Third Conference on Hydrometeorology, Bogota, Colombia, August 20–24*, pp. 67–73, Am. Meteorol. Soc., Boston, Mass.

Frederick, R. H., V. A. Myers, and E. P. Auciello (1977a), Storm depth-area relations from digitized radar returns, *Water Resour. Res.*, *13*(3), 675–679.

Frederick, R. H., V. A. Myers, and E. P. Auciello (1977b), Five- to 60-minute precipitation frequency for the eastern and central United States, *NOAA Tech. Memo. NWS HYDRO-35*, Natl. Weather Serv., Silver Spring, Md.

Haestad Methods (2002), *Computer Applications in Hydraulic Engineering*, Haestad Press, Waterbury, Conn.

Hershfield, D. M. (1961), Rainfall frequency atlas of the United States for durations from 30 minutes to 24 hours and return periods from 1 to 100 years, *Tech. Pap. 40*, U.S. Weather Bur., Washington, D. C.

Hosking, J. R. M., and J. R. Wallis (1997), *Regional Frequency Analysis*, Cambridge Univ. Press, Cambridge, Mass.

Huff, F. A., and J. R. Angel (1992), Rainfall frequency atlas of the Midwest, *Bull. 71 (MCC Res. Rep. 92-03)*, Midwestern Clim. Cent., Champaign, Ill.

Kite, G. W. (1977), *Frequency and Risk Analyses in Hydrology*Water Resour. Publ., Littleton, Colo.

Langousis, A., and D. Veneziano (2007), Intensity-duration-frequency curves from scaling representations of rainfall, *Water Resour. Res.*, *43*, W02422, doi:10.1029/2006WR005245.

McCuen, R. H., and R. E. Beighley (2003), Seasonal flow frequency analysis, *J. Hydrol.*, *279*, 43–56.

McKay, M., and D. S. Wilks (1995), Atlas of short-duration precipitation extremes for the northeastern United States and southeastern Canada, *Publ. RR 95-1*, Northeast Reg. Clim. Cent., Ithaca, N. Y.

Miller, J. F. (1963), Probable maximum precipitation and rainfall-frequency data for Alaska, *Tech. Pap. 47*, U.S. Weather Bur., Washington, D. C.

Miller, J. F. (1964), Two- to ten-day precipitation for return periods of 2 to 100 years in the contiguous United States, *Tech. Pap. 49*, U.S. Weather Bur., Washington, D. C.

Miller, J. F. (1965), Two- to ten-day precipitation for return periods of 2 to 100 years in Alaska, *Tech. Pap. 52*, U.S. Weather Bur., Washington, D. C.

Miller, J. F., R. H. Frederick, and R. J. Tracey (1973), *Precipitation-Frequency Atlas of the Coterminous Western United States, NOAA Atlas 2*, 11 vols., Natl. Weather Serv., Silver Spring, Md.

Myers, V. A., and R. M. Zehr (1980), A methodology for point-to-area rainfall frequency ratios, *NOAA Tech. Rep. NWS 24*, Natl. Weather Serv., Silver Spring, Md.

Naghavi, B., and F. X. Yu (1995), Regional frequency analysis of extreme precipitation in Louisiana, *J. Hydraul. Eng.*, *121*(11), 819–827.

Olivera, F., J. Choi, D. Kim, and M.-H. Li (2007), Estimation of average rainfall areal reduction factors in Texas using NEXRAD data, *J. Hydrol. Eng.*, *13*(6), 438–448.

Oliveria, J. T. D. (1975), Bivariate extremes: Extensions, *Bull. Int. Stat. Inst.*, *46*(2), 241–251.

Omolayo, A. S. (1993), On the transposition of areal reduction factors for rainfall frequency estimation, *J. Hydrol.*, *145*, 191–205.

Osborn, H. B., L. J. Lane, and V. A. Myers (1980), Rainfall/watershed relationships for southwestern thunderstorms, *Trans. ASAE*, *23*, 82–87, 91.

Ouarda, T. B. M. J., J. M. Cunderlik, A. St-Hilaire, M. Barbet, P. Bruneau, and B. Bobée (2006), Data-based comparison of seasonality-based regional flood frequency methods, *J. Hydrol.*, *330*, 329–339.

Parrett, C. (1997), Regional analyses of annual precipitation maxima in Montana, *U.S. Geol. Surv. Water Resour. Invest. Rep.*, *97-4004*.

Perica, S., et al. (2009), *Precipitation-Frequency Atlas of the United States*, vol. *4*, *NOAA Atlas*, vol. *14*, NOAA, Silver Spring, Md.

Rasmussen, P. F., and D. Rosbjerg (1989), Risk estimation in partial duration series, *Water Resour. Res.*, *25*(11), 2319–2330.

Roche, M. (1963), *Hydrologie de Surface,* Gauthier-Villars, Paris.

Rodriguez-Iturbe, I., and J. M. Mejía (1974), On the transformation of point rainfall to areal rainfall, *Water Resour. Res.*, *10*(4), 729–735.

Rosbjerg, D. (1985), Estimation in partial duration series with independent and dependent peak values, *J. Hydrol.*, *76*, 183–196.

Schaefer, M. G. (1990), Regional analyses of precipitation annual maxima in Washington state, *Water Resour. Res.*, *26*(1), 119–131.

Singh, V. P., and W. G. Strupczewski (Eds) (2007), Copulas in hydrology, *J. Hydrol. Eng.*, *12*(4), 345–439.

Siriwardena, L., and P. E. Weinmann (1996), Development and testing of methodology to derive areal reduction factors for long duration rainfalls, *Working Doc. 96/4*, Coop. Res. Cent. for Catchment Hydrol., Monash Univ., Clayton, Victoria, Australia.

Sivapalan, M., and G. Blöschl (1998), Transformation of point rainfall to areal rainfall: Intensity-duration-frequency curves, *J. Hydrol.*, *204*, 150–167.

Srikathan, R. (1995), A review of the methods for estimating areal reduction factors for design rainfalls, *Rep. 95/3*, Coop. Res. Cent. for Catchment Hydrol., Monash Univ., Clayton, Victoria, Australia.

Stedinger, J. R., R. M. Vogel, and E. Foufoula-Georgiou (1993), Frequency analysis of extreme events, in *Handbook of Hydrology*, edited by D. R. Maidment, chap. 18, pp. 18.1–18.66, McGraw-Hill, New York.

Strupczewski, W. G., and K. Kochanek (2008), On the rationale of seasonal approach to flood frequency analysis, *Publ. Inst. Geophys. Pol. Acad. Sci. D*, *E-10*(406), 161–174.

Strupczewski, W. G., K. Kochanek, W. Feluch, E. Bogdanowicz, and V. P. Singh (2009), On seasonal approach to nonstationary flood frequency analysis, *Phys. Chem. Earth, Parts A/B/C*, *34*, 612–618.

Tortorelli, R. L., A. Rea, and W. H. Asquith (1999), Depth-duration frequency of precipitation for Oklahoma, *U.S. Geol. Surv. Water Resour. Invest. Rep.*, *99-4232*.

Veneziano, D., and P. Furcolo (2002), Multifractality of rainfall and scaling of intensity-duration-frequency curves, *Water Resour. Res.*, *38*(12), 1306, doi:10.1029/2001WR000372.

Wanielista, M., R. Eaglin, and L. Eaglin (1996a), Intensity-duration frequency curves for the State of Florida, report, Fla. Dep. of Transp., Tallahassee.

Wanielista, M., R. Eaglin, and L. Eaglin (1996b), Isopluvial contour curves for long duration storms in Florida, report, Fla. Dep. of Transp., Tallahassee.

Weather Bureau (1953), Rainfall intensities for local drainage design in the United States for durations of 5 to 240 minutes and 2-, 5- and 10-year return periods, Part I: West of the 115th meridian, *Tech. Pap. 24*, Washington, D. C.

Weather Bureau (1954), Rainfall intensities for local drainage design in the United States for durations of 5 to 240 minutes and 2-, 5- and 10-year return periods. Part II: Between 105°W and 115°W, *Tech. Pap. 24*, Washington, D. C.

Weather Bureau (1955), Rainfall intensity-duration-frequency curves for selected stations in the United States, Alaska, Hawaiian Islands, and Puerto Rico, *Tech. Pap. 25*, Washington, D. C.

Weather Bureau (1956), Rainfall intensities for local drainage design in western United States, for durations of 20 minutes to 24 hours and 1- to 100-year return periods, *Tech. Pap. 28*, Washington, D. C.

Weather Bureau (1957), Rainfall intensity-frequency regime. Part I: The Ohio Valley, *Tech. Pap. 29*, Washington, D. C.

Weather Bureau (1958a), Rainfall intensity-frequency regime. Part II: Southeastern United States, *Tech. Pap. 29*, Washington, D. C.

Weather Bureau (1958b), Rainfall intensity-frequency regime. Part III: The Middle Atlantic Region, *Tech. Pap. 29*, Washington, D. C.

Weather Bureau (1959), Rainfall intensity-frequency regime. Part IV: Northeastern United States, *Tech. Pap. 29*, Washington, D. C.

Weather Bureau (1960), Rainfall intensity-frequency regime. Part V: Great Lakes region, *Tech. Pap. 29*, Washington, D. C.

Weiss, L. L. (1964), Ratio of true to fixed-interval maximum rainfall, *J. Hydraul. Div. Am. Soc. Civ. Eng.*, *90*(HY1), 77–82.

Wilks, D. S., and R. P. Cember (1993), Atlas of precipitation extremes for the northeastern United States and southeastern Canada, *Publ. RR 93-5*, Northeast Reg. Clim. Cent., Ithaca, N. Y.

Xu, Y.-P., and Y.-K. Tung (2009), Constrained scaling approach for design rainfall estimation, *Stochastic Environ. Res. Risk Assess.*, *23*(6), 697–705.

Yarnell, D. L. (1935), Rainfall intensity-frequency data, *Misc. Publ. 204*, U.S. Dep. of Agric., Washington, D. C..

Yue, S. (2000), The Gumbel mixed model applied to storm frequency analysis, *Water Resour. Manage.*, *14*, 377–389.

S. R. Durrans, Department of Civil, Construction, and Environmental Engineering, University of Alabama, Box 870205, Tuscaloosa, AL 35487, USA. (rdurrans@eng.ua.edu)

Frequency Analysis of Extreme Rainfall Events

Salaheddine El Adlouni

Institut National de Statistique et d'Économie Appliquée, INSEA, Rabat, Morocco

Taha B. M. J. Ouarda

Institut National de la Recherche Scientifique Eau, Terre et Environnement, Québec City, Québec, Canada

The study of the rainfall probability distributions is important to estimate large events and their probability of occurrence. This paper presents the usefulness of extreme hydrological frequency analysis when the occurrences are independent in time to describe the likelihood of extreme events over the time horizon. Classical and recent developments are presented with illustrative examples.

1. INTRODUCTION

The occurrence of many extreme events in hydrology cannot be forecasted on the basis of deterministic information with sufficient skill and lead time. In such cases, a probabilistic approach is required to incorporate the effects of such phenomena into decisions. If the occurrences can be assumed to be independent in time, i.e., the timing and magnitude of an event bears no relation to preceding events, extreme hydrologic frequency analysis (HFA) can be used to describe the likelihood of any one or a combination of events over the time horizon of a decision. HFA is useful for a variety of engineering applications including hydraulic and municipal structure design (culverts, storm sewers) and landslide hazard evaluation [see, e.g., *Crosta*, 1998].

In probabilistic analysis, such as HFA, a series is a convenient sequence of data, hourly, daily, seasonal, or annual observations of hydrological variables. If the series contains all the observed values, it is called complete-duration series (CDS). When the record contains only events that exceed a fixed threshold, the series is called a partial-duration series (PDS). A series that contains only the event with largest magnitude that occurred in each year is called an annual-maximum series (AMS). The use of AMS is very common in probabilistic analysis because of the availability of the data and the theoretical basis for extrapolating beyond the range of the observations. A reason for the absence of statistical theory for the PDS is the lack of independence of consecutive events. More arguments in favor of either of these techniques are well described in the literature [*National Research Council of Canada*, 1989; *Stedinger et al.*, 1993]. Due to its simpler structure, the AMS-based method is more popular in practice and will be detailed in the present chapter in the case of stationary and nonstationary data sets.

The three main steps involved in the AMS approach are (1) selection of a sample in the form of a data series, which satisfies certain statistical criteria, (2) fitting of the best theoretical probability distribution to represent this sample using the best fitting technique available for this distribution, and (3) use of this fitted distribution to make statistical inferences about the underlying population.

2. DATA REQUIREMENTS FOR HFA OF RAINFALL DATA

2.1. Data Validation

As for any statistical analysis studies, both quantity and quality of the data used are important. The precipitation data should be collected for a long period of time. A sufficiently

Rainfall: State of the Science
Geophysical Monograph Series 191

10.1029/2010GM000976

long record of precipitation data provides a reliable basis for frequency analysis. It is known that a data sample of size n, in the absence of a priori distributional assumptions, can provide information only about exceedance probabilities greater than approximately $1/n$. It is a common rule of thumb to restrict extrapolation of at-site quantile estimates to return periods (years) of up to twice as long as the record length [*Stedinger et al.*, 1993]. Hence, long-term precipitation data are extremely valuable to enable determination of statistically based rainfall estimates of reasonable reliability, especially for extreme rainfalls with high return periods (e.g., greater than 100 years).

The quality of precipitation data may affect its usability and proper interpretation in frequency analysis studies. Precipitation measurements are subject to random and systematic errors [*Sevruk*, 1985]. The random error is due to irregularities of topography and microclimatical variations around the gauged site. The random error arises also due to inadequate network density to account for the natural spatial variability of rainfall. The systematic error in point precipitation measurements is, however, believed to be the most important source of error. The largest systematic error component is considered to be the loss due to wind field deformation above the orifice of elevated precipitation gages. Other sources of systematic error are wetting and evaporation losses of water that adheres to the funnel and measurement container, and rain splash [*Sevruk*, 1985].

For frequency analysis studies, it is necessary to check precipitation data for outliers and consistency. An outlier is an observation that departs significantly from the general trend of the remaining data. Procedures for treating outliers require hydrologic and mathematical judgment [*Stedinger et al.*, 1993]. In the context of regional analysis of precipitation, the outliers could provide critical information for describing the upper tail of the rainfall distribution. Hence, high outliers are considered as historical data if sufficient information is available to indicate that these outlying observations are not due to measurement errors. However, when the existence of outlier is confirmed, it can affect all the stages of frequency analysis: model identification, estimation, and forecasting. Thus, the detection of outliers and correction are vital in hydrological frequency analysis [*Tolvi*, 2000; *Battagila and Orfei*, 2002]. The main purpose of outlier detection is to identify outlier and proceed to their correction before any estimation, testing, or inference. Outliers may have significant impact on the results of standard methodology; therefore, it is important to detect them, estimate their effects, and undertake the appropriate corrective actions.

In spite of clear importance of this topic, published works on the detection of outliers in time series was slow to appear. *Barnett and Lewis* [1994] pointed out that the major difficulty in detecting outliers is that they are not necessarily extreme values. In addition, almost all outlier tests, such as Grubbs, Rosner, Dixon-Thompson, Anscombe were developed for normally distributed data [*Bradu and Hawkins*, 1995; *Rorabacher*, 1990; *Yu et al.*, 2004] and may fall when applied to extreme hydrological time series. Recently, classical outlier tests have been adapted to take in account the existence of extremes in the data set by considering nonnormal skewed probability distributions [*Moreau et al.*, 2008].

Changes in gauging instruments or station environment may lead to the presence of inhomogeneities in precipitation time series [*Beaulieu et al.*, 2008]. Data from the gages located in forest areas may not be compatible with those measured in open areas. Measurements in the valley and mountain stations and at various altitudes cannot be compared with one another. Therefore, care must be used in applying and combining precipitation data. Homogenization techniques can be applied to correct artificial inhomogeneities in precipitation data series [see *Beaulieu et al.*, 2009].

2.2. Statistical Criteria and Tests

Before carrying out a frequency analysis for any variable, the data series must meet certain statistical criteria such as independence, homogeneity, and stationarity. Independence means that no significant autocorrelation exists in the data series. Likewise, according to *Hershfield and Kohler* [1960], the elements of annual series of short-duration rainfall may, in practice, be assumed to be independent. In some cases, however, there may be significant dependence even between annual maximum values. Homogeneity implies that all the elements of the data series originate from a single population. Stationarity means that, excluding random fluctuations, the data series is invariant with respect to time. Types of nonstationarity include trends, jumps, and cycles. In rainfall frequency analysis, jumps are normally due to a change in station location, while trends and cycles may be associated with long-term climatic fluctuations.

Appropriate statistical tests are indicated [*Anderson*, 1941; *Wald and Wolfowitz*, 1943; *Mann and Whitney*, 1947; *Terry*, 1952]; and summarized here. A more complete description of these tests can be found in the work of *Siegel* [1956]. A summary of these tests is presented hereafter.

2.3. Wald and Wolfowitz (W-W) Test for Independence

For a sample of size N, $x_1, \ldots, x_N$ the W-W test considers the statistic R such that $R = \sum_{i=1}^{N-1} x_i x_{i+1} + x_1 x_N$ to test the hypothesis "H0: The observations are independent" against the alternative "H1: The observations are dependent" [*Wald and Wolfowitz*, 1943]. When H0 is true, the statistic R follows a normal distribution with mean and variance given, respectively, by

$$\bar{R} = \left(s_1^2 - s_2\right)/(N-1) \text{ and}$$

$$\mathrm{Var}(R) = \left(\frac{s_2^2 - s_4}{(N-1)}\right) + \left(\frac{s_1^4 - 4s_1^2 s_2 + 4s_1 s_3 + s_2^2 - 2s_4}{(N-1)(N-2)}\right) - \bar{R}^2,$$

where $s_r = Nm_r'$ and m_r' is the no-centered rth moment of the sample.

The quantity $u = (R - \bar{R})/(\mathrm{Var}(R))^{1/2}$ follows a standardized normal distribution (with mean 0 and variance 1) and can be used to test at the level α the hypothesis of independence, by comparing $|u|$ with the standard normal deviate $u_{\alpha/2}$ corresponding to a probability of exceedance $\alpha/2$.

2.4. Kendall Test for Stationarity

The Kendall test compares the following hypotheses: "H_0: The mean of the random variables is constant in time" and the alternative "H_1: The mean of the random variables is not constant in time." Let n random variables $X_1, X_2, X_3, \ldots, X_n$ be ordered chronologically. The S statistic of the Kendall test [*Kendall*, 1975] is $S = \sum_{i=1}^{n-1} \sum_{j=i+1}^{n} \mathrm{sgn}(X_j - X_i)$ where sign (D) is equal to 1 if D is positive and -1 if D is negative.

Under the null hypothesis, i.e., when random variables are stationary, S is asymptotically normally distributed with mean 0 and variance given by

$$\mathrm{Var}\{S\} = \frac{1}{18}\left[n(n-1)(2n+5) - \sum_t t(t-1)(2t+5)\right],$$

where t is the size of a tied group. By adding a continuity correction, the standardized statistic

$$K = \begin{cases} \dfrac{S-1}{\sqrt{\mathrm{Var}\{S\}}} & \text{if } S > 0 \\ 0 & \text{if } S = 0 \\ \dfrac{S+1}{\sqrt{\mathrm{Var}\{S\}}} & \text{if } S < 0 \end{cases}$$

is asymptotically normally distributed with mean zero and variance 1. The observations are stationary if S is close to its mean, i.e., 0. Thus, the null hypothesis is rejected for large values of $|K|$, calculated from the observations $x_1, x_2, \ldots, x_n$. Then, the critical region of the Kendall test at a significance level α is $\{|K| > z_{\alpha/2}\}$, where $z_{\alpha/2}$ is the standardized normal quantile corresponding to a probability of exceedance $\alpha/2$.

The decision rule of this test at a given significance level α is the following: If $|K| > z_{\alpha/2}$, H_0 is rejected, the observations cannot be considered as stationary. If not, H_0 is not rejected.

2.5. Mann-Whitney (M-W) Test for Homogeneity and Stationarity (Jump)

For two samples of size p and q (with $p \leq q$), the combined set of size $N = p + q$ is ranked in increasing order [*Mann-Whitney*, 1947]. The M-W test considers the quantities $V = R - p(p+1)/2$ and $W = pq - V$, where R is the sum of the ranks of the elements of the first sample (of size p) in the combined series, and V and W are calculated from R, p, and q. V represents the number of times that an item in sample 1 follows in the ranking of an item in sample 2; W can also be computed in a similar way for sample 2 following sample 1.

The M-W statistic, U, is defined by the smaller of V and W. When $N > 20$, and $p, q > 3$, and under the null hypothesis that the two samples come from the same population, U is approximately normally distributed with mean $\bar{U} = pq/2$ and variance

$$\mathrm{Var}(U) = \left[\frac{pq}{N(N-1)}\right]\left[\frac{N^3 - N}{12} - \sum T\right] \text{ with } T = (J^3 - J)/12,$$

where J is the number of observations tied at a given rank. The summation $\sum T$ is over all groups of tied observations in both samples of size p and q. For a test at a level of significance α, the quantity $|u| = |(U - \bar{U})/[\mathrm{Var}(U)]^{1/2}|$ is compared with the standardized normal quantile $u_{\alpha/2}$ corresponding to a probability of exceedance $\alpha/2$.

3. STATISTICAL DISTRIBUTIONS USED IN RAINFALL FREQUENCY ANALYSIS AND THEIR PROPERTIES

Common distributions that have been applied to the analysis of rainfall extremes include the Gumbel [*Gumbel*, 1942; *Koutsoyiannis*, 2004], generalized extreme value (GEV) [*Natural Environment Research Council*, 1975], lognormal (LN) [*Pilgrim*, 1998], log-Pearson type 3 [*Niemczynowicz*, 1982; *Pilgrim*, 1998], Halphen [*El Adlouni et al.*, 2008], and generalized logistic (GLO) [*Fitzgerald*, 2005] distributions. Among these distributions, the GEV and its special form, the Gumbel distribution, have been the most often used in modeling annual maximum rainfall series. The Gumbel distribution was found, however, to underestimate the extreme precipitation amounts [*Wilks*, 1993]. *Brooks and Carruthers* [1953] indicated that the Gumbel distribution, which is commonly used in hydrological frequency analysis, tends to underestimate the magnitude of the rarest rainfall events.

Bernier [1959] suggested the log-Gumbel distribution for hydrological series. The log-Gumbel distribution is also called Fréchet distribution [*Fréchet*, 1927], which is a special case of the GEV distribution. The two-component extreme value distribution has also been used for the frequency analysis of rainfall, mainly to alleviate the concerns over some of the theoretical restrictions of the GEV. Halphen distributions (Halphen type A (HA), type B, and inverse type B, HIB) have been introduced to fit a large variety of data sets [*Halphen*, 1941; *Morlat*, 1956]. They constitute with their limiting forms, the gamma and inverse gamma distributions, a complete system to model hydrological variables. Indeed, the (δ_1,δ_2) diagram [*Morlat*, 1956; *El Adlouni and Bobée*, 2007] makes it possible to represent this family of distributions and their limiting cases in a plane the same way as the well-known Pearson system with the (β_1,β_2) diagram corresponding to the skewness and kurtosis coefficients. For any sample, the corresponding (δ_1,δ_2) point is associated to one and only one member of the Halphen family or their limiting cases.

In order to select the most adequate distribution, tests and information criteria are considered, and empirical comparisons are commonly used for a given region. A comparison of the LN, gamma, Gumbel, Fréchet, and log-Pearson type 3 fits was given by *Benson* [1968] for 10 U.S. streamflow stations. Based on this study, the United States adopted a uniform approach to hydrological frequency estimation, which consists in fitting the annual maximum magnitude to the log-Pearson type 3 (LP3) distribution [*U.S. Water Resources Council*, 1981]. Although many countries, such as Australia, have adopted the LP3 distribution, others have selected different distributions, such as the GLO distribution in the United Kingdom and the LN in China, [*Bobée*, 1999; *Robson and Reed*, 1999]. Discussions and reviews of the application of these and other statistical distributions to flood frequency analysis are given by *Stedinger et al.* [1993], *Bobée and Rasmussen* [1995], and *Rao and Hamed* [2000]. A comparison study for some distributions, given by *Koutsoyiannis* [2004], shows that the Fréchet distribution is more adequate to represent extreme rainfall series.

Studies using rainfall data from many climatic regions [*Wilks*, 1993; *Zalina et al.*, 2002] suggest also that a three-parameter distribution can provide sufficient flexibility to represent extreme precipitation data. In particular, the GEV distribution has been found to be the most convenient, since it requires a simpler method of parameter estimation, and it is more suitable for regional estimation of extreme rainfalls at sites with limited or without data [*Nguyen et al.*, 2002; *Leclerc and Ouarda*, 2007]. When the return periods associated with frequency-based rainfall estimates greatly exceed the length of record available, discrepancies between commonly used distributions tend to increase. Distributions commonly used for rainfall frequency analysis are summarized hereafter.

Malamud and Turcotte [2006] showed that the most commonly used frequency-magnitude distributions in hydrology can be divided into four groups: the normal family (normal and LN), the generalized extreme value (GEV) family (GEV, Gumbel, Fréchet, and reverse Weibull), the Pearson type 3 family (Gamma, Pearson type 3, and log-Pearson type 3), and the generalized Pareto distribution. These groupings were deduced from the historical development of the distributions that compose them. However, no discussion of the tail behavior of these classes was given in their work. In practice, almost all these models are fitted to data and compared using conventional goodness-of-fit tests (see section 5). However, these procedures essentially test the adequacy of the model to the central range of the sample, and the adequacy should be tested for the extreme range of observations in order to estimate high return period events. *Ouarda et al.* [1994] presented an analysis of the tail behavior of these distributions. Recently, *El Adlouni et al.* [2008] presented a classification of these models based on their tail behavior.

3.1. GEV Distribution

The generalized extreme value distribution encompasses the three standard extreme value distributions: Fréchet, Weibull, and Gumbel. The cumulative distribution function, as presented by *Von Mises* [1954] and *Jenkinson* [1955] is written

$$F_{\text{GEV}}(x) = \begin{cases} \exp\left[-\left(1-\frac{\kappa}{\alpha}(x-\mu)\right)^{1/\kappa}\right] & \kappa \neq 0 \quad \text{(EV2 and EV3)} \\ \exp\left[-exp\left(-\frac{(x-\mu)}{\alpha}\right)\right] & \kappa = 0 \quad \text{(EV1)} \end{cases},$$

whenever $1-\kappa\frac{(x-u)}{\alpha} \geq 0$.

The mean and the variance of a random variable X that follows the GEV distribution are, respectively, given by

$$E(X) = \mu + \frac{\alpha}{\kappa}(1-\Gamma(1+\kappa))$$

and

$$\text{Var}(X) = \frac{\alpha^2}{\kappa^2}(\Gamma(1+2\kappa)-\Gamma^2(1+\kappa)).$$

In practice, in order to identify the appropriate form of the GEV distribution (EV1, EV2, or EV3), the sign of the shape parameter should be determined. Therefore, the efficiency of the estimation procedure of the GEV parameters is of great importance.

3.2. *The Three-Parameter LN Distribution*

The three-parameter LN distribution differs from the two-parameter LN distribution by the introduction of a lower bound x_0, so that if X follows the LN distribution, $Z = \log(X - x_0)$ is normally distributed with parameters μ and σ^2. Thus, it is a three-parameter distribution for which x_0 is the location parameter, while μ and σ^2 control the scale and shape, respectively. It can be shown that the probability density function (pdf) of x is

$$f(x) = \frac{1}{(x-x_0)\sigma\sqrt{2\pi}} e^{-\frac{1}{2}[(\log(x-x_0)-\mu)/\sigma]^2}.$$

The moments of X may be obtained as functions of those of the corresponding normal distribution. The mean and the variance are given by

$$E(X) = x_0 + \exp\left(\mu + \frac{1}{2}\sigma^2\right)$$

and

$$\mathrm{Var}(X) = \exp(2\mu + \sigma^2)(e^{\sigma^2} - 1).$$

The coefficient of variation and the coefficient of skewness are

$$\mathrm{Cv} = (e^{\sigma^2}-1)^{\frac{1}{2}} \quad \text{and} \quad \mathrm{Cs} = (e^{\sigma^2}-1)^{\frac{1}{2}}(e^{\sigma^2}+2).$$

3.3. *HA Distribution*

Halphen [1941] first suggested what he called the harmonic distribution:

$$f(x) = \frac{1}{2x\, k_0(2\alpha)} \exp\left[-\alpha\left(\frac{x}{m} + \frac{m}{x}\right)\right], \quad x > 0,$$

where $m > 0$ is a scale parameter, $\alpha > 0$ is a shape parameter, and $k_0(.)$ is the modified Bessel function of the second kind of order zero [*Perreault et al.*, 1999]. Being a two-parameter distribution, the shape of the harmonic distribution is entirely determined by the shape parameter α. To obtain additional flexibility, Halphen generalized the harmonic distribution by adding a second shape parameter, ν, giving rise to the HA distribution:

$$f_A(x) = \frac{1}{2m^\nu\, k_\nu(2\alpha)} x^{\nu-1} \exp\left[-\alpha\left(\frac{x}{m} + \frac{m}{x}\right)\right], \quad x > 0,$$

where $m > 0$ is a scale parameter, and $\alpha > 0$ and $\nu \in \mathbb{R}$ are shape parameters. $k_\nu(.)$ is the modified Bessel function of the second kind of order ν. The HA distribution may be recognized as a re-parameterized form of the generalized inverse Gaussian distribution (GIG). The noncentered r moments (r is an integer) of a random variable following HA distribution are given by

$$\mu'_r = E\left[X^r\right] = \frac{m^r k_{\nu+r}(2\alpha)}{k_\nu(2\alpha)}.$$

Thus, the mean and the variance are given by

$$E[X] = m\frac{k_{\nu+1}(2\alpha)}{k_\nu(2\alpha)}$$

and

$$\mathrm{Var}[X] = \frac{m^2}{k_\nu^2(2\alpha)}\left(k_\nu(2\alpha)k_{\nu+2}(2\alpha) - k_{\nu+1}^2(2\alpha)\right).$$

3.4. *HIB Distribution*

The pdf of the HIB distribution is given by [*Perreault et al.*, 1999]

$$f_{\mathrm{HIB}}(y; m, \alpha, \nu) = \frac{2}{m^{-2\nu} ef_\nu(\alpha)} y^{-2\nu-1} \exp\left[-\left(\frac{m}{y}\right)^2 + \alpha\left(\frac{m}{y}\right)\right], \quad y > 0,$$

where $m > 0$ is a scale parameter and $\alpha \in \mathbb{R}$ and $\nu > 0$ are shape parameters.

The noncentral moment expression is given by

$$[\mu'_r]_{\mathrm{HIB}} = E(Y^r) = \frac{m^r ef_{\nu-r/2}(\alpha)}{ef_\nu(\alpha)},$$

where $[\mu_r']_{\mathrm{HIB}}$ exists provided that $\nu - r/2 > 0$, and the mean and the variance are

$$E[X] = m\frac{ef_{\nu-1/2}(\alpha)}{ef_\nu(\alpha)}$$

and

$$\mathrm{Var}[X] = \frac{m^2}{ef_\nu^2(\alpha)}\left(ef_\nu(\alpha)ef_{\nu-1}(\alpha) - ef_{\nu-1/2}^2(\alpha)\right).$$

More details on the Halphen system of distributions are given by *Chebana et al.* [2010] and *El Adlouni and Bobée* [2007].

3.5. *The Pearson Type 3 Distribution*

The Pearson Type 3 distribution has three parameters, which can be denoted by x_0, β, and γ. One special case occurs when $x_0 = 0$ and gives the Gamma distribution. Another arises when $\gamma = 1$ leading to the exponential distribution. For the Pearson type 3 distribution, the pdf and cdf are

$$f(x) = \frac{(x-x_0)^{\gamma-1} e^{-(x-x_0)/\beta}}{\beta^\gamma \Gamma(\gamma)}$$

and

$$F(x) = \int_{x_0}^{x} \frac{(x-x_0)^{\gamma-1} e^{-(x-x_0)/\beta}}{\beta^\gamma \Gamma(\gamma)} dx.$$

If on the right-hand side of the cdf the variable $z = x - x_0$ is introduced, this becomes

$$\int_0^z \frac{z^{\gamma-1} e^{-z/\beta}}{\beta^\gamma \Gamma(\gamma)} dz$$

showing that z has a gamma distribution with parameters β and γ. This means that a Pearson type 3 distribution can be considered to be a gamma distribution with parameters β and γ and whose variate has its origin at $x = x_0$.

The mean, the variance, and the skewness are, respectively, given by mean $= \mu' = E(x) = x_0 + \beta\gamma$, variance $= \mu_2 = E(x - E(x))^2 = \beta^2\gamma$, and skewness $= g = \frac{2}{\sqrt{\gamma}}$.

3.6. GLO Distribution

The cumulative distribution function of the GLO distribution with parameters ξ (location), α (scale), and κ (shape) is [*Hosking and Wallis*, 1997]

$$F(x) = \frac{1}{1+e^{-y}},$$

where

$$y = -\frac{ln\left(1-\frac{\kappa(x-\xi)}{\alpha}\right)}{\kappa}, \quad \kappa \neq 0.$$

and

$$y = \frac{x-\xi}{\alpha}, \quad \kappa = 0.$$

The distribution is bounded at $\xi + \frac{\alpha}{\kappa}$ from the right (left) if $\kappa > 0$ (if $\kappa < 0$). The first L moments are defined for $-1 < \kappa < 1$ and are given by

$$\lambda_1 = \xi + \alpha\left(\frac{1}{\kappa} - \frac{\pi}{\sin \kappa\pi}\right), \lambda_2 = -\frac{\alpha\,\kappa\,\pi}{\sin \kappa\pi}, \text{ and}$$
$$\lambda_3 = -\kappa\,\lambda_2.$$

The L moment estimators of the GLO parameters are given by matching the theoretical expression of L moments and their corresponding sample estimates l_1, l_2, and l_3. The L moments estimators are

$$\hat{\kappa} = -\frac{l_3}{l_2}, \hat{\alpha} = \frac{l_2 \sin \kappa\pi}{\kappa\pi} \text{ and } \hat{\xi} = l_1 - \alpha\left(\frac{1}{\kappa} - \frac{\pi}{\sin \kappa\pi}\right).$$

4. PARAMETER ESTIMATING METHODS

Several methods for the estimation of the distribution parameters are available in the hydrological and statistical literature. The simplest method is the method of moments that provides parameter estimates such that the theoretical moments are equal to the computed sample moments. An alternative method for estimating parameters is based on the sample L moments [*Hosking*, 1990]. Sample L moments are found to be less biased than traditional moment estimators and thus are better suited for use with small sample sizes. The L moment method has proven quite effective in the estimation of the GEV distribution parameters [*Stedinger et al.*, 1993]. Another method is the method of maximum likelihood. The maximum likelihood method provides estimators that maximize the likelihood function with very good statistical properties for large samples. However, these estimators are often not available in closed form and thus must be computed using an iterative numerical method.

The GEV is among the most frequently used distributions for extreme value analysis [*Stedinger et al.*, 1993; *Ouarda et al.*, 2001; *Katz et al.*, 2002] in hydrology and climatology. A particular attention was given to estimate the parameters of the GEV distribution in statistical literature as well as in hydrological studies. Several approaches were proposed to avoid the computational problems related to the maximum likelihood approach especially for small samples. The method of maximum likelihood is efficient when the sample size is sufficiently large. Because of the complexity of the likelihood function, the solution of the maximum likelihood (ML) method can only be obtained through numerical methods. The ML method may diverge when the sample size is small. To resolve the problems of divergence occurring in the numerical techniques used for ML, *Martins and Stedinger* [2000] suggested the use of a "prior" distribution for the shape parameter of the GEV model such that the most probable values of the parameter are included. This approach was also investigated by *Park and Sohn* [2005] who studied the optimal selection of the hyperparameters. The generalized maximum likelihood (GML) approach is similar to the quasi-Bayesian maximum likelihood estimator of *Hamilton* [1991], which considers prior information about the parameters to eliminate the singularities associated with the ML method. The latter approach was successfully applied by *Venkataraman* [1997] for the solution of problems in finance.

Morrison and Smith [2003] introduced a new method, named "mixed L moments: maximum likelihood," to resolve the same singularity problem and to obtain unbiased estimators such as those produced by the ML method. This method consists of solving the equations of the maximum likelihood estimation method under the constraints given by the first L moment or the first two L moments. Through a simulation experiment, *Morrison and Smith* [2003] have shown that the estimators obtained with this method conserve the property of being unbiased and are characterized by a low variance. *Dose and Menzel* [2004] used a Bayesian approach to build nonparametric models when studying climate change effects in phenology. Recently, *Zhang* [2007] proposed the likelihood moment estimator, which is computationally simple and possesses asymptotic efficiency.

5. PROBABILITY PLOTS, GOODNESS-OF-FIT TESTS, AND CRITERIA

5.1. *Plotting Positions and Probability Plots*

Initial evaluation of the adequacy of a fitted probability distribution is best done by generating a probability plot of the observations. When the sorted observations are plotted against an appropriate probability scale, except for sampling fluctuation, they would fall approximately on a straight line. Such a plot serves both as an informative visual display of the data and a check as to whether the fitted distribution is actually consistent with the data.

Such plots can be generated with special probability papers for some distributions (including the normal, two-parameter LN, and Gumbel distributions, all of which have a fixed shape). However, with modern software, it is generally easier to generate such plots without use of special papers [*Bobée and Ashkar*, 1991; *Stedinger et al.*, 1993]. The trick is that the *i*th largest observed flood $x_{(i)}$ is plotted versus the estimated quantile associated with the exceedance probability, or probability-plotting position q_i, assigned to each ranked observed value $x_{(i)}$; $x_{(1)} \geq x_{(2)} \geq \cdots \geq x_{(n)}$. The exceedance probability of the *i*th largest flood $x_{(i)}$ can be estimated by any of several reasonable formulas. Three commonly used formulas are the Weibull formula with $p_i = \frac{i}{(n+1)}$, the Cunnane formula with $p_i = \frac{i-0.4}{(n+0.2)}$, and the Hazen formula with $p_i = \frac{i-0.5}{n}$. *Aldenberg and Jaworska* [2000] and *Stedinger et al.* [1993] provide a discussion of the plotting position issue.

5.2. *Goodness-of-Fit Tests*

Several rigorous statistical tests are available and are useful in climatology and hydrology to determine whether the selected model gives a reasonable fit to a given set of observations. The tests allow one to identify if the sample was drawn from a particular family of distributions [*Stedinger et al.*, 1993, *Singh and Prevert*, 2004; *Kidson and Richards*, 2005; *Katz et al.*, 2002; *El Adlouni et al.*, 2008]. The Kolmogorov-Smirnovtest provides bounds within which every observation on a probability plot should lie if the sample is actually drawn from the assumed distribution [*Kottegoda and Rosso*, 1997]. The probability plot correlation test is a more powerful test of whether a sample has been drawn from a postulated distribution [*Filliben*, 1975; *Vogel*, 1986, 1987; *Vogel and Kroll*, 1989; *Chowdhury et al.*, 1991]. Discussion of the development and interpretation of probability plots is provided by *Haan* [1977], *Stedinger et al.*, [1993], *Kottegoda and Rosso* [1997], *Beirlant et al.* [2004], and *El Adlouni et al.* [2008].

5.3. *Information Criteria*

Several approaches have been suggested for the comparison of the distributions of extreme rainfall. Goodness-of-fit tests have been applied to assess the suitability of different probability distributions for describing annual maximum precipitation series (or to simulated samples in the case of simulation studies). These tests establish which distributions are, in general, the most appropriate for extreme rainfall modeling. To assess the quality of a fitted model, *Akaike* [1974] introduced an information criterion called Akaike information criterion (AIC), which can be adapted to a large number of different situations. It consists in minimizing an information measure that is defined as: $\mathrm{AIC}(f) = -2\log L(\hat{\theta}, x) + 2k$, where $L(\hat{\theta}, x)$ is the likelihood function, and k is the number of parameters. According to *Akaike* [1974], the model that better explains the data with the least number of parameters is the one with the lowest AIC. To select an appropriate model, some compromises between the goodness of fit and the complexity of the model must be done. Alone, the Akaike information criterion is not appropriate for model selection.

A Bayesian extension of the minimum AIC concept is the Bayesian information criterion called BIC. It is defined as $\mathrm{BIC}(f) = -2\log L(\hat{\theta}, x) + k\log(n)$, where $L(\hat{\theta}, x)$ is the likelihood function, k is the number of parameters, and n is the sample size. BIC is also a parsimony criterion. Between several models, the model with the lowest BIC is considered to be the best one. The *Schwarz* method [1978] is often used to obtain the BIC. However, this method can also be used only to get an asymptotic approximation of a Bayes factor.

Furthermore, it can be combined to an a priori probability distribution to obtain the a posteriori probability for each distribution of a given set of statistical distributions.

6. NONSTATIONARY RAINFALL FREQUENCY ANALYSIS

6.1. Nonstationarity of Hydrological Series

There are two fundamental assumptions for the classical frequency analysis to provide useful engineering design values. The proper estimation of design values requires that the data series from which the probability distribution parameters are to be estimated come from independent and identically distributed observations. The proper assessment of risk factors for an engineering structure requires that the statistical inference has also to be valid during the projected life span of the structure. This requires that the conditions (e.g., climate) under which the inferences are made will remain the same in the future. There is, however, mounting evidence suggesting that such assumptions can hardly be met in reality. On one hand, observed historical extreme events are hardly nonstationary. In fact, statistically significant trends have been identified in extreme values of different hydroclimatological series [*IPCC*, 2001] in different parts of the world. On the other hand, the anthropogenic influence on the climate system caused by the increase in the emission of greenhouse gases into the atmosphere has a potential to make future climate very different from what it is today. Climate extremes will likely change in the future [*Jain and Lall*, 2001; *Wang and Swail*, 2004; *Wang et al.*, 2004; *Kharin and Zwiers*, 2005]. The reality of nonstationary hydrometeorological extremes needs to be properly addressed, since the GEV model with constant parameters may no longer be valid under nonstationary conditions [*Leadbetter et al.*, 1983].

Nonstationarity of extreme values may be detected by identifying trends in the extreme values [*Zhang et al.*, 2001, 2004; *Clarke*, 2002]. However, trend estimation has limited usefulness when return values are required. In this situation, "covariates" may be introduced into the probability distribution when modeling extreme values [*Smith*, 1989]. *Scarf* [1992] introduced a trend in the position parameter for the GEV model. *Coles* [2001] provided a general description of the covariate approach to the modeling of extreme values. An example of a hydrological application of the covariate approach can be found in the work of *Katz et al.* [2002]. *Sankarasubramanian and Lall* [2003] studied quantile regression with climate indicators to estimate quantiles under climate change conditions. The covariate approach has also found applications in climate studies. Using Monte Carlo simulations, *Zhang et al.* [2004] compared several methods for detecting trends in the magnitude of extreme values. They found that methods that specifically model trend in the parameters of extreme value distributions provide the highest power of detection of statistically significant trends. *Wang and Swail* [2004] and *Wang et al.* [2004] used covariates in their analyses of projected extreme ocean wave heights for the end of the 21st century. *Kharin and Zwiers* [2005] applied a similar model to general climate model simulated extremes to estimate the impact of anthropogenic climate change on climate extremes. All of these studies used the ML method for parameter estimation.

The majority of the models that consider dependence of parameters on covariates are based on the normality assumption. This assumption is not always verified, especially in the case of extreme values. Meanwhile, even when trends or any other causes of nonstationarity are eliminated, the resulting residual series is not necessarily normal. Hydroclimatic extreme value variables are often characterized by a strong skewness.

6.2. Nonstationary GEV Model

The distributions of extreme values, introduced by *Fisher and Tippett* [1928], include three families: Gumbel, Fréchet, and Weibull. *Jenkinson* [1955] combined the three families into the GEV distribution with a cumulative distribution function F_{GEV} (see section 3.1), where, $\mu + \alpha/\kappa \le x < +\infty$ when $\kappa < 0$ (Fréchet), $-\infty < x < +\infty$ when $\kappa = 0$ (Gumbel), and $-\infty < x \le \mu + \alpha/\kappa$ when $\kappa > 0$ (Weibull). Here μ ($\in\mathbb{R}$), α (>0), and κ ($\in\mathbb{R}$) are the location, the scale, and the shape parameters, respectively.

In the case of the model with covariates, the parameters depend on other variables such as time: GEV $(\mu_t, \alpha_t, \kappa_t)$ [*Coles*, 2001]. To ensure a positive value for the scale parameter, a transformation such that $\varphi_t = \log(\alpha_t)$ is used when estimating the parameters. We assume that the location parameter μ_t is a function of n_μ covariates $U = (U_1\, U_2 \ldots U_{n_\mu})'$. Let $\beta = (\beta_1\, \beta_2 \ldots \beta_{n_\mu})'$ be the vector of corresponding parameters. In the case of linear dependence, we have

$$\mu_t = U'(t)\beta = \sum_{i=1}^{n_\mu} \beta_i U_i(t).$$

For the scale parameter α_t, let $V = (V_1\, V_2 \ldots V_{n_\alpha})'$ be the vector of covariates. We have

$$\varphi_t = \log(\alpha_t) = V'(t).\delta = \sum_{i=1}^{n_\alpha} \delta_i V_i(t),$$

where $\delta = (\delta_1\, \delta_2 \ldots \delta_{n_\alpha})'$ are the corresponding parameters. The same applies to the shape parameter

$$\kappa_t = W'(t).\gamma = \sum_{i=1}^{n_\kappa} \gamma_i W_i(t),$$

where $W = (W_1\; W_2 \ldots W_{n_\kappa})'$ are the covariates, and $\gamma = (\gamma_1\, \gamma_2 \ldots \gamma_{n_\kappa})'$ are the corresponding parameters.

For the GEV model with covariates, the likelihood function for a given sample $\underline{x}'_n = \{x_1, \ldots, x_n\}$ is $L_n = \prod_{t=1}^{n} f(x_t|\mu_t, \varphi_t, \kappa_t)$, where f is the pdf of the GEV distribution.

In general, in hydrological studies, the shape parameter is taken to be constant ($\kappa_t = \kappa$) because it corresponds to the physical phenomenon. For the location and scale parameters, dependence is linear, and the number of covariates is restricted to $1 \le n_\mu \le n_\mu(\max)$ and $1 \le n_\alpha \le n_\alpha(\max)$, in order to limit the number of parameters to be estimated. However, several models can still be considered given particular values of n_μ and n_α. Here are some of these models:

1. $\mathrm{GEV}_{1,1}$ (μ,α,κ) is the classic model with all parameters being constant: $\mu_t = \mu$, $\alpha_t = \alpha$ and $\kappa_t = \kappa$. In this case, $n_\mu = n_\alpha = 1$.
2. $\mathrm{GEV}_{2,1}$ ($\mu_t = \beta_1 + \beta_2 Y_t$,$\alpha$,$\kappa$) is the homoscedastic model, and the location parameter is a linear function of one covariate Y_t. (Here $n_\mu = 2$, $U(t) = (U_1(t) = 1 \;\; U_2(t) = Y_t)$, $n_\alpha = 1$, and $V_1 = 1$).
3. $\mathrm{GEV}_{2,2}$ ($\mu_t = \beta_1 + \beta_2 Y_t$,$\alpha_t = \exp(\delta_1 + \delta_2 Y_t)$,$\kappa$) is the model in which the location and scale parameters are function of the covariate Y_t. This model is recommended when the covariate is time $Y_t = t$, since trends are usually observed at the same time in the location and scale parameters ($n_\mu = 2$, $U(t) = (U_1(t) = 1 \;\; U_2(t) = Y_t)$, $n_\alpha = 2$, $V(t) = (V_1(t) = 1 \;\; V_2(t) = Y_t)$).
4. $\mathrm{GEV}_{3,2}$ ($\mu_t = \beta_1 + \beta_2 Y_t + \beta_3 Y_t^2$,$\alpha_t = \exp(\delta_1 + \delta_2 Y_t)$,$\kappa$) is the model in which the location is a quadratic function of the covariate Y_t, and the scale parameter is a linear function of the same covariate ($n_\mu = 3$, $U(t) = (U_1(t) = 1 \;\; U_2(t) = Y_t U_3(t) = Y_t^2)$, $n_\alpha = 2$, $V(t) = (V_1(t) = 1 \;\; V_2(t) = Y_t)$).

In the same manner, model $\mathrm{GEV}_{1,2}$ can be defined as the model with a constant location parameter and a scale parameter that is a linear function of the covariate. $\mathrm{GEV}_{3,1}$ is the model with a location expressed as a quadratic function of the covariate and a constant scale parameter. Other models can be defined using a vector of covariates.

6.3. *Parameter Estimators for the Model with Covariates*

To estimate the parameters of the GEV model with covariates, the maximum likelihood method is the most commonly used approach [*Coles*, 2001]. More recently, the GML method, developed originally for the classical model (GEV), was extended to the case with covariates [*El Adlouni et al.*, 2007]. The GML method is based on the same principle as the ML Method with an additional constraint on the shape parameter to eliminate potentially invalid values of this parameter. A prior distribution of κ, in the case of the hydrometeorological series, was introduced by *Martins and Stedinger* [2000] based on practical considerations (see section 4). The GML method improves considerably the ML method and avoids some computational problems related to the ML maximization [*El Adlouni and Ouarda*, 2008].

6.4. *Nonstationary Model Selection*

The most general model ($\mathrm{GEV}_{3,2}$) is usually the best model to represent data variance. However, when the difference between two models is not evident, it is preferable to use the simplest model in order to respect the parsimony principle. A simple method to compare the validity of a model M_1 against another model M_0, such as $M_0 \subset M_1$, is to use the deviance statistic defined by *Coles* [2001], $D = 2\{l_n^*(M_1) - l_n^*(M_0)\}$, where $l_n^*(M)$ is the maximized log-likelihood function of model M. Large values of D indicate that model M_1 is more adequate and explains more of the data variation than model M_0. The statistic D is distributed according to a chi-square (χ_ν^2) distribution. The parameter ν is the difference between the dimensions (number of parameters) of the M_1 and M_0 models. Values of D greater than the quantiles of the χ_ν^2 distribution for a particular confidence level are considered significant, thus model M_1 is better than model M_0.

Recently, *El Adlouni and Ouarda* [2009] proposed the use of the Reversible Jump Markov Chain Monte Carlo

Figure 1. Geographic location of the Randsburg station (latitude, 35.3700; longitude, 117.650) in California.

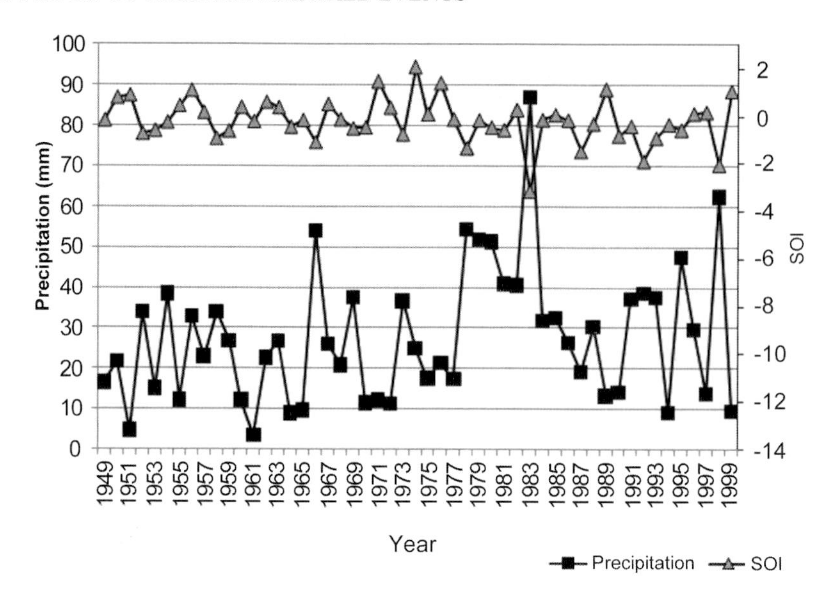

Figure 2. Observed annual maximal precipitation and Southern Oscillation Index (SOI) series.

algorithm for parameter estimation and model selection of the GEV model with covariates.

7. CASE STUDY: ANNUAL MAXIMUM PRECIPITATIONS IN THE RANDSBURG STATION (CALIFORNIA) AND THE SOUTHERN OSCILLATION INDEX (SOI)

This section illustrates the use of frequency analysis of the series $\underline{X} = (x_1, \ldots, x_n)$ of annual maximum precipitation (mm), recorded at the Randsburg station in California (Station 047253) with 51 years of record. The latitude of the station is 35.3700, its longitude 117.650, and the period of record is 1949–1999. Figure 1 illustrates the geographic location of the Randsburg station. Located in the southern part of California, precipitation in this station should be strongly affected by the El Niño phenomenon.

The use of the GEV models with covariables allows studying the frequency of the annual maximal precipitation (X) conditionally to the effect of climate variability represented by the SOI. The correlation coefficient between the annual maximal precipitations X and the SOI is $\rho = -0.60$ (Figures 2 and 3).

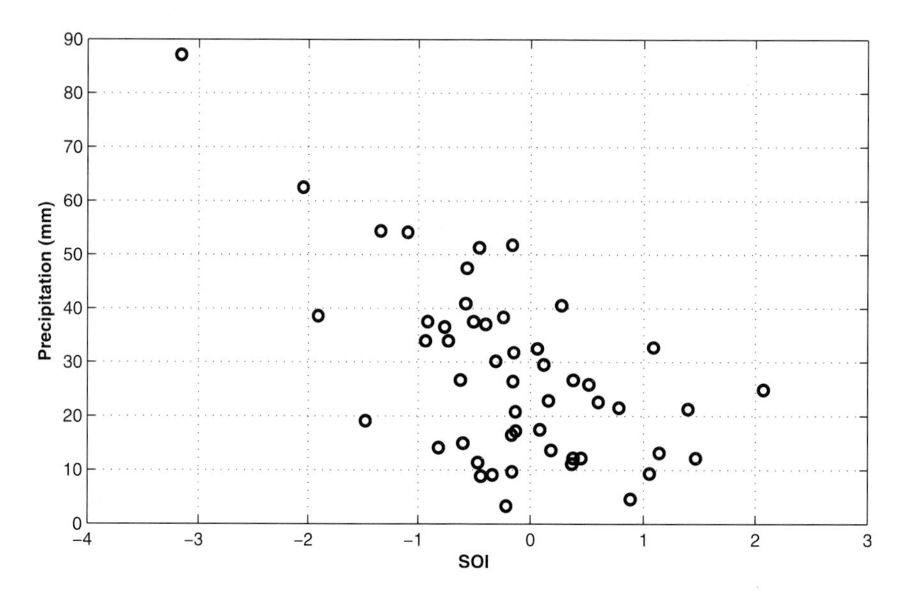

Figure 3. Observed annual maximal precipitation and corresponding SOI values.

Table 1. Maximized Log-Likelihood Function and Generalized Maximum Likelihood (GML) Parameter Estimators for Each Model

	l_n^*	β_1	β_2	β_3	α	κ
Generalized extreme value (GEV)$_{1,1}$	−209.96	19.52	–	–	12.36	−0.05
GEV$_{2,1}$	−206.86	18.89	−9.92	–	12.21	−0.07
GEV$_{3,1}$	−204.71	16.57	−10.61	3.03	12.14	−0.06

We observe a significant negative correlation between X and the SOI. Figure 3 shows that the extreme precipitation values correspond to low SOI values. Three GEV models with covariates were compared: GEV$_{1,1}$, GEV$_{2,1}$, and GEV$_{3,1}$. The parameters of these models were estimated by the GML and are given in Table 1.

The deviance statistic-based test shows that a significant difference exists between the GEV$_{1,1}$ and the GEV$_{2,1}$ models since $D = 6.2$ is bigger than the 0.95-quantile of the χ_1^2 distribution ($\Pr(\chi_1^2 \leq 6.2) = 0.9872$). The GEV$_{2,1}$ model led to a maximized log-likelihood value of $l_n^*(\text{GEV}_1) = -206.86$. In the case of quadratic dependence in the location parameter μ, the maximized log-likelihood becomes $l_n^*(\text{GEV}_{3,1}) = -204.71$. The deviance statistic for comparing these two models is therefore $D = 4.3$. This value is large when compared to the χ_1^2 distribution ($\Pr(\chi_1^2 \leq 4.3) = 0.9619$), implying that the quadratic model (GEV$_{3,1}$) explains a substantial amount of the variation in the data and is likely to best represent the dependence between the annual maximal precipitation and the SOI index.

The difference between the three models can also be demonstrated by comparing the quantiles estimated by each model. Figure 4 illustrates the conditional medians, for various values of the SOI, estimated for the GEV$_{1,1}$, GEV$_{2,1}$, and the GEV$_{3,1}$ models.

Figure 5 represents the quantiles estimated by the selected model, conditional to the SOI covariate. Note that the use of the common notion of "return period" is no longer appropriate in a nonstationary framework. The return period associated to any extreme event value depends on the covariate. Risk assessment should then be carried out through integrating the risk level throughout the lifetime of a structure or by considering the worst case scenario, which may occur toward the end of the lifetime of the structure.

Table 2 presents, in more detail, the exact GML estimator values of the median and the 95% confidence intervals for the following values of the SOI: −3.16, 0.04, and 2.04. These values correspond to the minimum observed value, a central value, and the maximum observed value of the SOI index.

Results show that the difference is greater for negative values of SOI, which correspond to extreme values in the annual maximal precipitation data. Indeed, the median estimated by the quadratic model, GEV$_{3,1}$, can be three times greater than that estimated by the classic model, GEV$_{1,1}$. Results of the comparison test with the deviance statistic

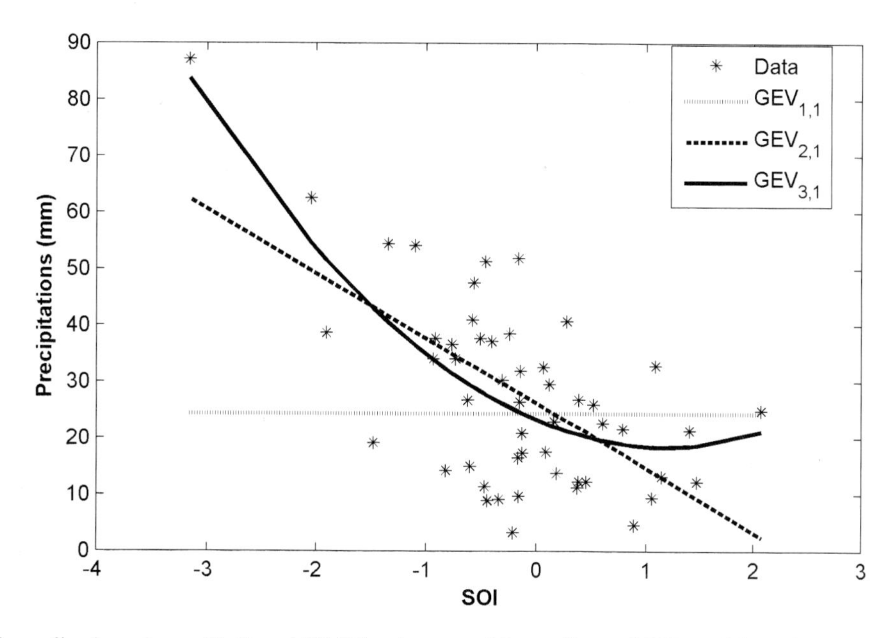

Figure 4. Generalized maximum likelihood (GML) estimators of the median and 95% confidence intervals conditional on values of the SOI, obtained by the generalized extreme value (GEV)$_{1,1}$, GEV$_{2,1}$, and the GEV$_{3,1}$ models.

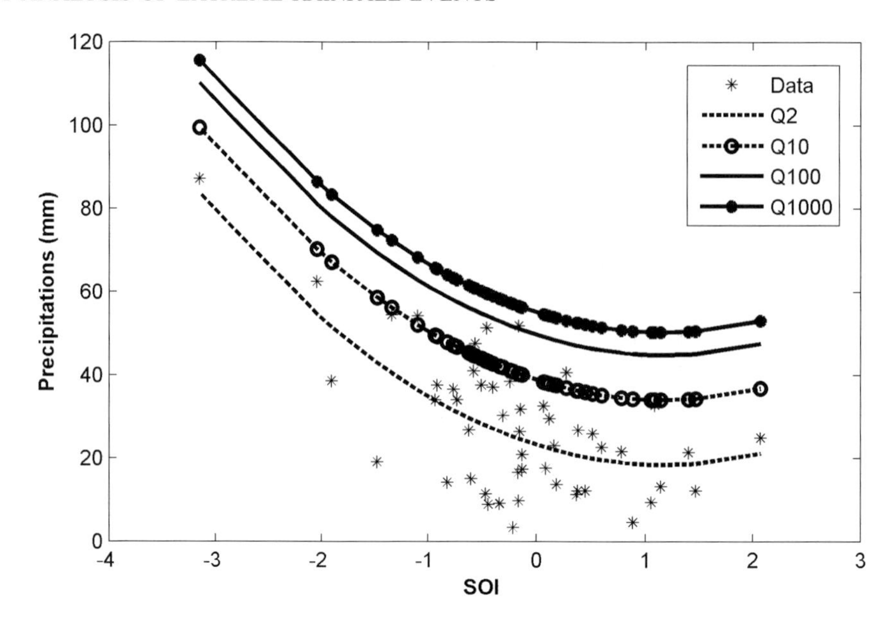

Figure 5. GML estimators of the quantiles by the selected model $GEV_{3,1}$ conditional on values of the SOI.

indicate that the $GEV_{3,1}$ model is more adequate to represent the precipitation data. Thus, the use of either simplified model ($GEV_{1,1}$ or $GEV_{2,1}$) could lead to a significant underestimation of the median in some cases. Indeed, the $GEV_{1,1}$ model leads to an underestimation of the median in the case of negative SOI values, while the $GEV_{2,1}$ leads to an underestimation of the median for large SOI values. On the other side, overestimation can result from the use of the $GEV_{1,1}$ model with large positive values of SOI.

It is important to recognize the practical significance of the climate indices modulation of flood risk. *Micevski et al.* [2006] show that the use of at-site flood data with an inadequate coverage of both IPO epochs (IPO− and IPO+) may result in biased estimates of long-run flood risk. For example, the use of flood data from an IPO+ period would likely lead to a large underestimate of the long-run (marginal) flood risk; whereas a relatively unbiased estimate should be obtainable when using just IPO+ data. In cases with inadequate coverage of one of the IPO epochs, the prospect of significant bias in long-run flood risk is high. It, therefore, may be necessary to use a regional flood frequency distribution that can been obtained from a data set containing sufficient samples from both IPO-positive and IPO-negative periods to augment the limited at-site data.

Table 2. GML Estimators of the Median, Conditional on Three Values of the Southern Oscillation Index (SOI), and 95% Confidence Intervals

	SOI = −3.16	SOI = 0.04	SOI = 2.04
$GEV_{1,1}$	24 (21–28)	24 (21–28)	24 (21–28)
$GEV_{2,1}$	54 (51–58)	23 (19–27)	4 (0.5–7)
$GEV_{3,1}$	77 (72–82)	21 (18–24)	17 (15–22)

8. BIVARIATE FREQUENCY ANALYSIS OF RAINFALL EVENTS

Many water resources projects require the joint probability distributions of rainfall characteristics (i.e., rainfall intensity, depth, and duration), which may or may not be correlated. *Cordova and Rodriguez-Iturbe* [1985] found that the correlation structure of rainfall intensity and duration had a significant effect on surface runoff. *Hashino* [1985] generalized the Freund bivariate exponential distribution [*Freund*, 1961] to represent the joint probability distribution of rainfall intensity and maximum storm surge in Osaka Bay, Japan.

The development of models, which take into account the dependencies between all rainfall characteristics of interest, involves the application of multivariate statistical methods. The mathematical theory of the univariate models for extreme events is well developed and applied in climatology and hydrology. However, for multivariate modeling, there are fewer distributions of interest, and the most used are the multivariate normal, bivariate gamma [*Yue et al.*, 2001], and bivariate extreme value distributions [*Adamson et al.*, 1999]. In most cases, the use of a multivariate normal distribution is not appropriate to model maximum discharges because marginal distributions are asymmetric and have a heavy tail. The dependence structure is generally different from the Gaussian case described by Pearson's correlation coefficient.

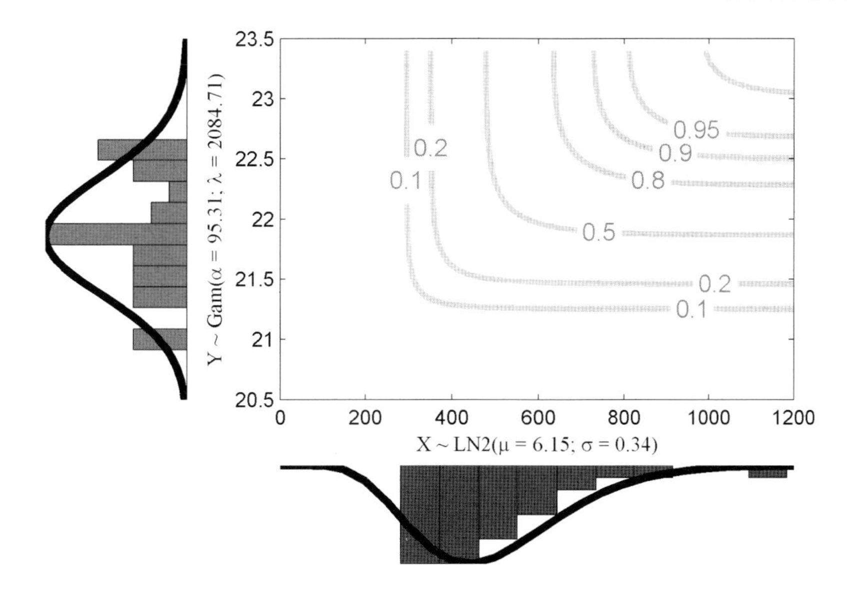

Figure 6. Bivariate probability of nonexceedance obtained from Frank copula, gamma, and lognormal distributions as marginals.

Furthermore, in the case of more complex marginal distributions, such as mixtures of distributions, it is not possible to use "classical" bivariate distributions (Figure 6).

A more generalized approach can be based on the notion of copulas [*Sklar*, 1959]. A copula is a distribution function whose margins are uniform. The multidimensional models of extreme values are the limit distributions of the joint distribution of margin maximum. If $F1$ and $F2$ are the marginal distributions of the extreme values, the associated limit distributions are of the form $C(F1(X1),F2(X2))$, where C is a copula (bivariate uniform distribution). A copula is useful to implement efficient algorithms for simulating joint distributions in a more realistic way. Indeed, copulas are able to model the dependence structure independently of the marginal distributions. Thus, it is possible to build multidimensional distributions with different margins, and the structure of dependence will be given by the copula.

The crucial step in the modeling process is the choice and the fit of the copula function which best fits the data. Copulas have been widely used in the financial domain and recently in climatology and hydrology to estimate the combined risk [see, e.g., *Salvadori and De Michele*, 2004; *Bárdossy*, 2006; *Singh and Zhang*, 2007; *El Adlouni and Ouarda*, 2008; *Gargouri-Ellouze and Chebchoub*, 2008; *Chebana and Ouarda*, 2010].

One of the most used families of copulas in hydrometeorological studies is the Archimedean family of copulas. *Genest and MacKay* [1986] defined bivariate Archimedean copulas as the following: $C(u_1,u_2) = \varphi^{-1}(\varphi(u_1) + \varphi(u_2))$, for $0 \le u_1,u_2 \le 1$, where φ is the generator of the copula. The most commonly used copulas are presented in Table 3, which shows that different choices of the generator yield several important bivariate families of copulas. A generator determines uniquely (up to a scalar multiplier) an Archimedean copula.

Figure 7 illustrates the cumulative distribution function and probability distribution function for the Clayton copula. As mentioned, Archimedean copulas allow representation of many dependence structures, and statistical inference for this family is well known in the literature.

Table 3. Examples of Archimedean Copulas and Their Generators

Copulas	Generator $\varphi(t)$	Parameter θ	Bivariate Copula $C_\varphi(u_1,u_2)$
Independent	$-\ln(t)$		u_1u_2
Clayton	$t^{-\theta} - 1$	$\theta > 0$	$(u_1^{-\theta} + u_2^{-\theta} - 1)^{-1/\theta}$
Frank	$-\ln\left(\frac{e^{-\theta t}-1}{e^{-\theta}-1}\right)$	$\theta \in \mathbb{R}$	$-\frac{1}{\theta} ln\left(1 + \frac{(e^{-\theta u_1}-1)(e^{-\theta u_2}-1)}{(e^{-\theta}-1)}\right)$

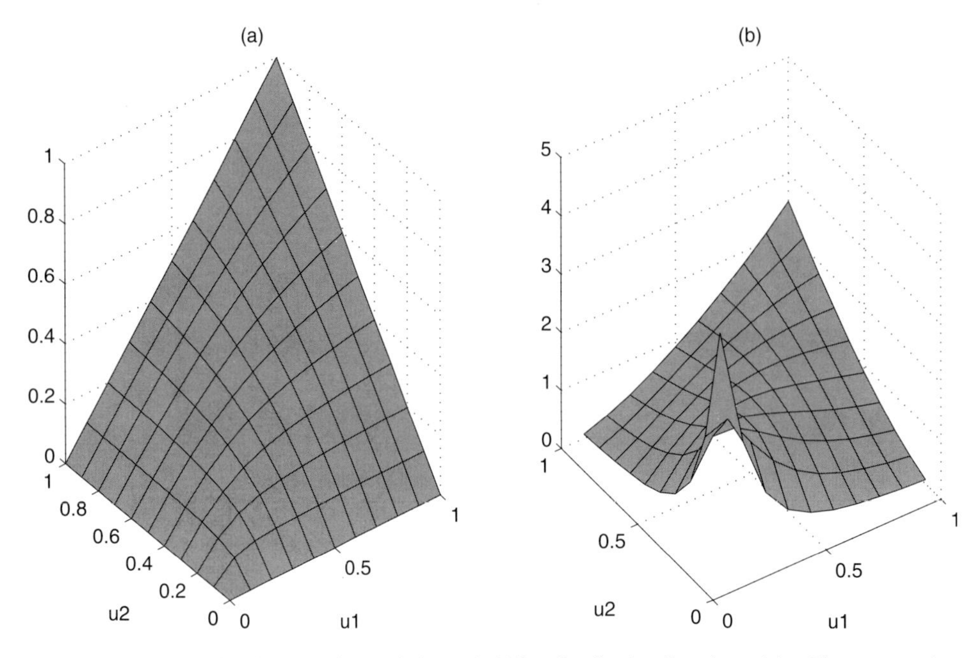

Figure 7. (a) Cumulative distribution function and (b) probability distribution function of the Clayton copula.

9. REGIONAL FREQUENCY ANALYSIS OF EXTREME RAINFALL EVENTS

Regional frequency analysis is commonly used for the estimation of extreme hydrological or meteorological events (such as floods or extreme rainfall events) at sites where little or no data are available [*Ouarda et al.*, 2001]. A number of regional frequency analysis procedures have been developed and used for the regionalization of rainfall events [*Baeriswyl and Rebetez*, 1997; *Fermindez Mills et al.*, 1994; *Fermindez Mills*, 1995; *Wotling et al.*, 2000; *Sveinsson et al.*, 2002; *Dinpashoha et al.*, 2004; *Durrans and Kirby*, 2004]. Most of these procedures are based on the use of multivariate statistical approaches such as canonical correlation analysis, principal component analysis, or factorial analysis. A literature review of regional precipitation frequency analysis procedures was presented in the work of *St-Hilaire et al.* [2003].

Onibon et al. [2004] presented a regional frequency analysis procedure for annual maximum daily precipitations in the province of Quebec, Canada. The procedure is composed of three steps. In the first step, homogeneous regions are delineated within the area of study, and the homogeneity within each region is tested. The delineation of homogenous regions is carried out based on L moment ratios. In the second step, the regional distribution is identified and its parameters estimated. The GEV distribution was identified as an appropriate regional distribution, and its parameters are estimated based on the computation of the regional L-CV, L-CS, and the mean of annual maximum daily precipitations. In the third step, precipitation quantiles corresponding to various return periods are estimated at the ungauged site. The procedure was shown to be robust and efficient with values of the root mean square error below 10% for a return period of 20 years and 20% for a return period of 100 years.

NOTATION

GEV	generalized extreme value distribution.
EV1	Gumbel distribution (or GUM).
EV2	Fréchet distribution.
HA	Halphen type A distribution.
HB	Halphen type B distribution.
HIB	Halphen type inverse B distribution.
G	gamma distribution.
IG	inverse gamma distribution.
f_D	probability density function (pdf) of the distribution D.
F_D	probability distribution of the distribution D.
A	arithmetic mean.
G	geometric mean.
H	harmonic mean.
δ_1	moment ratio ln(A/G).
δ_2	moment ratio ln(G/H).
μ_r	rth central moment.
$[\mu_r']_D$	rth noncentral moment of the distribution D.
Cv	coefficient of variation ($\sqrt{\mu_2}/\mu_1$).
Cs	coefficient of skewness ($\mu_3/\mu_2^{3/2}$).

QT quantile of return period T years
QT (D1/D2) QT estimated by fitting the distribution D1 for sample generated from the distribution D2.
n sample size.
RB relative bias.

REFERENCES

Adamson, P. T., A. V. Metcalfe, and B. Parmentier (1999), Bivariate extreme value distributions: An application of the Gibbs sampler to the analysis of floods, *Water Resour. Res.*, *35*, 2825–2832.

Akaike, H. (1974), A new look at the statistical model identification, *IEEE Trans. Autom. Control*, *19*(6), 716–723.

Aldenberg, T., and J. S. Jaworska (2000), Estimation of the hazardous concentration and fraction affected for normally distributed species sensitivity distributions, *Ecotoxicol. Environ. Saf.*, *46*, 1–18.

Anderson, R. L. (1941), Distribution of the serial correlation coefficients, *Ann. Math. Stat.*, *8*(1), 1–13.

Baeriswyl, P.-A., and M. Rebetez (1997), Regionalization of precipitation in Switzerland by means of principal component analysis, *Theor. Appl. Climatol.*, *58*, 31–41.

Bárdossy, A. (2006), Copula-based geostatistical models for groundwater quality parameters, *Water Resour. Res.*, *42*, W11416, doi:10.1029/2005WR004754.

Barnett, V., and T. Lewis (1994), *Outliers in Statistical Data*, 3rd ed., John Wiley, New York.

Battagila, F., and L. Orfei (2002), Outlier detection and estimation in nonlinear time series, *J. Time Ser. Anal.*, *26*, 107–121.

Beaulieu, C., O. Seidou, T. B. M. J. Ouarda, X. Zhang, G. Boulet, and A. Yagouti (2008), Intercomparison of homogenization techniques for precipitations data, *Water Resour. Res.*, *44*, W02425, doi:10.1029/2006WR005615.

Beaulieu, C., O. Seidou, T. B. M. J. Ouarda, and X. Zhang (2009), Intercomparison of homogenization techniques for precipitation data continued: Comparison of two recent Bayesian change point models, *Water Resour. Res.*, *45*, W08410, doi:10.1029/2008WR007501.

Beirlant, J., Y. Goegebeur, J. Segers, and J. Teugels (2004), *Statistics of Extremes: Theory and Applications*, 490 pp., Wiley, Chichester, U. K.

Benson, M. A. (1968), Uniform flood-frequency estimating methods for federal agencies, *Water Resour. Res.*, *4*, 891–908.

Bernier, J. (1959), Comparaison des lois de Gumbel et de Fréchet sur l'estimation des débits maxima de crue, *La Houille Blanche (SHF)*, *1*, 681.

Bobée, B. (1999), Extreme flood events valuation using frequency analysis: A critical review, *Houille Blanche*, *54*(7–8), 100–105.

Bobée, B., and F. Ashkar (1991), *The Gamma Family and Derived Distributions Applied in Hydrology*, 217 pp., Water Resour. Publ., Littleton, Colo.

Bobée, B., and P. F. Rasmussen (1995), Recent advances in flood frequency analysis, *Rev. Geophys.*, *33*(S1), 1111–1116.

Bradu, D., and D. M. Hawkins (1995), An Anscombe type robust regression statistic, *Comput. Stat. Data Anal.*, *20*, 355–386.

Brooks, C. E. P., and N. Carruthers (1953), *Handbook of Statistical Methods in Meteorology*, 85–111, Her Majesty's Stationery Office, London.

Chebana, F., and T. B. M. J. Ouarda (2010), Multivariate quantiles in hydrological frequency analysis, Environmetrics, doi:10.1002/env.1027.

Chebana, F., S. El Adlouni, and B. Bobée (2010), Mixed estimation methods for Halphen distributions with applications in extreme hydrologic events, *Stochastic Environ. Res. Risk Assess.*, *24*(3), 359–376.

Chowdhury, J. U., J. R. Stedinger, and L. H. Lu (1991), Goodness-of-fit tests for regional generalized extreme value flood distribution, *Water Resour. Res.*, *27*(7), 1765–1777.

Clarke, R. T. (2002), Estimating trends in data from the Weibull and a generalized extreme value distribution, *Water Resour. Res.*, *38* (6), 1089, doi:10.1029/2001WR000575.

Coles, G. S. (2001), *An Introduction to Statistical Modeling of Extreme Values*, 208 pp., Springer, London.

Cordova, J. R., and I. Rodriguez-Iturbe (1985), On the probabilistic structure of storm surface runoff, *Water Resour. Res.*, *21*(5), 755–763.

Crosta, G. (1998), Regionalization of rainfall thresholds: An aid to landslide hazard evaluation, *Environ. Geol.*, *35*(2–3), 131–145.

Dinpashoha, Y., A. Fakheri-Farda, A. Moghaddamb, S. Jahanbakhshc, and M. Mirnia (2004), Selection of variables for the purpose of regionalization of Iran's precipitation climate using multivariate methods, *J. Hydrol.*, *297*, 109–123.

Dose, V., and A. Menzel (2004), Bayesian analysis of climate change impacts in phenology, *Global Change Biol.*, *10*, 259–272, doi:10.1111/j.1529-8817.2003.00731.x.

Durrans, S. R., and J. T. Kirby (2004), Regionalization of extreme precipitation estimates for the Alabama rainfall atlas, *J. Hydrol.*, *295*, 101–107.

El Adlouni, S., and B. Bobée (2007), Sampling techniques for Halphen distributions, *J. Hydrol. Eng.*, *12*(6), 592–604.

El Adlouni, S., and T. B. M. J. Ouarda (2008), Étude de la loi conjointe débit-niveau par les copules: Cas de la rivière Châteauguay, *Can. J. Civ. Eng.*, *35*(10), 1128–1137, doi:10.1139/L08-054.

El Adlouni, S., and T. B. M. J. Ouarda (2009), Joint Bayesian model selection and parameter estimation of the GEV model with covariates using birth-death Markov chain Monte Carlo, *Water Resour. Res.*, *45*, W06403, doi:10.1029/2007WR006427.

El Adlouni, S., T. B. M. J. Ouarda, X. Zhang, R. Roy, and B. Bobée (2007), Generalized maximum likelihood estimators for the non-stationary generalized extreme value model, *Water Resour. Res.*, *43*, W03410, doi:10.1029/2005WR004545.

El Adlouni, S., B. Bobée, and T. B. M. J. Ouarda (2008), On the tails of extreme event distributions, *J. Hydrol.*, *355*, 16–33.

Fermindez Mills, G. (1995), Principal component analysis of precipitation and rainfall regionalization in Spain, *Theor. Appl. Climatol.*, *50*, 169–183.

Fermindez Mills, G., X. Lana, and C. Serra (1994), Catalonian precipitation patterns: Principal component analysis and automated regionalization, *Theor. Appl. Climatol.*, *49*, 201–212.

Filliben, J. J. (1975), The probability plot correlation coefficient test for normality, *Technometrics*, *17*(1), 111–117.

Fisher, R. A., and L. H. C. Tippett (1928), Limiting forms of the frequency distribution of the largest or smallest member of a sample, *Proc. Cambridge Philos. Soc.*, *24*, 180–190.

Fitzgerald, D. L. (2005), Analysis of extreme rainfall using the log-logistic distribution, *Stochastic Environ. Res. Risk Assess.*, *19*, 249–257.

Fréchet, M. (1927), Sur la loi de Probabilité de l'Écart Maximum, *Ann. Soc. Pol. Math*, *6*, 93–116.

Freund, J. E. (1961), A bivariate extension of the exponential distribution, *J. Am. Stat. Assoc.*, *56*, 971–977.

Gargouri-Ellouze, E., and A. Chebchoub (2008), Modélisation de la structure de dépendance hauteur-durée d'événements pluvieux par la copule de Gumbel, *Hydrol. Sci. J.*, *53*(4), 802–817.

Genest, C., and R. J. MacKay (1986), Copules archimédiennes et familles de lois bidimensionnelles dont les marges sont données, *Can. J. Stat.*, *14*, 145–159.

Gumbel, E. J. (1942), On the frequency distribution of extreme values in meteorological data, *Bull. Am. Meteorol. Soc.*, *23*, 95–104.

Haan, C. T. (1977), *Statistical Methods in Hydrology*, 378 pp., Iowa State Univ. Press, Ames, Iowa.

Halphen, E. (1941), Sur un nouveau type de courbe de fréquence, C. R. Acad. Sci., 213, 633–635. (Due to war constraints, published under the name "Dugué")

Hamilton, J. (1991), A quasi-Bayesian approach to estimating parameters for mixtures of normal distributions, *J. Bus. Econ. Stat.*, *9*(1), 27–39.

Hashino, M. (1985), Formulation of the joint return period of two hydrologie variâtes associated with a Poisson process, *J. Hydrosci. Hydraul. Eng.*, *3*(2), 73–84.

Hershfield, D. M., and M. A. Kohler (1960), An empirical appraisal of the Gumbel extreme-value procedure, *J. Geophys. Res.*, *65*(6), 1737–1746.

Hosking, J. R. M. (1990), L-moments: Analysis and estimation of distributions using linear combinations of order statistics, *J. R. Stat. Soc., Ser. B*, *52*, 105–124.

Hosking, J. R. M., and J. R. Wallis (1997), *Regional Frequency Analysis—An Approach Based on L-Moments*, 240 pp., Cambridge Univ. Press, New York.

Intergovernmental Panel for Climate Change [IPCC] (2001), *Climate Change 2001: Impacts, Adaptation and Vulnerability*, 1042 pp., Cambridge Univ. Press, New York.

Jain, S., and U. Lall (2001), Floods in a changing climate: Does the past represent the future?, *Water Resour. Res.*, *37*(12), 3193–3205.

Jenkinson, A. F. (1955), The frequency distribution of the annual maximum (or minimum) of meteorological elements, *Q. J. R. Meteorol. Soc.*, *81*, 158–171.

Katz, R. W., M. B. Parlange, and P. Naveau (2002), Statistics of extremes in hydrology, *Adv. Water Resour.*, *25*, 1287–1304.

Kendall, M. G. (1975), *Rank Correlation Methods*, 272 pp., Charles Griffin, London.

Kharin, V. V., and F. W. Zwiers (2005), Estimating extremes in transient climate change simulations, *J. Clim.*, *18*, 1156–1173.

Kidson, R. L., and K. S. Richards (2005), Flood frequency analysis: Assumptions and alternatives, *Prog. Phys. Geogr.*, *29*(3), 392–410.

Kottegoda, N. T., and R. Rosso (1997), *Probability, Statistics, and Reliability for Civil and Environmental Engineers*, 768 pp., McGraw-Hill, New York.

Koutsoyiannis, D. (2004), On thc appropriateness of the Gumbel distribution for modelling extreme rainfall, in *Hydrological Risk: Recent Advances in Peak River Flow Modelling, Prediction and Real-time Forecasting. Assessment of the Impacts of Land-use and Climate Changes*, edited by A. Brath, A. Montanari, and E. Toth, pp. 303–319, Editoriale Bios, Castrolibero, Bologna, Italy.

Leadbetter, M. R., G. Lindgren, and H. Rootzen (1983), *Extremes and Related Properties of Random Sequences and Processes*, 336 pp., Springer, New York.

Leclerc, M., and T. B. M. J. Ouarda (2007), Non-stationary regional flood frequency analysis at ungauged sites, *J. Hydrol.*, *343*(3–4), 254–265.

Malamud, B. D., and D. L. Turcotte (2006), The applicability of power-law frequency statistics to floods, *J. Hydrol.*, *322*, 168–180.

Mann, H. B., and D. R. Whitney (1947), On a test of whether one of two random variables is stochastically larger than the other, *Ann. Math. Stat.*, *18*, 50–60.

Martins, E. S., and J. R. Stedinger (2000), Generalized maximum-likelihood generalized extreme-value quantile estimators for hydrologic data, *Water Resour. Res.*, *36*(3), 737–744.

Micevski, T., S. W. Franks, and G. Kuczera (2006), Multidecadal variability in coastal eastern Australian flood data, *J. Hydrol.*, *327*, 219–225.

Moreau, I., S. El Adlouni, T. B. M. J. Ouarda, P. Bruneau, L. Roy, M. Barbet, and B. Bobée (2008), Détection et identification des valeurs aberrantes DIVA: Guide pour l'interface Matlab, *Rapp. Rech. R-957*, 32 pp., INRS-ETE, Québec, Que., Canada.

Morlat, G. (1956), Les lois de probabilité de Halphen, *Rev. Stat. Appl.*, *3*, 21–43.

Morrison, J. E., and J. A. Smith (2003), Stochastic modeling of flood peaks using the generalized extreme value distribution, *Water Resour. Res.*, *38*(12), 1305, doi:10.1029/2001WR000502.

National Research Council of Canada (1989), *Hydrology of Floods in Canada: A Guide to Planning and Design* edited by W. E. Watt, 245 pp., NRCC, Ottawa.

Natural Environment Research Council (1975), Flood studies report, *vol. 1*, London.

Nguyen, V. T. V., T. D. Nguyen, and F. Ashkar (2002), Regional frequency analysis of extreme rainfalls, *Water Sci. Technol.*, *45*, 75–81.

Niemczynowicz, J. (1982), Areal intensity-duration-frequency curves for short term rainfall events, *Nord. Hydrol.*, *13*(4), 193–204.

Onibon, H., T. B. M. J. Ouarda, M. Barbet, A. St-Hilaire, B. Bobée, and P. Brunveau (2004), Analyse fréquentielle régionale des précipitations journalières maximales annuelles au Québec, Canada, *Hydrol. Sci. J.*, *49*(4), 717–735.

Ouarda, T. B. M. J., F. Ashkar, E. Bensaid, and I. Hourani (1994), *Statistical Distributions Used in Hydrology. Transformations and Asymptotic Properties, Scientific Report*, 31 pp., Department of Mathematics, Univ. of Moncton, New Brunswick.

Ouarda, T. B. M. J., C. Girard, G. Cavadias, and B. Bobée (2001), Regional flood frequency estimation with canonical correlation analysis, *J. Hydrol.*, *254*(1–4), 157–173.

Park, H. W., and H. Sohn (2005), Parameter estimation of the generalized extreme value distribution for structural health monitoring, *Probab. Eng. Mech.*, *21*(4), 366–376.

Perreault, L., B. Bobée, and P. F. Rasmussen (1999), Halphen distribution system. I: Mathematical and statistical properties, *J. Hydrol. Eng.*, *4*, 189–199.

Pilgrim, D. H. (Ed.) (1998), Australian Rainfall and Runoff: A Guide to Flood Estimation, Volumes I and II, Institution of Engineers Australia, Canberra.

Rao, A. R., and K. H. Hamed (2000), *Flood Frequency Analysis*, 350 pp., CRC Press, Boca Raton, Fla.

Robson, A., and D. W. Reed (1999), *Flood Estimation Handbook*, vol. 3, *Statistical Procedures for Flood Frequency Estimation*, p. 108, Centre for Ecology and Hydrology, Wallingford, U. K.

Rorabacher, D. B. (1990), Statistical treatment for rejection of deviant values: Critical values of Dixon's "Q" parameter and related subrange ratios at the 95% confidence level, *Anal. Chem.*, *63(2), 139–146.*

Salvadori, G., and C. De Michele (2004), Frequency analysis via copulas: Theoretical aspects and applications to hydrological events, *Water Resour. Res.*, *40*, W12511, doi:10.1029/2004WR003133.

Sankarasubramanian, A., and U. Lall (2003), Flood quantiles in a changing climate: Seasonal forecasts and causal relations, *Water Resour. Res.*, *39*(5), 1134, doi:10.1029/2002WR001593.

Scarf, P. A. (1992), Estimation for a four parameter generalized extreme value distribution, *Commun. Stat. Theory Methods*, *21*, 2185–2201.

Schwarz, G. (1978), Estimating the dimension of a model, *Ann. Stat.*, *6*, 461–464.

Sevruk, B. (1985), Correction of monthly precipitation for wetting losses, *Instrum. Obs. Methods Rep. 22*, 7–12, World Meteorol. Organ., Geneva, Switzerland.

Siegel, S. (1956), *Nonparametric Statistics for the Behavioural Sciences*, 312 pp., McGraw-Hill, New York.

Singh, V. P., and D. K. Prevert (2004), Review of mathematical models of large watershed hydrology, *J. Hydrol. Eng.*, *130*(1), 89–90.

Singh, V. P., and L. Zhang (2007), IDF curves using the Frank Archimedean copula, *J. Hydrol. Eng.*, *12*, 651–662, doi:10.1061/(ASCE)1084-0699(2007)12:6(651).

Sklar, A. (1959), Fonctions de répartition á n dimensions et leurs marges, *Publ. Inst. Stat. Univ. Paris*, *8*, 229–231.

Smith, R. L. (1989), Extreme value analysis of environmental time series: An application to trend detection in ground-level ozone (with discussion), *Stat. Sci.*, *4*, 367–393.

Stedinger, J. R., R. M. Vogel, and E. Foufoula-Georgio (1993), Frequency analysis of extreme events, in *Handbook of Hydrology*, edited by D. R. Maidment, pp. 18.1–18.66, McGraw-Hill, New York.

St-Hilaire, A., T. B. M. J. Ouarda, M. Lachance, B. Bobée, M. Barbet, and P. Bruneau (2003), Régionalisation des précipitacipitations: Une revue bibliographique des développements récents, *Rev. Sci. Eau*, *16*(1), 27–54.

Sveinsson, O. G. B., J. D. Salas, and D. C. Boes (2002), Regional frequency analysis of extreme precipitation in northeastern Colorado and Fort Collins flood of 1997, *J. Hydrol. Eng.*, *7*, 49–63, doi:10.1061/(ASCE)1084-0699(2002)7:1(49).

Terry, M. E. (1952), Some rank order tests which are most powerful against specific parametric alternatives, *Ann. Math. Stat.*, *23*, 346–366.

Tolvi, J. (2000), The effects of outliers on two nonlinearity tests, *Commun. Stat. Simulation Comput.*, *29*, 897–918.

U. S. Water Resources Council (1981), Guideline for Determining Flood Flow Frequency, Revised Bulletin 17B of the Hydrology Committee, September.

Venkataraman, S. (1997), Value at risk for a mixture of normal distributions: The use of quasi-Bayesian estimation technique, *Econ. Perspect.*, *21*, 2–13.

Vogel, R. M. (1986), The probability plot correlation coefficient test for the normal, lognormal and Gumbel distributional hypotheses, *Water Resour. Res.*, *22*(4), 587–590.

Vogel, R. M. (1987), Correction of the probability plot correlation coefficient test for the normal, lognormal and Gumbel distributional hypotheses, *Water Resour. Res.*, *23*(10), 2013.

Vogel, R. M., and C. N. Kroll (1989), Low flow frequency analysis using probability plot correlation coefficients, *J. Water Resour. Plann. Manage.*, *115*(3), 338–357.

Von Mises, R. (1954), La distribution de la plus grande de n valeurs, in *Selected Papers*, vol. II, pp. 271–294, American Mathematical Society, Providence, R. I.

Wald, A., and J. Wolfowitz (1943), An exact test for randomness in the non-parametric case based on serial correlation, *Ann. Math. Stat.*, *14*(4), 378–388.

Wang, X. L., and V. R. Swail (2004), Historical and possible future changes of wave heights in Northern Hemisphere oceans, in *Atmosphere Ocean Interactions*, vol. 2, edited by W. Perrie, Wessex Inst. of Technol., Southampton, U. K.

Wang, X. L., F. W. Zwiers, and V. Swail (2004), North Atlantic Ocean wave climate scenarios for the 21st century, *J. Clim.*, *17*, 2368–2383.

Wilks, D. S. (1993), Comparison of three-parameter probability distributions for representing annual extreme and partial duration precipitation series, *Water Resour. Res.*, *29*(10), 3543–3549.

Wotling, G., C. Bouvier, J. Danloux, and J.-M. Fritsch (2000), Regionalization of extreme precipitation distribution using the principal components of the topographical environment, *J. Hydrol.*, *233*, 86–101.

Yu, R. C., H. W. Teh, P. A. Jaques, C. Sioutas, and J. R. Froines (2004), Quality control of semi-continuous mobility size-fractionated particle number concentration data, *Atmos. Environ.*, *38*, 3341–3348.

Yue, S., T. B. M. J. Ouarda, and B. Bobée (2001), A review of bivariate gamma distribution for hydrological application, *J. Hydrol.*, *246*, 1–18.

Zalina, M. D., M. N. Desa, V. T. V. Nguyen, and M. K. Hashim (2002), Selecting a probability distribution for extreme rainfall series in Malaysia, *Water Sci. Technol.*, *45*, 63–68.

Zhang, J. (2007), Likelihood moment estimation for the generalized Pareto distribution, *Aust. N. Z. J. Stat.*, *49*(1), 69–77.

Zhang, X., K. D. Harvey, W. D. Hogg, and T. R. Yuzyk (2001), Trends in Canadian streamflow, *Water Resour. Res.*, *37*(4), 987–999.

Zhang, X., F. W. Zweirs, and G. Li (2004), Monte Carlo experiments on the detection of trends in extreme values, *J. Clim.*, *17*, 1945–1952.

S. El Adlouni, Institut National de Statistique et d'Économie Appliquée, INSEA, Rabat 10100, Morocco. (el_adlouni@yahoo.com)

T. B. M. J. Ouarda, Institut National de la Recherche Scientifique Eau, Terre et Environnement, Québec City, QC G1K 9A9, Canada.

Methods and Data Sources for Spatial Prediction of Rainfall

T. Hengl

Institute for Biodiversity and Ecosystem Dynamics, University of Amsterdam, Amsterdam, Netherlands

A. AghaKouchak

Department of Civil and Environmental Engineering, University of California, Irvine, California, USA

M. Perčec Tadić

Meteorological and Hydrological Service of Croatia, Zagreb, Croatia

This chapter reviews both rain gauge (point) data sources and remote-sensing (visible, thermal IR and microwave (MW)) imagery sources used for producing precipitation maps, and then shows "in action" a number of mechanical and stochastic spatial prediction methods that can be used to generate maps of rainfall intensity. Special focus was put on using geostatistical techniques implemented in the `R` environment for statistical computing (via `stats`, `gstat`, and `geoR` packages). The spatial prediction methods are illustrated using a small case study (97 points obtained from the National Climatic Data Center Global Summary of Day) covering the scanning radius of the Bilogora weather radar in Croatia (366 daily images) and the national rain gauge network in Italy (1901 stations). The results show that the rainfall estimated using ground-based radar can be of variable accuracy. The radar images can contain many artifacts, especially at high distances from the ground radar, so that the correlation with ground measurements is often marginal. Daily rainfall intensity is commonly skewed toward small values with many zeros, thus rainfall intensity estimates at shorter time intervals are suitable for modeling using zero-inflated regression models. The chapter contains code snippets showing how to implement various prediction techniques from local trend surfaces to ordinary kriging, zero-inflated regression models, and regression-kriging in `R`.

1. INTRODUCTION

One of the main objectives of atmospheric sciences is to generate maps of meteorological variables from ground-based station records. Rainfall is one such meteorological variable, spatial distribution of which is of interest for various climatic and ecological models: from hydrological runoff models to soil-plant response and pollutant transport models. An accurate map of rainfall is often crucial for the success of such applied models; hence, there is an increasing interest in procedures that can generate reliable maps of rainfall using limited point and remote-sensing data.

The most common techniques used to generate maps of rainfall range from various versions of splines, then various

Rainfall: State of the Science
Geophysical Monograph Series 191

10.1029/2010GM000999

types of regression (local and global), kriging, and their combination(s). Discussion is still ongoing about which method is optimal, although it is very likely that various methods will perform differently under different input conditions, different sampling designs, spatial and temporal support sizes; they probably also perform with various success in different regions in the world.

This chapter will guide you through the various analysis and spatial prediction steps using "state-of-the-art" geostatistical techniques. The emphasis is put on procedures and software tools and on how to generate maps for your own case study. Before analyzing real data, we present a review of basic geostatistics and list some important statistical aspects of rainfall data (section 1.2). We then focus on predicting rainfall for a smaller case study for which a time-series of radar images is also available (section 3) and then demonstrate use of an extensive national network of rain gauges (section 4). We introduce and demonstrate techniques ranging from mechanical interpolators to geostatistical techniques, and then compare and interpret results of predictions. To follow these exercises on your own, consider obtaining the R script and input data from the contact author's website. The code snippets presented in further text are shown for demonstration purposes only. They are also specific to the MS Window OS.

1.1. Spatial Prediction

Spatial prediction is a process of estimating values (and associated uncertainty) of some target variable at unvisited locations using field samples and auxiliary information [*Pebesma*, 2006; *Hengl*, 2009]. Generation of maps representing spatial patterns of features of interest is one of the main objectives of geostatistics. Note that, from the point of statistics, the term "spatial prediction" is preferred to "spatial interpolation," although the latter is more widely used in literature.

We can denote a set of observations of rainfall intensity (Z in mm Δ^{-1}) as $z(\mathbf{s}_1)$, $z(\mathbf{s}_2)$,. . ., $z(\mathbf{s}_n)$, where $\mathbf{s}_i = (x_i, y_i)$ is a location and x_i and y_i are the coordinates (primary locations) in geographical space, and n is the number of observations. The field samples of rainfall are usually based on ground-based rain gauge networks and/or climatic stations and are measured at point support. The geographical domain of interest (area, land surface, object) can be denoted as $\mathbb{A}$. The objective of spatial prediction is to determine values of the target variable at some new location $\mathbf{s}_0 \in \mathbb{A}$ using an analytical spatial prediction model.

According to *Lanza et al.* [2001], there are only two main groups of spatial prediction models: "data-driven" (stochastic) and "physically based" (atmospheric) models. In this chapter, we will mainly focus on the data-driven models, although we will also discuss some possibilities to build hybrid models that rely both on physical modeling of precipitation and ground data.

Data-driven spatial prediction models are based on generic mathematical models (e.g., splines, kriging), whose parameters are adjusted to local data using some optimization criteria. In most cases, the model parameters are set to pass "through" ground data, so that the resulting map also closely reflects the spatial pattern in the data. Depending on whether the model parameters are estimated objectively or are adjusted by the user, statisticians also tend to distinguish between "mechanical" and "stochastic" spatial predictors. For example, splines and inverse weighted distance interpolators are mechanical interpolation models because the smoothness and similar parameters can be set subjectively by the user.

In the case of splines with tension, values at new location can be determined by using

$$\hat{z}(\mathbf{s}_0) = a_1 + \sum_{i=1}^{n} w_i \cdot R(\upsilon_i) \tag{1}$$

where the a_1 is a constant, w_i are the weights, and $R(\upsilon_i)$ is the radial basis function determined using [*Mitasova and Mitas*, 1993]

$$R(\upsilon_i) = -[E_1(\upsilon_i) + ln(\upsilon_i) + C_E] \tag{2}$$

$$\upsilon_i = \left[\varphi \cdot \frac{\mathbf{h}_0}{2}\right]^2, \tag{3}$$

where $E_1(\upsilon_i)$ is the exponential integral function, C_E = 0.577215 is the Euler constant, φ is the generalized tension parameter, and $\mathbf{h}_0$ is the distance between the new and the interpolation point ($d\{h_i, h_0\}$). The coefficients a_1 and w_i are obtained by solving the system:

$$\sum_{i=1}^{n} w_i = 0 \tag{4}$$

$$a_1 + \sum_{i=1}^{n} w_i \cdot \left[R(\upsilon_i) + \delta_{ij} \cdot \frac{\varpi_0}{\varpi_i}\right] = z(\mathbf{s}_i);\ j = 1, . . , n, \tag{5}$$

where ϖ_0/ϖ_i are positive weighting factors representing a smoothing parameter at each given point $\mathbf{s}_i$. The tension parameter φ controls the distance over which the given points influence the resulting surface, while the smoothing parameter controls the vertical deviation of the surface from the points. By using an appropriate combination of tension and smoothing, one can produce a surface that accurately fits the empirical knowledge about the expected variation

[*Mitasova et al.*, 2005]. Again, these parameters are typically set by the user, although various cross-validation procedures exist that can be used to estimate them more objectively [*Mitasova et al.*, 2005].

Splines without external predictors are, in a way, very similar to local polynomial regression fitting and universal kriging with spatial coordinates as predictors. These three techniques typically produce maps that (at least visually) do not differ significantly [*Durbrule*, 1983; *Mitasova et al.*, 2005].

Sharples and Hutchinson [2004] proposed an extension of the splines with tension model to additive regression splines:

$$z(\mathbf{s}) = \beta_0(\mathbf{s}) + \sum_{j=1}^{p} \beta_j \cdot q_j(\mathbf{s}) + \varepsilon, \tag{6}$$

where q are external predictors, also known as "covariates" or explanatory variables (need to be available at any location within $\mathbb{A}$), and ε is the error term. The functions $f[q_0,\ldots,q_p]$ may be estimated by minimizing

$$\sum_{i=1}^{n} \left(\mathbf{z} - \beta_0(\mathbf{s}_i) - \sum_{j=1}^{p} \beta_{ij} \cdot q_j(\mathbf{s}_i) \right)^2 + \sum_{j=0}^{p} \rho_j \cdot J_d^m(q_j) \tag{7}$$

where $J_d^m(q)$ is the m-order roughness penalty [*Sharples and Hutchinson*, 2004]. This means that the data model can also be simply thought of as spatially varying linear regression with the constant parameters found in linear regression models replaced by the multivariate functions $q_0,\ldots,q_p$ (additive components). Thin-plate smoothing spline interpolation has been implemented in the ANUSPLIN (http://fennerschool.anu.edu.au/publications/software/anusplin.php) software [*Hutchinson*, 1995], and has been used for example by *Hijmans et al.* [2005] to generate the 1 km world map of precipitation.

Stochastic spatial prediction models aim to obtain unbiased, minimum variance estimates of rainfall at new locations. Unlike mechanical predictors, stochastic prediction models also typically provide estimates of uncertainty, which is increasingly used in decision making and for error assessment. In mathematical terms, a stochastic spatial prediction model draws realizations, either the most probable or a set of equiprobable realizations, of the feature of interest given a list of inputs:

$$\hat{z}(\mathbf{s}_0) = E\{Z|z(\mathbf{s}_i), \quad q_k(\mathbf{s}_0), \quad \gamma(\mathbf{h}), \quad \mathbf{s} \in \mathbb{A}\}, \tag{8}$$

where $z(\mathbf{s}_i)$ is the input point data set, $\gamma(\mathbf{h})$ is the covariance model defining the spatial autocorrelation structure, and $q_k(\mathbf{s}_0)$ is the list of deterministic predictors. Stochastic spatial prediction models also assume that we deal with only one reality (samples $z(\mathbf{s}_n)$), which is a realization of a process ($\mathbf{Z} = \{Z(\mathbf{s}), \forall \mathbf{s} \in \mathbb{A}\}$) that could have produced many realities.

All versions of linear statistical predictors share the same objective of minimizing the estimation error variance $\hat{\sigma}_E^2(\mathbf{s}_0)$ under the constraint of unbiasedness [*Goovaerts*, 1997]. In mathematical terms, the estimation error:

$$\hat{\sigma}^2(\mathbf{s}_0) = E\left\{ (\hat{z}(\mathbf{s}_0) - z(\mathbf{s}_0)) \cdot (\hat{z}(\mathbf{s}_0) - z(\mathbf{s}_0))^T \right\} \tag{9}$$

is minimized under the (unbiasedness) constraint that

$$E\{\hat{z}(\mathbf{s}_0) - z(\mathbf{s}_0)\} = 0, \tag{10}$$

which leads to a generalized linear regression model [*Goovaerts*, 1997]:

$$z(\mathbf{s}) = \mathbf{q}^{\mathbf{T}} \cdot \beta + \varepsilon(\mathbf{s}) \tag{11}$$

$$E\{\varepsilon(\mathbf{s})\} = 0 \tag{12}$$

$$E\{\varepsilon(\mathbf{s}) \cdot \varepsilon^{\mathbf{T}}(\mathbf{s})\} = \mathbf{C}, \tag{13}$$

where β is the vector of regression coefficients, ε is the residual variation, and $\mathbf{C}$ is the $n \times n$ positive-definite variance-covariance matrix of residuals. This model can be read as follows: (1) the information signal is a function of deterministic and residual parts, (2) the best estimate of the residuals is 0, and (3) the best estimate of the correlation structure of residuals is the variance-covariance matrix. The best linear unbiased prediction of the target variable is then

$$\hat{z}(\mathbf{s}_0) = \hat{\delta}_0^{\mathbf{T}} \cdot \mathbf{z} = \sum_{i=1}^{n} w_i(\mathbf{s}_0) \cdot z(\mathbf{s}_i), \tag{14}$$

which is the general form used to represent almost any interpolation technique. In most simple terms, spatial prediction is always a sort of a weighted average of sampled values; the issue is only how to determine the weights.

The parameters of the Best Linear Unbiased Predictor (BLUP) model can be obtained by solving the following system [*Goovaerts*, 1997]:

$$\begin{bmatrix} \mathbf{C} & \mathbf{q} \\ \mathbf{q}^{\mathbf{T}} & 0 \end{bmatrix} \cdot \begin{bmatrix} \delta \\ \phi \end{bmatrix} = \begin{bmatrix} \mathbf{c}_0 \\ \mathbf{q}_0 \end{bmatrix}, \tag{15}$$

where $\mathbf{c}_0$ is the vector of $n \times 1$ covariances at a new location, $\mathbf{q}_0$ is the vector of $p \times 1$ predictors at a new location (e.g., elevation, distance to the coast line, atmospheric images), and ϕ is a vector of Lagrange multipliers. It can be further shown that, by solving equation (15), we get the following model [*Christensen*, 2001, p. 277]:

$$\hat{z}(\mathbf{s}_0) = \mathbf{q}_0^{\mathbf{T}} \cdot \hat{\beta} + \hat{\lambda}_0^{\mathbf{T}} \cdot (\mathbf{z} - \mathbf{q} \cdot \hat{\beta})$$

$$\hat{\beta} = (\mathbf{q}^{\mathbf{T}} \cdot \mathbf{C}^{-1} \cdot \mathbf{q})^{-1} \cdot \mathbf{q}^{\mathbf{T}} \cdot \mathbf{C}^{-1} \cdot \mathbf{z} \tag{16}$$

$$\hat{\lambda}_0 = \mathbf{C}^{-1} \cdot \mathbf{c}_0,$$

which is the regression-kriging spatial prediction model, known in the literature under the names "Universal Kriging" and "Kriging with External Drift" [*Hengl*, 2009].

Under the assumption of the first order stationarity, i.e., constant trend (or nonexistent trend):

$$E\{z(\mathbf{s})\} = \mu \qquad \forall \mathbf{s} \in \mathbb{A} \tag{17}$$

the regression-kriging (equation 16) modifies to

$$\hat{z}(\mathbf{s}_0) = \mu + \hat{\lambda}_0^{\mathbf{T}} \cdot (\mathbf{z} - \mu)$$

$$\hat{\lambda}_0 = \mathbf{C}^{-1} \cdot \mathbf{c}_0,$$

i.e., to ordinary kriging [*Matheron*, 1969], which is still one of the most widely used techniques in geostatistics. Note that the covariances in equation (15) are determined by fitting a variogram model $\gamma(\mathbf{h})$, e.g., the exponential model:

$$\gamma(\mathbf{h}) = \begin{cases} 0 & \text{if} \quad |\mathbf{h}| = 0 \\ C_0 + C_1 \cdot \left[1 - e^{-\left(\frac{\mathbf{h}}{R}\right)}\right] & \text{if} \quad |\mathbf{h}| > 0, \end{cases} \tag{18}$$

where $\mathbf{h}$ is the distance between the point pairs and C_0, C_1, and R are nugget, sill, and range parameters. A variogram of the target variable can be estimated by deriving and plotting the so-called semivariances, differences between the pairs of values:

$$\gamma(\mathbf{h}) = \frac{1}{2} E\left[\left(z(\mathbf{s}_i) - z(\mathbf{s}_i + \mathbf{h})\right)^2\right], \tag{19}$$

where $z(\mathbf{s}_i)$ is the value of a target variable at some sampled location and $z(\mathbf{s}_i + \mathbf{h})$ is the value of target variable at distance $\mathbf{s}_i + \mathbf{h}$ (see further Figure 6).

Kriging can be used to produce both predictions and simulations, i.e., equiprobable realization of a rainfall intensity. Use of simulations of rainfall intensity is especially important for the assessment of the propagated error affecting, e.g., hydrological models [*Teo and Grimes*, 2007; *AghaKouchak et al.*, 2010].

The results from numerous case studies have shown that, by incorporating elevation and similar topographic variables in the interpolation of rainfall measurements through regression-kriging, it is possible to produce more accurate maps than if pure geographical techniques are used [*Goovaerts*, 2000; *Kyriakidis et al.*, 2001; *Lloyd*, 2005]. Hence, hybrid geostatistical techniques such as regression-kriging are now, in general, preferred methods for geostatistical mapping of climatic variables.

1.2. Specific Properties of Rainfall Data

Before we start generating spatial predictions of rainfall, it is first necessary to be clear about its intrinsic properties. First of all, the term "rainfall" (or more accurately "precipitation") can mean various things to various research groups. The focus of geostatistical mapping is usually prediction of the rainfall intensity (i.e., volume or mass of rainfall per time unit), although other rainfall-type variables can also be of interest, e.g., rainfall status (yes/no indicator variable), rainfall duration, precipitation type (snow, ice, rain type), and others. Modeling the spatial distribution of snow or rain duration typically requires different inputs and even different mathematical models than those described in this chapter. In this chapter, we also mainly focus on mapping rainfall intensity expressed in mm d^{-1} or mm $month^{-1}$.

Rainfall intensity data has several specific features that make it different from various similar meteorological and environmental data. Most importantly, daily rainfall measurements contain many zeros, even longer periods with absolutely no rainfall within the domain of interest. The distribution of the rainfall data is typically heavily skewed toward small values or even "Poissonic," so that one needs to consider fitting Generalized Linear Models and/or General Additive Models with a zero-inflated distribution or similar, instead of using a simple linear regression modeling [*Kamarianakis et al.*, 2008].

Zero-inflated models are capable of dealing with excess zero counts [*Ridout et al.*, 1998; *Zeilies et al.*, 2008]. For example, zero-inflated Poisson (ZIP) distribution can be used to model count data for which the portion of counts is greater than expected, based on the nonzero counts. The ZIP model is an adjustment of the Poisson regression model:

$$\log[E(z_i|\mathbf{q}_i)] = \log(\mu_i) = \mathbf{q}^{\mathbf{T}} \cdot \beta, \tag{20}$$

with different probability distributions for zero and positive values [*Ridout et al.*, 1998]:

$$\Pr(Z = z) = \begin{cases} \omega + (1 - \omega) \cdot \exp(-v) & z = 0 \\ (1 - \omega) \cdot \exp(-v) \cdot v^z / z! & z > 0 \end{cases}, \tag{21}$$

where $\log(\mu_i)$ is the target variable in transformed space. The basic ZIP model is

$$\log(v) = \mathbf{q} \times \beta \quad \text{and} \quad \log\left(\frac{\omega}{1-\omega}\right) = \tau \cdot \mathbf{q} \times \beta, \tag{22}$$

where τ is the scalar parameter ($\omega = (1 + \nu^{-\tau})^{-1}$), ν is the Poisson parameter, β is the vector of regression coefficients, and μ is the link function, in this case:

$$E\{z\} = \mu = (1 - \omega) \cdot \nu. \tag{23}$$

Zero-inflated models have been shown to produce fits that are more significant than regression models that ignore that the zero values have higher probability [*Zeilies et al.*, 2008]. Especially if we aim at interpolating daily measurements of rainfall, we need to consider using such models to account for dominant zero values. Note also that rainfall measurements are not really "count-type" data, but can be easily converted to integers, expected by the Poisson model, by dividing the values in original scale with measurement error (0.1 mm for gauge readings); after the regression modeling, the integers can be back-transformed to the original scale.

The regression-kriging model in equation (16) permits a combination of ZIP and geostatistics because we can, instead of using a list of linear predictors, first fit a ZIP model, then use the predicted trend in the regression-kriging framework:

$$\hat{\mu}(\mathbf{s}_0) = \hat{\beta}_0 + \hat{\beta}_1 \cdot \mu^{ZIP}(\mathbf{s}_0) + \hat{\lambda}_0^{\mathbf{T}} \cdot e^{ZIP}(\mathbf{s}_0) \tag{24}$$

$$\hat{z}(\mathbf{s}_0) = \exp(\hat{\mu}(\mathbf{s}_0)) \tag{25}$$

where μ^{ZIP} is the link-variable fitted using ZIP regression, and e^{ZIP} is the ZIP residual in transformed space. *Agarwal et al.* [2002] further show how to estimate ZIP regression parameters for spatial count data by using a Bayesian framework. Note that this procedure is, in general, computationally demanding and hence not recommended for large data sets.

An alternative to using ZIP models to predict daily rainfall is separation of prediction of rainfall for areas of positive and no rainfall. For example, *Sun et al.* [2003] propose a double kriging procedure where first indicator kriging is used to delineate areas of positive rainfall, then the actual rainfall intensity values can be predicted for the areas of positive rainfall. A similar concept is, in fact, implemented via the `zeroinfl` model as implemented in the `pscl` package. This method fits a Poisson regression model to the positive values and then model positive and zero values using a binomial model with logit link function [*Zeilies et al.*, 2008].

Second, rainfall is very dynamic, meaning that it has a high temporal variation. Probability of rain is determined by physical factors such as topography and atmospheric conditions, but rain often follows a chaotic behavior. The relation between topography and previous rainfall term is not so obvious for rainfall intensities of short duration [*Kebaili Bargaoui and Chebbi*, 2009]. Even excessive rain can happen over the domain of interest that has shown no relation with local topography or geographic location. Rain can be also very short. The rain reaches the ground within 30 min from the beginning of convective cloud formation and in 1–3 h in case of stratiform clouds [*Houze*, 1993], so that meteorological images of clouds need to be collected for very short time intervals. By increasing the temporal support of rainfall measurements, from days to month to years, spatial patterns of rainfall will become more and more distinct (see further Figure 1). *Chen et al.* [2002] show, using global data,

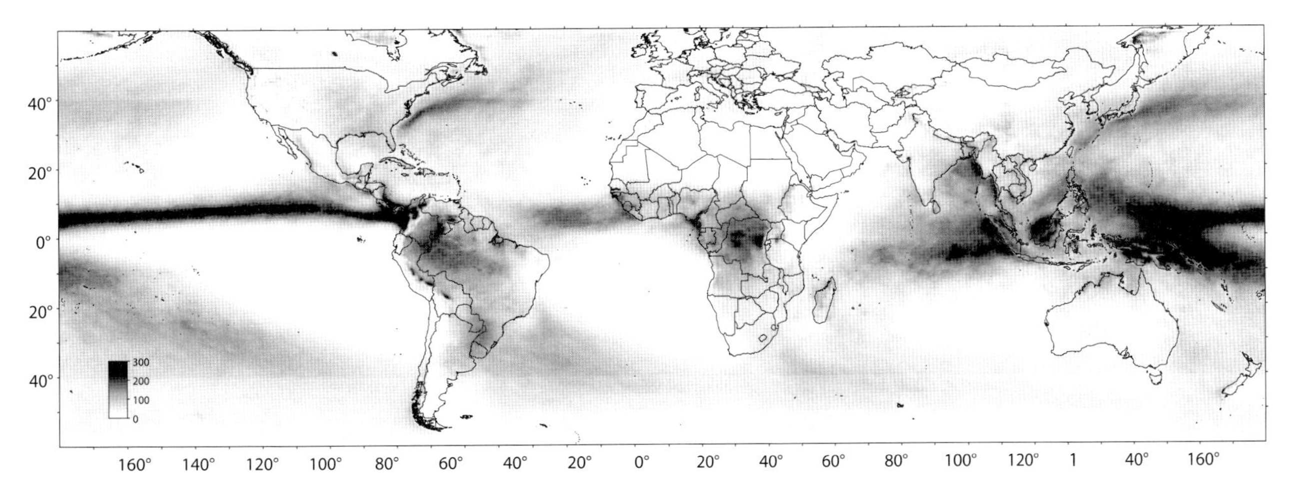

Figure 1. Global long-term precipitation pattern (mm month^{-1}) based on the GSMaP images (2003–2006). Compare with the 1-arcdegree map of global precipitation for 1979–2001 by *Yin et al.* [2004]. For more details about the GSMaP project see the work of *Kubota et al.* [2007].

that the models to explain spatial distribution of rainfall will become more significant if one increases the temporal support: to produce "stable" patterns of rainfall, one needs at least 10 years of observations. Modeling of rainfall patterns for shorter periods (days, weeks, and months) is also possible, but one needs to be aware that these models will often be less significant or even contradict with our knowledge of the physical processes.

Third, atmospheric images of clouds have a high potential of explaining rainfall patterns. One might anticipate that, in the near future, meteorological images of rainfall will be so accurate that we will not need to interpolate point data any more. However, even if meteorological images are of very high quality (very fine spatial resolution, temporal coverage of less than few minutes, etc.), we still require ground measurements of rainfall to calibrate those models. So our assumption is that both data-driven and process-driven models will remain important for geostatistical analysis of rainfall, and a hybrid model (equation 24) is probably the best approach to spatial prediction of rainfall.

1.3. Software: `R+OSGeo`

To demonstrate spatiotemporal modeling of rainfall values, we use a small case study containing time-series of hourly radar images of rainfall for a whole year. We process these data using a combination of open source packages developed for the `R` environment for statistical computing [*R Development Core Team*, 2009] and open source GIS packages `SAGA GIS` (http://saga-gis.org) and `FWTools` (http://fwtools.maptools.org). The combination of `R` and open source GIS packages will be in further text referred to as `R+OSGeo`. For more information about `R+OSGeo`, please refer to *Bivand et al.* [2008] and/or *Hengl* [2009].

`R` offers a wide family of spatial prediction models, available either via the contributed or basic packages. For example, geostatistical methods are available via `gstat` (http://cran.r-project.org/web/packages/gstat/), `geoR` (http://cran.r-project.org/web/packages/geoR/), and `fields` packages, thin-plate spline interpolators are available via `MBA` (http://cran.r-project.org/web/packages/MBA/) and `fields` (http://cran.r-project.org/web/packages/fields/). *Pebesmat et al.*, [2009] has recently launched a wrapper package called `intamap` (http://www.intamap.org) that automates filtering of outliers, fitting of variograms, and selection of spatial predictors based on the inherent properties of the input data.

Because the possibilities for quantitative modeling data in `R` are constantly increasing, `R` syntax is now widely considered to be a lingua franca for postgraduate students. With the extensive support for analysis of spatial data and time series, this software is also increasingly attractive for climatological mapping. The advantage of using `R` is that, both model fitting and spatial predictions can typically be implemented in a few lines. Furthermore, the link between GIS and `R` operations is established via the `RSAGA` (http://cran.r-project.org/web/packages/RSAGA/) library [*Brenning*, 2008]. This library basically sends commands from `R` to `SAGA` command line and then reads the results of analysis back in `R`. `SAGA` has been primarily used for manipulation and calculations within spatial grids, while `R` is the central programming environment which stands "on top" of the external GIS applications. This means that the complete data analysis shown in this article is available in a single script. This script and the data set can be obtained from the corresponding author's website.

2. DATA SOURCES

2.1. Rainfall Measurements (Stations)

Rainfall intensity measurements are historically collected through networks of rain gauges, marked tubes that collect raindrops. Ground measurement of rainfall typically refers to a point support size and to a temporal support size of 24 h (mm/24 h). Readings are done usually early in the morning. Even though the precision of the rainfall reading is in tenths of millimeter, all rain collectors suffer from errors that are related to splash-out, in-splash, evaporative losses, and errors due to the gauge not being leveled accurately [*Thompson and Perry*, 1997]. The most important are errors due to wind that can carry away some of the raindrops that should have fallen into the gauge. This error can be up to 30% or more in exposed conditions [*Thompson and Perry*, 1997], which needs to be taken into account when evaluating the results of analysis.

Monitoring of rainfall is inexpensive: a rain gauge is easy to set up, readings are usually done manually by an observer, although automated systems are now also used increasingly. Consequently, the density of rain gauges in the world is high. First, there are many local or regional projects that are usually connected with various hydrological or land management applications. Almost every country in the world has its own National Meteorological Agency that daily collects data from a network of rain gauges. These data are increasingly being integrated to produce continental scale or even global data sets. The best known global meteorological data sets are the following [*Hijmans et al.*, 2005]:

1. The Global Historical Climatology Network (GHCN) contains monthly values for the 1950–2000 period for precipitation (20,590 stations), mean temperature (7280 stations), and minimum and maximum temperature (4966

stations). It has global coverage for precipitation and mean temperature, but there are large gaps in the geographic distribution of stations with minimum and maximum temperature data.

2. The WMO climatological normals database includes filtered monthly mean (3084 stations), minimum, and maximum (both 2504 stations) temperature and precipitation (4261 stations) measurements.

3. FAOCLIM, the FAO's global climate database, contains monthly values for precipitation (27,372 stations), mean temperature (20,825 stations), and minimum and maximum temperature (11,543 stations). It includes long-term averages (1960–1990) as well as time series data for temperature and precipitation.

4. The National Climatic Data Center (NCDC) Global Summary of Day (GSOD) provides daily meteorological measurements from over 9000 stations prepared by the NCDC in Asheville, North Carolina. This is now one of the most extensive global meteorological databases.

5. The Global Precipitation Climatology Centre is now the largest monthly precipitation station database of the world with data from more than 70,000 stations.

If all records from different projects/countries are put together, there is plenty of ground data in the world to accurately model rainfall in both space and time. *Chen et al.* [2002], for example, used gauge observations from over 17,000 stations (GHCN) to generate mean monthly and annual precipitation maps of the world. *Hijmans et al.* [2005] use meteorological data from a variety of sources (total of 47,554 locations) to produce a global 1 km resolution map of precipitation.

A problem is that the coverage of the rain gauges in GHCN and similar global data sets is heavily biased toward developed countries. Tropics, dislocated islands, and regions toward the poles are represented with few rain gauges for thousands of square kilometers. This means that the global spatial patterns of rainfall presented in, e.g., the works of *Chen et al.* [2002] and/or *Hijmans et al.* [2005] could be much improved using the atmospheric remote-sensing products.

2.2. Satellite Imagery (Precipitation Products)

As mentioned previously, conventional maps of precipitation and other climatic parameters are made by collecting ground observations locally at meteorological stations, which are then interpolated over large areas to produce complete maps. The focus of meteorology today is to "measure" rather than to "model." During the last two decades, satellite instruments have been designed and used that directly estimate parameters such as temperature, aerosols, cloud properties, rainfall intensity, snowfall, lightning intensity, and similar [*Levizzani et al.*, 2007]. Remote-sensing observations (images) at different wavelengths (visible, thermal IR, and MWs), in passive and active modes (MW radiometers and precipitation radar), and from low and geostationary orbits, are now analyzed and coupled to produce global estimates of meteorological parameters [*Prigent*, 2010].

In principle, there are four main sources of remote-sensing-based precipitation estimates: (1) visible/IR satellite images, (2) passive MW satellite images, (3) active MW satellite images, and (4) ground-based radar images. For a more detailed review of the remote-sensing systems used for precipitation monitoring, refer also to the work of *Kidd et al.* [this volume]. Here we review the RS data sources in the context of geostatistical mapping only.

One source of high-resolution visible/IR meteorological images is the Spinning Enhanced Visible and Infrared Instrument (SEVIRI) aboard the Meteosat Second Generation geostationary satellite. This instrument collects IR and thermal images at intervals of 15 to 30 min. The SEVIRI IR images can be used to derive precipitation rate based on the estimated cloud top temperature. In general, IR images have been a major source of satellite-based precipitation estimation, mainly due to their fine spatial and temporal resolution. Previous studies have shown that IR-based precipitation retrieval algorithms perform reasonably well where convective systems are dominant [*Arkin and Xie*, 1994]. A disadvantage of using visible/IR imagery for precipitation monitoring, in general, is that it records only cloud properties from the "top." In reality, clouds come in layers which can have different properties. For example, a cold precipitating cloud located below a warm nonprecipitating cloud may not be detected by IR sensors. Furthermore, an IR sensor may detect a cold nonprecipitating cloud as a precipitating system due to its cold temperature.

Another source of precipitation estimation is based on MW remote sensing, which is currently recognized as the most accurate satellite-based estimate of precipitation rate [*Adler et al.*, 2001]. The most used MW instruments focused on precipitation monitoring include the following: Tropical Rainfall Measuring Mission Microwave Imager; Advanced Microwave Scanning Radiometer on Aqua satellite; Advanced Microwave Sounding Unit-B sensors on NOAA polar-orbiting operational meteorological satellites (NOAA 15, 16, 17); Microwave Humidity Sounder on NOAA 18, 19, and MetOp Polar orbiting satellites; and Special Sensor Microwave Imager (SSM/I) on U.S. Defense Meteorological Satellite Program satellite F-13.

One limitation of MW-based precipitation retrieval is poor spatial and temporal coverage of MW sensors. To account for this problem, most of the current precipitation retrieval algorithms rely on combined data from IR and MW instruments.

Imagery collected from these sensors is processed using various algorithms and then converted to precipitation products. There are many research and operational precipitation products in the world, some of which are listed below:

1. Tropical Rainfall Measuring Mission (TRMM) Multi-Satellite Precipitation Analysis (TMPA) is available in real-time and adjusted versions. This product is MW-based and uses IR precipitation data to fill the gaps of MW precipitation estimates [*Huffman et al.*, 2007].
2. CPC Morphing Technique (CMORPH) relies on MW precipitation estimates advected in space and time based on IR data [*Joyce et al.*, 2004].
3. Precipitation Estimation from Remotely Sensed Information using Artificial Neural Networks (PERSIANN) is a multispectral IR-based MW calibrated precipitation data set [*Hsu et al.*, 1997; *Sorooshian et al.*, 2000].
4. Global Precipitation Climatology Project (GPCP) is based on the gauge data provided by the Global Precipitation Climatology Centre [*Huffman et al.*, 2001; *Adler et al.*, 2003].
5. PRO-OBS-5 is a SEVIRI-based estimate of daily precipitation.
6. PERSIANN-CCS provides IR-based MW-calibrated precipitation estimates based on a cloud classification system [*Hong et al.*, 2004].
7. Global Mapping of Precipitation (GSMaP) [*Kubota et al.*, 2007].
8. NRL-GEO [*Turk et al.*, 2000].

Not all of these are available globally and publicly; for example, the SEVIRI PR-OBS-5 Estimated Total Precipitation expressed in mm d^{-1} of cumulative precipitation [*Italian Meteorological Service*, 2008; *Van de Vyver and Roulin*, 2009]. The original images come in the WMO General Regularly distributed Information Binary form (GRIB) files and are projected in the Geostationary projection system. The coverage of the SEVIRI data is limited to Europe and Africa (between 25° and 75°N latitudes, 25°W and 45°E longitudes) and is only available since 2007. From the above listed products, only TMPA, CMORPH, and PERSIANN are operationally available in near real-time with global coverage for the past 9–12 years. Note also that, while the most common spatial resolution of merged precipitation products is 0.25 arcdegrees (cca 25 km), CMORPH and PERSIANN-CCS are available in 0.08 (cca 8 km) and 0.04 (cca 4 km), respectively.

The GPCP, established by the World Climate Research Programme, includes multiple merged precipitation data sources such as GPCP Version 2 monthly Satellite-Gauge *Adler et al.* [2003], One-Degree Daily *Huffman et al.* [2001] and Pentad *Xie et al.* [2003] combined precipitation data. All GPCP products are produced by optimally merging satellite IR, MW, and sounder data, and precipitation gauge analyzes. They are probably the most accurate precipitation products in the world, but then come at coarse spatial resolution. Read more about this topic in the chapter by *Kidd et al.* [this volume].

A high-resolution publicly available source of mosaicked precipitation images is the Global Satellite Mapping of Precipitation (GSMaP) project, sponsored by the Core Research for Evolutional Science and Technology of the Japan Science and Technology Agency [*Kubota et al.*, 2007]. The rainfall images are derived from passive microwave remote-sensing bands using the precipitation retrieval algorithm explained in the work of *Aonashi et al.* [2009]. The 0.1 arcdegree (~10 km) monthly images of rainfall for the period 2003–2006 can be obtained from the project website (http://sharaku.eorc.jaxa.jp/GSMaP_crest/). At the moment, the data set consists of 49 images of monthly precipitation at a resolution of 0.1 arcdegrees (or cca 10 km). In order to be able to process satellite data using the `R+OSGeo` toolbox, one needs to convert each image to `SAGA` GIS format first:

```
# list all binary files in dir:
> pr.list <- dir(path=getwd(), pattern=glob2rx("*.v484"))
> for(j in 1:length(pr.list)){
# make a short name using year/month designation e.g. "PRE200612":
> pr.name <- paste("PRE", substr(pr.list[j], nchar(pr.list[j])-25, nchar(pr.list[j])-20), sep="")
# convert to SAGA format:
> rsaga.geoprocessor(lib="io_grid", module=4, param=list(GRID=set.file.extension(pr.name, ".sgrd"),
+       FILE_DATA=pr.list[j], NX=3600, NY=1200, DXY=0.1, XMIN=0.05, YMIN=-59.95, NODATA=-999,
+       DATA_OFFSET=0, LINE_OFFSET=0, DATA_TYPE=6, BYTEORDER_BIG=0, TOPDOWN=1))
}
```

where `rsaga.geoprocessor` is the generic command to send an operation from R to SAGA, NX=3600, NY=1200" is the size of the map and XMIN=0.05, YMIN=−59.95 are the coordinates of the origin. Now one can derive the mean (PRECm) and standard deviation (PRECs) for precipitation values:

```
# list all transformed files:
> prg.list <- dir(path=getwd(), pattern=glob2rx("*.sgrd"))
> rsaga.geoprocessor(lib="geostatistics_grid", module=5, param=list(GRIDS=paste(prg.list, collapse=";"),
+        MEAN="PRECm.sgrd", MIN="tmp.sgrd", MAX="tmp.sgrd", VAR="tmp.sgrd", STDDEV="PRECs.sgrd",
+        STDDEVLO="tmp.sgrd", STDDEVHI="tmp.sgrd"))
```

By zooming into various regions (Figure 1), you will notice a number of obvious artificial ("ghost") lines in various parts of the world map. This is because the GSMaP images are produced by merging a large number of scenes, which are difficult to harmonize, a problem typical for many similar remote-sensing products. Nevertheless, it seems that the pattern in the rainfall values is clear: rainfall is high between 4° and 8° latitudes and toward eastern coasts of large continental masses as an effect of Earth rotation.

We can focus on the European continent to model the distribution of rainfall more accurately, i.e., by using maps of physical parameters such as global elevation model, global map of sea temperature, and distance from the sea. (The maps can be obtained from http://spatial-analyst.net/worldmaps/.) The correlation plots (see Figure 2) show that, indeed, spatial distribution of long-term precipitation is systematic and can be largely explained by geophysical parameters.

From the plot in Figure 2, it seems that Moderate Resolution Imaging Spectroradiometer (MODIS) daily mean land surface temperature is the most significant predictor of rainfall: it explains 32% of variability over the European continent (cooler areas receive more rainfall in this case). Another significant predictor is elevation, which is positively correlated with log-transformed rainfall intensity. Note also that, by correlating the map in Figure 1 with other environmental layers, we discovered that long-term precipitation is an

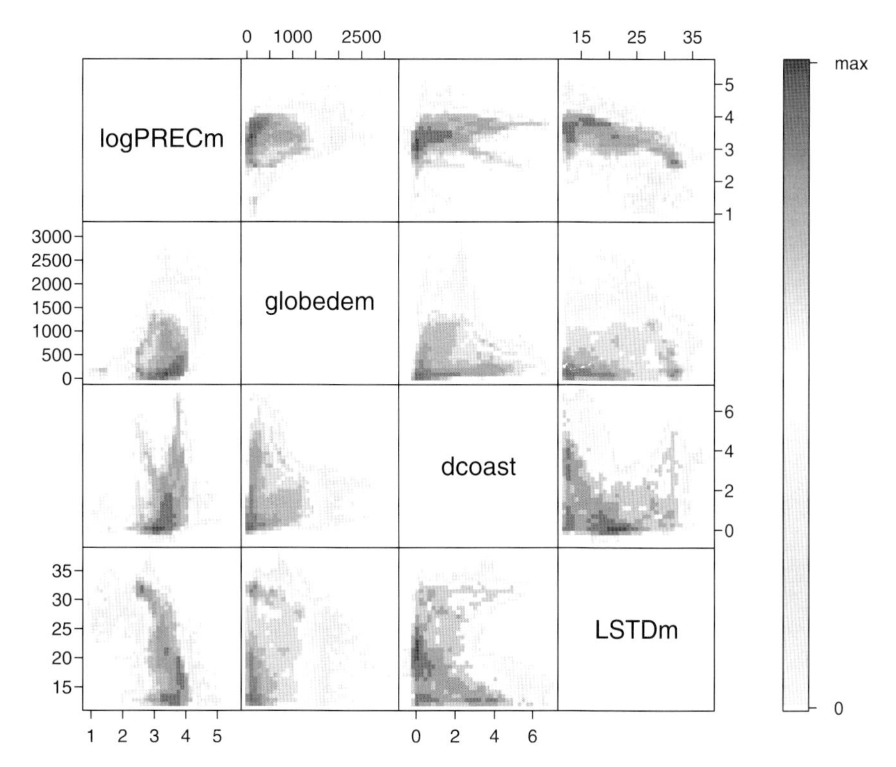

Figure 2. Correlation between mean precipitation (PRECm) and geophysical parameters: elevation (globedem), distance from coastline (dcoast), and Moderate Resolution Imaging Spectroradiometer-estimated mean land surface temperature (LSTDm). Plots refer to the European continent only.

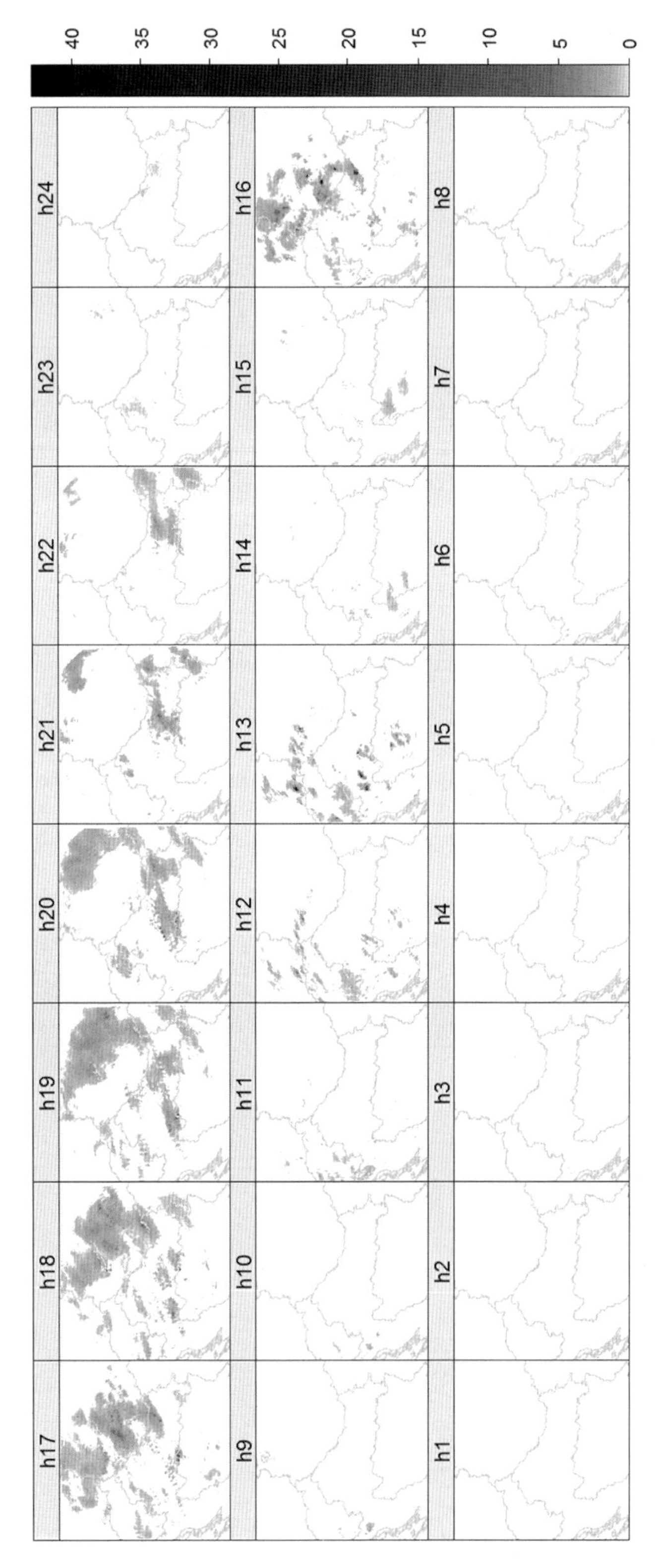

Figure 3. An example of a radar estimate of rainfall intensity (in mm) for 1 May 2008. See further section 3.

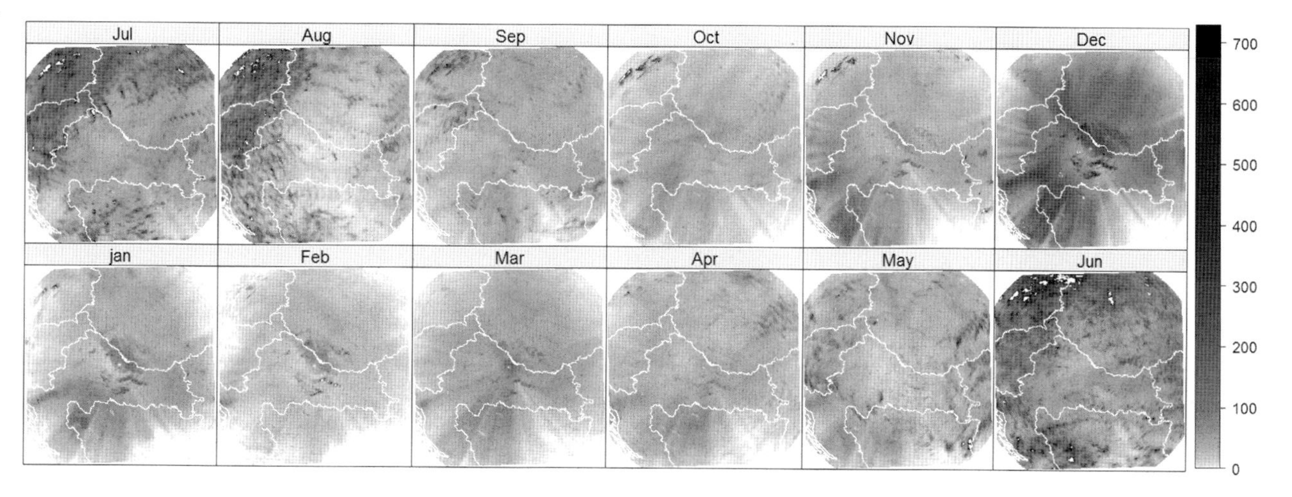

Figure 4. Radar-based monthly estimates of rainfall in mm month^{-1} for 2008. Notice the artificial radial features (not visible in the hourly images), illustrating the limitations of this technology.

excellent predictor of the global long-term leaf area index (LAI) (MODIS estimated). Precipitation explains 51% of variability in the LAI values (globally), and this relationship is close to linear. This proves the assumption that precipitation is also the most important predictor of biodiversity on the planet [*Kreft and Jetz*, 2007] and, hence, definitively worth monitoring globally.

2.3. Ground-Based Radar Images

Possibly the most accurate (and definitively the most detailed in both space and time) images of precipitation are produced by ground-based weather radars. Ground radars can provide estimates of total precipitation, along with estimates of the base and composite reflectivities, mean radial velocity, vertically integrated liquid water, wind profile, and more. Some of these parameters can be directly linked to cloud dynamics and are now increasingly used to study clouds.

The most commonly used radars for rainfall estimation are the Doppler-type weather radars that are capable of detecting the motion of rain droplets in addition to intensity of the precipitation [*Bringi and Chandrasekar*, 2001]. Continents such as Europe and North America (Map of ground radars for the United States of America is available at http://radar.weather.gov/; for the rest of the world at http://weather.org/radar.htm) are already densely covered by weather ground radars, which allows creation of radar

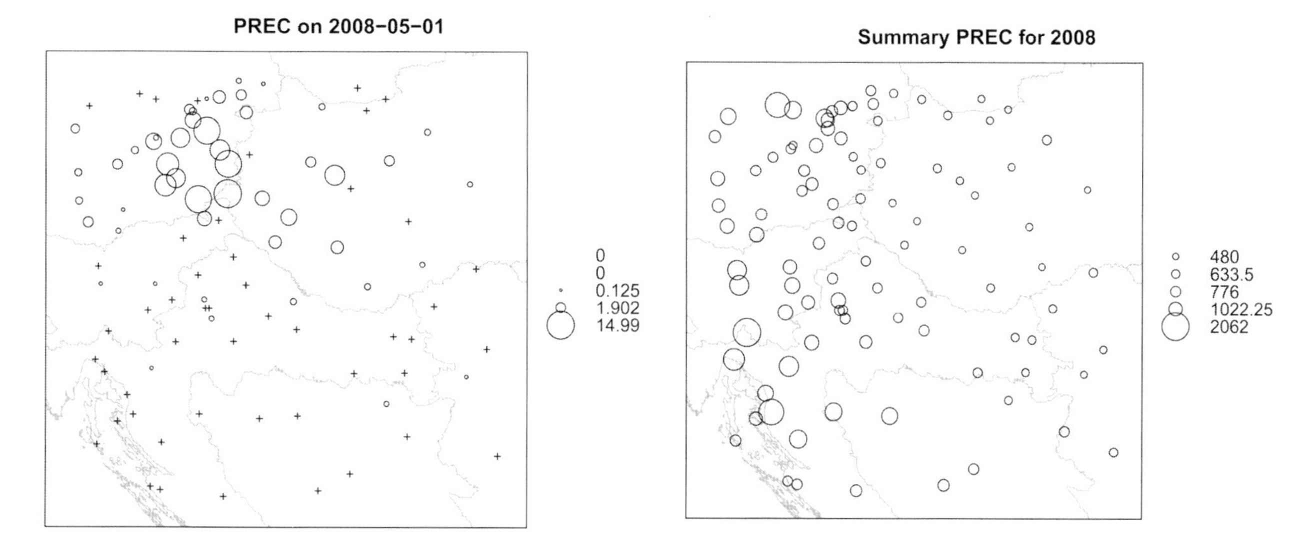

Figure 5. Rainfall intensity in mm: (left) values measured on 1 May 2008 and (right) summary values for the whole year.

mosaics with complete coverage for large areas. The next generation of weather radars are the polarimetric radars, also referred to as dual-polarization radars that can transmit both horizontal and vertical radio wave pulses. The horizontal pulses provide a measure of the horizontal dimension precipitation, while the vertical pulses give an estimate of the corresponding vertical dimension. The U.S. National Weather Service has planned to upgrade the Next Generation Weather Radar network to polarimetric radar technology by 2012.

Weather radars produce a series of images at short intervals that can then be aggregated to longer time periods. One such series of hourly radar images for the Bilogora station in Croatia, part of the European network of weather radars, is shown in Figure 3. This radar produces radial images showing estimated rainfall (noncalibrated) in mm h^{-1}. Accuracy of the radar images decreases exponentially with distance from the radar location [*Bringi and Chandrasekar*, 2001], and estimates of rainfall for distances beyond 240 km from the radar position at this station are not recommended.

A simple way to represent the rainfall dynamics using time series of radar estimated rainfall is to generate the summary rainfall for longer periods (e.g., months, Figure 4). In theory, one could also derive number of motion parameters using such time series of radar images: cloud motion vectors, average speed, average size of rainy clouds, etc. These are not trivial image processing tasks because clouds are highly dynamic and fuzzy entities, so that it can often be impossible to match the same pixels in two sequential cloud images. In the future, we can anticipate that algorithms that derive such quantitative measures of cloud dynamics (Figure 5) will be more and more sophisticated, which should also improve accuracy of spatiotemporal models for rainfall. For the moment, we will focus on using the daily and monthly sums only.

3. CASE STUDY: BILOGORA

For the purpose of this exercise, we use the NCDC GSOD data sets, which is one of the highest quality meteorological data set in the world (see previous section 2.1). It contains daily aggregates of a number of meteorological parameters: mean temperature (0.1°F), mean sea level pressure (0.1 mb), mean station pressure (0.1 mb), mean visibility (0.1 mi), mean wind speed (0.1 knots), maximum sustained wind speed (0.1 knots), precipitation amount (0.01 inches), snow depth (0.1 inches), and similar. The station data can be obtained directly via the NCDC's FTP server (ftp://ftp.ncdc.noaa.gov/pub/data/gsod/). The records are sorted per year (1929–2009), then per station. This data is still not available in some queryable database format so that you will need to download the complete data for the period of interest, then import it to `R`, trim empty spaces, and finally merge the table values with coordinates of the stations (cca 9000). The sorted values for the Bilogora case study can be obtained from the contact author's website (http://spatial-analyst.net/book/Bilogora). The sorted and subset data set for the case study has the following structure (daily measurements for the whole year 2008):

```
> GSOD.2008.bilogora <- read.table("GSOD_2008_bilogora.csv", sep=";", header=TRUE)
> str(GSOD.2008.bilogora)

 'data.frame' :  38889 obs. of 13 variables:
 $ STNID        : Factor w/ 111 levels "110162-99999",..: 1 1 1 1 1 1 1 1 1 1 ...
 $ STN          : int 110162 110162 110162 110162 110162 110162 110162 110162 110162 110162 ...
 $ WBAN         : int 99999 99999 99999 99999 99999 99999 99999 99999 99999 99999 ...
 $ YEARMODA     : int 20080802 20080804 20080806 20080808 20080805 20080810 20080801 ...
 $ TEMPC        : num 22.7 27.3 24.7 23.1 21.5 22.8 26.8 25.5 18.4 25 ...
 $ TEMP.count   : int 11 11 12 12 12 12 12 12 12 11 ...
 $ PREC         : num NA 0 0 NA NA 0 0 0 0 NA ...
 $ PREC.flag    : Factor w/ 8 levels "A","B","E","F",..: 8 7 7 8 8 7 7 7 7 8 ...
 $ DATE         : Factor w/ 366 levels "2008-01-01","2008-01-02",..: 215 217 219 221 218 ...
 $ STATION.NAME : Factor w/ 107 levels "AGARD","AIGEN IM ENNSTAL",..: 46 46 46 46 46 46 ...
 $ LAT          : num 48 48 48 48 48 ...
 $ LON          : num 16.3 16.3 16.3 16.3 16.3 ...
 $ ELEV..1M.    : int 233 233 233 233 233 233 233 233 233 233 ...
```

`STN` is the WMO unique station number, `WBAN` is the historical Weather Bureau Air Force Navy number, `TEMPC` is the cumulative daily temperature (now in Celsius degrees), `TEMP.count` is the number of observations used in calculating mean temperature, `PREC` is the total daily precipitation sum (now in mm d^{-1}), `LAT` and `LON` are latitude and longitude in the WGS84 system. In this case, the target variable is the total daily precipitation (rain and/or melted snow) reported during a continuous period (PREC), but we also want to look at the precipitation Flag value (6-h, 12-h, 24-h, or precipitation occurred but not reported).

In addition to the GSOD point data, we will also use a set of auxiliary predictors that can potentially explain variability in the ground-sampled rainfall values: (1) digital elevation model (dem), (2) DEM-derived topographic wetness index (twi) and (3) radar-derived estimates of the cumulative daily rainfall (RN1). These radar images were obtained from the Meteorological and Hydrological Service of Croatia (http://vrijeme.hr/aktpod.php?id=bradar). The radar station *Bilogora* is located at latitude = 45.88356°N and longitude = 17.20565°E, in central/north Croatia. This is a Doppler-type radar (DWSR 88 S) that creates radar images in intervals of 15 min; the total "volume scan" takes about 8 min. The original images are distributed in the Azimuthal Equidistant projection system; for the purpose of this exercises, all images have been converted to UTM coordinates. Note that an important aspect of the radar images is that, unlike other remote-sensing images, spatial accuracy of radar images decreases radially from the position of the radar (see also Figure 4).

3.1. Spatial Prediction of Daily Rainfall Using Plain Geostatistics

We can begin making predictions of rainfall by fitting a polynomial surface determined by geographical coordinates LON and LAT. For this, we can use the loess method as implemented in the stats package [*Venables and Ripley*, 2002, p. 423–424]:

```
# fit the model:
> PREC.loess <- loess(log1p(PREC)~LON+LAT, data=GSOD.20080501.ov@data, span=0.4)
> summary(PREC.loess)

  Call:
  loess(formula = log1p(PREC) ~ LON + LAT, data = GSOD.20080501.ov@data,
      span = 0.4)

  Number of Observations: 75
  Equivalent Number of Parameters: 16.24
  Residual Standard Error: 0.5332
  Trace of smoother matrix: 19.39

  Control settings:
    normalize: TRUE
    span     : 0.4
    degree   : 2
    family   : gaussian
    surface  : interpolate        cell = 0.2

# new locations:
> str(grids2km@data)

 'data.frame':  166464 obs. of  4 variables:
 $ LAT           : num  47.7 47.7 47.7 47.7 47.7 ...
 $ LON           : num  14.6 14.6 14.6 14.6 14.6 ...

# generate predictions:
> PREC.ls <- predict(PREC.loess, grids2km@data, se=TRUE)
> str(PREC.ls)

  List of 4
  $ fit           : num[1:166464] 0.0254 0.0262 0.0271 0.0281 0.0292 ...
  $ se.fit        : num[1:166464] 0.1037 0.1024 0.1012 0.1 0.0988 ...
  $ residual.scale: num 0.194
  $ df            : num 77.1
```

this will fit a model and save it as an R object PREC.loess, and predict method will generate predictions for all new locations defined in grids2km@data (Figure 7). The loess function fits a statistical (regression) model, and this method will, in fact, produce very similar results to spline interpolators. By setting the span parameter, we define the amount of smoothing (smaller values will smooth less; in this case, we used 0.4, which is an arbitrary value), by setting se=TRUE, we specify that we also want to estimate the prediction errors.

We can next try to fit a variogram for this data. Because the rainfall is a variable that is typically skewed, i.e., non-Gaussian, a suitable package for modeling such feature is geoR [*Diggle and Ribeiro*, 2007]:

```
--------------------------------------------------------------
Analysis of geostatistical data
For an Introduction to geoR go to http://www.leg.ufpr.br/geoR
geoR version 1.6-27 (built on 2009-10-15) is now loaded
--------------------------------------------------------------
```

We first convert the sp-type data to geoR compatible data and then fit a variogram using the default method (maximum likelihood):

```
> PREC.geo <- as.geodata(GSOD.20080501.XY["PREC"])
> PREC.geo$data <- PREC.geo$data+0.1 # add measurement error otherwise transformation fails!
> PREC.svar2 <- variog(PREC.geo, lambda=0, max.dist=200000, messages=FALSE)
# fit a variogram:
> PREC.vgm2 <- likfit(PREC.geo, lambda=0, messages=FALSE,
+         ini=c(PREC.vgm$psill[2],PREC.vgm$range[2]), cov.model="exponential")
> PREC.vgm2

  likfit: estimated model parameters:
          beta          tausq      sigmasq             phi
  "   -1.2819" "    0.5087" "    1.6777" "44099.1503"
  Practical Range with cor=0.05 for asymptotic range: 132109.2

  likfit: maximised log-likelihood = -82.58
```

where likfit is the generic variogram fitting method, ini is the initial variogram, tausq and sigmasq corresponds to nugget and sill parameters, phi is the range parameter, and lambda=0 indicates log-transformation. The resulting variogram is visible in Figure 6. In this case, the variogram indicates that the values are strongly spatially autocorrelated up to a distance of cca 150 km.

Note that geoR implements the Box-Cox transformation family [*Diggle and Ribeiro*, 2007, p. 61], which is somewhat more sophisticated than simple log() transformation. From Figure 6 (right), we can also notice that the variogram fitted using this method does not really go through all points (compare with Figure 6, left). This is because the maximum likelihood method discounts the potentially wayward influence of sample variogram at large interpoint distances [*Diggle and Ribeiro*, 2007]. Note also that the confidence bands in Figure 6 (right) also confirm that the variability of the empirical variogram is significant, which is not unusual considering that we deal with a relatively small point data set (98 points).

Once we have fitted a variogram, we can produce predictions (or simulations) by using

```
# prepare prediction locations:
> locs <- pred_grid(c(467000, 874000), c(4879000, 5286000), by=1000)
# ordinary kriging:
> PREC.ok2 <- krige.conv(PREC.geo, locations=locs, krige=krige.control(obj.m=PREC.vgm2))
```

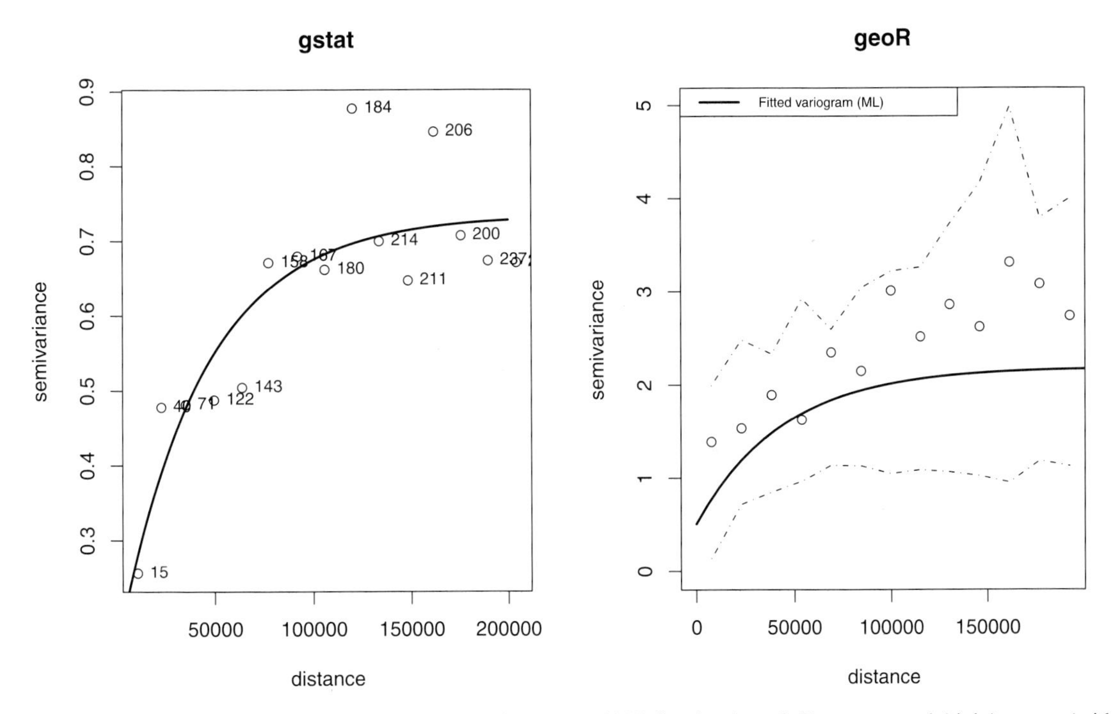

Figure 6. Fitted variogram models for precipitation for 1 May 2008 fitted using: (left) `gstat` and (right) `geoR` (with confidence bands).

```
# geoR will automatically back-transform the values!

  krige.conv: model with constant mean
  krige.conv: performing the Box-Cox data transformation
  krige.conv: back-transforming the predicted mean and variance
  krige.conv: Kriging performed using global neighbourhood
```

where `krige.conv` is the generic name used in `geoR` for "kriging conventional," i.e., any version of linear kriging. The resulting map is shown in Figure 7. The advantage of using `geoR` is that is contains "state-of-the-art" statistical fitting and prediction techniques and can deal with non-normal data sets. On the other hand, operations in `geoR` are computationally very demanding and can take significant amount of time even with smaller data sets; hence, we will subsequently use `gstat` to generate predictions.

3.2. Spatial Prediction of Daily Rainfall Using Auxiliary Predictors

In the next phase, we consider a number of auxiliary predictors that have the potential to explain the spatial variation of values. The most important predictor is the ground-based radar estimate of the daily rainfall available from the Bilogora station for this day (Figure 3). Recall from section 2.2, we know the long-term rainfall is strongly controlled by topography; hence, we can also try to include this information to improve the spatial prediction of rainfall. Recall also from section 1.2 that the daily rainfall data is typically skewed toward 0 values. Hence, we use the ZIP regression analysis as implemented in the package `pscl` (http://cran.r-project.org/web/packages/pscl/) via the method `zeroinfl`. For example, to fit a regression model that explains spatial distribution of rainfall using the elevation data, TWI and radar estimated rainfall, we can run:

```
> library(pscl)

  Loading required package: MASS
  Loading required package: mvtnorm
  Loading required package: coda
  Classes and Methods for R developed in the
  Political Science Computational Laboratory
  Department of Political Science
  Stanford University
  Simon Jackman
  hurdle and zeroinfl functions by Achim Zeileis

# convert real numbers to integers (based on the precision):
> GSOD.20080501.ov$PREC.i <- as.integer(GSOD.20080501$PREC/0.1)
# fit a zero inflated regression model:
> PREC.zip <- zeroinfl(PREC.i~dem+twi+RN10501, data=GSOD.20080501.ov@data)
> summary(PREC.zip)

  Call:
  zeroinfl(formula = PREC.i ~ dem + twi + RN10501, data = GSOD.20080501.ov@data)

  Pearson residuals:
        Min        1Q    Median        3Q      Max
  -5.42743 -0.93482 -0.63437 -0.03149  9.63001

  Count model coefficients (poisson with log link):
              Estimate Std. Error z value Pr(>|z|)
  (Intercept) 0.5556445 0.5851274   0.950 0.342309
  dem         0.0017665 0.0001477  11.956  < 2e-16 ***
  twi         0.0888628 0.0254466   3.492 0.000479 ***
  RN10501     0.3438259 0.0178854  19.224  < 2e-16 ***

  Zero-inflation model coefficients (binomial with logit link):
              Estimate Std. Error z value Pr(>|z|)
  (Intercept) -4.542187  5.703863 -0.796  0.4258
  dem         -0.002617  0.001604 -1.631  0.1029
  twi          0.267920  0.259094  1.034  0.3011
  RN10501     -0.376759  0.198612 -1.897  0.0578 .
  ---
  Signif. codes: 0'***' 0.001'**' 0.01'*' 0.05'.' 0.1' ' 1

  Number of iterations in BFGS optimization: 14
  Log-likelihood: -723.2 on 8 Df

# make predictions:
> PREC20080501.zip <- predict(PREC.zip, grids2km@data, type="response")
> grids2km$PREC.reg <- PREC20080501.zip*0.1
```

which shows that rainfall intensity (positive values) can be explained by dem and RN10501 (radar estimated rainfall), but the model has problems distinguishing between 0 and positive areas. In fact, it seems that these predictors cannot help us distinguish between areas where the rain actually fell and did not. In summary, the regression models are not distinct, and it is very likely that the output pattern will reflect mainly the residual part of variation (see further Figure 7). Now that we have estimated the trend model, we can proceed with making predictions using equation (24). We first need to estimate a variogram model for the residuals:

```
> GSOD.20080501.ov <- overlay(grids2km, GSOD.20080501.XY)
> GSOD.20080501.ov@data <- cbind(GSOD.20080501@data, GSOD.20080501.ov@data)
# mask NA values:
```

```
> sel <- !is.na(GSOD.20080501.ov$PREC.reg)
# fit a variogram:
> PREC.rvgm <- fit.variogram(variogram(log1p(PREC)~log1p(PREC.reg), GSOD.20080501.ov[sel,]),
+       model=vgm(nugget=0.2, model="Exp", range=100000, psill=0.8))
> PREC.rvgm

   model     psill     range
 1   Nug 0.3628927       0.0
 2   Exp 1.2135109 429981.7
```

In this case, we have used the exponential model to fit the variogram. This is one of the many standard variogram models available in the `gstat` package:

```
# print all possible variogram models in gstat:
> vgm()

   short                                      long
 1   Nug                             Nug (nugget)
 2   Exp                        Exp (exponential)
 3   Sph                          Sph (spherical)
 4   Gau                           Gau (gaussian)
 5   Exc              Exclass (Exponential class)
 6   Mat                             Mat (Matern)
 7   Ste Mat (Matern, M. Stein' s parameterization)
 8   Cir                           Cir (circular)
 9   Lin                             Lin (linear)
 10  Bes                             Bes (bessel)
 11  Pen                   Pen (pentaspherical)
 12  Per                           Per (periodic)
 13  Hol                               Hol (hole)
 14  Log                        Log (logarithmic)
 15  Pow                              Pow (power)
 16  Spl                             Spl (spline)
 17  Leg                           Leg (Legendre)
 18  Err                  Err (Measurement error)
 19  Int                         Int  (Intercept)
```

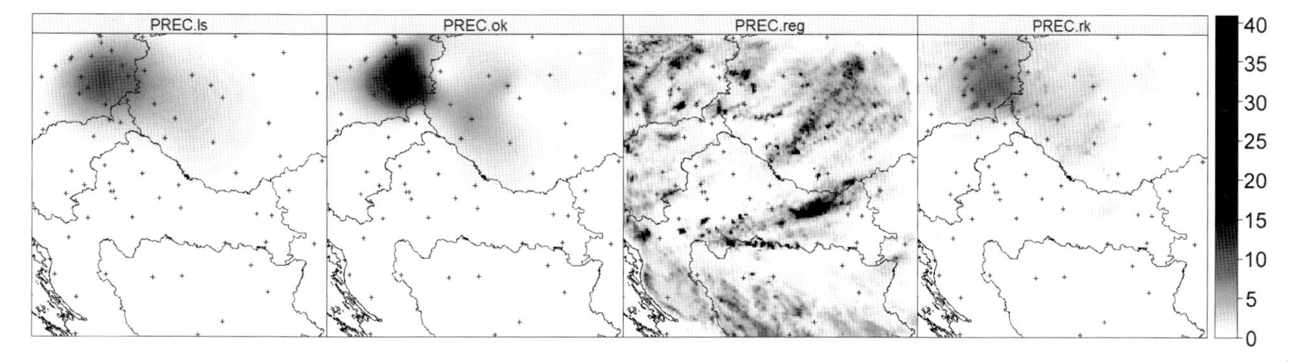

Figure 7. Precipitation (in mm) for the Bilogora case study on 1 May 2008 predicted using: (`PREC.ls`) polynomial surface, (`PREC.ok`) ordinary kriging in `geoR`, (`PREC.reg`) regression with zero-inflated Poisson model, and (`PREC.rk`) regression-kriging. Based on the point data in Figure 5, left.

To generate predictions we use the `krige` method as implemented in the `gstat` package:

```
# make predictions and back-transform:
> grids2km$PREC.rk <- expm1(krige(log1p(PREC)~log1p(PREC.reg), locations=GSOD.20080501.ov[sel,],
+     newdata=grids2km, model=PREC.rvgm)$var1.pred)

 [using universal kriging]
```

Results of prediction using various methods are shown in Figure 7. The OK and RK maps look, in fact, fairly different, which demonstrates the difficulties of modeling daily rainfall. The general spatial trend of relatively higher rainfall intensities in both ordinary kriging and regression-kriging, in general, corresponds. The issues for discussion are the small patches where the weather radar image indicates excessive rain, but there were not enough rain gauges to prove this. Another issue is that the regression-kriging model can predict much higher values in rainfall intensities than what was measured at the meteorological stations. For example, on 1 May 2008, the highest measured precipitation values on the ground are about 18 mm, while the weather radar predicts values up to 65 mm. It is very possible that those excessive rains did occur, but because they do not spatially overlap with the rain gauges, the model needs to extrapolate.

4. CASE STUDY: ITALY

In the next case study, we use a more extensive data set consisting of 1950 rain gauge measurements for the whole of Italy, obtained from the Consiglio Nazionale delle Ricerche (CNR) research Institute for Geo-Hydrological Hazard Assessment (http://geomorphology.irpi.cnr.it). This rainfall data is used for landslide and flooding hazard prediction [*Marchesini et al.*, 2010]. A formatted version of this data set can also be obtained from the contact author's website (http://spatial-analyst.net/book/ITrain2010). Note that this is now a much larger data set; hence modeling and spatial prediction might become more computationally demanding. The data contains 329,550 observations of daily rainfall for the period 2 January until 25 June 2010 (a total of 168 days). As auxiliary predictors, we now only use the coarse images obtained from the same repository of the 5-km global data.

4.1. Filtering of Gross Errors and Automated Interpolation

We start with importing the data to `R`:

```
> rain_it <- read.table("rain_it.csv", sep=";", header=TRUE)
> str(rain_it)  # this is a large file!

 'data.frame' :  329550 obs. of 11 variables:
 $ PK_Station : int  364300 2444 247000 2504 367800 450300 79300 351600 283500 2611 ...
 $ name_stat  : Factor w/ 1901 levels "Abbadia S.Salvatore",..: 26 1599 1475 1609 1611 1612 ...
 $ location   : Factor w/ 1331 levels  "Abbasanta","Abbiategrasso",..: NA NA NA NA NA NA  ...
 $ region     : Factor w/ 21 levels "ABRUZZO","BASILCIATA",..: NA NA NA NA NA NA NA NA  ...
 $ x          : num  10.97 8.44 13.83 12.51 11.24 ...
 $ y          : num  44.7 44.3 41.4 43.8 44.4 ...
 $ PK_Sensor  : int  4851 -1254 8536 -602 4997 9584 468 3509 6358 -1238 ...
 $ descra     : Factor w/ 1 level "P": 1 1 1 1 1 1 1 1 1 1 ...
 $ date       : Factor w/ 168 levels "2010-01-02","2010-01-03",..: 168 168 168 168 168 168 ...
 $ fk_thresh  : int.1 1 1 1 1 1 1 1 1 1 ...
 $ rain_24    : num  0 0 0 0 0 0 0 0 0.2 0 ...
```

which shows that the there are 1901 unique stations in 21 regions. Our interest is the `rain_24` variable, which is the measured cumulative daily rainfall (mm d^{-1}) at station. We can focus on a specific data, e.g., 23 January 2010:

```
> rain_it$DATE <- as.Date(rain_it$date)
> it.20100123 <- subset(rain_it, rain_it$DATE==as.Date("2010-01-23")&!is.na(rain_it$rain_24))
> str(it.20100123)  # 1950 points

 'data.frame' :   1950 obs. of 12 variables:
 $ PK_Station : int 2452 242400 364800 198100 57800 278900 352500 202500 145100 351400 ...
 $ name_stat  : Factor w/ 1901 levels "Abbadia S.Salvatore",..: 626 323s 321 312 310 306 300 297 287 278 ...
 $ location   : Factor w/ 1331 levels "Abbasanta","Abbiategrasso",..: 526 NA NA NA NA NA NA NA NA NA ...
 $ region     : Factor w/ 21 levels "ABRUZZO","BASILCIATA",..:  9 NA NA NA NA NA NA NA NA NA ...
 $ x          : num 8.71 7.89 12.28 14.01 14.63 ...
 $ y          : num 44.5 44.5 44.1 42.2 41.9 ...
 $ PK_Sensor  : int -245 6329 4859 5852 9837 4472 3518 4793 934 3508  ...
 $ descra     : Factor w/ 1 level "P": 1 1 1 1 1 1 1 1 1 1 ...
 $ date       : Factor w/ 168 levels "2010-01-02","2010-01-03",..: 22 22 22 22 22 22 22 22 22 22 ...
 $ fk_thresh  : int 1 1 1 1 1 1 1 1 1 1  ...
 $ rain_24    : num 0 0 0 0 0.2 2.6 0 0 0 0 ...
 $ DATE       :Class 'Date'  num[1:1950] 14632 14632 14632 14632 14632 ...

> coordinates(it.20100123) <- ~x+y
> proj4string(it.20100123) <- CRS("+proj=longlat +datum=WGS84")
> it.20100123.XY <- spTransform(it.20100123, CRS(utm33))
> hist(it.20100123.XY$rain_24, breaks=45, col="grey", main="Histogram", ylim = c(0, 100))
# values are heavily skewed!
```

The `it.20100123.XY` point map is shown in Figure 8, with the 5-km resolution global topographic model in the background. Notice in the map in Figure 8 that there are a few isolated locations with high values of rainfall, which are assumed to be gross errors, i.e., mistyped numbers. A way to select and filter these locations automatically is to use the `krige.cv` function available in the `gstat` package. This will run the leave-one-out cross-validation using all points and then compare the predicted values versus the measured values [*Bivand et al.*, 2008, p. 221–226]. Obviously, if the difference between the measured and predicted value is very large and improbable ($P < 0.001$), we can assume that this number need to be checked or removed. Cross-validation can be run via the following:

```
# run Leave-One-Out cross-validation to detect suspicious points:
> PREC_it.cv <- krige.cv(log1p(rain_24)~1, it.20100123.XY, nmax=20, PREC_it.vgm)

  | =========================================== | 100%
```

We can now determine the local (positive) outliers by using the `zscore` (normalized error with mean zero and standard deviation one). If, e.g., `zscore` > 7, we already deal with very improbable values:

```
> it.20100123.XY[PREC_it.cv$zscore>7, c("location", "rain_24")]

              coordinates        location rain_24
 287287 (162294, 4905630)         Sestola     7.2
 288586 (521406, 4503080) Contursi Terme    22.2

# Select suspicious points based on the cross-validation error:
> sel <- PREC_it.cv$zscore<q99
```

which successfully detects isolated points also clearly visible in Figure 8.

We can now test the automated interpolation method available via the `intamap` package. This is an automated mapping function, which means we only need to specify the input point data set and domain of interest, and the package will select the most suitable method from a variety of methods:

```
# add measurement error:
> it.20100123.XY$value <- it.20100123.XY$rain_24+0.1
# predict values automatically:
> output <- interpolate(observations=it.20100123.XY[sel,], predictionLocations=mask.xy)

 R 2010-07-26 21:20:44 interpolating 1901 observations, 12043 prediction locations
 [Time models loaded...]
 [1] "estimated time for copula 8166.67338767378"
 Checking object ... OK
```

What is not immediately evident from this output is that intamap has automatically checked some properties of the data and then chosen between: (1) kriging, (2) copula methods, (3) inverse distance interpolation, projected spatial Gaussian process methods in the psgp package, and (4) transGaussian kriging or Yamamoto interpolation [*Pebesma et al.*, 2009]. By printing the summary of the output object, we notice that, in this case, intamap has selected the transGaussian kriging as the most suitable technique.

If you visualize the results of automated interpolation using intamap package for various dates, you will notice that the predictions can be critically out of range in many areas, and there can be many artifacts in the final map. This is possibly because the variogram fit is poor, or there are still some outliers in the data, which illustrates some possible limitations of automated mapping for spatial prediction of skewed and noisy variables such as daily precipitation.

Indeed, the variogram for rainfall for 23 January 2010 cannot be fitted automatically (Figure 9). Instead, we need to try to visually determine the variogram parameters, and then, after a few iterations produce a satisfactory result. In this case, it seems that the spatial autocorrelation can be best described with the Matérn variogam (adjusted using the M. Stein's parameterization):

```
> PREC_it.svar <- variogram(log1p(rain_24)~1, it.20100123.XY)
# initial variogram:
> PREC_it.ivgm <- vgm(nugget=0.1, model="Ste", range=sqrt(diff(grids.it@bbox["x",])^2 +
+      diff(grids.it@bbox["y",])^2)/5, psill=var(log1p(it.20100123.XY$rain_24))*1.4, kappa=5)
> PREC_it.vgm <- fit.variogram(PREC_it.svar, model=PREC_it.ivgm)
> PREC_it.vgm

  model     psill    range  kappa
 1  Nug 0.1021922      0.0      0
 2  Ste 1.6987252 831597.9      5
```

4.2. *Spatial Prediction of Daily Rainfall Using Auxiliary Predictors*

We next focus on modeling daily rainfall using auxiliary predictors: elevation, distance to the coastline, and the long-term precipitation pattern derived in section 2.2. We again fit a ZIP model using auxiliary predictors:

```
> it.20100123.XY.ov$PREC.i <- as.integer(it.20100123.XY.ov$rain_24*10)
> PREC_it.zip <- zeroinfl(PREC.i~globedem+dcoast+PRECm, data=it.20100123.XY.ov@data[sel,])
> summary(PREC_it.zip)

 Call:
 zeroinfl(formula = PREC.i ~ globedem + dcoast + PRECm, data =  it.20100123.XY.ov@data[ sel, ])

 Pearson residuals:
   Min      1Q  Median      3Q     Max
 -0.8954 -0.6628 -0.4627 -0.2063 56.8036
```

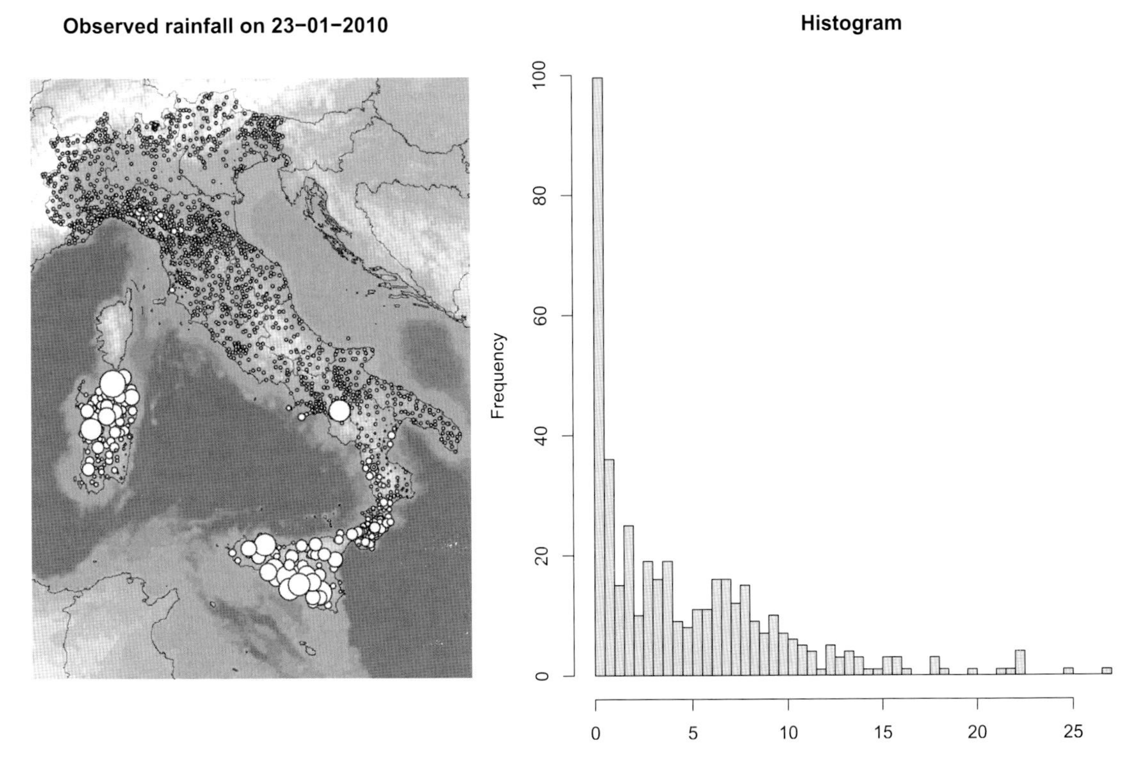

Figure 8. Rainfall observed at 1950 locations in Italy on 23 January 2010 and histogram of values for that day. Notice the few isolated points with high (improbable) values.

```
Count model coefficients (poisson with log link):
              Estimate Std. Error z value Pr(>|z|)
(Intercept) 5.205e+00  3.471e-02  149.97   <2e-16 ***
globedem    8.283e-04  2.089e-05   39.65   <2e-16 ***
dcoast     -1.454e+00  4.306e-02  -33.77   <2e-16 ***
PRECm      -4.873e-02  1.316e-03  -37.02   <2e-16 ***

Zero-inflation model coefficients (binomial with logit link):
              Estimate Std. Error z value Pr(>|z|)
(Intercept) -0.0202399  0.1795416 -0.113   0.9102
globedem    -0.0002214  0.0001510 -1.466   0.1425
dcoast       1.4262280  0.1887880  7.555  4.2e-14 ***
PRECm        0.0119877  0.0060757  1.973   0.0485 *
---
Signif. codes:  0'***'  0.001'**' 0.01'*'  0.05'.' 0.1' ' 1

Number of iterations in BFGS optimization: 27
Log-likelihood: -1.119e+04 on 8 Df
```

which shows that the rainfall areas on 23 January are mainly controlled by distance to the coastline (binomial with logit link), while all three auxiliary maps are significant predictors of the positive values (compare with the Bilogora case study). This is also visible from the predicted ZIP values shown in Figure 10. Note that this model is specific to this day only, so that explanation of why, for example, PRECm (long-term precipitation trend) is negatively correlated with daily rainfall on 23 January is not trivial. These relationships would probably differ for each subsequent day, although in

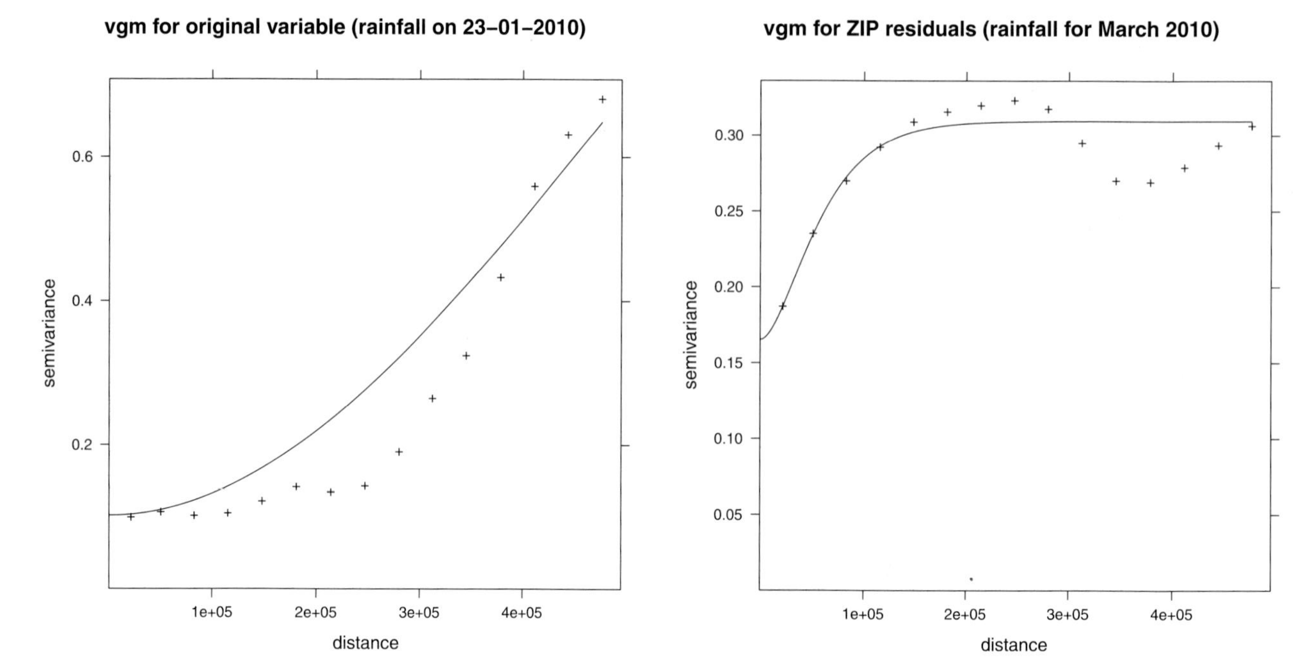

Figure 9. Variogram model fitted for rainfall values observed (left) on 23 January 2010 and (right) for regression residuals.

most cases, we would expect to see a positive correlation between, e.g., long-term precipitation, elevation, and daily rainfall.

The final prediction map (Figure 10, right) shows that the predictions now include both the pattern of the topography/coastline and the local high spots. Local differences between points are smoothed out by spatial and topographic trends in the data.

4.3. Spatial Prediction and Simulation of Aggregated Rainfall

In the last exercise, we focus on making predictions and simulations for a single month (March 2010). We need to aggregate the values first:

```
> it.2010March <- subset(rain_it, rain_it$DATE>as.Date("2010-02-28")&rain_it$DATE<as.Date("2010-03-31"))
> rain_it.2010March <- aggregate(it.2010March[!is.na(it.2010March$rain_24),c("rain_24","x","y")],
+        by=list(it.2010March$name_stat[!is.na(it.2010March$rain_24)]),FUN=mean)
> coordinates(rain_it.2010March) <- ~x+y
> proj4string(rain_it.2010March) <- CRS("+proj=longlat +datum=WGS84")
> rain_it.2010March <- remove.duplicates(rain_it.2010March)
> it.2010March.XY <- spTransform(rain_it.2010March, CRS(utm33))
```

Next, we can fit a variogram and prepare the data for interpolation. The variogram fitted in `gstat` is:

```
> PREC_it_March.rvgm

  model     psill     range kappa
1   Nug 0.1652977      0.00   0.0
2   Mat 0.1441707  31104.79   1.5
```

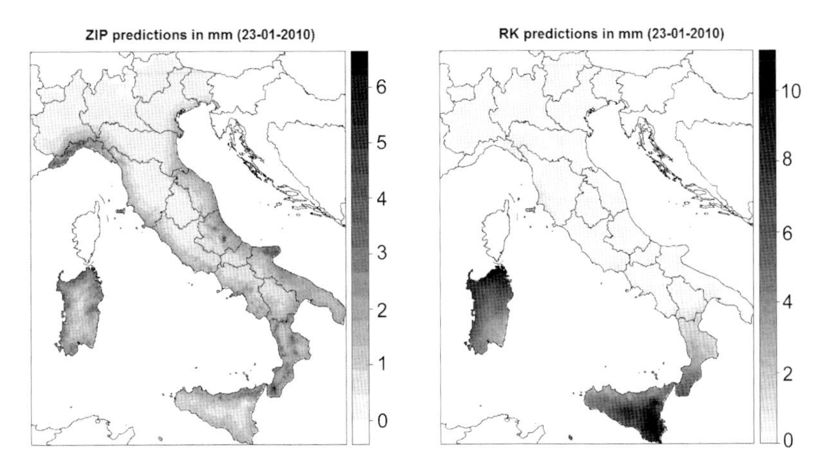

Figure 10. Rainfall values predicted (left) using zero-inflated model and (right) final prediction using regression-kriging. Compare with Figure 8.

Unlike for daily measurements, the variogram for monthly averaged values is bounded, but again contains a significant nugget (Figure 9). This high nugget is then reflected in the simulations that locally show larger differences between values. This indicates that spatial prediction of daily/monthly rainfall is not trivial, and the possible impacts of the mapping error need to be considered seriously (Figure 11).

5. DISCUSSION AND CONCLUSIONS

In this chapter, we have focused on using geostatistical techniques with emphasis on regression-kriging, a technique which many geostatisticians now consider to be the preferred Best Linear Unbiased Prediction model for spatial data [*Gotway Crawford and Young*, 2008; *Hengl*, 2009]. Other geostatistical techniques, such as copula-based interpolation [*Bárdossy and Li*, 2008], or probabilistic simulation approaches for mapping precipitation fields and their uncertainties [*AghaKouchak et al.*, 2010], are equally valid alternatives to simple ordinary kriging or spline interpolators. Even more complex approaches of (stochastic) spatial prediction of rainfall values include neural networks [*Karmakar et al.*, 2009], various Bayesian and maximum entropy techniques, and others [*Lanza et al.*, 2001]. These techniques were not considered in this chapter, but are also available through various R packages (or their combination).

The main problem of mapping rainfall remains validation of produced maps [*Lanza et al.*, 2001]. There is no guarantee that the radar images shown in Figure 3 are true images of rainfall intensity. In addition, rain gauges are on point support

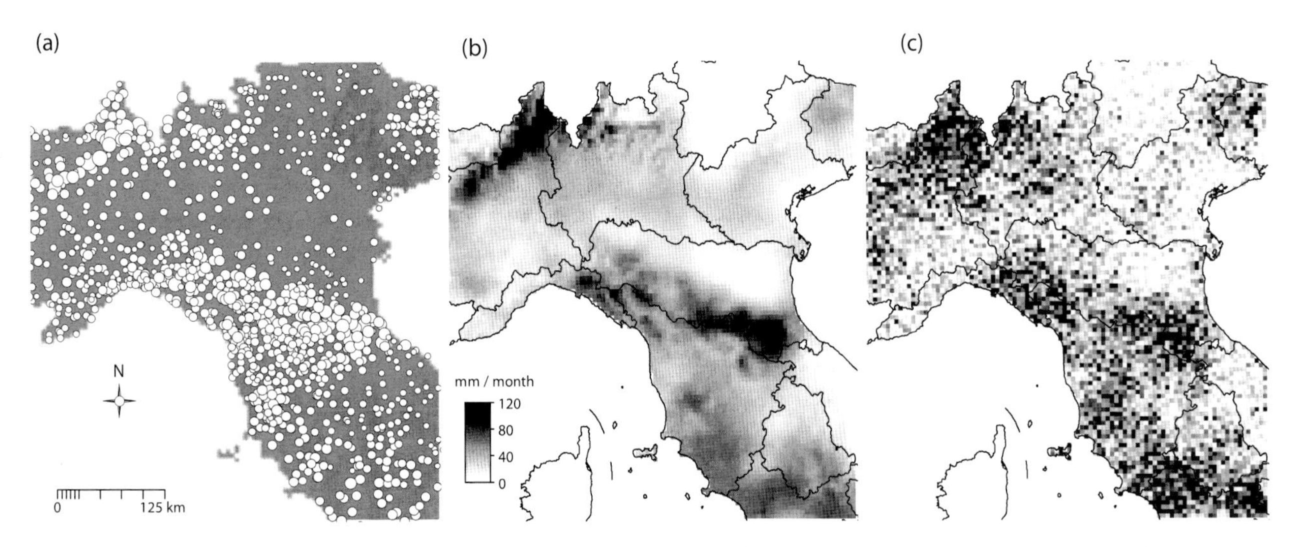

Figure 11. Rainfall values (a) observed for March 2010, (b) predicted using regression-kriging, and (c) simulations using the same model.

so that it is difficult to validate aerial (block support) estimates of rainfall produced by, e.g., radar images of 1 km.

The usability of radar images investigated in this chapter for spatial prediction of rainfall is somewhat disappointing, but this might change in the coming years. Several new pending space missions will improve what we know about rainfall dynamics and patterns. One such mission is the joint European/Japanese EarthCARE (http://www.esa.int/esaLP/LPearthcare.html) mission that will include similar (three-dimensional) radar satellite sensors to CloudSat but with additional capabilities for providing Doppler motion measurements: high spectral resolution (HSR) Lidar (atmospheric lidar) and equipped with a HSR receiver and the Cloud Profiling Radar [*Lajas et al.*, 2008]. Another pending mission is the joint NASA/Japan Aerospace Exploration Agency Global Precipitation Measurement (http://gpm.gsfc.nasa.gov) mission, which is planned for 2013. This satellite will carry both a dual frequency precipitation radar and a high-resolution, multichannel passive MW rain radiometer [*Levizzani et al.*, 2007, p.611–653]. These recent evaluation and verification studies suggest multisensor estimates (gauge-radar or gauge-satellite) are preferred over single sensor data.

We foresee that the most anticipated development of the field will be connected with using space-time interpolation techniques and a combination of process-based models and data-driven techniques. Although `R` contains a variety of packages suitable for spatial prediction of rainfall, our impression is that they often experience problems: `geoR` is computationally demanding and is not suited for data sets with $\gg 10^3$ points, zero-inflated models can lead to serious over-estimates of the rainfall intensity. Automation of geostatistical mapping (e.g., `intamap` package) of daily rainfall will need to mature before it becomes a reliable method. At least, the results have to be constantly cross-checked by the analyst for artifacts caused by poorly fitted models.

There is also much room to improve modeling of rainfall at global level. Once we are able to produce a stable estimate of global precipitation pattern using remote-sensing imagery, we will also be able to build a global regression model that explains the majority of long-term spatial variation of precipitation. We anticipate that this model will be complex: nonlinear and possibly area specific. The relationships shown in Figure 2, for example, are valid only for the European continent (land mass) and cannot be used for other continents. These relationships will certainly be different for sea and land masses, and for tropical areas. Once information about which parameters to use and in which form (i.e., the mathematical model) has been defined at a global scale, we should also be able to model local changes in rainfall for any given area in the world with much higher accuracy than achieved in the two case studies.

Based on the results of testing various spatial prediction methods in this chapter, following the cited literature source, we can conclude the following:

1. Daily rainfall is a highly variable, dynamic feature (with chaotic properties). The only way to predict it accurately is to collect enough ground and remote-sensing data at fine(r) time intervals (e.g., every few minutes).

2. Daily rainfall from rain gauge networks usually contains many zero values; hence, rainfall data at shorter time intervals need to be modeled using zero-inflated regression models.

3. Rainfall sampled at rain gauges is typically spatially autocorrelated (up to distances of >100 km), which makes it suitable for various types of kriging.

4. Rainfall estimated using ground-based radar can be of variable accuracy. The radar images can carry many artifacts, especially at high distances from the ground radar. The accuracy of estimated rainfall is critically poor for low-intensity rainfalls.

5. The rain gauge data should remain one of the main "hard data" sources for rainfall mapping.

6. The significance of correlation between rainfall intensity and radar images, and other auxiliary data, gradually increases with larger temporal support.

7. There is a need for a global geophysical model that can explain long-term variability in precipitation distribution (as a function of sea temperature and evapotranspiration, elevation, Earth rotation, and proximity to land masses).

Acknowledgments. The authors would like to thank Zvonko Komerički from the Meteorological and Hydrological Service of Croatia for providing the weather radar images for year 2008, Ivan Marchesini and Mauro Rossi from the CNR research Institute for Geo-Hydrological Hazard Assessment in Perugia for providing the rainfall ground data for Italy, and the National Climatic Data Center/NOAA for providing a public access to the global surface summary of day (GSOD) data.

REFERENCES

Adler, R. F., C. Kidd, G. Petty, M. Morrissey, and H. M. Goodman (2001), Intercomparison of global precipitation products: The Third Precipitation Intercomparison Project (PIP-3), *Bull. Am. Meteorol. Soc.*, *82*, 1377–1396.

Adler, R. F., et al. (2003), The Version 2 Global Precipitation Climatology Project (GPCP) monthly precipitation analysis (1979–present, *J. Hydrometeorol.*, *4*, 1147–1167.

Agarwal, D. K., A. E. Gelfand, and S. Citron-Pousty (2002), Zero-inflated models with application to spatial count data, *Environ. Ecol. Stat.*, *9*(4), 341–355.

AghaKouchak, A., A. Bárdossy, and E. Habib (2010), Conditional simulation of remote sensed rainfall data using a non-gaussian v-transformed copula, *Adv. Water Resour.*, *33*(6), 624–634.

Aonashi, K., et al. (2009), GSMaP passive, microwave precipitation retrieval algorithm: Algorithm description and validation, *J. Meteorol. Soc. Jpn.*, *87A*, 119–136.

Arkin, P. A., and P. Xie (1994), The global precipitation climatology project–1st algorithm intercomparison project, *Bull. Am. Meteorol. Soc.*, *75*, 401–419.

Bárdossy, A., and J. Li (2008), Geostatistical interpolation using copulas, *Water Resour. Res.*, *44*, W07412, doi:10.1029/2007WR006115.

Bivand, R., E. Pebesma, and V. Rubio (2008), *Applied Spatial Data Analysis With R,* Use R! Series, 400 pp., Springer, Heidelberg, Germany.

Brenning, A. (2008), Statistical Geocomputing combining R and SAGA: The example of landslide susceptibility analysis with generalized additive models, in *SAGA–Seconds Out*, vol. 19, edited by J. Böhner, T. Blaschke, and L. Montanarella, pp. 23–32, Hamburger Beitr. zur Phys. Geogr. und Landschaftsökol, Hamburg, Germany.

Bringi, V. N., and V. Chandrasekar (2001), *Polarimetric Doppler Weather Radar: Principles and Applications*, 636 pp., Cambridge Univ. Press, New York.

Chen, M., P. Xie, J. E. Janowiak, and P. A. Arkin (2002), Global land precipitation: A 50-yr monthly analysis based on gauge observations, *J. Hydrometeorol.*, *3*(3), 249–266.

Christensen, R. (2001), *Linear Models for Multivariate, Time Series, and Spatial Data*, 2nd ed., 393 pp., Springer, New York

Diggle, P. J., and P. J. Ribeiro Jr. (2007), *Model-Based Geostatistics, Springer Series in Statistics*, 288 pp., Springer, New York.

Dubrule, O. (1983), Two methods with different objectives: Splines and kriging, *Math. Geol.*, *15*(2), 245–257.

Goovaerts, P. (1997), *Geostatistics for Natural Resources Evaluation (Applied Geostatistics)*, 496 pp., Oxford Univ. Press, New York.

Goovaerts, P. (2000), Geostatistical approaches for incorporating elevation into the spatial interpolation of rainfall, *J. Hydrol.*, *228*, 113–129.

Gotway Crawford, C. A., and L. J. Young (2008), Geostatistics: What's hot, what's not, and other food for thought, *Proceedings of the 8th International Symposium on Spatial Accuracy Assessment in Natural Resources and Environmental Sciences (Accuracy 2008)*, edited by Y. Wan et al., pp. 8–16, World Academic Union, Shanghai, China.

Hengl, T. (2009), *A Practical Guide to Geostatistical Mapping*, 2nd ed., 291 pp., Univ. of Amsterdam, Amsterdam, Netherlands.

Hijmans, R. J., S. E. Cameron, J. L. Parra, P. G. Jones, and A. Jarvis (2005), Very high resolution interpolated climate surfaces for global land areas, *Int. J. Climatol.*, *25*, 1965–1978.

Hong, Y., K. Hsu, X. Gao, and S. Sorooshian (2004), Precipitation estimation from remotely sensed imagery using artificial neural network-cloud classification system, *J. Appl. Meteorol.*, *43*, 1834–1853.

Houze, R. A. J. (1993), *Cloud Dynamics*, *International Geophysics Series*, *53*, 573 pp., Academic, San Diego, Calif.

Hsu, K., X. Gao, S. Sorooshian, and H. Gupta (1997), Precipitation estimation from remotely sensed information using artificial neural networks, *J. Appl. Meteorol.*, *36*, 1176–1190.

Huffman, G. J., R. F. Adler, M. Morrissey, D. T. Bolvin, S. Curtis, R. Joyce, B. McGavock, and J. Susskind (2001), Global precipitation at one-degree daily resolution from multi-satellite observations, *J. Hydrometeorol.*, *2*, 36–50.

Huffman, G. J., R. F. Adler, D. T. Bolvin, G. Gu, E. J. Nelkin, K. P. Bowman, Y. Hong, E. F. Stocker, and D. B. Wolff (2007), The TRMM Multisatellite Precipitation Analysis (TMPA): Quasi-global, multiyear, combined-sensor precipitation estimates at fine scales, *J. Hydrometeorol.*, *8*, 38–55.

Hutchinson, M. F. (1995), Interpolation of mean rainfall using thin plate smoothing splines, *Int. J. Geogr. Inf. Syst.*, *9*, 385–403.

Italian Meteorological Service (2008), *Satellite Application Facility on Support to Operational Hydrology and Water Management (H-SAF): Product User Manual*, 98 pp., Italian Meteorol. Serv., Rome.

Joyce, R. J., J. E. Janowiak, P. A. Arkin, and P. Xie (2004), CMORPH: A method that produces global precipitation estimates from passive microwave and infrared data at high spatial and temporal resolutions, *J. Hydrometeorol.*, *5*, 487–503.

Kamarianakis, Y., H. Feidas, G. Kokolatos, N. Chrysoulakis, and V. Karatzias (2008), Evaluating remotely sensed rainfall estimates using nonlinear mixed models and geographically weighted regression, *Environ. Modell. Software*, *23*(12), 1438–1447.

Karmakar, S., M. K. Kowar, and P. Guhathakurta (2009), Spatial interpolation of rainfall variables using artificial neural network, in *ICAC3 '09: Proceedings of the International Conference on Advances in Computing, Communication and Control*, pp. 547–552, ACM, New York.

Kebaili Bargaoui, Z., and A. Chebbi (2009), Comparison of two kriging interpolation methods applied to spatiotemporal rainfall, *J. Hydrol.*, *365*, 56–73.

Kidd, C., V. Levizzani, and S. Laviola (2010), Quantitative precipitation estimation from Earth observation satellites, in *Rainfall: State of the Science*, *Geophys. Monogr. Ser.*, doi:10.1029/2009GM000920, this volume.

Kreft, H., and W. Jetz (2007), Global patterns and determinants of vascular plant diversity, *Proc. Natl. Acad. Sci. U. S. A.*, *104*, 5925–5930.

Kubota, T., et al. (2007), Global precipitation map using satellite-borne microwave radiometers by the GSMaP Project: Production and validation, *IEEE Trans. Geosci. Remote Sens.*, *45*(7), 2259–2275.

Kyriakidis, P. C., J. Kim, and N. L. Miller (2001), Geostatistical mapping of precipitation from rain gauge data using atmospheric and terrain characteristics, *J. Appl. Meteorol.*, *40*, 1855–1877.

Lajas, D., T. Wehr, M. Eisinger, and A. Lefebre (2008), An overview of the EarthCARE mission and end-to-end simulator, *Proceedings of SPIE, the International Society for Optical Engineering*, pp. 7107, Soc. of Photo-Opt. Instrum. Eng., Bellingham, Wash.

Lanza, L. G., J. A. Ramírez, and E. Todini (2001), Stochastic rainfall interpolation and downscaling, *Hydrol. Earth Syst. Sci.*, *5*(2), 139–143.

Levizzani, V., P. Bauer, and F. J. Turk (2007), *Measuring Precipitation From Space: EURAINSAT and the Future*, *Advances in Global Change Research*, *28*, 722 pp., Springer, Dodrecht, Netherlands.

Lloyd, C. D. (2005), Assessing the effect of integrating elevation data into the estimation of monthly precipitation in Great Britain, *J. Hydrol.*, *308*(1–4), 128–150.

Marchesini, I., V. Balducci, G. Tonelli, M. Rossi, and F. Guzzetti (2010), *Geospatial Information on Landslides and Floods in Italy*, Consiglio Naz. delle Ric., Turin, Italy.

Matheron, G. (1969), *Le Krigeage Universel*, vol. 1, 83 pp., Cahiers du Cent. de Morphol. Math., École des Mines de Paris, Fontainebleau, France.

Mitasova, H., and L. Mitas (1993), Interpolation by regularized spline with tension, I Theory and implementation, *Math. Geol.*, *25*, 641–655.

Mitasova, H., L. Mitas, and R. Harmon (2005), Simultaneous spline approximation and topographic analysis for lidar elevation data in open-source GIS, *IEEE Geosci. Remote Sens. Lett.*, *2*, 375–379.

Pebesma, E. (2006), The role of external variables and GIS databases in geostatistical analysis, *Trans. GIS*, *10*(4), 615–632.

Pebesma, E., D. Cornford, G. Dubois, G. B. M. Heuvelink, D. Hristopoulos, J. Pilz, U. Stoehlker, and J. Skoien (2009), INTAMAP: An interoperable automated interpolation web service, in *StatGIS Conference Proceedings, Milos, Greece*, edited by D. Cornford et al., pp. 1–6.

Prigent, C. (2010), Precipitation retrieval from space: An overview, *C. R. Geosci.*, *342*(4–5), 380–389.

R Development Core Team (2009), *R: A Language and Environment for Statistical Computing*, 409 pp., R Found. for Stat. Comput., Vienna.

Ridout, M. S., C. G. B. Demetrio, and J. P. Hinde (1998), Models for counts data with many zeros, paper presented at XIXth International Biometric Conference, Int. Biometric Soc., Cape Town, South Africa.

Sharples, J. J., and M. F. Hutchinson (2004), Multivariate spatial smoothing using additive regression splines, *ANZIAM J.*, *45*, suppl., C676–C692.

Sorooshian, S., K. Hsu, X. Gao, H. V. Gupta, B. Imam, and D. Braithwaite (2000), Evolution of the PERSIANN system satellite-based estimates of tropical rainfall, *Bull. Am. Meteorol. Soc.*, *81* (9), 2035–2046.

Sun, X., M. J. Manton, and E. E. Ebert (2003), Regional rainfall estimation using double-kriging of raingauge and satellite observations, *BMRC Res. Rep. 94*, 46 pp., Bur. of Meteorol., Melbourne, Victoria, Australia.

Teo, C. K., and D. I. F. Grimes (2007), Stochastic Modeling of rainfall from satellite data, *J. Hydrol.*, *346*, 33–50.

Thompson, R. D., and A. H. Perry (1997), *Applied Climatology: Principles and Practice*, 388 pp., Routledge, London, U. K.

Turk, F. J., G. D. Rohaly, J. Hawkins, E. A. Smith, F. S. Marzano, A. Mugnai, and V. Levizzani (2000), Meteorological applications of precipitation estimation from combined SSM/I, TRMM and infrared geostationary satellite data, in *Microwave Radiometry and Remote Sensing of the Earth's Surface and Atmosphere*, edited by P. Pampaloni and S. Paloscia, pp. 353–363, VSP Int. Sci. Publ., Zeist, Netherlands.

Van de Vyver, H., and E. Roulin (2009), Belgian contribution to the validation of the precipitation products of the Hydrology-SAF: methodology developed and preliminary results, *Precipitation Products Users Training Course*, pp. 1–8, EUMETSAT, Rome.

Venables, W. N., and B. D. Ripley (2002), *Modern Applied Statistics with S*, 4th ed., 481 pp., Springer, New York.

Xie, P., J. E. Janowiak, P. A. Arkin, R. F. Adler, A. Gruber, R. R. Ferraro, G. J. Huffman, and S. Curtis (2003), GPCP pentad precipitation analyses: An experimental dataset based on gauge observations and satellite estimates, *J. Clim.*, *16*, 2197–2214.

Yin, X., A. Gruber, and P. Arkin (2004), Comparison of the GPCP and CMAP merged gauge-satellite monthly precipitation products for the period 1979–2001, *J. Hydrometeorol.*, *5*(6), 1207–1222, doi:10.1175/JHM-392.1.

Zeileis, A., C. Kleiber, and S. Jackman (2008), Regression models for count data in R, *J. Stat. Software*, *27*(8), 1–25.

A. AghaKouchak, Department of Civil and Environmental Engineering, University of California, Irvine, Irvine, CA 92697, USA. (amir.a@uci.edu)

T. Hengl, Institute for Biodiversity and Ecosystem Dynamics, University of Amsterdam, Nieuwe Achtergracht 166, NL-1018 WV Amsterdam, Netherlands. (T.Hengl@uva.nl)

M. Perčec Tadić, Meteorological and Hydrological Service of Croatia, Grič 3, 10000 Zagreb, Croatia. (melita.percec.tadic@cirus.dhz.hr)

Rainfall Generation

Ashish Sharma and Raj Mehrotra

School of Civil and Environmental Engineering, University of New South Wales, Sydney, New South Wales, Australia

This chapter presents an overview of methods for stochastic generation of rainfall at annual to subdaily time scales, at single- to multiple-point locations, and in a changing climatic regime. Stochastic rainfall generators are used to provide inputs for risk assessment of natural or engineering systems that can undergo failure under sustained (high or low) extremes. As a result, generation of rainfall has evolved to provide options that adequately represent such conditions, leading to sequences that exhibit low-frequency variability of a nature similar to the observed rainfall. The chapter consists of three key sections: the first two outlining approaches for rainfall generation using endogenous predictor variables and the third highlighting approaches for generation using exogenous predictors often simulated to represent future climatic conditions. The first section presents approaches for generation of annual and seasonal rainfall and daily rainfall, both at single-point locations and multiple sites, with an emphasis on alternatives that ensure appropriate representation of low-frequency variability in the generated rainfall sequences. The second section highlights advancements in the subdaily rainfall generation procedures including commonly used approaches for daily to subdaily rainfall generation. The final section (generation using exogenous predictors) presents a range of alternatives for stochastic downscaling of rainfall for climate change impact assessments of natural and engineering systems. We conclude the chapter by outlining some of the key challenges that remain to be addressed, especially in generation under climate change conditions, with an emphasis on the importance of incorporating uncertainty present in both measurements and models, in the rainfall sequences that are generated.

1. INTRODUCTION

Stochastic generation of rainfall is of considerable interest for a probabilistic failure assessment of natural or man-made systems where rain is an important input. Such systems include most Civil Engineering infrastructure, such as buildings or bridges, which need to satisfy safety requirements to withstand varying levels of floods, water storages such as reservoirs, detention basins, or even rainwater tanks, which need to maintain supply under specified conditions of reliability, or agricultural and environmental systems, where both the magnitude and the sustenance of the rainfall incident on the system is of interest.

Stochastic generation, in general, requires characterization of the full probability distribution of the response (rainfall) and its associated predictors, so as to derive the conditional probability distribution from which the generation proceeds. In the context of rainfall, this characterization is often specific to the time scale being modeled. For instance, annual or monthly rainfall assume rain to be a continuous variable where the occurrence of a discrete zero state is often an anomaly to be disregarded, whereas daily or subdaily rainfall

Rainfall: State of the Science
Geophysical Monograph Series 191

10.1029/2010GM000973

needs to be modeled as a mixed distribution that accounts for the rainfall occurrence and nonoccurrence as distinct states. Furthermore, generation of subdaily and sometimes daily rainfall, especially for extreme events that are not sampled well in observed records, requires the formulation of intelligent disaggregation approaches that can transverse coarse scales to finer ones. Additionally, if the intent is to use the generated rainfall for an assessment of flood characteristics, issues related to persistence become less important. If, however, the generated rainfall is to be used to assess storage, long-term persistence, or low-frequency variability, often not well characterized by the default predictors that are identified by search algorithms, it needs to be carefully incorporated to avoid the problem of "overdispersion" that leads to biases in the storage attributes simulated. The above problems are compounded when the generation is to proceed at multiple locations or a regional grid, or when multiple variables such as temperature, wind speed, and evaporation are also required to be generated.

This chapter presents an overview of stochastic generation of rainfall, with a focus on daily and subdaily rainfall generation at point and multiple locations, for the current climate assuming climatic stationarity, as well as for future climates using exogenous inputs simulated using general circulation models under assumed greenhouse gas emission scenarios. The chapter is organized into three main sections, the first two dealing with stochastic generation of rainfall in a stationary climate, while the third on stochastic downscaling of rainfall for future climates affected by global warming. These sections also include a summary of rainfall generation procedures currently in use and an outline of the many challenges that remain to be addressed and form the basis for research in these areas in the coming years. Throughout the chapter, our aim is to summarize the current practice, the associated advantages and limitations, and our assessment of the research needs to move the discipline further. An outline of the subsections that constitute the main sections is provided at the outset of each of the sections.

2. STOCHASTIC RAINFALL GENERATION

This section summarizes many approaches that have been developed for stochastic generation of rainfall at annual, daily, and subdaily time scales, for single (point) or multiple locations, and for the case where long-term persistence or low-frequency variability assumes importance. Before getting into the various approaches that are discussed, it is useful to briefly visit the theory stochastic generation is based on.

Let $J(t)$ represents a binary random variable (assuming values of 0 or 1) at time t, $t = 1, \ldots, T$, where T is the length of the observed time series to be used to develop the stochastic model. In the context of rainfall, $J(t)$ represents the rainfall occurrence/nonoccurrence process over time, where

$$J(t) = 0$$

if no rain occurs during time interval t, and

$$J(t) = 1$$

if time interval t has rain.

Assume that the variable evolves as a function of its state at the previous time step, or that it exhibits order-1 Markovian dependence:

$$\begin{aligned} P[J(t) &= 1 | J(t-1), J(t-2), \ldots, J(1)] \\ &= P[J(t) = 1 | J(t-1)]. \end{aligned} \tag{1}$$

Stochastic generation of $J(t)$ proceeds through the specification of the conditional or transition probability in equation (1). Since the variable being modeled exists in two states, the conditional probability consists of a transition probability matrix of size 2×2:

$$\begin{bmatrix} p_{00} = P[J(t) = 0 | J(t-1) = 0] & p_{01} = (1 - p_{11}) = P[J(t) = 0 | J(t-1) = 1] \\ p_{10} = (1 - p_{00}) = P[J(t) = 1 | J(t-1) = 0] & p_{11} = P[J(t) = 1 | J(t-1) = 1] \end{bmatrix}. \tag{2}$$

Hence, the conditional probability distribution can be defined by the specification of the transition probabilities p_{00} and p_{01}, which can be estimated simply as the relative frequency with which the respective transitions are recorded in the observed record.

The general steps to generate a stochastic series from the model described through equations (1) and (2) are (algorithm 1):

Algorithm 1: Markov order-1 process-based generation

1. Set $t = 0$. Randomly specify a value of 0 or 1 to $J(0)$.
2. Increment t by 1.
3. Generate a uniform random number (u) between 0 and 1.

4. Sample $J(t)$ conditional to $J(t-1)$ as follows:

$$\begin{aligned}
&\text{If} \quad J(t-1)=0\\
&\qquad J(t)=0 \quad \text{if} \quad u \le p_{00}\\
&\qquad J(t)=1 \quad \text{if} \quad u > p_{00}\\
&\text{If} \quad J(t-1)=1\\
&\qquad J(t)=0 \quad \text{if} \quad u \le p_{01}\\
&\qquad J(t)=1 \quad \text{if} \quad u > p_{01}.
\end{aligned}$$

5. Repeat steps 2–4 ($N_w + N$) times where N is the length of the desired time series, and N_w is a "warm-up" period, the first N_w generated values being discarded to account for the random initialization used.

It is important to note that the above logic involves two specifications: (a) the choice of the predictors used, in the above case $J(t-1)$, and (b) the choice of the full or conditional probability distributions, in this case the conditional probability mass function defined by (p_{00}, p_{01}). In general, $J(t)$ could represent a nondiscrete response (such as seasonal or annual rainfall) and $J(t-1)$ be a predictor vector that includes functions of prior lags of $J(t)$ as well as exogenous predictors that originate from other processes or models (such as pressure and temperature simulated using a general circulation model if the aim is to generate rainfall for future climates). If this were so, all that is needed is a different formulation of the full and conditional probability distributions that define the relationship between the response and predictors, distributions that may be simpler to specify if they belong to a certain family and are characterizable through a handful of parameters, or alternately by the full historical record, assuming it is vast enough to reasonably characterize the full probability distribution of the variables being modeled. These two approaches are referred to as the parametric and the nonparametric alternatives for stochastic generation of rainfall. While each have their merits and demerits, in general, parametric alternatives are favored when data is limited, and models need to be applied at ungauged locations and nonparametric alternatives favored when data is plenty and available at the location being modeled.

It should also be noted that while the above example is specific to a univariate response, extension to multiple responses is often a matter of representing the pertinent equations in the form of matrices and vectors. What becomes a problem often is having sufficient data to specify these matrices in a reliable fashion. In what follows, readers will be exposed to a variety of modeling alternatives for both univariate and multivariate responses that belong to both the parametric and nonparametric categories. Our presentation goes from coarser to finer time scales (with an increasing focus on the latter), with the ensuing sections dealing with seasonal/annual rainfall generation, daily rainfall generation, and subdaily rainfall generation, respectively.

2.1. *Seasonal to Annual Stochastic Rainfall Generation*

In the context of the generation algorithm presented previously, generation of annual rainfall at a single-point location, a Markov order-1 dependence process, or, dependence of $J(t)$ does not extend beyond $J(t-1)$. Additionally, the joint probability distribution of $(J(t), J(t-1))$ is assumed to be a bivariate normal distribution, which often requires the use of normalizing transformations for skewed annual rainfall, the most common transformation used being a "log" transform (often with a specified offset parameter) and less commonly a "Box-Cox" transformation.

A multisite version of the above generation model can be written as

$$\boldsymbol{J}_t = \mathbf{A}\,\boldsymbol{J}_{t-1} + \mathbf{B}\,\boldsymbol{e}_t, \tag{3}$$

where $\boldsymbol{J}_t$ and $\boldsymbol{J}_{t-1}$ represent standardized random variables, $\mathbf{A}$ and $\mathbf{B}$ are parameter matrices specified using order-0 and order-1 moments of the historical data (for details see chapters on multivariate autoregressive models in the work of *Bras and Rodriguez-Iturbe* [1985] and *Salas et al.* [1980]), and $\boldsymbol{e}_t$ is a vector of $N(0,1)$ random variates of the same length as $\boldsymbol{J}_t$ or $\boldsymbol{J}_{t-1}$ or the number of sites being modeled. While the above formulation requires the specification of $2 \times d \times d$ parameters, where d is the number of sites modeled, there are alternatives [*Hipel and McLeod*, 1994] that reduce the number of modeled parameters by invoking assumptions on the strength of the lagged cross-correlations between elements of $\boldsymbol{J}_t$ and $\boldsymbol{J}_{t-1}$.

The above representation has been criticized as being inadequate at representing sustained periods of droughts and high flows in data, leading to the development of alternatives such as the hidden state Markov (HSM) model [*Frost et al.*, 2007; *Thyer and Kuczera*, 2000] and the moving block bootstrap (MBB) approach [*Srinivas and Srinivasan*, 2006]. The HSM approach deviates from the usual assumption of Markovian dependence and instead assumes that the process being modeled depends on a "state" that is not explicitly defined (hence the term "hidden"), or,

$$\boldsymbol{J}_t = \boldsymbol{J}_t | S_t, \tag{4a}$$

where

$$S_t = S_t | S_{(t-1)}. \tag{4b}$$

Furthermore, S_t is assumed to be binary, representing a discrete "wet" or a "dry" state, allowing modeling of the conditional relationship in equation (4b) by a simple order-1 transitional probability matrix. Consequently, the generation of $\boldsymbol{J}_t$ proceeds through an appropriately specified unconditional multivariate probability distribution (assumed as a multivariate normal by *Thyer and Kuczera* [2000]). The advantage of using the hidden state structure is the ability to model sustained dry and wet states explicitly, often noted in annual hydrologic data and found to impact reservoir storage and related attributes of interest.

In contrast to the HSM, the MBB or its variant, the hybrid MBB [*Srinivas and Srinivasan*, 2006] conditionally "resamples" an entire block of observed data that has been standardized and whitened that is subsequently transformed back to the original variables. Unlike the alternatives in equations (3) and (4), this approach is nonparametric and assumes that the observed residuals (obtained through the standardization and whitening steps on the historical data) have the necessary persistence attributes that allow for the generation of the sustained droughts that are needed to qualify the generated series as a "representative" version of the historical record.

Seasonal rainfall generation proceeds using alternatives not too dissimilar to those discussed above, the main difference being that the parameters (or the data segments being resampled in the nonparametric equivalents) are a function of season and, hence, are estimated based on the seasonal and lagged seasonal data, the model for any given season corresponds to. Without going into details, readers are referred to chapters on monthly or seasonal rainfall/streamflow generation in stochastic generation texts such as those of *Bras and Rodriguez-Iturbe* [1985], *Salas et al.* [1980], and *Hipel and McLeod* [1994].

A key drawback of seasonal (and to a greater extent, daily) generation models is "overdispersion," which occurs when generated sequences are aggregated to annual and longer time scales and are found to not exhibit the persistence and variability that is present in the historical record. This occurs because the stochastic model assumes season-to-season persistence, with the impact of the first season becoming progressively smaller as one proceeds to the next year and beyond. This drawback is important when sequences are used for storage assessment where low-frequency variability or long-term persistence needs to be well simulated. Corrective options include using higher-order Markovian dependence in the conditional generation process [*Coe and Stern*, 1982; *Gates and Tong*, 1976; *Eidsvik*, 1980; *Pegram*, 1980; *Stern and Coe*, 1984; *Wilks*, 1999b], using aggregate predictors that convey the long-term behavior to the response [*Sharma and O'Neill*, 2002], and disaggregation of annual data, which has been separately generated ensuring an appropriate representation of persistence [*Koutsoyiannis*, 2001; *Tarboton et al.*, 1998; *Valencia and Schaake*, 1973]. The next section presents options for stochastic generation at finer time steps, where the overdispersion problem becomes more acute.

2.2. Daily Rainfall Generation

2.2.1. Background. The modeling of daily rainfall process has attracted a lot of interest in the past decades. Various daily rainfall models have been developed. Unlike annual or seasonal rainfall, daily rainfall cannot be characterized by a continuous probability distribution and is usually modeled using a binary (wet or dry) rainfall occurrence following a conventional Markov process and a continuous rainfall amount for the generated wet days. Consequently, models for daily rainfall generation follow a two-step procedure: rainfall occurrence generation and generation of rainfall amounts on wet days. Rainfall occurrence is modeled as a Markov process, the state (wet or dry) of the current and a few prior days deciding the state of the day that follows, while amounts on wet days are modeled through appropriately specified probability distributions. To reduce the number of parameters of the higher-order Markov models of rainfall, *Pegram* [1980] devised a Markov chain modeling structure, which allows the specification of a lag-k dependence structure with $k + 1$ parameters rather than $2k$. Similarly, *Stern and Coe* [1984] have suggested Markov chains of hybrid order requiring only $k + 1$ rather than $2k$ parameters. Although the Markov model is inappropriate in some areas due to event clustering or other phenomena [*Srikanthan and McMahon*, 2001], it has been found useful, in general.

While rainfall has traditionally been modeled using parametric probability distributions to characterize the conditional or transition probabilities needed, there has been a surge in nonparametric equivalents that are based on bootstrap and kernel density nonparametric approaches [*Lall and Sharma*, 1996; *Lall et al.*, 1996; *Rajagopalan et al.*, 1996; *Rajagopalan and Lall*, 1999]. The bootstrap-based approaches have a limitation on the data extrapolation beyond the range of historical records and the estimation of high-dimensional density function [*Silverman*, 1986; *Scott*, 1992; *Srikanthan and McMahon*, 2001].

Stochastic models of daily time scale do not capture the wet and dry spell-related rainfall statistics very well. Spells can be of great importance for applications such as agriculture and water resources. Modeling the process as wet and dry spells can be an attractive alternative. Another category of the rainfall generation models consist of approaches that represent the length of the current dry or wet spell as a random variable and model it as an alternating renewal process using parametric and nonparametric approaches [*Buishand*, 1978; *Foufoula-Georgiou and Lettenmaier*, 1987; *Woolhiser and Roldan*, 1982; *Lall et al.*, 1996; *Sharma and Lall*, 1999].

Seasonality in rainfall simulation procedure is typically incorporated by fitting different rainfall occurrence (transition probabilities) and amounts (parameters of gamma or exponential distribution) models for each month. This may result in sharp transition from 1 month to another. As an alternative, *Rajagopalan et al.* [1996] proposed a nonhomogeneous Markov chain, wherein the transition probabilities are estimated for each day of the year using information within an optimal sliding window. The rainfall amounts (parameters) are also simulated using the data available within this window.

The final category of models includes all other approaches to generate rainfalls and perhaps illustrates recent developments of generating hydroclimatic time series. Models such as a hidden Markov model [*Thyer and Kuczera*, 2000], a Markov switching model [*Shami and Forbes*, 2000], semiparametric models [*Kim and Valdés*, 2005; *Mehrotra and Sharma*, 2007a], and modified Markov model [*Mehrotra and Sharma*, 2007b] fall under this class of models. Generalized linear modeling (GLM)-based weather generators seem to provide an attractive and complementary alternative to traditional rainfall and weather simulation and are gaining popularity [*Chandler and Wheater*, 2002; *Yang et al.*, 2005; *Furrer and Katz*, 2007]. The GLM approach is flexible in that it can easily be incorporated in a hierarchical modeling framework. These recent approaches enable us to incorporate other characteristics of rainfall, such as measures of nonlinearity, nonstationarity, and long-term persistence, respectively. A summary of the various modeling alternatives for daily rainfall occurrence and amount generation is presented in Table 1.

A key drawback in conventional models for daily rainfall generation is the problem of "overdispersion," discussed before in the context of seasonal rainfall generation. Overdispersion becomes more of an issue when modeling daily rainfall, with variances at annual and longer time scales exhibiting significant (greater than 50%) biases. To address this issue, approaches have been suggested in the past to enhance the variability in the generated rainfall series, like conditioning on covariates or other continuous variables [*Wilks*, 1989; *Katz and Zheng*, 1999; *Sharma and O'Neill*, 2002; *Mehrotra and Sharma*, 2007a, 2007b; *Bárdossy and Plate*, 1992; *Woolhiser*, 1992; *Woolhiser et al.*, 1993; *Hughes and Guttorp*, 1994; *Katz and Parlange*, 1993, 1998] and variance inflation [*Klein et al.*, 1959; *Srikanthan*, 2004] that increases variability by multiplying by a suitable factor. Some of the daily rainfall generation studies that have explicitly attempted to mitigate overdispersion are outlined in Table 2.

Another important issue in rainfall simulation, more specifically at daily time scale, is related to proper simulation of rainfall extremes. It has been noted that parametric and nonparametric rainfall generation procedures perform reasonably well in reproducing averages, however, simulate poorly the extremes, especially high precipitation amounts [*Sharif and Burn*, 2006; *Furrer and Katz*, 2008; *Wilks*, 1999b]. Parametric weather generators do not produce a heavy enough upper tail for the distribution of daily precipitation amount, whereas those based on resampling have inherent limitations in representing extremes. Some recent studies have attempted to address this issue, majority of them are based on consideration of some kind of mixture distribution for rainfall amount generation, for example, *Furrer and Katz* [2008] considered gamma distribution for low to moderate values and a generalized Pareto (GP) distribution for high values; *Johnson et al.* [1996] and *Wilks* [1999b] considered a mixture of two exponential distributions; *Cameron et al.* [2001] used the GP, instead of the exponential, for the distribution of high intensities from a rain cell. *Semenov* [2008] used kernel density smoothing for simulation of precipitation extremes. *Wilson and Toumi* [2005] argued that the distribution of high precipitation intensity can be approximated as stretched exponential. *Vrac and Naveau* [2007] used a dynamic mixture with gamma and GP distributions. The dynamic mixture avoids the need of a threshold for switching from one distribution to other as it is designed to assign more weight to the gamma distribution for low rainfall values, while the GP distribution gets more weight for high rainfall values.

The following sections present details of four daily rainfall modeling alternatives that are used commonly for stochastic generation. While the first approach does not specifically aim at mitigating overdispersion (posttreatment is possible using a nesting logic and correction factor), the next three employ a range of strategies to reduce overdispersion in generated sequences. While the first two fall in the parametric category, the latter two employ a mix of parametric and nonparametric alternatives to generate the rainfall sequences. The section concludes with a short discussion on the model nesting procedure that aims to reduce the over dispersion.

2.2.2. TPM approach. The transition probability matrix (TPM) [*Allen and Haan*, 1975; *Carey and Haan*, 1978; *Srikanthan and McMahon*, 1983] approach-based models simulate both precipitation occurrence and amounts, by defining different ranges of precipitation amounts as constituting distinct states. Thus, the TPM model is a multistate first-order Markov chain where daily rainfall amounts are divided into a number of states, with state 1 as dry (no rainfall) and the other states as wet. The day to day probabilities of transition from one state to another are calculated from historical data. The seasonality in occurrence and magnitude of daily rainfall can be incorporated by considering

Table 1. A Summary of Alternatives for Stochastic Generation of Daily Rainfall

Model	Description/Advantages/Drawbacks	References
	Daily Rainfall Occurrence Generation	
Low-order Markov chain models	Based on wet days probabilities. For some regions generates rainfall series with too few long dry spells.	*Gabriel and Neumann* [1962], *Buishand* [1977, 1978], *Racsko et al.* [1991], *Wilks* [1998], *Caskey* [1963], *Weiss* [1964], *Hopkins and Robillard* [1964], *Feyerherm and Bark* [1965, 1967], *Lowry and Guthrie* [1968], *Selvalingam and Miura* [1978], *Stern* [1980a, 1980b], *Garbutt et al.* [1981], *Richardson* [1981], *Stern and Coe* [1984]
Higher-order Markov chain models	Based on wet day probabilities of few consecutive days. The approach increases the length of the Markov model's "memory" of antecedent wet and dry days. The number of parameters (i.e., transition probabilities) required increases exponentially as the order increases, being 2^k for akth-order chain. These models improve the representation of observed interannual variance in the simulations but still fell short of observed climatic variability on average.	*Dennett et al.* [1983], *Singh and Kripalani* [1986], *Jones and Thornton* [1997], *Chin* [1977], *Coe and Stern* [1982], *Gates and Tong* [1976], *Eidsvik* [1980], *Pegram* [1980], *Singh et al.* [1981]
"Hybrid-order" Markov models	The Markov "memory" extends further back in time for the dry spells only.	*Stern and Coe* [1984], *Wilks* [1999b]
Alternating renewal process based models	These spell-length models operate by fitting probability distributions to observed relative frequencies of wet and dry spell lengths. The approach may not be suited in arid regions or in cases with less than 25 years of observations.	*Williams* [1947], *Green* [1964], *Buishand* [1977, 1978], *Roldan and Woolhiser* [1982], *Wilks* [1999b], *Racsko et al.* [1991]
	Nonparametric wet-dry spell length models	*Lall et al.* [1996], *Sharma and Lall* [1997, 1999]
	Daily Rainfall Amount Generation	
Parametric precipitation amounts models	Based on some distribution like a two-parameter gamma distribution, exponential, and mixed exponential distribution. These models assume that precipitation amounts on wet days are independent and follow the same distribution.	*Jones et al.* [1972], *Goodspeed and Pierrehumbert* [1975], *Coe and Stern* [1982], *Richardson* [1981], *Todorovic and Woolhiser* [1975], *Woolhiser and Pegram* [1979], *Woolhiser and Roldan* [1982, 1986], *Wilks* [1999c]
Wet spell-based precipitation amount models	These models allow different probability distributions for precipitation amounts depending on that day's position in a wet spell (separate models for start, mid, and end of a wet spell).	*Cole and Sherriff* [1972], *Buishand* [1977, 1978], *Katz* [1977], *Chin and Miller* [1980], *Wilks* [1999c]
Nonparametric precipitation amount models	A nonparametric kernel density estimation-based procedure is used to simulate the rainfall conditional on previous time step value of rainfall and/or other variables	*Harrold et al.* [2003b], *Mehrotra and Sharma* [2006], *Mehrotra et al.* [2006]
Multistate Markov models	These Markov models simulate both precipitation occurrence and amounts, by defining different ranges of precipitation amounts as constituting distinct states. The outcome of this approach depends on the choice of the number of states, their ranges, and on the distributions used for wet-day amounts in any given state. These models involve comparatively large numbers of parameters, and thus require quite long data records in order to be estimated well.	*Gregory et al.* [1993], *Haan et al.* [1976], *Srikanthan and McMahon* [1983, 1985], *Boughton* [1999]
Cluster-based point processes models	Rainfall process is described using cluster of rectangular pulses. In the approach, storms arrive according to a Poisson process and are represented by clusters of rainfall cells temporally displaced from the storm center.	*Neyman and Scott* [1958], *Kavvas and Delleur* [1981], *Rodriguez-Iturbe et al.* [1984, 1987, 1988], *Onof and Wheater* [1993, 1994], *Waymire and Gupta* [1981a, 1981b, 1981c], *Ramírez and Bras* [1985]
Multifractal simulation techniques	These models characterize rainfall by scale invariant (scaling) and fractal properties.	*Marshak et al.* [1994], *Olsson* [1996], *Menabde et al.* [1997]
Time series models	Time series models similar to streamflow data generation are used to generate daily rainfall data.	*Adamowski and Smith* [1972]

Table 2. Alternatives for Daily Rainfall Generation Aimed at Reducing Overdispersion

Model	Description/Advantages/Drawbacks	References
Conditioning on covariates	Monthly statistics of rainfall, long-range forecasts of the monthly statistics, random numbers or a "hidden" mixture approach to capture some interannual variability.	*Wilks* [1989], *Briggs and Wilks* [1996], *Jones and Thornton* [1997], *Katz and Zheng* [1999]
Conditioning on previous time history of simulated rainfall	Rainfall occurrences and amounts are simulated conditional on the recent past rainfall behavior.	*Sharma and O'Neill* [2002], *Harrold et al.* [2003a, 2003b], *Mehrotra and Sharma* [2007a, 2007b]
Conditioning on some aspect of large-scale atmospheric circulation	Using the Lamb Weather Type weather classification, monthly Southern Oscillation Index, North Atlantic Oscillation Index, North Atlantic sea surface temperature anomalies and other atmospheric predictors.	*Hay et al.* [1991], *Bárdossy and Plate* [1992], *Woolhiser* [1992], *Woolhiser et al.* [1993], *Hughes and Guttorp* [1994], *Katz and Parlange* [1993, 1998], *Wallis and Griffiths* [1997], *Kiely et al.* [1998], *Wilby* [1998]
Model nesting at multiple time scales	Rainfall amounts are adjusted at monthly/seasonal and annual time scales to maintain the desired variability at higher time scales	*Boughton* [1999], *Wang and Nathan* [2002], *Srikanthan* [2004]

each season/month separately. The rainfall values in the intermediate states are modeled either by a uniform [*Haan et al.*, 1976] or by a linear distribution [*Srikanthan and McMahon*, 1985]. Similarly, for the last state, an exponential distribution [*Haan et al.*, 1976] or a Box-Cox transformation as suggested by *Srikanthan and McMahon* [1985] can be used. The following describes a stepwise rainfall generation procedure using TPM model (algorithm 2).

Algorithm 2: Stepwise daily rainfall generation procedure using TPM model.

1. Estimate the daily transition probabilities using

$$p_{i,j}(k) = \frac{f_{i,j}(k)}{\sum_{j=1}^{C} f\chi_{i,j}(k)} \qquad i,j = 1, \ldots, C; k = 1, \ldots S,$$

where $p_{i,j}(k)$ probability of transition from state i to state j within season k, $f_{i,j}(k)$ = historical frequency of transition from state i to state j within season k, C = the maximum number of states, and S = number of seasons.

2. Assume that the initial state is dry (i.e., state one).
3. Generate a uniformly distributed random number U between 0 and 1. Using the appropriate TPM for the season, determine the state of the next day.
4. If the state is wet, go to step 5. Otherwise, set the rainfall depth to zero and go to step 3.
5. Calculate the rainfall depth by using a uniform/linear distribution for the intermediate states and a power/Box-Cox transformation for the largest state.
6. Repeat steps 3 to 5 until the required length of daily rainfall data is achieved.

To improve the annual variability of rainfall, *Boughton* [1999] proposed an empirical adjustment factor (F) in the TPM procedure to match the generated standard deviation of the annual rainfall with the observed value. The generated daily rainfall in each year is multiplied by the following ratio:

$$I_i = \frac{\{M + (T_i - M)F\}}{T_i}, \tag{5}$$

where M is the generated mean annual rainfall, T_i is the generated annual rainfall for year I, and F is an adjustment factor that is defined as

$$F = \frac{s_h}{s_g}, \tag{6}$$

where s_g and s_h are standard deviations of the generated and historical annual rainfall series, respectively.

2.2.3. ROG-RAG model. Harrold et al. [2003a, 2003b] presented a nonparametric model for generating single-site daily rainfall occurrence and amounts (ROG-RAG) with the aim of reproducing longer-term variability and low-frequency features such as drought and sustained wet periods in the simulations, while still reproducing characteristics at daily time scales (see Figure 1 for the model results). This was achieved within a Markovian framework by using "aggregate" predictor variables that describe how wet it has been over a period of time.

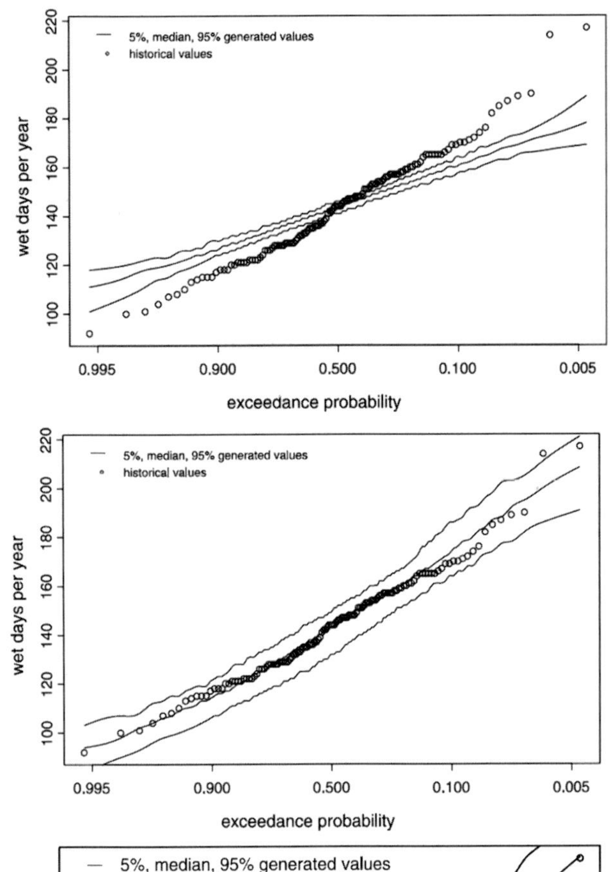

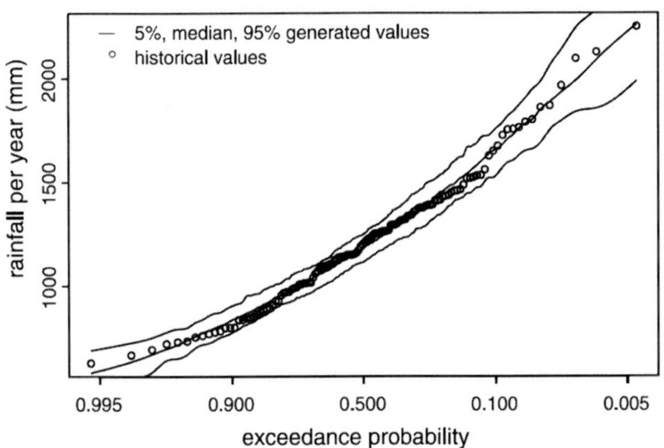

Figure 1. Distribution of wet days and rainfall amounts per year for Sydney using (top) (1) Markov order-1 model, (middle) ROG(4) (three discrete wetness states) model, and (bottom) ROG(4)-RAG(2) model. From *Harrold et al.* [2003a, 2003b].

The procedure used by them for rainfall occurrence generation involves the following steps (algorithm 3).

Algorithm 3: Stepwise daily rainfall generation procedure using ROG-RAG model

At the preprocessing stage of the algorithm, for each day t, calculate the values of the predictor set (J_{t-1}; $b_{m_i t}$), where J_{t-1} is previous time step rainfall state, $b_{m_i t}$ is ith "wetness index" (i varies from 1 to ns, ns being the number of discrete states considered) and is a transformation of the previous time period wetness state into a small number of discrete indices for all historical days. (ROG(4) model of *Harrold et al.* [2003a, 2003b] consisted of the following four predictors: (1) rainfall occurrence on the previous day, (2) 90-day wetness index, (3) 365-day wetness index and, (4) 1825-day wetness index). Also calculate the scaling weights for each predictor using the following:

$$w_i = 1/s_i,$$

where w_i denotes the scaling weight for predictor variable i, where i ranges from 1 to np, with np being the number of predictor variables ($ns + 1$), and s_i is the estimated local standard deviation, which is seasonal and calculated using the moving window.

1. Select a short part of the historical sequence at random, to use for the initial specification of the predictor variables J_t and $b_{m_i t}$. The first day in the generated sequence is the day immediately after the end of this startup sequence. The main algorithm can then be implemented as follows:
2. Calculate the values of the predictors for the current day in the generated sequence. Assume some appropriate value of the number of nearest neighbors (k) to be considered.
3. Randomly choose a value for i using, $p(i) = 1/i \Big/ \sum_{j=1}^{k} 1/j$.
4. Formulate an l-day moving window centered on the current date.
5. Select the ith historical nearest neighbor from the moving window using the following Euclidean distance measure:

$$E_d = \sqrt{(w_0(J_{t-1} - J_{d-1}))^2 + \sum_{i=1}^{3} (w_i(b_{m_i t} - b_{m_i d}))^2}.$$

The ith nearest neighbor has the ith smallest value of E_d. Whenever ties occur, randomly choose from all dates in the moving window with E_d equal to the ith smallest value.

6. Put the chosen historical value J_d (i.e., the ith nearest neighbor) into the current position in the generated sequence, i.e., set J_t equal to J_d.
7. Move on to the next date in the generated sequence.
8. Repeat steps 3–8 until the desired length of generated sequence is obtained.

2.2.4. Modified Markov model. *Mehrotra and Sharma* [2007b] proposed an algorithm to simulate rainfall at individual locations using separate models for rainfall occurrences and rainfall amounts on the simulated wet days. The rainfall occurrence model (Modified Markov model (MMM)) is based on a modification of the transition probabilities of the traditional Markov model through an analytically derived factor that represents the influence of rainfall aggregated over long time periods (higher time scale variables) in an attempt to incorporate low-frequency variability in simulations. The rainfall amounts on the wet days then can be simulated using either parametric or nonparametric conditional simulation approach [as adopted in *Mehrotra and Sharma*, 2007b]. Figure 2 presents the modeled results for Sydney using the model, while algorithm 4 describes a stepwise procedure to be followed for simulation of rainfall occurrences.

Algorithm 4: Stepwise daily rainfall generation procedure using MMM.

1. For all calendar days of the year, calculate the transition probabilities of the standard first-order Markov model using the observations falling within the moving window of 31 days centered on each day. Denote these transition probabilities as p_{11} for previous day being wet and p_{10} for previous day being dry.
2. Also estimate the means, variances, and covariances of the higher time scale predictor variables separately for occasions when current day is wet/day and previous day is wet/dry. (*Mehrotra and Sharma* [2007b], identified two variables, namely, previous 30 and 365 days wetness state.)
3. Consider a day. Ascertain appropriate critical transition probability to the day t based on previous day's rainfall state of the generated series. If previous day is wet, assign critical probability p as p_{11}; otherwise, assign p_{10}.
4. Calculate the values of the 30 and 365 days wetness state for the day t and the available generated sequence (Jo). To have values of wetness state in the beginning of the simulation, randomly pick up a year from the historical record and calculate values of 30 and 365 days wetness states.
5. Modify the critical transition probability p of step 3 using the following equation, and conditional means, variances, covariances, and tth day value of higher time scale predictors for the generated day t. Denote the modified transition probability as $\hat{p}$.

$$\hat{p} = p_{1i} \frac{\dfrac{1}{\det(\mathbf{V}_{1,i})^{1/2}} \exp\left\{ -\frac{1}{2}(\mathbf{X}_t - \boldsymbol{\mu}_{1,i})\mathbf{V}_{1,i}^{-1}(\mathbf{X}_t - \boldsymbol{\mu}_{1,i})' \right\}}{\left[\dfrac{1}{\det(\mathbf{V}_{1,i})^{1/2}} \exp\left\{ -\frac{1}{2}(\mathbf{X}_t - \boldsymbol{\mu}_{1,i})\mathbf{V}_{1,i}^{-1}(\mathbf{X}_t - \boldsymbol{\mu}_{1,i})' \right\} p_{1i}\right] + \left[\dfrac{1}{\det(\mathbf{V}_{0,i})^{1/2}} \exp\left\{ -\frac{1}{2}(\mathbf{X}_t - \boldsymbol{\mu}_{0,i})\mathbf{V}_{0,i}^{-1}(\mathbf{X}_t - \boldsymbol{\mu}_{0,i})' \right\}(1 - p_{1i})\right]},$$

where $\mathbf{X}_t$ is the predictor set at time t, the $\boldsymbol{\mu}_{1,i}$ parameters represent the mean $E(X_t | J_t = 1, J_{t-1} = i)$, and $\mathbf{V}_{1,i}$ is the corresponding variance-covariance matrix. Similarly, $\boldsymbol{\mu}_{0,i}$ and $\mathbf{V}_{0,i}$ represent, respectively, the mean vector and the variance-covariance matrix of $\mathbf{X}$ when $(J_{t-1} = i)$ and $(J_t = 0)$. The p_{1i} parameters represent the baseline transition probabilities of the first-order Markov model defined by $P(J_t = 1 | J_{t-1} = i)$, and det() represents the determinant operation.
6. Compare $\hat{p}$ with the uniform random variate $u_t(k)$ for station k. If $u_t(k)$ is $\leq \hat{p}$, assign rainfall occurrence, Jo_t for the day t as 1; otherwise, zero.
7. Move to the next date in the generated sequence and repeat steps 2–5 until the desired length of generated sequence is obtained.

2.2.5. Model nesting at multiple time scales. As mentioned before, the commonly used daily rainfall generation models preserve the daily rainfall characteristics; however, they undersimulate the monthly and annual characteristics. By nesting the daily model in monthly and annual models, the characteristics of rainfall at daily, monthly, and annual

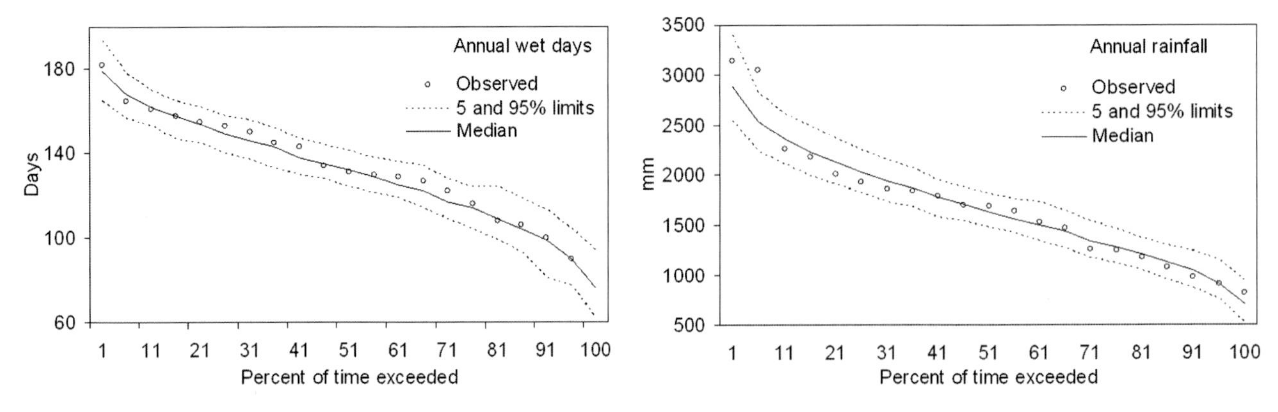

Figure 2. Distribution of wet days and rainfall amount per year for Sydney using modified Markov model.

levels are preserved. *Wang and Nathan* [2002] proposed a daily monthly mixed algorithm to preserve the monthly rainfall characteristics explicitly. *Srikanthan* [2004] proposed a nested daily rainfall model to preserve the daily, monthly, and annual characteristics. The steps involved in formulating the nested model used by *Srikanthan* [2004] are described in the next box (algorithm 5).

Algorithm 5: Stepwise daily rainfall generation procedure using nested model.

1. Let x be the series of observed daily rainfall at a location. Calculate the two transition probabilities of a first-order Markov chain: p_{10}, the conditional probability of a wet day given that the previous day was dry; p_{11}, the conditional probability of a wet day given that the previous day was wet.
2. For wet days, obtain the rainfall depth using a Gamma distribution whose probability density function is given by

$$f(x) = \frac{(x/\beta)^{\alpha-1}\exp(-x/\beta)}{\beta\Gamma(\alpha)},$$

where α is the shape parameter and β the scale parameter. The mean and variance of the Gamma distribution are given as: $\mu(x) = \alpha\beta$ and $\sigma^2(x) = \alpha\beta^2$.
3. The seasonality in daily rainfall is taken into account by considering each month separately. Once the daily rainfall is generated for a month, the monthly rainfall is obtained by summing the daily rainfall values. The generated monthly rainfall value, $\tilde{X}_i$, is modified by using the Thomas-Fiering monthly model to preserve the monthly characteristics.

$$\frac{X_i - \mu(X_i)}{\sigma(X_i)} = \rho_{i,i-1}\frac{X_{i-1} - \mu(X_{i-1})}{\sigma(X_{i-1})} + (1 - \rho_{i,i-1}^2)^{1/2}\frac{\tilde{X}_i - \mu'(X_i)}{\sigma'(X_i)},$$

where $\rho_{i,i-1}$ is the correlation coefficient between months i and $i-1$., $\mu(X)$ and $\sigma(X)$ are mean and standard deviation of the observed data for each month, and superscript' is used for the generated data.

The generated daily rainfall data is multiplied by the ratio $X_i/\tilde{X}_i$. Once the values for the 12 months of a year (k) have been generated, the generated monthly values are aggregated to obtain the annual value. The aggregated annual value, $\tilde{Z}_k$, is modified by using a lag one autoregressive model to preserve the annual characteristics.

$$\frac{Z_k - \mu(Z)}{\sigma(Z)} = \rho(Z)\frac{Z_{k-1} - \mu(Z)}{\sigma(Z)} + [1 - \rho^2(Z)]^{1/2}\frac{\tilde{Z}_k - \mu'(Z)}{\sigma'(Z)},$$

where ρ is the lag one autocorrelation coefficient. If the annual rainfall data exhibits significant skewness, then the noise term in the above expression is modified by using the Wilson-Hilferty transformation (1931). The theoretical values of the mean and variance of the aggregated annual rainfall are given by

$$\mu(Z) = \sum_{j=1}^{12} \mu(X_j)$$

$$\sigma^2(Z) \approx \sum_{j=1}^{12} \sigma^2(X_j) + 2\sum_{j=2}^{12} \sigma(X_j)\sigma(X_{j-1})\rho_{j,j-1} + 2\sum_{j=3}^{12} \sigma(X_j)\sigma(X_{j-2})\rho_{j,j-1}\rho_{j-1,j-2} + 2\sum_{j=4}^{12} \sigma(X_j)\sigma(X_{j-3})\rho_{j,j-1}\rho_{j-1,j-2}\rho_{j-2,j-3}.$$

The generated monthly rainfall value is multiplied by the ratio $Z_k/\tilde{Z}_k$. This will preserve the annual characteristics. The modified monthly rainfall values are used to adjust the daily rainfall values. Rather than adjusting the daily rainfall values twice, the adjustment to the daily rainfall values can be carried out in one step by multiplying the generated daily rainfall values for each month (i) by the ratio $X_i Z_k / \tilde{X}_i \tilde{Z}_k$.

2.2.6. Extension to multisite generation. The generation of daily rainfall occurrence or amount in previous sections has been presented as a point generation problem. Most of the approaches discussed have varying sets of options to enable generation at multiple point locations. *Srikanthan and McMahon* [2001] provide an extensive review of rainfall models, including multisite network modeling. There has been continued interest in the subject, and some of the

Table 3. Multisite Daily Rainfall Generation Alternatives

Model	Description/Advantages/Drawbacks	References
Weather state-based Hidden Markov model for the occurrence/ nonoccurrence of rainfall.	Multisite rainfall is simulated conditional on the weather states and/or atmospheric circulation patterns.	*Zucchini and Guttorp* [1991], *Bárdossy and Plate* [1991, 1992], *Wilson et al.* [1992], *Hughes and Guttorp* [1994], *Hughes et al.* [1999], *Mehrotra and Sharma* [2005, 2006], *Pegram and Seed* [1998], *Thyer and Kuczera* [2000]
Multisite Markov models	Individual occurrence and amount models are fitted to each of the sites, and spatial dependence is introduced by making use of spatially correlated random numbers.	*Wilks* [1998], *Mehrotra and Sharma* [2006], *Srikanthan and Pegram* [2009], *Bárdossy and Pegram* [2009]
Nonparametric multisite models	k-nearest neighbor approach is adopted for simultaneous simulation of rainfall at multiple locations	*Buishand and Brandsma* [2001], *Zorita et al.* [1995], *Zorita and von Storch* [1999], *Beersma and Buishand* [2003]
Reshuffling approach-based models	Rainfall is generated at individual site using any appropriate approach, and temporal and spatial dependence is introduced by shuffling the generated records across realizations.	*Clark et al.* [2004a, 2004b], *Mehrotra and Sharma* [2009]
Random cascade models	A Markov chain model generates a daily time series of the regionally averaged rainfall and a spatial model based on a nonhomogeneous random cascade process disaggregates the regionally averaged rainfall to produce spatial patterns of daily rainfall.	*Jothityangkoon et al.* [2000]
Generalized Linear models (GLM)	These models are based on an extension of linear regression models and can represent the spatial and temporal non-stationarities of multisite daily rainfall.	*Chandler and Wheater* [2002], *Yang et al.* [2005], *Furrer and Katz* [2007]

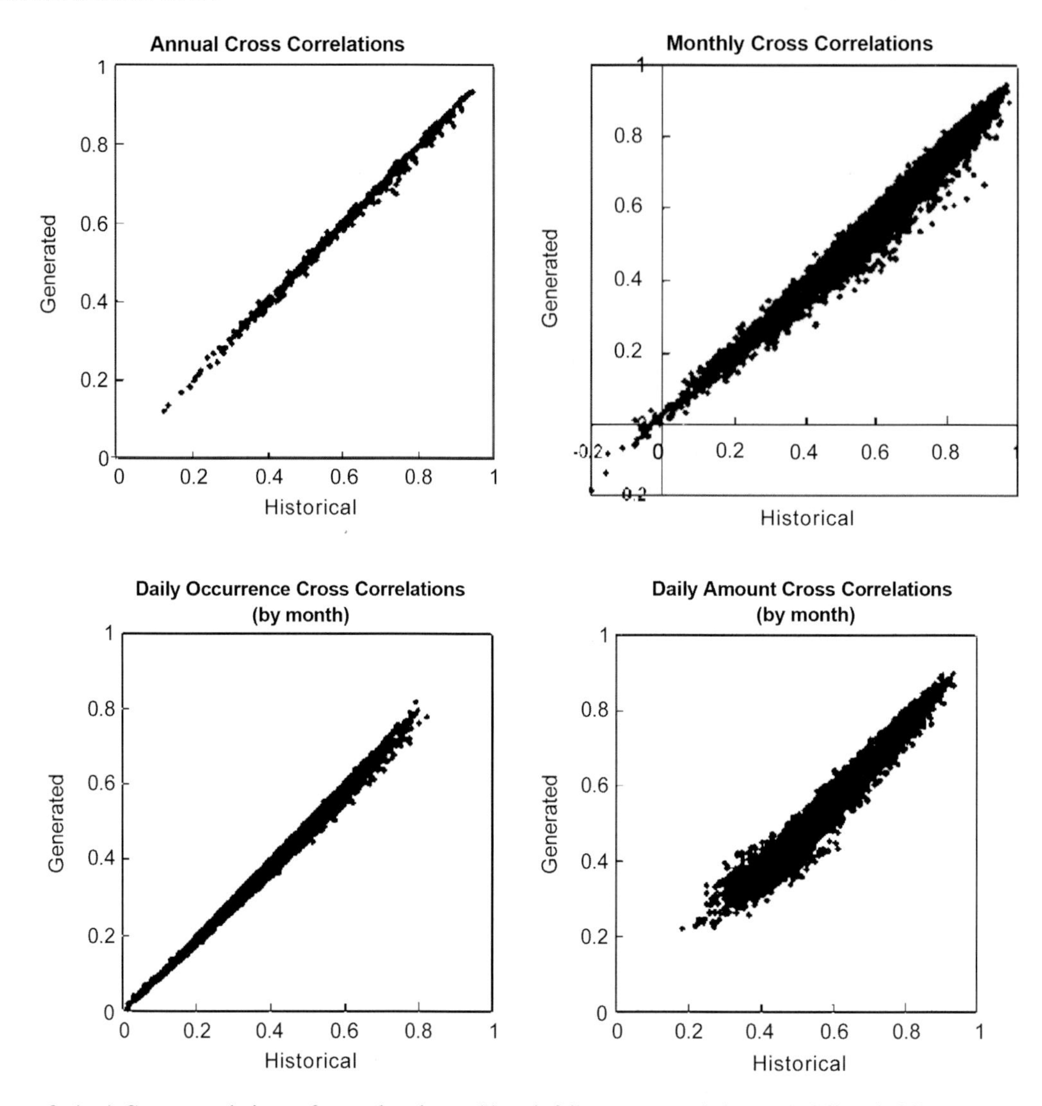

Figure 3. (top) Cross-correlations of annual and monthly rainfall amounts and (bottom) daily rainfall occurrences and amounts across 30 rain gauge stations over Sydney region. There are 435 points in the annual cross-correlations and 5220 in each of the remaining plots. Reprinted from *Srikanthan and Pegram* [2009], with permission from Elsevier.

commonly used approaches (this is not an exhaustive list) are mentioned in Table 3. Readers are referred to the papers cited for implementation details of the methods that are described.

The issue of spatial dependence of precipitation has been addressed in a number of space-time parametric stochastic models [e.g., *Bras and Rodriguez-Iturbe*, 1976; *Waymire et al.*, 1984; *Cox and Isham*, 1988; *Smith and Karr*, 1985; *Wilks*, 1999a, 1999b; *Qian et al.*, 2002; *Mehrotra et al.*, 2004], nonparametric stochastic models [*Buishand and Brandsma*, 2001; *Mehrotra et al.*, 2004], and other parametric, semiparametric, and nonparametric variations including reshuffling and GLM [*Mehrotra and Sharma*, 2007a, 2007b; *Srikanthan*, 2005; *Apipattanavis et al.*, 2007; *Srikanthan and Pegram*, 2009; *Clark et al.*, 2004a, 2004b; *Bárdossy and Pegram*, 2009; *Chandler and Wheater*, 2002; *Yang et al.*, 2005].

As an example, Figure 3 compares the observed and model simulated cross correlations of the multisite model of *Srikanthan and Pegram* [2009]. The model captures well the observed cross correlation of occurrences and amounts at daily as well as annual time scales in the simulations.

2.2.7. Rainfall generation on a fine spatial grid. Most frequently, time series models, e.g., multivariate autoregressive (AR) models [*Bras and Rodríguez-Iturbe*, 1985] or generalized linear models [*Chandler and Wheater*, 2002] are limited to simulations only at rain gauge locations for which historic data exist [*Wilks*, 1998; *von Storch and Zwiers*, 1999]. However, the spatial density of available records may not be adequate for some applications, such as hydrological simulation or high-resolution crop modeling, or in any application requiring rainfall time series at locations for which no or limited observations are available. Transfer of information from the nearby meteorologically similar locations, spatial interpolation of the parameters of stochastic model, use of output of physically based models (although, obtaining multiple realizations may not be feasible), or use of remote sensing data could be possible solutions to this problem of inadequate resolution of the observing stations. Recently, time series approaches (in the form of weather generators) have been generalized to a continuous spatial domain, and maps of precipitation levels are constructed at any arbitrary location via interpolation of time series model parameters [*Hutchinson*, 1995; *Johnson et al.*, 2000; *Kyriakidis and Journel*, 2001; *Wilks*, 2008].

2.2.8. Stochastic disaggregation of rainfall. Temporal or spatial disaggregation techniques are used to enhance the time or space resolution of hydrologic variables, while trying to maintain the multiple scale representation of the stochastic nature of hydrologic processes. A pioneer work has been done in this area by *Valencia and Schaake* [1972, 1973] by introducing linear disaggregation model. This has been followed by the contributions from other researchers [for example, *Mejia and Rousselle*, 1976; *Tao and Delleur*, 1976; *Curry and Bras*, 1978; *Hoshi and Burges*, 1979; *Lane*, 1979, 1982; *Salas et al.*, 1980; *Todini*, 1980; *Loucks et al.*, 1981; *Stedinger and Vogel*, 1984; *Pereira et al.*, 1984; *Stedinger et al.*, 1985; *Oliveira et al.*, 1988; *Grygier and Stedinger*, 1987, 1988, 1990; *Lane and Frevert*, 1988, 1990; *Lin*, 1990a, 1990b; *Santos and Salas*, 1992]. The disaggregation approach of Valencia and Schaake has been the most widely acknowledged scheme for stochastic disaggregation problems in hydrological applications for disaggregation of annual or seasonal rainfall, runoff, and other hydrologic variables in space and time.

At a finer time scale of less than a month, this kind of approaches has limited success as rainfall has skewed distribution and may also contain many zeros [*Valencia and Schaake*, 1972]. Keeping these limitations in mind, other disaggregation models have been proposed and used for the disaggregation of finer time scale rainfall [*Woolhiser and Osborn*, 1985; *Marien and Vandewiele*, 1986; *Hershenhorn and Woolhiser*, 1987; *Koutsoyiannis and Xanthopoulos*, 1990; *Koutsoyiannis and Foufoula-Georgiou*, 1993; *Koutsoyiannis*, 1994; *Glasbey et al.*, 1995; *Olsson and Berdtsson*, 1997; *Connolly et al.*, 1998; *Gyasi-Agyei*, 1999; *Koutsoyiannis and Onof*, 2000, 2001].

2.2.9. Remarks. There has been a lot of research on the generation of daily rainfall from individual sites to multiple sites. A shortcoming of the many of the existing models is the consistent underestimation of the variances of the simulated monthly and annual totals. The daily rainfall adjusting procedures such as proposed by *Srikanthan and Pegram* [2009] and *Mehrotra and Sharma* [2007b] appear to address this issue well in single as well as multisite situations.

Continuous simulation of the daily rainfall-runoff process in space and time has some potential advantages. The dependence of runoff response on antecedent conditions is explicitly represented, and the spatial interaction of runoff process is clearly addressed. The procedure, proposed by *Wilks* [1998], provides an easy way of extending the single-site models to multiple sites. The reshuffling approach of *Clark et al.* [2004a, 2004b] also performs well; however, it is of limited use when there are many zeros in the record.

2.3. Subdaily Rainfall Generation

2.3.1. Background. The importance of high-resolution temporal (and spatial) rainfall data for planning, design, and management of water and environmental systems has been increasingly realized recently. Occurrence of flash floods and associated storm water runoff and pollutant load in densely populated urban areas shows the vital roles high-resolution rainfall data can play in devising effective short-term emergency measures as well as long-term management strategies. Despite their obvious importance, the existence of continuous rainfall data in reality is scarce due to the fact that its measurements are costly and time-consuming.

With advancement in computing resources, many complex tools for simulating subdaily rainfall time series at a given point have been proposed. These models are based on a variety of structures, depending on the choice of variables used for describing the rainfall series and the intermittent nature of rainfall. The family of models developed by *Rodriguez-Iturbe et al.* [1987], *Entekhabi et al.* [1989], *Istok and Boersma* [1989], *Cowpertwait* [1991], and *Onof et al.* [1995] is based on the Neyman-Scott [*Neyman and Scott*, 1958] aggregation process. The starting assumption in this category of models is that storms can be considered as aggregating according to a group hierarchy described by a Poisson's law. These types of models involve many parameters.

Models based on the nondimensional description of storm intensity patterns [*Huff*, 1967; *Bonta and Rao*, 1987] can also be used to generate storms [*Garcia-Gazman and Aranda-Olivier*, 1993; *Koutsoyiannis and Foufoula-Georgiou*, 1993]. Models based on multiplicative cascade processes make use of the scale invariance properties of multifractals [*Schertzer and Lovejoy*, 1987; *Over and Gupta*, 1994; *Olsson and

Table 4. Commonly Used Subdaily Rainfall Generation Models

Model	Description/Advantages/Drawbacks	References
Poisson cluster process based models	Represents rainfall events as clusters of rain cells, where each cell is considered a pulse with a random duration and random intensity. A Rainfall generation model, however, can also be used for rainfall disaggregation.	*Rodriguez-Iturbe et al.* [1987], *Cowpertwait* [1991], *Onof et al.* [1995], *Bo et al.* [1994], *Gyasi-Agyei* [2005], *Glasbey et al.* [1995], *Onof and Koutsoyannis* [2001]
Scale invariance theory-based models	Utilizes the moment scaling function and an appropriate probability distribution for the weights.	*Schertzer and Lovejoy* [1987], *Over and Gupta* [1996], *Gupta and Waymire* [1993], *Lovejoy and Schertzer* [1990], *Hubert et al.* [1993], *Olsson et al.* [1993], *Tessier et al.* [1993, 1996], *de Lima and Grasman* [1999], *Menabde et al.* [1997, 1999], *Deidda et al.* [1999], *Olsson* [1998], *Molnar and Burlando* [2005], *Olsson and Berndtsson* [1998], *Sivakumar et al.* [2001]
Parametric and nonparametric stochastic disaggregation models	Based on disaggregation of daily rainfall based on distribution of subdaily rainfall statistics/rainfall values.	*Hershenhorn and Woolhiser* [1987], *Arnold and Williams* [1989], *Econopouly et al.* [1990], *Cowpertwait et al.* [1996], *Connolly et al.* [1998], *Sharma and Srikanthan* [2006]

Berndtsson, 1998; *Schmitt et al.*, 1998; *Menabde et al.*, 1999]. These models combined an underlying hypothesis of cascade-type scaling with empirically observed features of temporal rainfall. So-called multiplicative random cascade models distribute rainfall mass on successive regular subdivisions of an interval in a multiplicative manner [*Molnar and Burlando*, 2005; *Olsson and Berndtsson*, 1998]. Other types of stochastic models [*Tourasse*, 1981; *Lebel*, 1984] are based on the use of a random-variable-based geometric description of storm intensity patterns.

Another category of models is based on a rather simpler approach of temporal disaggregation procedures [*Woolhiser and Osborn*, 1985; *Econopouly et al.*, 1990; *Koutsoyiannis*, 1994; *Sivakumar et al.*, 2001; *Sivakumar and Sharma*, 2008] and the nonparametric method of fragment-based approach [*Sharma and Srikanthan*, 2006]. This option is attractive, since daily data are widely available, and the records are usually much longer in comparison to high-resolution rainfall data. As this is less expensive, research in this direction has been an exciting topic in hydrology during the past two decades. Table 4 presents a summary of the various modeling alternatives for subdaily rainfall generation. A few commonly used subdaily rainfall generation models are discussed further in the following paragraphs. This is followed by a short discussion on the multisite subdaily rainfall generation alternatives and a brief conclusion.

2.3.2. Poisson cluster process based models. In this category of models, storms arrive randomly according to a Poisson process, each storm consisting of a random number of cells and each cell causing rain for a random period. The arrivals of cells form a stochastic series of points in time subject to clustering. Two models have been used in the literature to represent such a clustered point process: the Neyman-Scott process and the Bartlett-Lewis process. In the Neyman-Scott and Bartlett-Lewis rectangular pulses models, each rainfall event is originated by a triggering mechanism, the origin of the event that primes several elementary rain cells. Both Neyman-Scott and Bartlett-Lewis models are characterized by three independent elementary stochastic processes responsible for (1) origin of the events, (2) number of rain cells generated by each event, and (3) origin of the cells. Each rain cell has a random duration and a random intensity. The difference between the Bartlett-Lewis and Neyman-Scott models lies in the variable that sets the origin of the cells. In the Neyman-Scott process, the arrival time of each cell is measured from the origin of the event, while in the Bartlett-Lewis process, it is the time interval between successive cells. Figures 4a and 4b provide schematic descriptions of these two models.

While the Neyman-Scott and Bartlett-Lewis models have been widely used as rainfall generators, these have also been utilized for rainfall disaggregation [*Glasbey et al.*, 1995; *Koutsoyannis and Onof*, 2001; *Cowpertwait et al.*, 1996; *Cowpertwait*, 2006].

2.3.3. Random multiplicative cascades. A cascade process repeatedly divides the available space (of any dimension) into smaller regions, in each step redistributing some associated quantity according to rules specified by the so-called cascade generator. In general, scaling may be defined as a log-log linear relationship between statistical moments of various orders and a scale parameter. This behavior is a generic feature

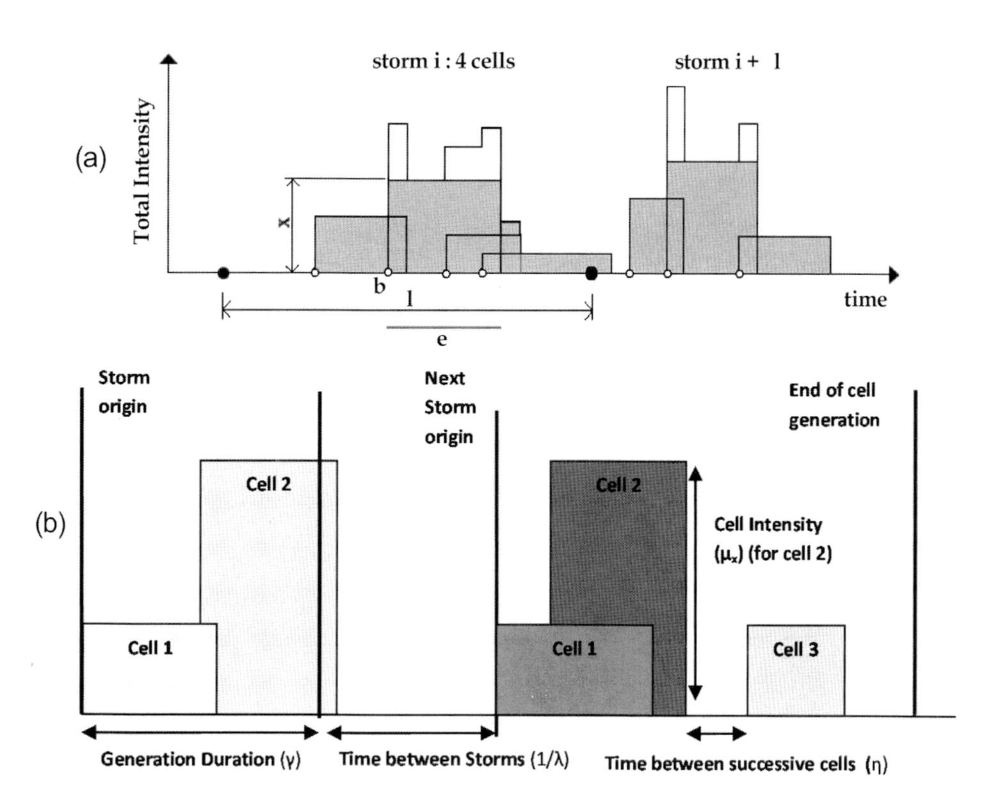

Figure 4. (a) Schematic showing the Neyman-Scott process. The total intensity at a given time is given by the sum of the cells intensities active at that time. Reprinted from *Favre et al.* [2004], with permission from Elsevier. (b) Schematic showing Bartlett-Lewis process. Storm arrivals follow Poisson process with mean of $1/\lambda$, with generation duration of each storm exponentially distributed with mean $1/\gamma$. Duration of each cell is exponentially distributed with mean of $1/\eta$. Each cell's intensity is exponentially distributed with a mean of μx.

of random cascades. According to general multifractal theory, once fluctuations at a given scale are understood, those at other scales are deduced from scale invariance (via connecting a common thread through moments at different scales) and need not be independently specified. The following summarizes the basic methodology of the models derived from previous studies [*Gupta and Waymire*, 1993; *Over and Gupta*, 1994, 1996; *Molnar and Burlando*, 2005].

2.3.3.1. The canonical RMC model. The canonical random multiplicative model distributes rainfall on successive subdivisions (see Figure 5) with b as the branching number. As such, the ith interval after n levels of subdivision is denoted as Δ^i_n. The dimensionless scale is defined as $\lambda_n = b^{-n}$. The distribution of mass then occurs via a multiplicative process through all levels, $1 \ldots n$ of the cascade, such that the mass, μ_n, in subdivision Δ^i_n is

$$\mu_{(\Delta_n)^i} = r_0 \lambda_n \prod_{j=1}^{n} W_j(i) \quad \text{for} \quad i = 1, 2, \ldots b^n, \tag{7}$$

where r_0 is the initial rainfall depth at $n = 0$, and $W_j(i)$ (hereafter denoted as just W for "weights") is a range of weights that essentially forms the cascade generator [*Molnar and Burlando*, 2005]. W is treated as an independent and identically distributed (iid) random variable, with the important condition that $E(W) = 0.5$, so that mass is conserved, on average through all levels of the cascade. Properties of W can be estimated from the moment scaling function behavior across all scales of interest. In particular, we

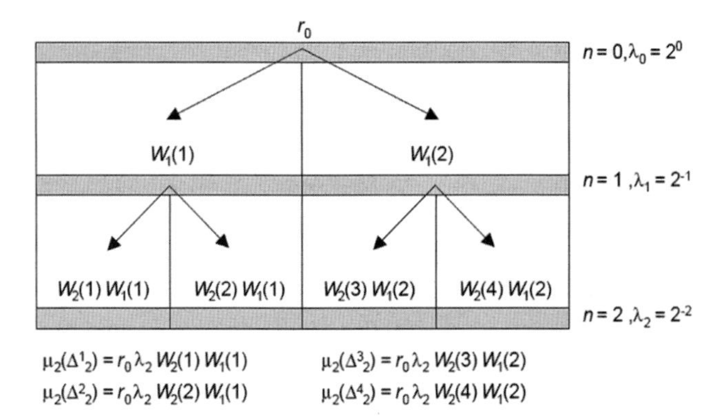

Figure 5. Framework of canonical random multiplicative with branching number $b = 2$ and cascade generator W for scales $n = 0$, 1, and 2. Here r_0 denotes rainfall amount at scale 0 (i.e., daily rainfall). Reprinted from *Molnar and Burlando* [2005], with permission from Elsevier.

need to obtain the rate of divergence/convergence of moments with scale. For a random cascade, the ensemble moments have been shown to be a log-log linear function of the scale of resolution. The slope of this scaling relationship is known as the Mahane Kahane Pierre function [*Mandelbrot*, 1974; *Kahane and Pierre*, 1976]. Having identified the moment scaling function, an appropriate probability distribution for the weights such as the Bernoulli distribution for intermittency and lognormal distribution for rainfall amounts can be chosen [*Over and Gupta*, 1994, 1996; *Molnar and Burlando*, 2005].

2.3.3.2. Microcanonical cascades. The microcanonical cascade model, by definition, conserves mass exactly at each cascade level. Hence, for a cascade with branching order (b) of two, the microcanonical weights (M) must either be equal to a combination of (0, 1) when intermittency arises (hereafter intermittent part), or (M, $1 - M$), where $0 < M < 1$ when there is no intermittency (hereafter variability part). The microcanonical cascade also differs from its canonical counterpart because its weights (M) are scale dependent and not independent and iid.

The intermittency parameter (p) and variability parameter (a) are estimated from the breakdown coefficients, which are defined as the ratio of rainfall of a random field averaged over different scales to account for decrease in variance with decrease in time scales [see *Menabde and Sivapalan*, 2000]. More specifically, p represents the probability that one of the intervals in disaggregation is dry for scales between n and $n + 1$ and, therefore, is simply estimated from historical rainfall. Here a is estimated from the single-parameter (symmetrical) "beta" distribution [*Molnar and Burlando*, 2005].

2.3.4. Multisite subdaily rainfall generation. The research on simulation of multisite subdaily rainfall is also expanding. *Cox and Isham* [1988] extended the single-site Poisson cluster methods to represent rainfall in continuous space and time. *Northrop* [1998] developed a spatial-temporal model based on the Bartlett-Lewis process for the temporal evolution of storm cells and the Neyman-Scott process for spatial arrivals. Similarly, *Cowpertwait et al.* [2002] proposed a spatial-temporal model based on the Neyman-Scott mechanism. *Kottegoda et al.* [2003] achieved the disaggregation of daily rainfalls into hourly values through dimensionless accumulated hourly amounts generated by a beta distribution with the assumption that the occurrence process of hourly rainfall had a geometric distribution conditioned on the total daily rainfall. *Segond et al.* [2006] presented a procedure that combined a multisite rainfall model at daily time scale and a disaggregation tool to generate the hourly rainfalls.

2.3.5. Discussion. Weather data such as rainfall at finer time scales are vital inputs to many watershed and crop simulation models. Yet, the limited availability of detailed temporal and spatial weather data in most areas exacerbates the problem of running such models. Only by use of temporally continuous (rather than discrete event) simulations the full effects of antecedent conditions can be properly represented, and the frequencies of the full range of output properties are defined. Some recent developments in the stochastic modeling of subdaily rainfall are reviewed covering a range of methodologies.

All point process-based subdaily rainfall simulation models require a delicate balance between mathematical complexity of analytical expressions and incorporation of more realistic features into the model. A useful cross-validation check would be the use of the simulated rainfall in the intended application, for example, generating streamflows using a rainfall-runoff model. The resulting streamflows will determine whether the rainfall model has done well in reproducing the desired statistics or not. Work on the spatial-temporal model involves elaborate data analysis and parameter estimation. Relating the storm types and model parameters with climatological variables may help in reducing the number of parameters and finding their meaningful estimates.

3. STOCHASTIC RAINFALL GENERATION FOR CLIMATE CHANGE CONDITIONS

3.1. Background

General circulation models (GCMs) are mathematical models of the general circulation of a planetary atmosphere or ocean and are widely used to simulate the present and future climate under assumed greenhouse gas emission scenarios,

Table 5. GCM Skill Score for Climate Variables (%) for a 20-Year Window Centered at 2030 for Two SRES Scenarios[a]

Variable	A2	B1
Surface pressure	97	99
Temperature	72	82
Shortwave radiation	68	69
Precipitable water	53	53
Specific humidity	53	51
Wind speed	42	50
Longwave radiation	24	24
Precipitation	7	7

[a]Higher skill score represents more consistent simulations for a variable across the GCMs.

both in space and time [e.g., *IPCC*, 2007; *Bergström et al.*, 2001; *Varis et al.*, 2004]. They attempt to represent the physical processes in the atmosphere, ocean, cryosphere, and land surface. Both atmospheric GCMs (AGCMs) and oceanic GCMs (OGCMs), with the addition of other components (such as a sea ice model or a model for evapotranspiration over land), form an atmosphere-ocean coupled general circulation model (CGCM or AOGCM). Within this structure, different variations can exist, and their varying response to climate change may be studied.

GCMs are currently the most credible tools available for simulating the response of the global climate system to increasing greenhouse gas concentrations and to provide estimates of climate variables (e.g., air temperature, precipitation, wind speed, pressure, etc.) on a global scale. GCMs demonstrate a significant skill at the continental and hemispheric spatial scales and incorporate a large proportion of the complexity of the global system; they are, however, inherently unable to represent local subgrid-scale features and dynamics. The spatial scale on which a GCM can operate (typically 125 to 500 km for Coupled Global Climate model) is very coarse compared to that of a hydrologic process (e.g., precipitation in a region, streamflow in a river, etc.), of interest in the climate change impact assessment studies, especially at a catchment scale [*IPCC*, 2007; *Charles et al.*, 2004; *Vicuna et al.*, 2007]. Additionally, accuracy of GCMs, in general, decreases from climate-related variables, such as wind, temperature, humidity, and air pressure to hydrologic variables such as precipitation, evapotranspiration, runoff, and soil moisture, which are also simulated by GCMs. These limitations of the GCMs restrict the direct use of their output in hydrology, water resources, and related fields.

There are uncertainties associated in the representation (parameterization) of climate processes in the climate models. For example, model representations of cloud physics, various effects of anthropogenic aerosols, chemical ozone and its interactions with climate, and carbon cycles are areas in which significant uncertainty exists [*Knutti*, 2008]. As a conse-

Table 6. Advantages and Disadvantages of Statistical and Dynamical Downscaling Techniques[a]

Statistical Downscaling	Dynamical Downscaling
Advantages	
Computationally efficient and easy to implement	Produces responses based on physically consistent processes
Capable of providing point-scale climate information from GCM-scale output	Produces finer resolution information from GCM-scale output that can resolve atmospheric processes on a smaller scale
Can be used to obtain information about variables not available from RCMs/GCMs	Can provide information at locations where no observed data is available
Easily transferable to other regions	
Based on established and accepted statistical/stochastic procedures	
Can incorporate observations into procedure	
Disadvantages	
Requires reliable observed data series of long length for calibration	Computationally intensive
Normally undersimulates climate variability at higher temporal or spatial scales	Limited number of scenario ensembles available
Heavily dependent upon choice of predictors	Strongly dependent on GCM boundary forcing and therefore affected by biases in underlying GCM
Climate system feedbacks not included	
Dependent on GCM output and therefore affected by biases in underlying GCM	
Domain size, climatic region, and season may influence the downscaling performance	

[a]Adapted and modified from *Fowler et al.* [2007]. Copyright Royal Meteorological Society, reprinted with permission.

quence the response from one model to another varies significantly, even for the same climate change emission scenario, and there is no objective way to place a higher confidence in a particular model.

Partially addressing this issue, *Johnson and Sharma* [2009] developed a skill score for climate variables across a number of GCMs (Table 5). The outputs of nine GCMs for eight different variables and two emission scenarios were examined in order to obtain a relative ranking of the variables averaged across Australia. The results (Table 5) indicate that the GCMs have lowest skill for simulating the precipitation and highest skill for surface pressure. This skill score is an important step in trying to evaluate the consistency of the predictions from a range of models in time and space for use in projects aiming to assess future climate change impacts at a regional level.

3.2. Downscaling Techniques

Downscaling techniques are used to transfer the GCM output from coarser spatial scales to local or regional scales for use in catchment-scale impact assessment studies. These

Table 7. Commonly Available Statistical/Stochastic Downscaling Methods

Model	Description/Advantages/Drawbacks	References
Scaling or delta-change approach	Simple to apply. Differences between the control and future GCM simulations are applied to current climate (baseline) observations by simply adding or scaling the each day value of the current climate by a factor. Some developments are made using other versions of scaling techniques such as using ranks or probability as a basis of scaling. The method assumes that GCMs more accurately simulate relative change than absolute values, and it only scales the mean, maxima, and minima of climatic variables, ignoring change in variability. For precipitation, the temporal sequence of wet days remains the same.	*Prudhomme et al.* [2002], *Harrold and Jones* [2003], *Ruosteenoja et al.* [2007], *Salathé* [2005], *Diaz-Nieto and Wilby* [2005]
Regression-based approaches	Offer a simple means of representing linear or nonlinear relationship between predictors and predictands. In the simplest form, multiple regression models are built using grid cell values of atmospheric variables or the principal components or singular value decomposition of pressure fields or geopotential heights as predictors for surface temperature and precipitation. Use of Canonical correlation analysis, artificial neural network, logistic regression, GLM, and partial regression is also attempted. Regression methods and some weather-typing approaches underpredict climate variability to varying degrees, i.e., underestimate variance and poorly represent extreme events.	*Hanssen-Bauer and Førland* [1998], *Hellström et al.* [2001], *Cubasch et al.* [1996], *Kidson and Thompson* [1998], *Hanssen-Bauer et al.* [2003], *Zorita and von Storch* [1999], *Karl et al.* [1990], *Wigley et al.* [1990], *von Storch et al.* [1993], *Busuioc et al.* [2001], *Huth* [1999], *von Storch and Zwiers* [1999], *Abaurrea and Asín* [2005], *Bergant and Kajfez-Bogataj* [2005], *Murphy* [1999]
Weather classification based-approaches	Based on the more traditional synoptic climatology concept (including analogs) and which relates a particular atmospheric state to a set of local climate variables. Weather classes may be defined synoptically, typically using empirical orthogonal functions, some indices, cluster analysis, or fuzzy rules applied to pressure fields. Weather classes can also be defined using ground rainfall distribution patterns. Analog methods and k-nearest neighbor-based approaches are also popular. These approaches assume that the characteristics of the weather classes will remain the same in future. Also, all weather patterns are defined using a few discrete classes, and within class variability is ignored.	*Hughes and Guttorp* [1994], *Conway et al.* [1996], *Fowler et al.* [2000, 2005], *Bárdossy et al.* [2002, 2005], *Bellone et al.* [2000], *Timbal et al.* [2003], *Timbal and McAvaney* [2001], *Mehrotra et al.* [2004], *Mehrotra and Sharma* [2005], *Vrac and Naveau* [2007], *Charles et al.* [2004]
Weather generators	Stochastic models, based on representation of daily precipitation occurrence via Markov process for wet/dry state or spell transitions using parametric/nonparametric approaches. Precipitation amount is drawn from a probability distribution using parametric/nonparametric approach. Models based on storm arrival time and mixture models also fall under this category. Parameter modification for future climate can lead to unanticipated results. Low Markov order models often underestimate the variability and persistence characteristics in the simulated series.	*Wilks* [1992], *Watts et al.* [2004], *Mehrotra and Sharma* [2010], *Rajagopalan and Lall* [1999], *Buishand and Brandsma* [2001], *Yates et al.* [2003], *Podesta et al.* [2009]

downscaling techniques can be classified into two categories: "dynamical downscaling" that uses regional climate models (RCMs) to simulate finer-scale physical processes [e.g., *Giorgi et al.*, 2001; *Mearns et al.*, 2003] and "statistical downscaling" that is based on developing statistical relationships between the regional climate and preidentified large-scale parameters [e.g., *Wilby et al.*, 2004; *Mehrotra and Sharma*, 2005; *Vrac and Naveau*, 2007; *Mehrotra and Sharma*, 2010]. There are limitations and assumptions involved in both techniques, which contribute to the uncertainty of results (Table 6) (see also *Yarnal et al.* [2001] and *Fowler et al.* [2007] for a complete discussion on the classifications, limitations, and assumptions of various downscaling techniques and *Wilby et al.* [2004] for guidance on the use of statistical and *Mearns et al.* [2003] on the dynamical downscaling methods).

3.2.1. Statistical/stochastic downscaling. The statistical downscaling techniques (Table 7) enable climate scenarios to be generated at a much lower computational cost than dynamical downscaling, particularly where large ensembles of integrations are required. The range of downscaling techniques and applications has increased significantly since the last few years, with most falling into a category where the responses (precipitation) are related to predictors (coarse scale atmospheric and local scale time-lagged variables) or into a category where the responses are related to a discrete or continuous state, which is modeled as a function of the atmospheric and local scale predictors [*Hewitson and Crane*, 1996; *Wilby and Wigley*, 1997; *Hughes et al.*, 1999; *Charles et al.*, 2004; *Bartholy et al.*, 1995; *Stehlík and Bárdossy*, 2002; *Mehrotra and Sharma*, 2005; *Vrac and Naveau*, 2007].

One of the more straightforward and popular approach for climate change impact assessment is the scaling approach, that applies "change factors," calculated as the multiplicative or additive difference between the control and future GCM simulations, to observations [*Prudhomme et al.*, 2002; *Wilby et al.*, 2004]. More refined statistical downscaling methods include those based on regression models [e.g., *Hellström et al.*, 2001], artificial neural networks [e.g., *Cavazos and Hewitson*, 2005], analog methods based on empirical orthogonal functions [e.g., *Zorita and von Storch*, 1999], weather typing schemes [e.g., *Hughes et al.*, 1999; *Charles et al.*, 2004; *Mehrotra and Sharma*, 2005; *Vrac and Naveau*, 2007]; resampling method-based approaches [*Rajagopalan and Lall*, 1999; *Buishand and Brandsma*, 2001; *Yates et al.*, 2003]; and stochastic methods, including weather generators [e.g., *Wilks*, 1992; *Mehrotra and Sharma*, 2010]. Stochastic downscaling offers the advantage of being able to model natural climatic variability at time scales from daily to subdecadal and produce multiple realizations of projections.

Comparative studies indicate that the skill of statistical downscaling techniques depends on the chosen application, study region, availability of data, ease of access to existing models, the nature of problem at hand, and time step of the analysis [*Wilby et al.*, 2004].

Multisite downscaling of rainfall is a maturing field with application of many recently proposed methods [e.g., *Fowler et al.*, 2005; *Haylock et al.*, 2006; *Vrac and Naveau*, 2007; *Wetterhall et al.*, 2006; *Charles et al.*, 2004; *Hope et al.*, 2006; *Mehrotra and Sharma*, 2006; *Timbal*, 2004, *Nguyen et al.*, 2008; *Mehrotra and Sharma*, 2010] with few papers providing evidence that the spatial dependence and intermittency structure of rainfall is reproduced. It is unclear from the literature whether these models adequately reproduce the total monthly variations evident throughout the year and from site to site, as may be required for hydrologic models reliant on such input.

3.2.2. Some statistical downscaling alternatives

3.2.2.1. Scaling method. In the empirical scaling method (hereafter referred to as the scaling method), the entire daily rainfall series in each of the sites in the calibration period is scaled by the relative difference between the distributions of the large-scale (2.5^0) daily rainfall series (for current and future climates) to obtain the daily rainfall series at sites for the prediction/verification period. Because this method is simple, variants of the simple scaling method (referred to also as constant scaling method, daily scaling method, delta method, or quantile-quantile mapping method) have been used in the majority of hydrological impact of climate change studies involving GCMs [e.g., *Nemec and Schaake*, 1982; *Lettenmaier and Gan*, 1990; *Xu*, 1999; *Chiew and McMahon*, 2002; *Sharma et al.*, 2007; *Chiew et al.*, 2008].

3.2.2.2. Analog method. The analog downscaling method (herein referred to as the analog) is one example of a statistical downscaling model based on weather classification methods in which predictands are chosen by matching previous (i.e., analogous situations) to the current weather state. The method has been used for daily temperature extremes [*Timbal and McAvaney*, 2001], rainfall occurrences [*Timbal et al.*, 2003], and amount [*Timbal*, 2004]. The choice of a single, best climatological analog on any given day (given a set of atmospheric predictors over a region) is based on a closest neighbor defined using a simple Euclidean metric as chosen in an analysis of various metrics.

3.2.2.3. GLIMCLIM method. The generalized linear model (GLM) for daily climate (GLIMCLIM) software package [*Chandler*, 2002] for daily time series provides an alternative conceptualization of the rainfall process and has been used to analyze and simulate spatial daily rainfall given natural climate variability influences in the United Kingdom [*Chandler and Wheater*, 2002; *Yang et al.*, 2005] and further to downscale multisite rainfall from larger spatial scale reanalysis simulations [*Frost*, 2007; *Frost et al.*, 2008]. GLIMCLIM relies on a linear regression like structure (inherent to GLMs) for rainfall occurrence (the occurrence of a "wet day" where rainfall is greater than 1.0 mm, rainfall less than 1.0 mm is set to 0.0 mm) and the amounts falling on these wet days. The methodology broadly follows a two-stage approach to modeling daily rainfall: occurrence and amount associated with wet days.

3.2.2.4. NHMM method. The nonhomogeneous hidden Markov model (NHMM) of *Hughes et al.* [1999] models multisite patterns of daily precipitation occurrence as a finite number of "hidden" (i.e., unobserved) weather states. The temporal evolution of these daily states is modeled as a first-order Markov process with state-to-state transition probabilities conditioned on a small number of synoptic-scale atmospheric predictors, such as sea-level pressure, geopotential heights, and measures of atmospheric moisture.

In its most general form, the NHMM is defined by the following assumptions:

$$P(\mathbf{J}_t|S_1^T, \mathbf{J}_1^{t-1}, \mathbf{X}_1^T) = P(\mathbf{J}_t|S_t) \tag{8a}$$

$$P(S_t|S_1^{t-1}, \mathbf{X}_1^T) = P(S_t|S_{t-1}, \mathbf{X}_t), \tag{8b}$$

where $\mathbf{J}_t$ is a multivariate vector giving precipitation occurrences at a network of n stations at time t, S_t is the weather state at time t, and $\mathbf{X}_t$ is the vector of atmospheric measures at time t for $1 \leq t \leq T$. The $\mathbf{X}_t$ will usually consist of one or more derived measures from the available atmospheric data (e.g., geopotential heights and mean sea level pressure). The notation $\mathbf{X}_1^T$ is used to indicate the sequence of atmospheric data from time 1 to T and similarly for $\mathbf{J}_1^T$ and S_1^T. Specific NHMMs are defined by the parameterizations chosen for $P(\mathbf{J}_t|S_t)$ and $P(S_t|S_{t-1},\mathbf{X}_t)$ [*Hughes et al.*, 1999]. The first assumption (equation (8a)) states that the precipitation process, $\mathbf{J}_t$, is conditionally independent given the current weather state. That is, all the temporal persistence in precipitation is captured by the persistence in the weather state described in equation (8a). Equation (8b) states that, given the history of the weather state up to time $t-1$ and the entire sequence of the atmospheric data (past and future), the weather state at time t depends only on the previous weather state and the current atmospheric data. In the absence of the atmospheric data, this is simply the Markov assumption applied to the hidden process. The atmospheric data, when included, modify the transition probabilities of the Markov process, hence the term "nonhomogeneous." The most likely weather state sequence is obtained from a fitted NHMM using the Viterbi algorithm to assign each day to its most probable state [*Forney*, 1978].

Charles et al. [1999a] used conditional multiple linear regression to simulate multisite daily precipitation amounts. The NHMM has been successfully applied at many places to simulate observed weather and for climate change studies [*Hughes et al.*, 1999; *Charles et al.*, 1999a, 2004; *Bates et al.*, 1998; *Charles et al.*, 1999b; *Mehrotra et al.*, 2004].

3.2.2.5. MMM-KDE method. MMM-KDE multistation stochastic downscaling framework [*Mehrotra and Sharma*, 2010] involves a MMM for downscaling of rainfall occurrence and a kernel probability density estimation (KDE) approach for downscaling of rainfall amounts at individual stations on days identified as wet by the occurrence model.

The MMM is characterized by the following:

$$P(J_t|J_{t-1}, \mathbf{X}_t) = \frac{P(J_t, J_{t-1})}{P(J_{t-1})} \times \frac{f(\mathbf{X}_t|J_t, J_{t-1})}{f(\mathbf{X}_t|J_{t-1})}, \tag{9}$$

where J_t denotes rainfall occurrence at a given site at time step t, and $\mathbf{X}_t$ represents a vector of atmospheric variables t (the daily chosen predictor variables in this case). This equation specifies that the probability of a transition from one state to another is altered at each time step by the vector of atmospheric predictors $\mathbf{X}_t$ (an example of a nonhomogeneous Markov model). The associated conditional probability density $f(\mathbf{X}_t|J_t,J_{t-1})$ is approximated as a multivariate normal.

The MMM downscaled series is generated by sequentially sampling the series of rainfall occurrences for a given site according to equation (9). A nonzero rainfall amount is simulated for each day and individual location that the MMM occurrence downscaling model simulates as wet using the KDE procedure. The spatial dependence in the downscaled rainfall occurrence and amount field is being induced by making use of spatially correlated and serially independent random numbers using a procedure outlined by *Wilks* [1998]. Readers are referred to *Mehrotra and Sharma* [2010] for the details on the MMM-KDE structure. Figure 6 presents the changes in number of wet days and rainfall amount in year 2070 over Sydney obtained using MMM-KDE model [*Mehrotra and Sharma*, 2010]. Slight increase in annual rainfall along the coast is projected.

3.2.2.6. SDSM method. Statistical DownScaling model (SDSM) is a freely available decision support tool for assessing local climate change impacts using a statistical downscaling technique. SDSM provides multiple single-site scenarios of daily surface weather variables under current and

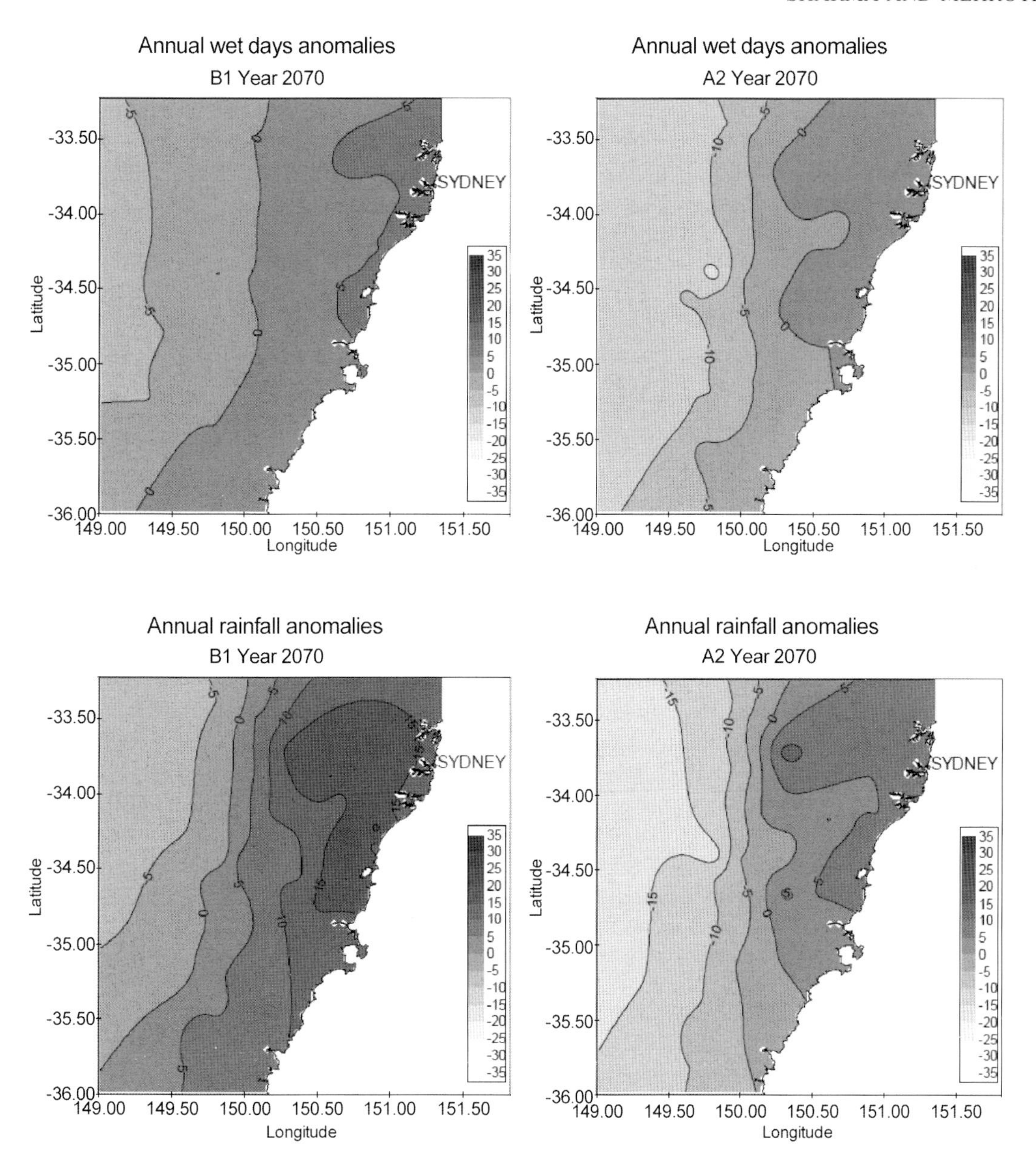

Figure 6. Annual wet days and rainfall amount anomalies expressed as a percentage difference of model simulated and current climate values for year 2070 over Sydney region. From *Mehrotra and Sharma* [2010].

future regional climate forcing using grid resolution GCM output. The model also performs additional tasks of predictor variable prescreening, model calibration, basic diagnostic testing, and statistical analyses of climate data. SDSM is a software package (with multiple versions available) and accompanying statistical downscaling methodology. The software allows five basic tasks: screening of potential predictor variables, model calibration, synthesis of current weather data, generation of future climate scenarios, diagnostic testing, and basic statistical analyses.

3.3. Discussion

GCMs perform reasonably well in simulating climatic variables at larger spatial scale (>10^4 km^2), but are of limited use at the smaller space and time scales relevant to regional impact analyses. This limitation has led to the development of downscaling techniques wherein large-scale GCM information is transferred to a finer spatial resolution. A number of papers provide a review of various downscaling alternatives, including the works of *Hewitson and Crane* [1996], *Wilby and*

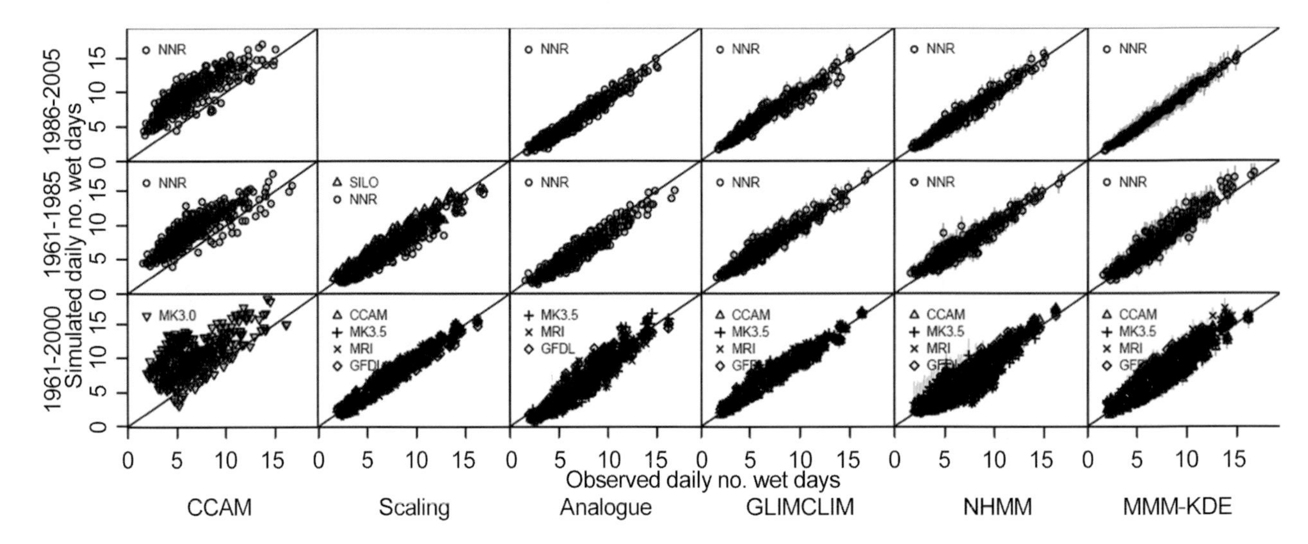

Figure 7. Number of wet days per month (days) for each model (column). Historical versus median simulated. Rows represent the simulation period: (a) 1986–2005 National Centers for Environmental Prediction/National Center for Atmospheric Research reanalysis data (NNR) calibration, (b) 1961–1985 NNR cross-verification, and (c) 1961–2000 using GCM data. From A. Frost, Bureau of Meteorology, Australia.

Wigley [1997], *Zorita and von Storch* [1997], *Xu* [1999], *Wilby et al.* [2004], *Hanssen-Bauer et al.* [2005], *Fowler et al.* [2007].

Studies comparing different statistical downscaling methods are now relatively common. However, there have been a few studies of the relative performance of dynamical and statistical methods in climate change impact assessment (see *Fowler et al.* [2007] for details). In a recent comparison study (unpublished work), six methods of downscaling large-scale simulations to multisite daily precipitation were applied to a set of 30 stations located within south-eastern Australia. These include a dynamic downscaling method, a simple empirical scaling method, a historical analog-based approach, and three stochastic downscaling methods (NHMM, MMM-KDE, and GLIMCLIM) were tested at reproducing a range of statistics important within hydrological studies. Figure 7 presents a comparison of observed and models simulated mean number of wet days per month for calibration, verification, and total time periods using National Centers for Environmental Prediction/National Center for Atmospheric Research reanalysis data and GCM data. GCM output of four models was considered including, (1) CSIRO Mk3.5, (2) CCAM (atmosphere far field nudged and SSTs from CSIRO Mk3.0), (3) GFDL-CM2.0, and (4) MRI-CGM2.3.2a (as described at the website of the Program for Climate Model Diagnosis and Intercomparison http://www-pcmdi.llnl.gov/). All models reproduced daily variability satisfactorily, and spatial statistics tended to suffer, while annual mean and standard deviation were slightly undersimulated.

As dynamical downscaling is computationally expensive, requires elaborate setup, and outputs from RCMs cannot be used in impact studies without a "bias correction" [*Fowler et al.*, 2007], statistical downscaling has found extensive use in various applications. Many recent studies have compared the performance of different statistical downscaling methods including statistical downscaling and dynamical downscaling methods. Simple statistical downscaling methods seem to perform well in reproducing the mean characteristics; however, for many applications, reproducing climate statistics such as extremes may be more important. In such cases, a selection of downscaling methods would be more appropriate. However, the choice of GCM, scenario, and predictor variables generally provide the largest source of uncertainty in downscaled results.

Despite of all the advancements in various downscaling approaches, only a few studies have looked into the topic of hydrological impact assessment and management decisions and adaptation. Multimodel downscaling approaches allowing for the uncertainty assessment in the downscaled results seem to offer the more prospective for advancement in this area.

3.4. *Uncertainty in Model Parameters*

Most of the methods described earlier only considered procedures, which rely on a single set of parameter estimates, and these estimated parameter values are assumed to be the "true" values free of the data sampling errors and model structure. In practice, there is considerable uncertainty with model parameters, even when all available historical data are efficiently used in the model establishment. The quantification of parameter uncertainty for multiple time series models presents a significant challenge. However, with recent advances, the Bayesian method has become a powerful technique for inferring complicated statistical models and dealing with parameter uncertainties [*Gelman et al.*, 1995; *Wang*, 2001; *Thyer and Kuczera*, 2003a, 2003b]. Bayesian approach provides a more coherent framework for keeping track of, and incorporating, the uncertainties involved in the prediction process. Computations for such models are intractable using conventional techniques, but are now almost routine using stochastic algorithms such as Markov chain Monte Carlo (MCMC).

Thyer and Kuczera [2003a] undertook a full Bayesian analysis to derive the posterior distribution of the hidden state Markov model and AR(1) model parameters, while *Mehrotra et al.* [2006] estimated the parameters of Hidden Markov model using Adaptive Metropolis sampling approach. Recently, *Wang* [2008] employed a full Bayesian analysis using MCMC simulations to derive the uncertainties in the parameters of a multisite annual data generation model. *Lima and Lall* [2009] employed a Bayesian approach to estimate the parameters of a multisite daily rainfall occurence model to define the onset and end of the rainfall season in northeast Brazil. *Chaouche and Parent* [1999] used a Bayesian approach to estimate the uncertainty in the parameters of a two-part daily rainfall model utilizing a Markov chain and a Gamma distribution. It has been argued that extremes and nonstationarity in the model due to seasonal effects can be better characterized through a Bayesian analysis [*Coles et al.*, 2003].

4. CONCLUDING REMARKS

Modeling precipitation is of importance to several disciplines in science and water engineering and in agricultural, ecosystem, and hydrological impact studies both as a means of infilling missing data and for providing long synthetic rainfall series from limited observed records. The amount and pattern of rainfall are among the most important weather characteristics, and they affect agriculture profoundly. In addition to the direct effects on water balance in soil, the rainfall is strongly related to other weather variables such as solar radiation, temperature, and humidity, which are also important factors affecting the growth and development of crops, pests, diseases, and weeds.

While the time scales modeled range from a year to a few minutes, simulation of daily precipitation has attracted the most attention. Estimation of water demand and the simulation of water supply systems generally require monthly data, while for rainfall-runoff and crop growth models, daily data are required.

This book chapter presented various alternatives for generation of rainfall across a range of scales in both space and time. The presentation was arranged to include approaches for stochastic generation of rainfall at annual, seasonal, daily, and subdaily time scales at single- and multiple-point locations, followed by approaches for stochastically downscaling rainfall conditional to simulations from general circulation models for future climates. Of special emphasis across the presentation was the representation of low-frequency variability in the generated rainfall sequences, so much of importance when the generated rainfall is used to assess the risk of failure of systems sensitive to sustained extremes. It was suggested that routine formulations of stochastic generation models, or even general circulation models, are not fully capable of simulating this low-frequency persistence, thus requiring additional enhancements in their formulation to impart this persistence through the use of exogenous slow-varying predictors or endogenous aggregate response indicators that convert the stochastic generator into a very high-order autoregressive process.

Stochastic generation cannot proceed without the specification of a full probability distribution relating the response(s) and predictors involved. Specifying such a model implicitly assumes that the conditional relationship represented can be classified as stationary (a more appropriate terminology here being "conditional stationarity" given the conditioning predictors that are used). The assumption of stationarity needed to generate stochastic sequences without the use of exogenous predictors is fast becoming untenable for future applications. Stochastic generation in our changing climate needs to proceed assuming conditional stationarity, with conditioning coming from sensibly selected exogenous climate predictors that are simulated using GCMs. Selection of these predictors poses a significant challenge for future applications. The basis for selecting the predictors needs to take into account whether the assumption of conditional stationarity that is invoked will be valid for future climates. In addition, considerations of the suitability of the predictor in formulating the conditional relationship (through measures of partial dependence such as the partial correlation or the partial mutual information) [*Sharma*, 2000], along with the stability with which the predictor is simulated by GCMs in future climatic settings (as measured by the variable convergence score) [*Johnson and Sharma*, 2009], need to be incorporated in the formulation of predictive models for the future.

An additional issue, often ignored in stochastic generation problems, involves the representation of errors in both historical and predicted response and exogenous variable records, along with uncertainty in the form of the stochastic model being used. Errors or uncertainty in historical data sets becomes especially important given the use of reanalysis products as the basis of forming climate inputs for the historical record. Care needs to be given in the selection of reanalysis variables and grid locations to ensure the associated input uncertainty is acceptably small. Care also needs to be given to the nonstationarity in the standard error associated with the variable over time, with procedures devised to take account for the additional uncertainty that would be invoked in any conditional relationship that is formulated. With the future, care needs to be given in the uncertainty of future projections, both due to the uncertainty associated with emissions, as well as the structural uncertainty due to model specification. Stochastic sequences for future climates should ideally be formulated as ensembles representing different GCMs, exogenous predictor sets, and historical data products, so as to encompass the full uncertainty present in the system being studied. Further research is needed to formalize the procedures for encompassing this uncertainty in stochastic generation of rainfall and other weather variables.

REFERENCES

Abaurrea, J., and J. Asín (2005), Forecasting local daily precipitation patterns in a climate change scenario, *Clim. Res.*, *28*, 183–197.

Adamowski, K., and A. F. Smith (1972), Stochastic generation of rainfall, *J. Hydraul. Div. Am. Soc. Civ. Eng.*, *98*(Hy 11), 1935–1945.

Allen, D. M., and C. T. Haan (1975), Stochastic simulation of daily rainfall, *Res. Rep. 82*, Water Resour. Inst., Univ. of Ky, Lexington.

Apipattanavis, S., G. Podestá, B. Rajagopalan, and R. W. Katz (2007), A semiparametric multivariate and multisite weather generator, *Water Resour. Res.*, *43*, W11401, doi:10.1029/2006WR005714.

Arnold, J. G., and J. R. Williams (1989), Stochastic generation of internal storm structure, *Trans. ASAE*, *32*(1), 161–166.

Bárdossy, A., and G. G. S. Pegram (2009), Copula based multisite model for daily precipitation simulation, *Hydrol. Earth Syst. Sci.*, *13*(12), 2299–2314.

Bárdossy, A., and E. J. Plate (1991), Modelling daily rainfall using a semi-Markov representation of circulation pattern occurrence, *J. Hydrol.*, *122*, 33–47.

Bárdossy, A., and E. J. Plate (1992), Space-time models for daily rainfall using atmospheric circulation patterns, *Water Resour. Res.*, *28*, 1247–1259.

Bárdossy, A., J. Stehlík, and H. J. Caspary (2002), Automated objective classification of daily circulation patterns for precipitation and temperature downscaling based on optimized fuzzy rules, *Clim. Res.*, *23*, 11–22.

Bárdossy, A., I. Bogardi, and I. Matyasovszky (2005), Fuzzy rule-based downscaling of precipitation, *Theor. Appl. Climatol.*, *82*, 119–129.

Bartholy, J., I. Bogardi, and I. Matyasovszky (1995), Effect of climate change on regional precipitation in Lake Balaton watershed, *Theor. Appl. Climatol.*, *51*, 237–250.

Bates, B. C., S. P. Charles, and J. P. Hughes (1998), Stochastic downscaling of numerical climate model simulations, *Environ. Modell. Software*, *13*, 325–331.

Beersma, J. J., and T. A. Buishand (2003), Multi-site simulation of daily precipitation and temperature conditional on the atmospheric circulation, *Clim. Res.*, *25*, 121–134.

Bellone, E., J. P. Hughes, and P. Guttorp (2000), A Hidden Markov model for downscaling synoptic atmospheric patterns to precipitation amounts, *Clim. Res.*, *15*, 1–12.

Bergant, K., and L. Kajfez-Bogataj (2005), N-PLS regression as empirical downscaling tool in climate change studies, *Theor. Appl. Climatol.*, *81*, 11–23.

Bergstrom, S., B. Carlsson, M. Gardelin, G. Lindstrom, A. Pettersson, and M. Rummukainen (2001), Climate change impacts on runoff in Sweden—Assessments by global climate models, dynamical downscaling and hydrological modelling, *Clim. Res.*, *16*, 101–112.

Bo, Z., S. Islam, and E. A. B. Eltahir (1994), Aggregation-disaggregation properties of a stochastic rainfall model, *Water Resour. Res.*, *30*, 3423–3435.

Bonta, J. V., and A. R. Rao (1987), Factors affecting development of Huff curves, *Trans. ASAE*, *30*, 1689–1693.

Boughton, W. C. (1999), A daily rainfall generating model for water yield and flood studies, *Report 99/9*, 21 pp., CRC for Catchment Hydrol., Monash Univ., Melbourne, Victoria, Australia.

Bras, R. L., and I. Rodriguez-Iturbe (1976), Rainfall generation: A nonstationary time varying multidimensional model, *Water Resour. Res.*, *12*, 450–456.

Bras, R. L., and I. Rodriguez-Iturbe (1985), *Random Functions and Hydrology*, 559 pp., Addison-Wesley, Reading, Mass.

Briggs, W. M., and D. S. Wilks (1996), Estimating monthly and seasonal distributions of temperature and precipitation using the new CPC long-range forecasts, *J. Clim.*, *9*, 818–826.

Buishand, T. A. (1977), Stochastic modelling daily rainfall sequences, *Meded. Landbouwhogesch. Wageningen, 77–3*, 211 pp.

Buishand, T. A. (1978), Some remarks on the use of daily rainfall models, *J. Hydrol.*, *36*, 295–308.

Buishand, T. A., and T. Brandsma (2001), Multisite simulation of daily precipitation and temperature in the Rhine basin by nearest-neighbor resampling, *Water. Resour. Res.*, *37*, 2761–2776.

Busuioc, A., D. Chen, and C. Hellström (2001), Performance of statistical downscaling models in GCM validation and regional climate change estimates: Application for Swedish precipitation, *Int. J. Climatol.*, *21*, 557–578.

Cameron, D., K. Bevin, and J. Tawn (2001), Modelling extreme rainfalls using a modified pulse Bartlett-Lewis stochastic rainfall model (with uncertainty), *Adv. Water Resour.*, *24*, 203–211.

Carey, D. I., and C. T. Haan (1978), Markov processes for simulating daily point rainfall, *J. Irrig. Drain. Div. Am. Soc. Civ. Eng.*, *104*(IR1), 111–125.

Caskey, J. E. (1963), A Markov Chain for the probability of precipitation occurrence in intervals of various lengths, *Mon. Weather Rev.*, *91*, 298–301.

Cavazos, T., and B. C. Hewitson (2005), Performance of NCEP-NCAR reanalysis variables in statistical downscaling of daily precipitation, *Clim. Res.*, *28*, 95–107.

Chandler, R. E. (2002), GLIMCLIM: Generalized linear modelling for daily climate time series (software and user guide), *Rep. 227*, Dep. of Stat. Sci., Univ. College London, London.

Chandler, R. E., and H. S. Wheater (2002), Analysis of rainfall variability using generalized linear models: A case study from the West of Ireland, *Water Resour. Res.*, *38*(10), 1192, doi:10.1029/2001WR000906.

Chaouche, A., and E. Parent (1999), Bayesian identification and validation of a daily rainfall model under monsoon conditions, in French*Hydrol. Sci. J.*, *44*, 199–220.

Charles, S. P., B. C. Bates, and J. P. Hughes (1999a), A spatio-temporal model for downscaling precipitation occurrence and amounts, *J. Geophys. Res.*, *104*, 31,657–31,669.

Charles, S. P., B. C. Bates, P. H. Whetton, and J. P. Hughes (1999b), Validation of downscaling models for changed climate conditions: Case study of southwestern Australia, *Clim. Res.*, *12*(1), 1–14.

Charles, S. P., B. C. Bates, I. N. Smith, and J. P. Hughes (2004), Statistical downscaling of daily precipitation from observed and modelled atmospheric fields, *Hydrol. Processes*, *18*(8), 1373–1394.

Chiew, F. H. S., and T. A. McMahon (2002), Modelling the impacts of climate change on Australian streamflow, *Hydrol. Processes*, *16*, 1235–1245.

Chiew, F. H. S., et al. (2008), Climate data for hydrologic scenario modelling across the Murray-Darling Basin: A report to the Australian Government from the CSIRO Murray-Darling Basin Sustainable Yields Project, *CSIRO Tech. Rep.*, 42 pp., CSIRO, Australia.

Chin, E. H. (1977), Modeling daily precipitation occurrence process with Markov chain, *Water Resour. Res.*, *13*, 949–956.

Chin, E. H., and J. F. Miller (1980), On the conditional distribution of precipitation amounts, *Mon. Weather Rev.*, *108*, 1462–1464.

Clark, M. P., S. Gangopadhyay, D. Brandon, K. Werner, L. Hay, B. Rajagopalan, and D. Yates (2004a), A resampling procedure for generating conditioned daily weather sequences, *Water Resour. Res.*, *40*, W04304, doi:10.1029/2003WR002747.

Clark, M. P., S. Gangopadhyay, L. E. Hay, B. Rajagopalan, and R. L. Wilby (2004b), The Schaake shuffle: A method for reconstructing space-time variability in forecasted precipitation and temperature fields, *J. Hydrometeorol.*, *5*, 243–262.

Coe, R., and R. D. Stern (1982), Fitting models to rainfall data, *J. Appl. Meteorol.*, *21*, 1024–1031.

Cole, J. A., and J. D. F. Sherriff (1972), Some single and multisite models of rainfall with discrete time increments, *J. Hydrol.*, *17*, 97–113.

Coles, S., L. R. Pericchi, and S. Sisson (2003), A fully probabilistic approach to extreme rainfall modeling, *J. Hydrol.*, *273*(1–4), 35–50.

Connolly, R. D., J. Schirmer, and P. K. Dunn (1998), A daily rainfall disaggregation model, *Agric. For. Meteorol.*, *92*(2), 105–117.

Conway, D., R. L. Wilby, and P. D. Jones (1996), Precipitation and air flow indices over the British Isles, *Clim. Res.*, *7*, 169–183.

Cowpertwait, P. S. P. (1991), Further developments of the Neyman Scott clustered point process for modeling rainfall, *Water Resour. Res.*, *27*, 1431–1438.

Cowpertwait, P. S. P. (2006), A spatial-temporal point process model of rainfall for the Thames catchment, UK, *J. Hydrol.*, *330*(3–4), 586–595.

Cowpertwait, P. S. P., P. E. O'Connell, A. V. Metcalfe, and J. A. Mawdsley (1996), Stochastic point process modelling of rainfall. II. Regionalisation and disaggregation, *J. Hydrol.*, *175*(1–4), 47–65.

Cowpertwait, P. S. P., C. G. Kilsby, and P. E. O'Connell (2002), A space–time Neyman–Scott model of rainfall: Empirical analysis of extremes, *Water Resour. Res.*, *38*(8), 1131, doi:10.1029/2001WR000709.

Cox, D. R., and V. Isham (1988), A simple spatial–temporal model of rainfall, *Proc. R. Soc. London, Ser. A*, *415*, 317–328.

Cubasch, U., H. von Storch, J. Waszkewitz, and E. Zorita (1996), Estimates of climate change in southern Europe using different downscaling techniques, *Clim. Res.*, *7*, 129–149.

Curry, K. D., and R. L. Bras (1978), Theory and applications of the multivariate broken line, disaggregation and monthly autoregressive generators to the Nile River, *Rep. 78-5*, 416 pp., Mass. Inst. of Technol., Cambridge, Mass.

de Lima, M. I. P., and J. Grasman (1999), Multifractal analysis of 15-min and daily rainfall from a semi-arid region in Portugal, *J. Hydrol.*, *220*, 1–11.

Deidda, R., R. Benzi, and F. Siccardi (1999), Multifractal modeling of anomalous scaling laws in rainfall, *Water Resour. Res.*, *35*, 1853–1867.

Dennett, M. D., J. A. Rodgers, and J. D. H. Keatinge (1983), Simulation of a rainfall record for a new site of a new agricultural development: An example from northern Syria, *Agric. Meteorol.*, *29*, 247–258.

Diaz-Nieto, J., and R. L. Wilby (2005), A comparison of statistical downscaling and climate change factor methods: Impacts on low flows in the River Thames, United Kingdom, *Clim. Change*, *69*, 245–268.

Econopouly, T. W., D. R. Davis, and D. A. Woolhiser (1990), Parameter transferability for a daily rainfall disaggregation model, *J. Hydrol.*, *118*, 209–228.

Eidsvik, K. J. (1980), Identification of models for some time series of atmospheric origin with Akaike's information criterion, *J. Appl. Meteorol.*, *19*, 357–369.

Entekhabi, D., I. Rodriguez-Iturbe, and P. Eagleson (1989), Probabilistic representation of the temporal rainfall process by a modified Neyman-Scott rectangular pulses model: Parameter estimation and validation, *Water Resour. Res.*, *25*, 295–302.

Favre, A., A. Musy, and S. Morgenthaler (2004), Unbiased parameter estimation of the Neyman-Scott model for rainfall simulation with related confidence interval, *J. Hydrol.*, *286*(1–4), 168–178.

Feyerherm, A. M., and L. D. Bark (1965), Statistical methods for persistent precipitation pattern, *J. Appl. Meteorol.*, *4*, 320–328.

Feyerherm, A. M., and L. D. Bark (1967), Goodness of fit of Markov chain model for sequences of wet and dry days, *J. Appl. Meteorol.*, *6*, 770–773.

Forney, G. D. (1978), The Viterbi algorithm, *Proc. IEEE*, *61*, 268–278.

Foufoula-Georgiou, E., and D. P. Lettenmaier (1987), A Markov renewal model for rainfall occurrences, *Water Resour. Res.*, *23*, 875–884.

Fowler, H. J., C. G. Kilsby, and P. E. O'Connell (2000), A stochastic rainfall model for the assessment of regional water resource systems under changed climatic conditions, *Hydrol. Earth Syst. Sci.*, *4*, 261–280.

Fowler, H. J., C. G. Kilsby, P. E. O'Connell, and A. Burton (2005), A weather type conditioned multi-site stochastic rainfall model for generation of scenarios of climatic variability and change, *J. Hydrol.*, *308*(1–4), 50–66.

Fowler, H. J., S. Blenkinsop, and C. Tebaldi (2007), Linking climate change modelling to impacts studies: Recent advances in downscaling techniques for hydrological modelling, *Int. J. Climatol.*, *27*(12), 1547–1578.

Frost, A. J. (2007), Australian application of a statistical downscaling technique for multi-site daily rainfall: GLIM CLIM, in *MODSIM 2007 International Congress on Modelling and Simulation* [CD-ROM], pp. 553–559, Modell. and Simul. Soc. of Aust. and New Zealand, Canberra, ACT, Australia. (Available at http://www.mssanz.org.au/MODSIM07/papers.htm).

Frost, A. J., M. A. Thyer, R. Srikanthan, and G. Kuczera (2007), A general Bayesian framework for calibrating and evaluating stochastic models of annual multi-site hydrological data, *J. Hydrol.*, *340*(3–4), 129–148.

Frost, A. J., R. Mehrotra, A. Sharma, and R. Srikanthan (2008), Comparison of statistical downscaling techniques for multi-site daily rainfall conditioned on atmospheric variables for the Sydney region. 29th Hydrology and Water Resources Symposium, Engineers AustraliaAdelaide, Australia.

Furrer, E. M., and R. W. Katz (2007), Generalized linear modeling approach to stochastic weather generators, *Clim. Res.*, *34*, 129–144.

Furrer, E. M., and R. W. Katz (2008), Improving the simulation of extreme precipitation events by stochastic weather generators, *Water Resour. Res.*, *44*, W12439, doi:10.1029/2008WR007316.

Gabriel, K. R., and J. Neumann (1962), A Markov chain model for daily rainfall occurrence at Tel Aviv, *Q. J. R. Meteorol. Soc.*, *88*, 90–95.

Garbutt, D. J., R. D. Stern, M. D. Dennet, and J. Elston (1981), A comparison of rainfall climate of eleven places in West Africa using a two-part model for daily rainfall, *Arch. Meteorol. Geophys. Bioklimatol., Ser. B*, *29*, 137–155.

Garcia-Guzman, A., and E. Aranda-Oliver (1993), A stochastic model of dimensionless hyetograph, *Water Resour. Res.*, *29*, 2363–2370.

Gates, P., and H. Tong (1976), On Markov chain modelling to some weather data, *J. Appl. Meteorol.*, *15*, 1145–1151.

Gelman, A., J. B. Carlin, H. S. Stern, and D. B. Rubin (1995), *Bayesian Data Analysis*, 526 pp., Chapman and Hall, London.

Giorgi, F., B. Hewitson, J. Christensen, M. Hulme, H. Von Storch, P. Whetton, R. Jones, L. Mearns, and C. Fu (2001), Regional climate information: Evaluation and projections, in *Climate Change 2001: The Scientific Basis, Contribution of Working Group I to the Third Assessment Report of the IPCC*, edited by J. T. Houghton, et al., pp. 739–768, Cambridge Univ. Press, Cambridge, U. K.

Glasbey, C. A., G. Cooper, and M. B. McGehan (1995), Disaggregation of daily rainfall by conditional simulation from a point-process model, *J. Hydrol.*, *165*, 1–9.

Goodspeed, M. J., and C. L. Pierrehumbert (1975), Synthetic input data time series for catchment model testing, in *Prediction in Catchment Hydrology*, edited by T. G. Chapman and F. X. Dunin, pp. 359–370, Aust. Acad. of Sci., Canberra, ACT, Australia.

Green, J. R. (1964), A model for rainfall occurrence, *J. R. Stat. Soc., Ser. B*, *26*, 345–353.

Gregory, J. M., T. M. L. Wigley, and P. D. Jones (1993), Application of Markov models to area-average daily precipitation series and interannual variability in seasonal totals, *Clim. Dyn.*, *8*, 299–310.

Grygier, J. C., and J. R. Stedinger (1987), *SPIGOT: A Stochastic Streamflow Generation Package, User's Manual, Version 2.0*, Dep. of Environ. Eng., Cornell Univ., Ithaca, N. Y.

Grygier, J. C., and J. R. Stedinger (1988), Condensed disaggregation procedures and conservation corrections for stochastic hydrology, *Water Resour. Res.*, *24*, 1574–1584.

Grygier, J. C., and J. R. Stedinger (1990), *SPIGOT, A Synthetic Streamflow Generation Software Package, Technical Description, Version 2.5*, Sch. of Civ. and Environ. Eng., Cornell Univ., Ithaca, N. Y.

Gupta, V. K., and E. C. Waymire (1993), A statistical analysis of mesoscale rainfall as a random cascade, *J. Appl. Meteorol.*, *32*, 251–267.

Gyasi-Agyei, Y. (1999), Identification of regional parameters of a stochastic model for rainfall disaggregation, *J. Hydrol.*, *223*(3–4), 148–163.

Gyasi-Agyei, Y. (2005), Stochastic disaggregation of daily rainfall into one-hour time scale, *J. Hydrol.*, *309*, 178–190.

Haan, C. T., D. M. Allen, and J. D. Street (1976), A Markov chain model of daily rainfall, *Water Resour. Res.*, *12*, 443–449.

Hanssen-Bauer, I., and E. J. Førland (1998), Long-term trends in precipitation and temperature in the Norwegian Arctic: Can they be explained by changes in atmospheric circulation patterns?, *Clim. Res.*, *10*, 143–153.

Hanssen-Bauer, I., E. J. Førland, J. E. Haugen, and O. E. Tveito (2003), Temperature and precipitation scenarios for Norway: Comparison of results from dynamical and empirical downscaling, *Clim. Res.*, *25*, 15–27.

Hanssen-Bauer, I., C. Achberger, R. E. Benestad, D. Chen, and E. J. Forland (2005), Statistical downscaling of climate scenarios over Scandinavia, *Clim. Res.*, *29*, 255–268.

Harrold, T. I., and R. N. Jones (2003), Generation of rainfall scenarios using daily patterns of change from GCMs, in *Water*

Resources Systems—Water Availability and Global Change, edited by S. Franks et al., *IAHS Publ.*, *280*, 165–174.

Harrold, T. I., A. Sharma, and S. J. Sheather (2003a), A nonparametric model for stochastic generation of daily rainfall occurrence, *Water Resour. Res.*, *39*(10), 1300, doi:10.1029/2003WR002182.

Harrold, T. I., A. Sharma, and S. J. Sheather (2003b), A nonparametric model for stochastic generation of daily rainfall amounts, *Water Resour. Res.*, *39*(12), 1343, doi:10.1029/2003WR002570.

Hay, L. E., G. J. McCabe, Jr., D. M. Wolock, and M. A. Ayers (1991), Simulation of precipitation by weather type analysis, *Water Resour. Res.*, *27*, 493–501.

Haylock, M. R., G. C. Cawley, C. Harpham, R. L. Wilby, and C. M. Goodess (2006), Downscaling heavy precipitation over the United Kingdom: A comparison of dynamical and statistical methods and their future scenarios, *Int. J. Climatol.*, *26*, 1397–1415.

Hellström, C., D. Chen, C. Achberger, and J. Raisanen (2001), Comparison of climate change scenarios for Sweden based on statistical and dynamical downscaling of monthly precipitation, *Clim. Res.*, *19*, 45–55.

Hershenhorn, J., and D. A. Woolhiser (1987), Disaggregation of daily rainfall, *J. Hydrol.*, *95*, 299–322.

Hewitson, B. C., and R. G. Crane (1996), Climate downscaling: Techniques and application, *Clim. Res.*, *7*, 85–95.

Hipel, K. W., and A. I. McLeod (1994), *Time Series Modelling of Water Resources and Environmental Systems*, 1013 pp., Elsevier, Amsterdam.

Hope, P., W. Drosdowsky, and N. Nicholls (2006), Shifts in the synoptic systems influencing southwest Western Australia, *Clim. Dyn.*, *26*, 751–764, doi:10.1007/s00382-006-0115-y.

Hopkins, J. W., and P. Robillard (1964), Some statistics of daily rainfall occurrences from the Canadian Prairie province, *J. Appl. Meteorol.*, *3*, 600–602.

Hoshi, K., and S. J. Burges (1979), Disaggregation of streamflow volumes, *J. Hydraul. Div. Am. Soc. Civ. Eng.*, *105*(HY1), 27–41.

Hubert, P., Y. Tessier, S. Lovejoy, D. Schertzer, F. Schmitt, P. Laddoy, J. P. Carbonnel, S. Violette, and I. Desurosne (1993), Multifractals and extreme rainfall events, *Geophys. Res. Lett.*, *20*, 931–934.

Huff, F. A. (1967), Time distributions of rainfall in heavy storms, *Water Resour. Res.*, *3*, 1007–1019.

Hughes, J. P., and P. Guttorp (1994), A class of stochastic models for relating synoptic atmospheric patterns to regional hydrologic phenomena, *Water Resour. Res.*, *30*, 1535–1546.

Hughes, J. P., P. Guttorp, and S. P. Charles (1999), A non-homogeneous Hidden Markov model for precipitation occurrence, *Appl. Stat.*, *48*(1), 15–30.

Hutchinson, M. F. (1995), Interpolating mean rainfall using thin plate smoothing splines, *Int. J. Geogr. Inf. Syst.*, *9*, 385–403.

Huth, R. (1999), Statistical downscaling in central Europe: Evaluation of methods and potential predictors, *Clim. Res.*, *13*, 91–101.

IPCC (2007), *Climate Change 2007: The Physical Science Basis. Contribution of Working Group I to the Fourth Assessment Report of the Intergovernmental Panel on Climate Change*, 996 pp., Cambridge Univ. Press, New York.

Istok, J. D., and L. Boersma (1989), A stochastic cluster model for hourly precipitation data, *J. Hydrol.*, *106*, 257–285.

Johnson, F., and A. Sharma (2009), Measurement of GCM skill in predicting variables relevant for hydroclimatological assessments, *J. Clim.*, *22*(16), 4373–4382.

Johnson, G. L., C. L. Hanson, S. P. Hardegree, and E. B. Ballard (1996), Stochastic weather simulation: Overview and analysis of two commonly used models, *J. Appl. Meteorol.*, *35*, 1878–1896.

Johnson, G. L., C. Daly, G. H. Taylor, and C. L. Hanson (2000), Spatial variability and interpolation of stochastic weather simulation model parameters, *J. Appl. Meteorol.*, *39*, 778–796.

Jones, J. W., R. E. Colwick, and E. D. Threadgill (1972), A simulated environmental model of temperature, evaporation, rainfall and soil moisture, *Trans. ASAE*, *15*, 366–372.

Jones, P. G., and P. K. Thornton (1997), Spatial and temporal variability of rainfall related to a third-order Markov model, *Agric. For. Meteorol.*, *86*, 127–138.

Jothityangkoon, C., M. Sivapalan, and N. R. Viney (2000), Tests of a space–time model of daily rainfall in southwestern Australia based on nonhomogeneous random cascades, *Water Resour. Res.*, *36*, 267–284.

Kahane, J. P., and J. Peyriere (1976), Sur certaines mertingales de Benoit Mandelbrot, *Adv. Math.*, *22*, 131–145.

Karl, T. R., W. C. Wang, M. E. Schlesinger, R. W. Knight, and D. Portman (1990), A method of relating general circulation model simulated climate to observed local climate. Part I: Seasonal statistics, *J. Clim.*, *3*, 1053–1079.

Katz, R. W. (1977), Precipitation as a chain-dependent process, *J. Appl. Meteorol.*, *16*, 671–676.

Katz, R. W., and M. B. Parlange (1993), Effects of an index of atmospheric circulation on stochastic properties of precipitation, *Water Resour. Res.*, *29*, 2335–2344.

Katz, R. W., and M. B. Parlange (1998), Overdispersion phenomenon in stochastic modeling of precipitation, *J. Clim.*, *11*, 591–601.

Katz, R. W., and X. Zheng (1999), Mixture model for overdispersion of precipitation, *J. Clim.*, *12*, 2528–2537.

Kavvas, M. L., and J. W. Delleur (1981), A stochastic cluster model of daily rainfall sequences, *Water Resour. Res.*, *17*, 1151–1160.

Kidson, J. W., and C. S. Thompson (1998), A comparison of statistical and model-based downscaling techniques for estimating local climate variations, *J. Clim.*, *11*, 735–753.

Kiely, G., J. D. Albertson, M. B. Parlange, and R. W. Katz (1998), Conditioning stochastic properties of daily precipitation on indices of atmospheric circulation, *Meteorol. Appl.*, *5*, 75–87.

Kim, T.-W., and J. B. Valdés (2005), Synthetic generation of hydrologic time series based on nonparametric random generation, *J. Hydrol. Eng.*, *10*(5), 395–404.

Klein, W. H., B. M. Lewis, and I. Enger (1959), Objective prediction of five-day mean temperatures during winter, *J. Meteorol.*, *16*, 672–682.

Knutti, R. (2008), Should we believe model predictions of future climate change?, *Philos. Trans. R. Soc. A*, *366*(1885), 4647–4664.

Kottegoda, N. T., L. Natale, and E. Raiteri (2003), A parsimonious approach to stochastic multisite modelling and disaggregation of daily rainfall, *J. Hydrol.*, *274*, 47–61.

Koutsoyiannis, D. (1994), A stochastic disaggregation method for design storm and flood synthesis, *J. Hydrol.*, *156*, 193–225.

Koutsoyiannis, D. (2001), Coupling stochastic models of different time scales, *Water Resour. Res.*, *37*, 379–392.

Koutsoyiannis, D., and E. Foufoula-Georgiou (1993), A scaling model of storm hyetograph, *Water Resour. Res.*, *29*, 2345–2361.

Koutsoyiannis, D., and C. Onof (2000), A computer program for temporal rainfall disaggregation using adjusting procedures, paper presented at General Assembly of European Geophysical Society, Eur. Geophys. Soc., Nice, France. (Available at http://www.itia.ntua.gr/e/docinfo/59/).

Koutsoyiannis, D., and C. Onof (2001), Rainfall disaggregation using adjusting procedures on a Poisson cluster model, *J. Hydrol.*, *246*, 109–122.

Koutsoyiannis, D., and T. Xanthopoulos (1990), A dynamic model for short-scale rainfall disaggregation, *Hydrol Sci. J.*, *35*(3), 303–321.

Kyriakidis, P. C., and A. G. Journel (2001), Stochastic modeling of atmospheric pollution, a spatial time series framework. Part I: Methodology, *Atmos. Environ.*, *35*(13), 2331–2337.

Lall, U., and A. Sharma (1996), A nearest neighbor bootstrap for time series resampling, *Water Resour. Res.*, *32*, 679–693.

Lall, U., B. Rajagopalan, and D. G. Tarboton (1996), A nonparametric wet/dry spell model for resampling daily precipitation, *Water Resour. Res.*, *32*, 2803–2823.

Lane, W. L. (1979), *Applied Stochastic Techniques, User's Manual*, Bur. of Reclam., Eng. and Res. Cent., Denver, Colo.

Lane, W. L. (1982), Corrected parameter estimates for disaggregation schemes, in *Statistical Analysis of Rainfall and Runoff*, edited by V. P. Singh, pp. 505–530, Water Resour. Publ., Littleton, Colo.

Lane, W. L., and D. K. Fervert (1988), *Applied Stochastic Techniques, User's Manual*, Bur. of Reclam., Eng. and Res. Cent., Denver, Colo.

Lane, W. L., and D. K. Frevert (1990), *Applied Stochastic Techniques, User's Manual*, Bur. of Reclam., Eng. and Res. Cent., Denver, Colo., Personal Computer Version.

Lebel, T. (1984), Moyenne spatiale de la pluie sur un bassin versant: Estimation optimale, génération stochastique et gradex des valeurs extrêmes, Ph.D. thesis, Inst. Natl. Polytech. de Grenoble, Grenoble, France.

Lettenmaier, D. P., and T. Y. Gan (1990), Hydrologic sensitivities of the Sacramento-San Joaquin River basin, California for global warming, *Water Resour. Res.*, *26*, 69–86.

Lima, C. H. R., and U. Lall (2009), Hierarchical Bayesian modeling of multisite daily rainfall occurrence: Rainy season onset, peak, and end, *Water Resour. Res.*, *45*, W07422, doi:10.1029/2008WR007485.

Lin, G. F. (1990a), Corrected parameter estimates for staged disaggregation, *J. Chin. Inst. Eng.*, *13*, 405–416.

Lin, G. F. (1990b), Parameter estimation for seasonal to subseasonal disaggregation, *J. Hydrol.*, *120*, 65–77.

Loucks, D. P., J. R. Stedinger, and D. A. Haith (1981), *Water Resources System Planning and Analysis*, 559 pp., Prentice-Hall, Englewood Cliffs, N. J.

Lovejoy, S., and D. Schertzer (1990), Fractals, raindrops, and resolution dependence of rain measurements, *J. Appl. Meteorol.*, *29*, 1167–1170.

Lowry, W. P., and D. Guthrie (1968), Markov chains of order greater than one, *Mon. Weather Rev.*, *96*, 798–801.

Mandelbrot, B. B. (1974), Intermittent turbulence in self-similar cascades: Divergence of high moments and dimension of the carrier, *J. Fluid Mech.*, *62*, 331–358.

Marien, J. L., and G. L. Vandewiele (1986), A point rainfall generator with internal storm structure, *Water Resour. Res.*, *22*, 475–482.

Marshak, A., A. Davis, R. Cahalan, and W. Wiscombe (1994), Bounded cascade models as nonstationary multifractals, *Phys. Rev. E*, *49*(1), 55–69.

Mearns, L. O., F. Giorgi, P. Whetton, M. Hulme, M. Lal, and D. Pabon (2003), Guidelines for the use of climate scenarios developed from regional climate model experiments, report, Intergov. Panel on Clim. Change, Geneva, Switzerland.

Mehrotra, R., and A. Sharma (2005), A nonparametric nonhomogeneous hidden Markov model for downscaling of multi-site daily rainfall occurrences, *J. Geophys. Res.*, *110*, D16108, doi:10.1029/2004JD005677.

Mehrotra, R., and A. Sharma (2006), A nonparametric stochastic downscaling framework for daily rainfall at multiple locations, *J. Geophys. Res.*, *111*, D15101, doi:10.1029/2005JD006637.

Mehrotra, R., and A. Sharma (2007a), A semi-parametric model for stochastic generation of multi-site daily rainfall exhibiting low frequency variability, *J. Hydrol.*, *335*, 180–193, doi:10.1016/j.jhydrol.2006.11.011.

Mehrotra, R., and A. Sharma (2007b), Preserving low-frequency variability in generated daily rainfall sequences, *J. Hydrol.*, *345*, 102–120, doi:10.1016/j.jhydrol.2007.08.003.

Mehrotra, R., and A. Sharma (2009), Evaluating spatio-temporal representations in daily rainfall sequences from three stochastic multi-site weather generation approaches, *Adv. Water Resour.*, *32*, 948–962.

Mehrotra, R., and A. Sharma (2010), Development and application of a multisite rainfall stochastic downscaling framework for climate change impact assessment, *Water Resour. Res.*, *46*, W07526, doi:10.1029/2009WR008423.

Mehrotra, R., A. Sharma, and I. Cordery (2004), Comparison of two approaches for downscaling synoptic atmospheric patterns to multisite precipitation occurrence, *J. Geophys. Res.*, *109*, D14107, doi:10.1029/2004JD004823.

Mehrotra, R., R. Srikanthan, and A. Sharma (2006), A comparison of three stochastic multi-site precipitation occurrence generators, *J. Hydrol.*, *331*, 280–292.

Mejia, J. M., and J. Rousselle (1976), Disaggregation models in hydrology revisited, *Water Resour. Res.*, *12*, 185–186.

Menabde, M., and M. Sivapalan (2000), Modeling of rainfall time series and extremes using the bounded random cascades and Levy-stable distributions, *Water Resour. Res.*, *36*, 3293–3300.

Menabde, M., D. Harris, A. Seed, G. Austin, and D. Stow (1997), Multiscaling properties of rainfall and bounded random cascades, *Water Resour. Res.*, *33*, 2823–2830.

Menabde, M., A. Seed, and G. Pegram (1999), A simple scaling model for extreme rainfall, *Water Resour. Res.*, *35*, 335–340, doi:10.1029/1998WR900012.

Molnar, P., and P. Burlando (2005), Preservation of rainfall properties in stochastic disaggregation by a simple random cascade model, *Atmos. Res.*, *77*, 137–151.

Murphy, J. (1999), An evaluation of statistical and dynamical techniques for downscaling local climate, *J. Clim.*, *12*, 2256–2284.

Nemec, J., and J. Schaake (1982), Sensitivity of water resources system to climate variation, *Hydrol. Sci. J.*, *2*, 327–343.

Neyman, J. E., and E. L. Scott (1958), A statistical approach to problems of cosmology, *J. R. Stat. Soc., Ser. B*, *20*, 1–43.

Nguyen, V. T. V., N. Desramaut, and T. D. Nguyen (2008), Estimation of urban design storms in consideration of GCM-based climate change scenarios, in *Proceedings International Conference on Water and Urban Development Paradigms: Towards an Integration of Engineering, Design and Management Approaches, Leuven, 15-17 September 2008*, edited by J. Feyen, K. Shannon, and M. Neville, pp. 347–356, CRC Press, Boca Raton, Fla.

Northrop, P. (1998), A clustered spatial–temporal model of rainfall, *Proc. R. Soc. London, Ser. A*, *454*, 1875–1888.

Oliveira, G. C., J. Kelman, M. V. F. Pereira, and J. R. Stedinger (1988), Representation of spatial cross-correlation in a seasonal streamflow model, *Water Resour. Res.*, *24*, 781–785.

Olsson, J. (1996), Scaling and fractal properties of rainfall, Ph.D. thesis, Univ. of Lund, Lund, Sweden.

Olsson, J. (1998), Evaluation of a cascade model for temporal rainfall disaggregation, *Hydrol. Earth Syst. Sci.*, *2*, 19–30.

Olsson, J., and R. Berndtsson (1997), Temporal rainfall disaggregation based on scaling properties, in *Third International Workshop on Rainfall in Urban Areas*, edited by R. Fankhauser, T. Einfalt, and K. Arnbjerg-Nielsen, *IHP-V, Technical Documents in Hydrology*, UNESCO, Paris.

Olsson, J., and R. Berndtsson (1998), Temporal rainfall disaggregation based on scaling properties, *Water Sci. Technol.*, *37*(11), 73–79.

Olsson, J., J. Niemczynowicz, and R. Berndtsson (1993), Fractal analysis of high-resolution rainfall time series, *J. Geophys. Res.*, *98*, 23,265–23,274.

Onof, C., and H. S. Wheater (1993), Modelling of British rainfall using a random parameter Bartlett–Lewis rectangular pulse model, *J. Hydrol.*, *149*, 67–95.

Onof, C., and H. S. Wheater (1994), Improvements to the modelling of British rainfall using a modified random parameter Bartlett–Lewis rectangular pulse model, *J. Hydrol.*, *157*, 177–195.

Onof, C., R. Chandler, A. Kakou, and P. Northrop (1995), Rainfall modelling using Poisson-cluster process, paper presented at International Conference in Honour of Jacques Bernire, UNESCO, Paris, 11–13 Sept.

Over, T. M., and V. K. Gupta (1994), Statistical analysis of mesoscale rainfall: Dependence of a random cascade generator on large-scale forcing, *J. Appl. Meteorol.*, *33*, 1526–1542.

Over, T. M., and V. K. Gupta (1996), A space–time theory of mesoscale rainfall using random cascades, *J. Geophys. Res.*, *101*, 26,319–26,332.

Pegram, G. G. S. (1980), An auto-regressive model for multi-lag Markov Chains, *J. Appl. Probab.*, *17*, 350–362.

Pegram, G. G. S., and A. W. Seed (1998), The feasibility of stochastically modelling the spatial and temporal distribution of rainfields, *WRC Rep. 550/1/98*, 94 pp., Dep. of Civ. Eng., Univ. of Natal, Durban, South Africa.

Pereira, M. V. F., G. C. Oliveira, C. C. G. Costa, and J. Kelman (1984), Stochastic streamflow models for hydroelectric systems, *Water Resour. Res.*, *20*, 379–390.

Podesta, G., F. Bert, B. Rajagopalan, S. Apipattanavis, C. Laciana, E. Weber, W. Easterling, R. Katz, D. Letson, and A. Menendez (2009), Inter-decadal climate variability in the Argentinean pampas: Regional impacts of plausible future climate scenarios on agricultural systems, *Clim. Res.*, *40*, 199–210.

Prudhomme, C., N. Reynard, and S. Crooks (2002), Downscaling of global climate models for flood frequency analysis: Where are we now?, *Hydrol. Processes*, *16*, 1137–1150.

Qian, B., J. Corte-Real, and H. Xu (2002), Multisite stochastic weather models for impact studies, *Int. J. Climatol.*, *22*, 1377–1397.

Racsko, P., L. Szeidl, and M. Semenov (1991), A serial approach to local stochastic weather models, *Ecol. Modell.*, *57*, 27–41.

Rajagopalan, B., and U. Lall (1999), A nearest neighbor bootstrap resampling scheme for resampling daily precipitation and other weather variables, *Water Resour. Res.*, *35*, 3089–3101.

Rajagopalan, B., U. Lall, and D. G. Tarboton (1996), A nonhomogeneous Markov model for daily precipitation simulation, *ASCE J. Hydrol. Eng.*, *1*(1), 33–40.

Ramírez, J. A., and R. Bras (1985), Conditional distributions of Neyman-Scott models for storm arrivals and their use in irrigation scheduling, *Water Resour. Res.*, *21*, 317–330.

Richardson, C. W. (1981), Stochastic simulation of daily precipitation, temperature and solar radiation, *Water Resour. Res.*, *17*, 182–190.

Rodriguez-Iturbe, I., V. K. Gupta, and E. Waymire (1984), Scale considerations in the modelling of temporal rainfall, *Water Resour. Res.*, *20*, 1611–1619.

Rodriguez-Iturbe, I., D. R. Cox, and V. Isham (1987), Some models for rainfall based on stochastic point processes, *Proc. R. Soc. London, Ser. A*, *410*, 269–298.

Rodriguez-Iturbe, I., D. R. Cox, and V. Isham (1988), A point process model for rainfall: Further developments, *Proc. R. Soc. London, Ser. A*, *417*, 283–298.

Roldan, J., and D. A. Woolhiser (1982), Stochastic daily precipitation models, 1. A comparison of occurrence processes, *Water Resour. Res.*, *18*, 1451–1459.

Ruosteenoja, K., H. Tuomenvirta, and K. Jylhä (2007), GCM-based regional temperature and precipitation change estimates for Europe under four SRES scenarios applying a super-ensemble pattern-scaling method, *Clim. Change*, *81*, 193–208, doi:10.1007/s10584-006-9222-3.

Salas, J. D., J. W. Delleur, V. Yevjevich, and W. L. Lane (1980), *Applied Modeling of Hydrologic Time Series*, 484 pp., Water Resour. Publ., Littleton, Colo.

Salathé, E. P. (2005), Downscaling simulations of future global climate with application to hydrologic modelling, *Int. J. Climatol.*, *25*, 419–436.

Santos, E. G., and J. D. Salas (1992), Stepwise disaggregation scheme for synthetic hydrology, *J. Hydraul. Eng.*, *118*, 765–784.

Schertzer, D., and S. Lovejoy (1987), Physical modeling and analysis of rain and clouds by anisotropic scaling multiplicative processes, *J. Geophys. Res.*, *92*, 9693–9714.

Schmitt, F., S. Vannitsem, and A. Barbosa (1998), Modeling of rainfall time series using two-state renewal processes and multifractals, *J. Geophys. Res.*, *103*, 23,181–23,193.

Scott, D. W. (1992), *Multivariate Density Estimation: Theory, Practice and Visualisation*, 376 pp., John Wiley, New York.

Segond, M. L., C. Onof, and H. S. Wheater (2006), Spatial-temporal disaggregation of daily rainfall from a generalized linear model, *J. Hydrol.*, *331*(3–4), 674–689.

Semenov, M. A. (2008), Simulation of extreme weather events by a stochastic weather generator, *Clim. Res.*, *35*, 203–212.

Selvalingam, S., and M. Miura (1978), Stochastic modelling of monthly and daily rainfall sequences, *Water Resour. Bull.*, *14*, 1105–1120.

Shami, R. G., and C. S. Forbes (2000), A structural time series model with Markov switching, working paper, Monash Univ., Melbourne, Victoria, Autralia.

Sharif, M., and D. H. Burn (2006), Simulating climate change scenarios using an improved k-nearest neighbor model, *J. Hydrol.*, *325*, 179–196.

Sharma, A. (2000), Seasonal to interannual rainfall probabilistic forecasts for improved water supply management: Part 1—A strategy for system predictor identification, *J. Hydrol.*, *239*, 232–239.

Sharma, A., and U. Lall (1997), A nearest neighbour conditional bootstrap for resampling daily rainfall, paper presented at 24th Hydrology and Water Resources Symposium, Inst. of Eng. Aust., Auckland, New Zealand.

Sharma, A., and U. Lall (1999), A nonparametric approach for daily rainfall simulation, *Math. Comput. Simulat.*, *48*, 361–371.

Sharma, A., and R. O'Neill (2002), A nonparametric approach for representing interannual dependence in monthly streamflow sequences, *Water Resour. Res.*, *38*(7), 1100, doi:10.1029/2001WR000953.

Sharma, A., and R. Srikanthan (2006), Continuous rainfall simulation: A nonparametric alternative, paper presented at 30th Hydrology and Water Resources Symposium, Inst. of Eng., Aust., Launceston, Tasmania, 4–7 Dec.

Sharma, D., A. D. Gupta, and M. S. Babel (2007), Spatial disaggregation of bias-corrected GCM precipitation for improved hydrologic simulation: Ping River Basin, Thailand, *Hydrol. Earth Syst. Sci. Discuss.*, *4*, 35–74.

Silverman, B. W. (1986), *Density Estimation for Statistics and Data Analysis*, 175 pp., Chapman and Hall, London.

Singh, S. V., and R. H. Kripalani (1986), Potential predictability of lower-tropospheric monsoon circulation and rainfall over India, *Mon. Weather Rev.*, *114*, 758–763.

Singh, S. V., R. H. Kripalani, P. Saha, P. M. M. Ismail, and S. D. Dahale (1981), Persistence in daily and 5-day summer monsoon rainfall over India, *Arch. Meteorol. Geophys. Bioklimatol., Ser. A*, *30*, 261–277.

Sivakumar, B., and A. Sharma (2008), A cascade approach to continuous rainfall data generation at point locations, *Stochastic Environ. Res. Risk Assess.*, *22*, 451–459.

Sivakumar, B., R. Berndtsson, and M. Persson (2001), Monthly runoff prediction using phase space reconstruction, *Hydrol. Sci. J.*, *46*(3), 377–387.

Smith, J. A., and A. F. Karr (1985), Parameter estimation for a model of space-time rainfall, *Water Resour. Res.*, *21*, 1251–1257.

Srikanthan, R. (2004), Stochastic generation of daily rainfall data using a nested model. 57th Canadian Water Resources Association Annual Congress, Montreal, Canada, 16–18 June, 2004.

Srikanthan, R. (2005), Stochastic generation of daily rainfall data at a number of sites, *Tech. Rep. 05/7*, 66 pp., CRC for Catchment Hydrolo., Monash Univ., Melbourne, Victoria, Australia.

Srikanthan, R., and T. A. McMahon (1983), Stochastic simulation of daily rainfall for Australian stations, *Trans. ASAE*, *26*, 754–766.

Srikanthan, R., and T. A. McMahon (1985), Stochastic generation of rainfall and evaporation data, *Aust. Water Resour. Counc., Tech. Pap. 84*, 301 pp., AGPS, Canberra, ACT, Australia.

Srikanthan, R., and T. A. McMahon (2001), Stochastic generation of annual, monthly and daily climate data: A review, *Hydrol. Earth Syst. Sci.*, *5*(4), 653–656.

Srikanthan, R., and G. G. S. Pegram (2009), A nested multisite daily rainfall stochastic generation model, *J. Hydrol.*, *371*(1–4), 142–153.

Srinivas, V. V., and K. Srinivasan (2006), Hybrid matched-block bootstrap for stochastic simulation of multiseason streamflows, *J. Hydrol.*, *329*(1–2), 1–15, doi:10.1016/j.jhydrol.2006.01.023.

Stedinger, J. R., and R. M. Vogel (1984), Disaggregation procedures for generating serially correlated flow vectors, *Water Resour. Res.*, *20*, 47–56.

Stedinger, J. R., D. Pei, and T. A. Cohn (1985), A condensed disaggregation model for incorporating parameter uncertainty into monthly reservoir simulations, *Water Resour. Res.*, *21*, 665–675.

Stehlík, J., and A. Bárdossy (2002), Multivariate stochastic downscaling model for generating daily precipitation series based on atmospheric circulation, *J. Hydrol.*, *256*, 120–141.

Stern, R. D. (1980a), Analysis of daily rainfall at Samaru, Nigeria, using a simple two-part model, *Arch. Meteorol. Geophys. Bioklimatol., Ser. B*, *28*, 123–135.

Stern, R. D. (1980b), Computing probability distribution for the start of the rains from a Markov chain model for precipitation, *J. Appl. Meteorol.*, *21*, 420–423.

Stern, R. D., and R. Coe (1984), A model fitting analysis of daily rainfall data, *J. R. Stat. Soc. Ser. A*, *147*, 1–34.

Tao, P. C., and J. W. Delleur (1976), Multistation, multiyear synthesis of hydrologic time series by disaggregation, *Water Resour. Res.*, *12*, 1303–1312.

Tarboton, D. G., A. Sharma, and U. Lall (1998), Disaggregation procedures for stochastic hydrology based on nonparametric density estimation, *Water Resour. Res.*, *34*, 107–119.

Tessier, Y., S. Lovejoy, and D. Schertzer (1993), Universal multifractals: Theory and observations for rain and clouds, *J. Appl. Meteorol.*, *32*, 223–250.

Tessier, Y., S. Lovejoy, P. Hubert, D. Schertzer, and S. Pecknold (1996), Multifractal analysis and modeling of rainfall and river flows and scaling, causal transfer functions, *J. Geophys. Res.*, *101*, 26,427–26,440.

Thyer, M., and G. Kuczera (2000), Modeling long-term persistence in hydroclimatic time series using a hidden state Markov model, *Water Resour. Res.*, *36*, 3301–3310.

Thyer, M., and G. Kuczera (2003a), A hidden Markov model for modelling long-term persistence in multi-site rainfall time series 1. Model calibration using a Bayesian approach, *J. Hydrol.*, *275*, 12–26.

Thyer, M., and G. Kuczera (2003b), A hidden Markov model for modelling long-term persistence in multi-site rainfall time series 2. Real data analysis, *J. Hydrol.*, *275*, 27–48.

Timbal, B. (2004), Southwest Australia past and future rainfall trends, *Clim. Res.*, *26*, 233–249.

Timbal, B., and B. McAvaney (2001), An analogue-based method to downscale surface air temperature: Application for Australia, *Clim. Dyn.*, *17*, 947–963.

Timbal, B., A. Dufour, and B. McAvaney (2003), An estimate of future climate change for Western France using a statistical downscaling technique, *Clim. Dyn.*, *20*, 807–823.

Todini, E. (1980), The preservation of skewness in linear disaggregation schemes, *J. Hydrol.*, *47*, 199–214.

Todorovic, P., and D. A. Woolhiser (1975), A stochastic model of n-day precipitation, *J. Appl. Meteorol.*, *14*, 17–24.

Tourasse, P. (1981), Analyses spatiales et temporelles de précipitations et utilisation opérationnelle dans un système de prévision des crues, Ph.D. thesis, Inst. Natl. Polytech. de Grenoble, Grenoble, France.

Valencia, D., and J. C. Schaake (1972), A disaggregation model for time series analysis and synthesis, *Rep. 149*, Ralph M. Parsons Lab. for Water Resour. and Hydrodyna., Mass. Inst. of Technol., Cambridge, Mass.

Valencia, D., and J. C. Schaake (1973), Disaggregation processes in stochastic hydrology, *Water Resour. Res.*, *9*, 580–585.

Varis, O., T. Kajander, and R. Lemmelä (2004), Climate and water: From climate models to water resources management and vice versa, *Clim. Change*, *66*, 321–344.

Vicuna, S., P. E. Murer, B. Joyce, J. A. Dracup, and D. Purkey (2007), The sensitivity of California water resources to climate change scenarios, *J. Am. Water Resour. Assoc.*, *43*, 482–499.

von Storch, H., and F. W. Zwiers (1999), *Statistical Analysis in Climate Research*, 494 pp., Cambridge Univ. Press, New York.

von Storch, H., E. Zorita, and U. Cubasch (1993), Downscaling of global climate change estimates to regional scales: An application to Iberian Rainfall in wintertime, *J. Clim.*, *6*, 1161–1171.

Vrac, M., and P. Naveau (2007), Stochastic downscaling of precipitation: From dry to heavy rainfalls, *Water Resour. Res.*, *43*, W07402, doi:10.1029/2006WR005308.

Wallis, T. W. R., and J. F. Griffiths (1997), Simulated meteorological input for agricultural models, *Agric. For. Meteorol.*, *88*, 241–258.

Wang, Q. J. (2001), A Bayesian joint probability approach for flood record augmentation, *Water Resour. Res.*, *37*, 1707–1712.

Wang, Q. J. (2008), A Bayesian method for multi-site stochastic data generation: Dealing with non-concurrent and missing data, variable transformation and parameter uncertainty, *Environ. Modell. Software*, *23*(4), 412–421, doi:10.1016/j.envsoft. 2007.04.013.

Wang, Q. J., and R. J. Nathan (2002), A daily and monthly mixed algorithm for stochastic generation of rainfall time series, paper presented at Hydrology and Water Resources Symposium, Inst. of Eng., Aust., Melbourne, Victoria, Australia, 20–23 May.

Watts, M., C. M. Goodess, and P. D. Jones (2004), The CRU Daily Weather Generator, Version 2, *BETWIXT Tech. Briefing Note 1*, Clim. Res. Unit, Univ. of East Anglia, Norwich, U. K.

Waymire, E., and V. K. Gupta (1981a), A mathematical structure of rainfall representations. 1. A review of the stochastic rainfall models, *Water Resour. Res.*, *17*, 1261–1272.

Waymire, E., and V. K. Gupta (1981b), A mathematical structure of rainfall representations, 2. A review of the theory of point processes, *Water Resour. Res.*, *17*, 1273–1285.

Waymire, E., and V. K. Gupta (1981c), A mathematical structure of rainfall representations, 3. Some applications of the point process theory to rainfall process, *Water Resour. Res.*, *17*, 1287–1294.

Waymire, E., V. K. Gupta, and I. Rodriguez-Iturbe (1984), A spectral theory of rainfall at the meso-b scale, *Water Resour. Res.*, *20*, 1453–1465.

Weiss, L. L. (1964), Sequences of wet and dry days described by a Markov chain model, *Mon. Weather Rev.*, *92*, 169–176.

Wetterhall, F., A. Bárdossy, D. Chen, S. Halldin, and C.-Y. Xu (2006), Daily precipitation-downscaling techniques in three Chinese regions, *Water Resour. Res.*, *42*, W11423, doi:10.1029/2005WR004573.

Wigley, T. M. L., P. D. Jones, K. R. Briffa, and G. Smith (1990), Obtaining subgrid scale information from coarse-resolution general circulation model output, *J. Geophys. Res.*, *95*, 1943–1953.

Wilby, R. L. (1998), Modeling low-frequency rainfall events using airflow indices, weather patterns and frontal frequencies, *J. Hydrol.*, *212–213*, 380–392.

Wilby, R. L., and T. M. L. Wigley (1997), Downscaling general circulation model output: A review of methods and limitations, *Prog. Phys. Geogr.*, *21*, 530–548.

Wilby, R. L., S. P. Charles, E. Zorita, B. Timbal, P. Whetton, and L. O. Mearns (2004), Guidelines for use of climate scenarios developed from statistical downscaling methods, technical report, 27 pp., IPCC Data Distrib. Cent.

Wilks, D. S. (1989), Conditioning stochastic daily precipitation models on total monthly precipitation, *Water Resour. Res.*, *23*, 1429–1439.

Wilks, D. S. (1992), Adapting stochastic weather generation algorithms for climate change studies, *Clim. Change*, *22*, 67–84.

Wilks, D. S. (1998), Multisite generalization of a daily stochastic precipitation generation model, *J. Hydrol.*, *210*, 178–191.

Wilks, D. S. (1999a), Simultaneous stochastic simulation of daily precipitation, temperature and solar radiation at multiple sites in complex terrain, *Agric. For. Meteorol.*, *96*, 85–101.

Wilks, D. S. (1999b), Interannual variability and extreme-value characteristics of several stochastic daily precipitation models, *Agric. For. Meteorol.*, *93*, 153–169.

Wilks, D. S. (1999c), Multisite downscaling of daily precipitation with a stochastic weather generator, *Clim. Res.*, *11*, 125–136.

Wilks, D. S. (2008), High-resolution spatial interpolation of weather generator parameters using local weighted regressions, *Agric. For. Meteorol.*, *148*, 111–120.

Williams, C. B. (1947), The log series and its applications to biological problems, *J. Ecol.*, *34*, 253–272.

Wilson, L. L., D. P. Lettenmaier, and E. Skyllingstad (1992), A hierarchical stochastic model of large-scale atmospheric circulation patterns and multiple station daily precipitation, *J. Geophys. Res.*, *97*, 2791–2809.

Wilson, P. S., and R. Toumi (2005), A fundamental probability distribution for heavy rainfall, *Geophys. Res. Lett.*, *32*, L14812, doi:10.1029/2005GL022465.

Woolhiser, D. A. (1992), Modelling daily precipitation—Progress and problems, in *Statistics in the Environmental and Earth Sciences*, edited by A. T. Walden, and P. Guttorp, pp. 71–89, Edward Arnold, London.

Woolhiser, D. A., and H. B. Osborn (1985), A stochastic model of dimensionless thunderstorm rainfall, *Water Resour. Res.*, *21*, 511–522.

Woolhiser, D. A., and G. G. S. Pegram (1979), Maximum likelihood estimation of Fourier coefficients to describe seasonal variation of parameters in stochastic daily precipitation models, *J. Appl. Meteorol.*, *18*, 34–42.

Woolhiser, D. A., and J. Roldan (1982), Stochastic daily precipitation models, 2. A comparison of distribution of amounts, *Water Resour. Res.*, *18*, 1461–1468.

Woolhiser, D. A., and J. Roldan (1986), Seasonal and regional variability of parameters for stochastic daily precipitation models, *Water Resour. Res.*, *22*, 965–978.

Woolhiser, D. A., T. O. Keefer, and K. T. Redmond (1993), Southern oscillation effects on daily precipitation in the southwestern United States, *Water Resour. Res.*, *29*, 1287–1295.

Xu, C. Y. (1999), From GCMs to river flow: A review of downscaling methods and hydrologic modelling approaches, *Prog. Phys. Geogr.*, *23*(2), 229–249.

Yang, C., R. E. Chandler, V. S. Isham, and H. S. Wheater (2005), Spatial-temporal rainfall simulation using generalized linear models, *Water Resour. Res.*, *41*, W11415, doi:10.1029/2004WR003739.

Yarnal, B., A. C. Comrie, B. Frakes, and D. P. Brown (2001), Developments and prospects in synoptic climatology, *Int. J. Climatol.*, *21*, 1923–1950.

Yates, D., S. Gangopadhyay, B. Rajagopalan, and K. Strzepek (2003), A technique for generating regional climate scenarios using a nearest-neighbor algorithm, *Water Resour. Res.*, *39*(7), 1199, doi:10.1029/2002WR001769.

Zorita, E., and H. von Storch (1997), A survey of statistical downscaling techniques, *GKSS Rep. 97/E/20*, GKSS Res. Cent., Geesthacht, Germany.

Zorita, E., and H. von Storch (1999), The analog method as a simple downscaling technique: Comparison with more complicated methods, *J. Clim.*, *12*, 2474–2489.

Zorita, E., J. Hughes, D. Lettenmaier, and H. von Storch (1995), Stochastic characterisation of regional circulation patterns for climate model diagnosis and estimation of local precipitation, *J. Clim.*, *8*, 1023–1042.

Zucchini, W., and P. Guttorp (1991), A hidden Markov model for space-time precipitation, *Water Resour. Res.*, *27*, 1917–1923.

R. Mehrotra and A. Sharma, School of Civil and Environmental Engineering, University of New South Wales, Sydney, NSW 2052, Australia. (a.sharma@unsw.edu.au)

Radar-Rainfall Error Models and Ensemble Generators

Pradeep V. Mandapaka and Urs Germann

MeteoSwiss, Locarno, Switzerland

The availability of radar-based rainfall products at high space-time resolutions and over continental scales has greatly advanced our understanding of the rainfall process and its interactions with other hydrological processes across a wide range of scales. However, it is well known that radars provide areal estimates of the rainfall, which are affected by systematic and random errors from various sources. Some of these errors are inherent to the observation system and unavoidable. Therefore, it is important to quantify the uncertainty associated with the radar-based rainfall products and to provide a strong basis for probabilistic quantitative precipitation estimation. Literature in the last four decades abounds with several studies comparing the radar estimates with estimates from other instruments, identifying the radar-rainfall error sources, minimizing the errors, and modeling the uncertainties. This chapter reviews key literature related to the characterization of errors for different space-time scales, rainfall regimes, and geographical settings. The emphasis is on the error models that can be utilized to represent the uncertainty in the form of ensembles. The chapter also lists a few open questions and challenges concerning the statistical structure of radar-rainfall errors and the generation of the ensembles.

1. INTRODUCTION

The characterization of spatial and temporal variability of the rainfall process across a range of scales is essential for a variety of applications in Earth sciences. Section 2 of this book volume discussed multiple aspects of rainfall measurement and estimation using such instruments as rain gauges, disdrometers, ground-based single- and dual-polarization weather radars, and spaceborne instruments. The purpose of this chapter is to present the state of the art of modeling the uncertainty in radar-rainfall products at different space and time scales. The increasing availability of radar-rainfall products at high resolution and over continental scales has greatly contributed toward our understanding of the space-time organization of the rainfall process [e.g., *Crane*, 1990; *Kumar and Foufoula-Georgiou*, 1994; *Steiner et al.*, 1996; *Onof and Wheater*, 1996; *Fan et al.*, 1996; *Venugopal et al.*, 1999; *Germann and Joss*, 2001; *Morin et al.*, 2006; *Zhang et al.*, 2009] and paved way for parsimonious models of rainfall variability [e.g., *Bell*, 1987; *Tessier et al.*, 1993; *Seed et al.*, 1999; *Pegram and Clothier*, 2001]. The radar-rainfall measurements are being increasingly employed in quantitative precipitation estimation and forecasting techniques and for better water management practices [e.g., *Wyss et al.*, 1990; *Pessoa et al.*, 1993; *Bell and Moore*, 1998; *Semperre-Torres et al.*, 1999; *Borga et al.*, 2000; *Li et al.*, 2004a, 2004b; *Bowler et al.*, 2006; *Germann et al.*, 2006a; *Rouault et al.*, 2008]. The information provided by the radar-rainfall products has also aided in better analysis and modeling of the complex interactions between the rainfall and other processes occurring on the landscape [e.g., *Smith et al.*, 1996a; *Morin et al.*, 2001; *Smith et al.*, 2004; *Hicks et al.*, 2005; *Mandapaka et al.*, 2009a; *Javier et al.*, 2010].

In parallel with the widespread acceptance of radar-based rainfall products, it has also been well recognized that these estimates are subject to large uncertainties [e.g., *Wilson and*

Rainfall: State of the Science
Geophysical Monograph Series 191

10.1029/2010GM001003

Brandes, 1979; *Zawadzki*, 1982; *Austin*, 1987; *Fabry et al.*, 1992; *Joss and Germann*, 2000; *Krajewski and Smith*, 2002; *Collier*, 2002; *Germann et al.*, 2006b; *Szturc et al.*, 2008; *Villarini and Krajewski*, 2010a; *Krajewski et al.*, 2010]. Over the years, several studies have identified various sources of radar-rainfall estimation uncertainties and proposed techniques to minimize the errors [e.g., *Zawadzki*, 1973; *Collier et al.*, 1983; *Austin*, 1987; *Kitchen et al.*, 1994; *Joss and Lee*, 1995; *Borga and Tonelli*, 2000; *Sanchez-Diezma et al.*, 2000; *Vignal et al.*, 2000; *Harrison et al.*, 2000; *Gabella*, 2004; *Wesson and Pegram*, 2006; *Chumchean et al.*, 2006a, 2006b]. Some of the major sources of uncertainty include variability of vertical profile of reflectivity (VPR), drop size distribution (DSD), radar signal attenuation, inappropriate reflectivity-rainfall (*Z-R*) transformation, beam shielding, ground clutter, beam smoothing, anomalous propagation, temporal sampling, hardware instabilities, and radar miscalibration. The relative importance of each uncertainty source depends on such factors as the orography (e.g., for beam shielding), prevailing meteorological conditions (attenuation and anomalous propagation) and the observation framework (temporal sampling and radar miscalibration). For a comprehensive review of different sources of uncertainty in radar-rainfall estimates, please refer to *Villarini and Krajewski* [2010a].

Although, the errors can be minimized by improving hardware capabilities, developing sophisticated estimation algorithms, and merging the radar estimates with those from other sources (e.g., rain gauges and satellites) (see the work of *Seo et al.* [this volume] for a review), the resulting rainfall products are still subject to uncertainties. A radar-rainfall product with an estimate of the uncertainty will help us to employ the products to their full potential. In this regard, the last three decades have witnessed several studies on statistical characteristics of errors [e.g., *Smith and Krajewski*, 1991; *Ciach and Krajewski*, 1999a, 1999b; *Seo et al.*, 2000; *Jordan et al.*, 2003; *Chumchean et al.*, 2003; *Ciach et al.*, 2007; *Villarini et al.*, 2008; *Germann et al.*, 2009; *Mandapaka et al.*, 2009b; *Villarini and Krajewski*, 2009a]. These studies presented models for such statistics as the bias, the error variance, space-time correlations, and the full error distribution.

This chapter presents an appraisal of empirically based radar-rainfall error models, ensemble generators, and summarizes recent developments in radar-rainfall uncertainty propagation. We do not discuss the uncertainty sources and the corresponding correction techniques as they are covered in an exhaustive manner by *Seo et al.* [this volume] and by *Villarini and Krajewski* [2010a]. Instead, we focus on the models for the residual errors and representing the radar-rainfall uncertainty in the form of an ensemble. The chapter is organized as follows: sections 2 and 3 provide a discussion on different approaches toward modeling radar-rainfall uncertainties. The error models can be employed to generate the rainfall ensemble, where each member is equally probable and is consistent with the radar-rainfall error structure. Section 4 reviews various techniques that have been employed to generate the ensembles. The impact of the residual errors depends on the specific application for which the radar-rainfall products are employed; it may not be significant for providing such information as the location and size of the storm as opposed to applications of quantitative nature such as rainfall forecasting and issuing flash flood warnings. The errors have to be propagated through the relevant hydrometeorologic models to assess their impact. Section 5 reviews recent literature on radar-rainfall uncertainty propagation. Besides advances in the characterization of errors and development of efficient generators, there are several open questions that need to be addressed. These challenges are discussed in section 6 followed by concluding remarks in section 7.

2. SOURCE-SPECIFIC ERROR MODELS

An ideal approach toward characterization of errors is to identify each source, model corresponding errors, and superimpose such "source-specific" models to obtain a model for the overall radar-rainfall uncertainty. Several studies were carried out that belong to this category [e.g., *Smith and Krajewski*, 1993; *Jordan et al.*, 2000, 2003; *Bellon et al.*, 2005; *Lee et al.*, 2007; *Berenguer and Zawadzki*, 2008]. In this section, we limit the discussion to studies that modeled errors from more than one source. Hereafter, the models are referred to by the names mentioned in Table 1.

2.1. Model JSA00

Jordan et al. [2000] presented a framework to model the uncertainties induced by variability in the VPR, spatial and

Table 1. List of Key Radar-Rainfall Error Models Discussed in This Chapter Along With the Type of the Model (source-specific versus total) and the corresponding references

Error Model	Reference	Errors
JSA00	*Jordan et al.* [2000]	*Z-R*, VPR, spatial and temporal sampling
JSW03	*Jordan et al.* [2003]	*Z-R*, VPR, spatial and temporal sampling
PED	*Ciach et al.* [2007]	total
BZ08	*Berenguer and Zawadzki* [2008]	*Z-R* and VPR
HAM08	*Habib et al.* [2008]	total
REAL	*Germann et al.* [2009]	total
AHB10	*Aghakouchak et al.* [2010]	total
KDBO10	*Kirstetter et al.* [2010]	total

temporal sampling of rainfall process by the radar, and insufficient radiometric resolution of the radar measurements. The study concluded that the variability in the VPR was the most significant source of uncertainty among the analyzed error sources. The authors compared the constant altitude plan position indicator (CAPPI) maps at elevations ranging from 2 to 5 km, with 1 km CAPPI map (assumed to be the ground reference) for different spatial and temporal scales. The VPR-related errors, defined as the difference between the reflectivity value for a given CAPPI map and the 1 km CAPPI map, were found to follow an exponential distribution, where the parameter is a function of height and spatial averaging. The space-time sampling errors were evaluated by using high-resolution (6–10 s in time and 100–150 m in space) X band radar measurements of a widespread rainfall event and an event characterized by a series of small showers. The 5 and 10 min accumulations were computed from the entire data set as well as using a sampling strategy at intervals ranging from 30 to 300 s. The root-mean-square errors between the original accumulations and those resulting from temporal sampling scheme were computed and expressed as a function of spatial averaging scale. As expected, the temporal sampling error increased with the sampling interval. For 10 min accumulations of widespread rainfall, the temporal sampling error for 1 km spatial scale ranged from ~2% for 30-s sampling scheme to ~27% for 5-min sampling scheme. For scattered showers, the same error ranged from ~5% to ~55%. In addition, the study reported that the errors due to the insufficient radiometric resolution were not as significant as the errors related to VPR and space-time sampling process.

2.2. Model JSW03

Jordan et al. [2003] proposed a stochastic model for radar-rainfall errors by separately modeling the errors caused by space-time sampling, variability in the reflectivity profile, and variability in the *Z-R* relationship. They proposed deterministic models for the spatial and temporal sampling errors and stochastic models for the errors due to variability in the reflectivity profile and *Z-R* relation. A multiplicative random cascade model was used to generate "true" space-time rainfall events at a high resolution of 1 km in space and 2.5 min in time. The parameters of the model were chosen so that the scaling properties of the simulated event matched those of the radar-rainfall observations. The "true" rainfall event was then sampled every 10 min to mimic the temporal sampling process by the radar. The authors then generated realizations of VPR sampling error fields and *Z-R* sampling error fields using the random cascade model. The parameters required for the random cascade model were derived using observed error fields. Assuming that the error from each source is unbiased, and independent from each other, the radar-observed fields were obtained by imposing the error fields on the "true" event. Rainfall accumulations from the "true" rainfall fields and the error-corrupted fields were then compared for different temporal and spatial scales. The study reported significant reduction in the errors due to spatial and temporal averaging. The study also showed that the sampling errors introduced by the variability of VPR had a dominant role in the overall sampling errors in the radar-rainfall estimates.

2.3. Model BZ08

Berenguer and Zawadzki [2008] developed a framework to model the uncertainties due to variability in the VPR and *Z-R* relation with an emphasis on assimilating radar-rainfall estimates into mesoscale numerical weather prediction models. The overall error for a pixel size of 15×15 km^2 was modeled as a combination of VPR related error, *Z-R* related error, and interaction between the aforementioned errors. The framework consists of modeling the bias, the variance, and the autocorrelation of errors from individual sources and the cross-correlation between the two error sources. Three dimensional reflectivity volume scans were simulated using the information from the S band radar observations over Montreal, Quebec. CAPPI maps corresponding to different heights were created from the simulated reflectivity scans and compared with the 1.3 km CAPPI map to characterize the VPR related error. The study reported significant effect of melting layer on the VPR related error. The autocorrelation in VPR related errors was estimated for different CAPPI heights and at different ranges from the radar. In general, the autocorrelation dropped rapidly with time (decorrelation scale $<$ 15 min), when the measurements were in the rain region. However, for the combination of lower CAPPIs and farther range, errors were significantly correlated (decorrelation scale ~2 h).

The errors due to variability in the *Z-R* relation were evaluated using collocated disdrometer and radar data. The *Z-R* related errors were found to be significantly correlated with a decorrelation scale around 2 h for 15×15 km^2 resolution. The cross-correlations between the VPR and *Z-R* related errors were negligible. However, the cross-correlations were significant in the region of bright band. The above results were then combined to obtain the characteristics of the overall error. The overall error was significantly correlated with decorrelation scale ranging from 40 to 80 min.

3. MODELS FOR TOTAL ERROR

Section 2 described studies that followed the source-specific approach toward modeling uncertainties. However, modeling source-specific errors and superposing them is quite challenging, as the interactions between each source of uncertainty are not fully understood. Therefore, many studies tried to model the residual error as one entity, where the error is defined as the discrepancy between the radar estimate and the corresponding ground reference, combining all uncertainty sources. In this section, we first present an overview of studies that proposed models for first- and second-order moments of radar-rainfall errors, and then proceed to more sophisticated models providing information on space-time dependence structure and the statistical distribution of errors. The treatment, therefore, is not necessarily in chronological order.

3.1. Models for First- and Second-Order Moments

The works of *Ahnert et al.* [1986] and *Smith and Krajewski* [1991] were some of the earliest studies that proposed rigorous statistical techniques to characterize the systematic errors in the radar-rainfall estimates. One such systematic error is the mean field bias defined as the ratio of average gauge rainfall to the average radar rainfall estimates at gauge locations. *Smith and Krajewski* [1991] modeled the mean field bias as a random process that varied hourly over the course of the storm. They applied the framework on radar and rain gauge data for the storm of 27 May 1987 over Norman, Oklahoma. *Anagnostou et al.* [1998] compared the performance of the aforementioned algorithms to estimate and correct the mean field bias. The study also quantified the sampling errors in the estimation of the mean field bias. *Seo et al.* [1999] adopted a similar definition for the mean field bias as in the work of *Smith and Krajewski* [1991] and proposed a recursive estimation technique for its estimation in real time. Since the radar-rainfall errors were shown to be dependent on rainfall estimate [e.g., *Ciach et al.*, 2000, 2007] and distance from the radar [e.g., *Smith et al.*, 1996b; *Young et al.*, 1999], the limited utility of uniform scaling factor was soon realized. *Seo and Breidenbach* [2002] proposed a methodology to estimate and correct the spatially nonuniform bias within the radar domain. The technique accounts for varying conditions of rain gauge network density and rainfall types. More recently, *Chumchaen et al.* [2006b] suggested correcting for the errors in *Z-R* conversion on a physical basis and then modeling the mean field bias. All the aforementioned techniques employed Kalman filter approach to estimate and correct the mean field bias in real time.

From the above studies, it can be noticed that rain gauges provide a valuable source of information for validating the radar-based estimates and also for improving the estimation algorithms, in general. However, it is also well acknowledged that the measurements by gauges are representative of an area of approximately 100 cm^2 and, therefore, caution should be exhibited while comparing these point measurements with the remotely sensed estimates, which are typically available at spatial scales ranging from 1 to 4 km^2 and fine temporal scales [e.g., *Kitchen and Blackall*, 1992; *Villarini and Krajewski*, 2008]. Adopting an additive definition of radar-rainfall error, *Ciach and Krajewski* [1999a] proposed a method to separate the area-point (or gauge representativeness) errors from the radar-gauge comparison. Referred to as error variance separation (EVS) method, the technique separates the variance of the area-point difference ($R_A - R_G$) from that of overall radar-gauge difference ($R_R - R_G$) to obtain the variance of radar error ($R_R - R_A$). The study applied EVS method to 20 days of radar and rain gauge data in Darwin, Australia and demonstrated that the variance of area-point difference can be as large as ~70% of the total radar-gauge variance for 4-km pixels and 5 min accumulations. It should be noted that the above result was obtained when measurement from a single rain gauge was approximated as a ground reference. With a larger number of rain gauges within the radar pixel, the representativeness errors would decrease. While formulating the EVS procedure, *Ciach and Krajewski* [1999a] assumed that the rainfall is second-order stationary and that the area-point differences and the radar errors are uncorrelated (zero-covariance assumption). *Ciach et al.* [2003] checked the validity of the zero-covariance assumption in the Little Washita watershed in Oklahoma covered by dense rain gauge network and reported that it was not fully satisfied. At the same time, they also mentioned that the EVS methodology provides better estimates of radar-rainfall error variance than the direct radar-gauge comparison.

In the last decade, many studies have employed the EVS method to characterize the errors in radar and satellite-derived rainfall products [e.g., *Krajewski et al.*, 2000; *Habib and Krajewski*, 2002; *Seo and Breidenbach*, 2002; *Gebremichael et al.*, 2003; *Zhang et al.*, 2007; *Mandapaka et al.*, 2009b]. *Anagnostou et al.* [1999] modified the EVS methodology for the multiplicative errors by working in the logarithmic domain. They applied the logarithmic EVS method on 2 months of summer radar-gauge data in Florida and reported that the area-point errors can contribute to as much as 60% of the radar-gauge errors in multiplicative form. Nevertheless, the zero-covariance assumption in the logarithmic domain remains to be verified. *Zhang et al.* [2007] extended the EVS method, where the pixel average rainfall can be obtained from the rain gauges using different techniques, such as block kriging, and the zero-covariance assumption was relaxed. *Mandapaka et al.* [2009b] further extended the EVS technique and proposed a framework to estimate the spatial correlation of radar-rainfall errors. According to their methodology, the spatial correlation of radar-rainfall errors can be obtained from the semivariograms of the radar-rainfall and the true pixel-averaged rainfall. While the semivariogram of the radar-rainfall can be estimated directly from the data, the semivariogram of the true pixel-scale rainfall can be obtained by averaging the spatial correlations from the gauge rainfall over the radar pixel. They estimated the radar-rainfall error spatial correlation for hourly rainfall products in Oklahoma using high-density rain gauge network (ARS Micronet, Figure 1). The errors were found to be spatially correlated with a correlation length of approximately 20 km (Figure 2).

The information regarding the first- and second-order statistics of the radar-rainfall errors for different radars, geographical regions, rainfall regimes, and space-time scales is extremely useful in representing the uncertainties in the radar-based rainfall products. However, recent literature has witnessed comprehensive error models being developed, with the main purpose of identifying the probability distribution and the space-time dependence of errors across various scales. We now proceed to description of such error models (Table 1).

3.2. Model PED

Ciach et al. [2007] proposed an empirically based product-error-driven (PED) framework to model radar-rainfall estimation errors for spatial resolution of 4 km and hourly and larger time scales. First, the study defined an overall multiplicative bias factor (B_o) as the ratio of pixel-scale true rainfall (R_a) accumulations to the radar-rainfall (R_r) accumulations for a certain spatial scale and time period. Assuming that R_a can be approximated by rain gauge measurements, they estimated B_o using 6 years of collocated radar-gauge data pairs in Oklahoma, United States. The study then separated the bivariate distribution of B_o-corrected radar-rainfall ($R_{br} = B_o R_r$) and R_a into two components: (1) an expectation function and (2) a random error component, both conditional on R_{br}. Therefore, the relation between R_a and R_{br} can be written as

$$R_a = h(R_{br}) \cdot \varepsilon(R_{br}), \tag{1}$$

where $h(\cdot)$ is the deterministic component that accounts for the conditional biases, and $\varepsilon(\cdot)$ is the random component that

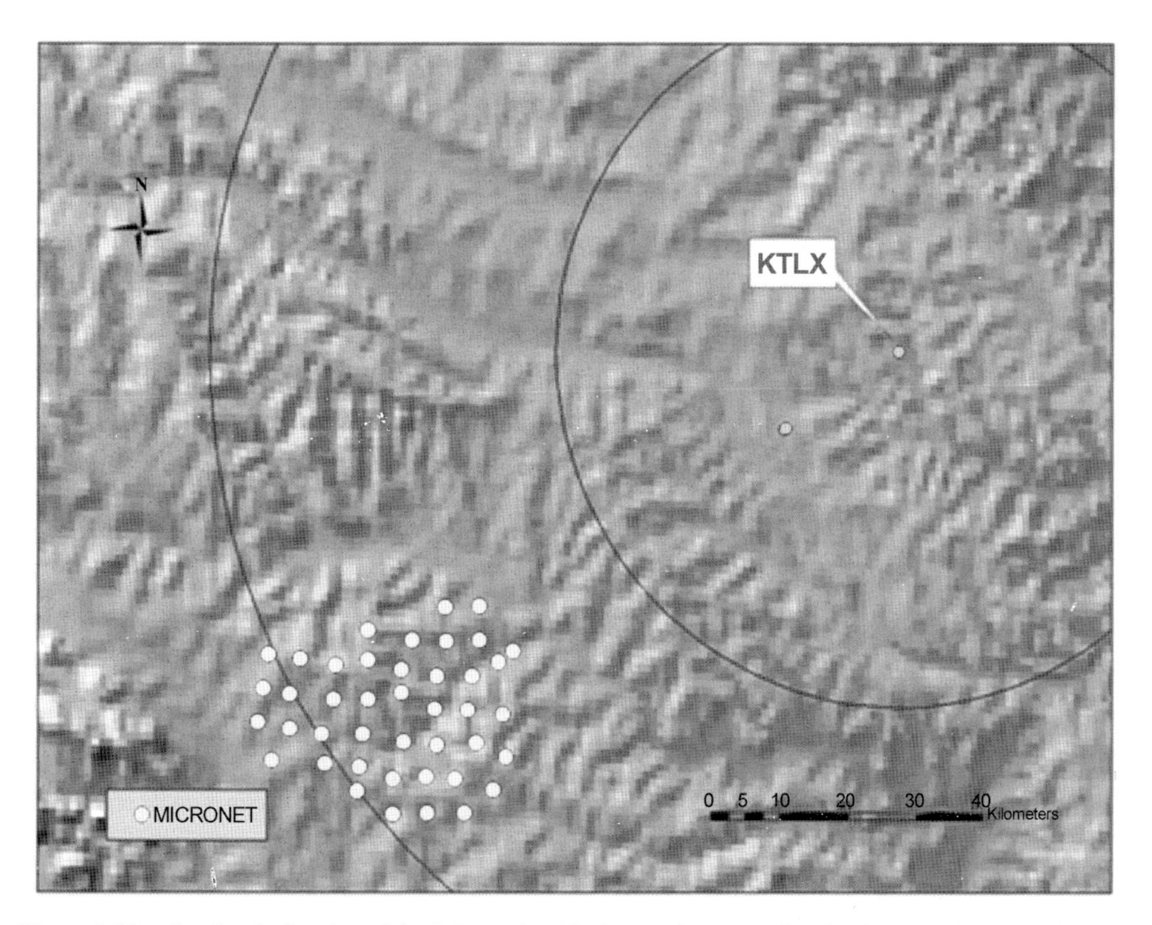

Figure 1. Map showing the location of the KTLX radar, Oklahoma Micronet, and regional topography. The radar rings shown are 50 km apart.

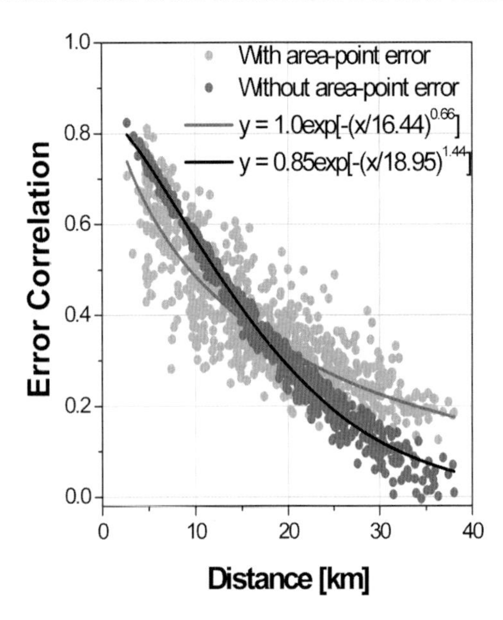

Figure 2. Radar-rainfall error spatial correlation estimated before and after accounting for the gauge representativeness errors for the 4-km resolution hourly rainfall accumulations over Oklahoma, United States. Adapted from the work of *Mandapaka et al.* [2009b].

accounts for all the remaining errors. The dependence of both the components on R_{br} was characterized using nonparametric techniques, but the study also provided simple parametric forms for modeling purposes. While $h(\cdot)$ was approximated by the following two-parameter power law,

$$h(r_r) = E\{R_a|R_{br} = r_r\} = a\cdot[r_r]^b; \quad (2)$$

$\varepsilon(\cdot)$ was found to roughly follow a Gaussian distribution with a mean of 1, and standard deviation was modeled as a rapidly decreasing hyperbolic function of radar estimate.

$$\sigma_\varepsilon^2(r_r) = E\left\{\left(\frac{R_a}{h(r_r)} - 1\right)^2 \middle| R_{br} = r_r\right\} = \sigma_{0\varepsilon} + \frac{c}{(r_r)^d}. \quad (3)$$

The availability of the high-density Micronet rain gauge network also allowed them to characterize the spatial and temporal correlation structure of residual errors with reasonable accuracy. The errors were significantly correlated with a correlation scale as large as 80 km in space and 145 min in time at hourly scales (Table 2). To account for the seasonality, and the range effect, the above framework was applied separately for three different seasons and five spatial zones. Furthermore, the parameters of the model were estimated at hourly, 3-hourly, 6-hourly, and daily time scales. Figure 3 show the parametric forms of conditional expectation function, error standard deviation, and space-time correlation of random errors for the warm season, hourly scales, and spatial zone of 50 to 100 km from the radar.

To verify if the structure of PED model is valid for a different geographic setting, *Villarini and Krajewski* [2009a] evaluated the radar-rainfall products from a C band radar in Wardon Hill, England using a high-density rain gauge network. The time scales ranged from 5 to 180 min, and the spatial resolution of the radar products was 2 km. The high density of the rain gauge network also allowed them to investigate the impact of area-point errors introduced by the assumption that the rain gauges represent the true areal rainfall accurately. The study found that the area-point sampling errors do not significantly affect the conditional bias. However, the Gaussian model for the random errors, as proposed by *Ciach et al.* [2007], was found to be valid only at hourly scale or larger. They attributed the lack of Gaussian error structure for subhourly scales to the presence of spatial sampling errors and the multiplicative nature of error.

Villarini and Krajewski [2010b] extended the PED framework to radar-rainfall products generated using different *Z-R* relations (Marshall-Palmer [*Marshall and Palmer*, 1948], NEXRAD [e.g., *Fulton et al.*, 1998], and Tropical [e.g., *Rosenfield et al.*, 1993]) and error definitions (multiplicative versus additive) for the hourly data from the Oklahoma region. The study reported that the NEXRAD *Z-R* relation results in the least conditionally biased rainfall estimates for the Oklahoma region. Regarding the random error structure, the multiplicative form was not particularly sensitive to the adopted *Z-R* equation, while the additive form showed significant sensitivity. The study further showed that the multiplicative (additive) random errors can be modeled using a Gaussian distribution with a mean of 1 (0) and a standard deviation, which is a decreasing (increasing) function of radar-rainfall estimate. Both the forms of random errors displayed significant correlation in space and time. In addition, the study investigated the effect of anomalous propagation (AP) correction on the parameters of the error model and found that the AP detection algorithm improved the comparison between radar and rain gauges in terms of overall bias [*Villarini and Krajewski*, 2010b].

3.3. *Model HAM08*

Habib et al. [2008] adopted a framework similar to that of PED model to characterize the uncertainties in the NEXRAD Stage III radar-rainfall products over 21 km^2 Goodwin Creek watershed in northern Mississippi, United States. A total of 12 rainfall events were considered, and the corresponding hourly radar-rainfall products with a spatial resolution of 4 km were evaluated using 30 rain gauges within the watershed. The study first estimated the overall multiplicative bias B_o, defined as in section 3.2 for each pixel containing the rain gauges. The value B_o displayed significant spatial and storm-to-storm

Table 2. Summary of Key Radar-Rainfall Error Models Discussed in Sections 2 and 3

Error Model	Location	Resolution		Random Error Characteristics
		Space	Time	
JSA00	radar data in Auckland, New Zealand, and Darwin, Australia	100–150 m; 1 km (Darwin)	6–10 s	VPR errors: exponential probability distribution
JSW03	Darwin, Australia; data-based simulation framework	1 km	2.5 min	assumed that the individual errors are lognormally distributed and uncorrelated
PED	Oklahoma, United States, and southwest England	4 km (U.S.), 2 km (U.K.)	1 h (U.S.), 5 min (U.K.)	multiplicative, Gaussian, and space-time correlated
PED (Additive)	Oklahoma, United States	4 km (U.S.)	1 h (U.S.)	additive, Gaussian, and space-time correlated
BZ08	Montreal, Quebec, Canada; data based simulation framework	15 km	–	correlated VPR, *Z-R* and combined errors
HAM08	Goodwin Creek Watershed, north central Mississippi, United States	4 km	1 h	multiplicative, lognormal, and space-time correlated
REAL	Southern Alps, Switzerland	2 km	1 h	multiplicative (expressed in dB), Gaussian, and space-time correlated
AHB10	Goodwin Creek Watershed, north central Mississippi, United States	1 km	15 min	multiplicative and additive error components, Gaussian, and uncorrelated
KDBO10	Cévennes-Vivarais region, France	1 km	1 h	additive, double exponential distribution, and space-time correlated

variability. The study then characterized the conditional bias $CB(r_r) = h(r_r)/r_r$, where $h(r_r)$ is the conditional expectation function defined in section 3.2. After correcting the radar estimates for the systematic errors, the study characterized the remaining errors considering only three pixels for which the true pixel-scale rainfall was obtained from rain gauges with reasonable level of accuracy. The results showed that the empirical distribution of random errors can be approximated with a lognormal distribution. While the mean of the log-transformed random errors was approximately equal to zero and independent of the radar estimate, the variance changed with radar-rainfall values. The study characterized the dependence of variance on the radar estimate as a power law of the form

$$\sigma_\varepsilon^2(r_r) = E\left\{\left(\frac{R_a}{r_r} - 1\right)^2 \middle| R_{br} = r_r\right\} = \frac{\alpha}{(r_r)^\beta}, \quad (4)$$

where α and β are the coefficients that were estimated from the data. In addition to overall bias, conditional bias, and random error probability distribution, the study also attempted to characterize the space-time dependence structure of random errors. The spatial correlation at a distance lag of 4 km and temporal correlation for time lags less than 2 h were found to be non-negligible. However, they did not include time correlation, while generating the rainfall ensemble stating that they "are rather low." Because of the limited size of the sample, they did not estimate the spatial correlations beyond the distance lag of 4 km.

3.4. Model REAL

Germann et al. [2006c, 2009] developed a model for radar-rainfall residual errors as part of their study to generate an ensemble of precipitation fields for a given radar-rainfall field. The study defined the error $\delta_{t,k}$ in radar estimates for the time t and pixel k as

$$\delta_{t,k} = 10 log\left(\frac{R_{g_{t,k}}}{R_{r_{t,k}}}\right), \quad (5)$$

where $R_{g_{t,k}}$ and $R_{r_{t,k}}$ are the rain gauge and radar estimates for the pixel k and time step t. Based on 6 months of data from 31 collocated radar-gauge pairs in a 2800-km^2 catchment in Southern Alps, the study estimated the characteristics of radar-rainfall errors at hourly scale. The errors were found to follow a Gaussian distribution reasonably well (Figure 4). The statistical characteristics of the errors were estimated as

$$\mu_k = \frac{1}{\sum_{t=1}^{Q} R_{r_{t,k}}} \sum_{t=1}^{Q} (R_{r_{t,k}} \cdot \delta_{t,k}) \quad (6)$$

$$C_{kl} = \frac{1}{\sum_{t=1}^{Q} (R_{r_{t,k}} \cdot R_{r_{t,l}})} \sum_{t=1}^{Q} R_{r_{t,k}} \cdot R_{r_{t,l}} \cdot (\delta_{t,k} - \mu_k) \cdot (\delta_{t,l} - \mu_l), \quad (7)$$

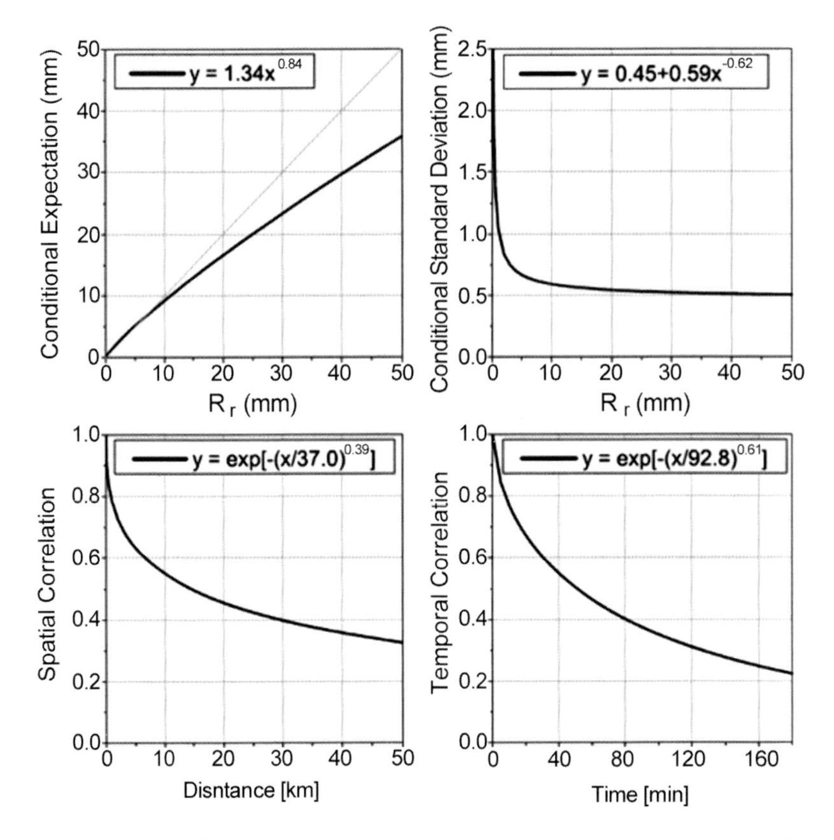

Figure 3. (top left) Average gauge rainfall and (top right) random error standard deviation conditional on the radar-rainfall value for hourly warm season rainfall products over Oklahoma, United States. (bottom) Random error spatial and temporal correlation functions. All the curves correspond to the region between 50 and 100 km from the radar Adapted from the work of *Ciach et al.* [2007] and *Villarini et al.* [2009b].

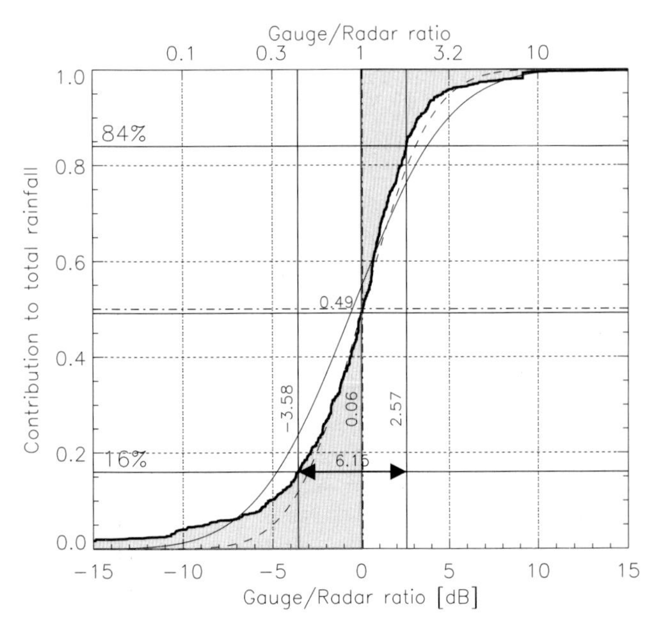

Figure 4. Gaussian error structure of log-transformed random errors. Adapted from *Germann et al.* [2009], reprinted with permission.

where Q is the number of time steps, μ_k is the average of the radar-rainfall errors over the pixel k, and C_{kl} is the spatial correlation of errors for the pixels k and l. In equations (6) and (7), weighting with the rainfall values was necessary to focus on the data with hydrologic relevance. Because of the strong dependence of the error structure on location, the study did not parameterize the behavior of the spatial correlation. The error structure for pixels without rain gauges was obtained by linear interpolation using Delaunay triangulation technique. The temporal evolution of the errors was characterized in terms of basin-averaged lag-1 and lag-2 autocorrelations in time. The decorrelation lengths in time and space were approximately 1 h and 10–40 km, respectively.

3.5. Model AHB10

Similar to PED model, *Aghakouchak et al.* [2010] separated the residual errors into two components: a purely random error and an error component that is proportional to the radar estimate. They defined the residual error as the difference between the radar estimate and the corresponding ground truth (approximated by the rain gauge measurement). According to the model,

$$R_g - \left(\frac{Z}{A}\right)^{1/B} = \left(\frac{Z}{A}\right)^{1/B} \cdot \varepsilon_1 + \varepsilon_2, \qquad (8)$$

where A is the coefficient in the Z-R relation, B is the exponent in the Z-R relation, ε_1 is the error that is proportional to the radar estimate, and ε_2 is the random error. They assumed that both the error components are Gaussian with a mean of zero and standard deviation of σ_1 and σ_2, respectively. The parameters of the model (A, B, σ_1, and σ_2) were estimated for four rainfall events over Goodwin Creek watershed, Mississippi, United States, using maximum likelihood method. The errors were characterized for the spatial resolution of 1 km and a time scale of 15 min. The model did not account for the spatiotemporal dependence of the errors and the area-point errors introduced by comparing radar estimates directly with the rain gauge measurements.

3.6. Model KDBO10

In a recent study, *Kirstetter et al.* [2010] followed an approach similar to PED framework [*Ciach et al.*, 2007] to evaluate radar data corresponding to five rainfall events over the Cevennes-Vivarais region, France. One of the five events is a severe mesoscale convective system that caused catastrophic flooding in the region. The residual error characteristics were estimated for the above severe event and the remaining four events separately, for the spatial scale of 1 km, time scales varying from 1 to 12 h, and range intervals of 0–50 and 50–100 km from the radar. However, there are two main differences between the PED model and the error model by *Kirstetter et al.* [2010]: (1) unlike PED model, the authors defined the residual error as the difference between the radar estimate and the ground reference, and (2) the study obtained ground reference values at the radar grid resolution by employing block kriging on the rain gauge measurements. The study also accounted for the strong anisotropy in the rainfall spatial structure, while performing block kriging. The distribution of the residual errors was found to closely follow a double exponential of the form

$$f(x) = \frac{1}{\sqrt{2}\beta} \exp\left(-\frac{\sqrt{2}|x-\alpha|}{\beta}\right), \qquad (9)$$

where the parameters α and β are the mean and the standard deviation, respectively. Both the parameters were found to be dependent on the radar estimate. The study suggested nonparametric fitting techniques for modeling the behavior of α and proposed a linear model for the dependence of β on R_r. The study then characterized the space-time correlation structure of the residual errors in terms of spherical variograms. The variogram range (distance at which the variogram reaches a constant value called sill) at hourly scale was found to be as large as 41 km indicating significant correlation in the residual errors. The residuals were also correlated in time with a variogram range of ~3 h for hourly scales. The error model was presented for multiple accumulation time scales, different range intervals, and rain events of different types.

Table 2 summarizes important characteristics of random error models discussed in sections 2 and 3.

4. ENSEMBLE GENERATION

The radar-rainfall ensemble generators developed in the last two decades can be broadly classified into (1) purely statistical techniques based on conceptual understanding of the error structure [e.g., *Krajewski and Georgakakos*, 1985; *Tadesse and Anagnostou*, 2005], (2) physical-statistical techniques that use stochastic generators of rainfall and mimic the physics of the radar observation process [e.g., *Krajewski et al.*, 1993; *Anagnostou and Krajewski*, 1997], and (3) the techniques that explicitly incorporate the empirical error structure, while generating the ensemble [e.g., *Germann et al.*, 2009; *Villarini et al.*, 2009b]. It should be noted that there is considerable overlap among the above categories, and the classification is only based on the dominant aspects of the generators.

4.1. Statistical Techniques

Krajewski and Georgakakos [1985] proposed a framework based on Turning Bands technique to generate an ensemble of rainfall fields by imposing nonstationary error fields on a given radar-rainfall field. The generator is capable of producing error fields with the desired spatial structure and conditional statistics. Since the radar-rainfall error characteristics were unknown at that time, the error fields were generated with arbitrary second-order statistics. *Tadesse and Anagnostou* [2005] proposed a technique based on generalized likelihood uncertainty estimation (GLUE) framework to obtain an ensemble of rainfall fields. According to the GLUE methodology, widely used in hydrologic modeling, several combinations of model parameters may be equally likely (equifinality) for a given performance measure [e.g., *Beven and Freer*, 2001]. *Tadesse and Anagnostou* [2005] used this concept to determine the equifinality range of three parameters in their radar-rainfall estimation algorithm: (1) the exponent in the Z-R relation, (2) a parameter that separates convective and stratiform rain regimes, and (3) an attenuation correction parameter. For a given Z field, they obtained a large number of rainfall fields using different sets of parameter values drawn randomly from a uniform distribution with certain upper and lower bounds. For each parameter set, the rainfall fields were compared with the observations, and the parameter sets that achieved a certain level of performance were retained as "behavioral." The retained parameter sets were then ranked according to their likelihood weights to form the cumulative distribution function from which uncertainty bounds (5% and 95% quantiles) were estimated.

An ensemble of rainfall fields can also be generated through stochastic rainfall models that replicate the space-time structure of the rainfall fields. Some of the examples are universal multifractal model by *Tessier et al.* [1993], String-of-Beads model by *Pegram and Clothier* [2001], and conditional stochastic rainfall generator by *Wojcik et al.* [2009]. The parameters for the models are obtained from the radar or multisensor rainfall maps. However, the rainfall estimation errors are not explicitly accounted for in the models that belong to this category.

4.2. Physical-Statistical Techniques

The generator developed by *Krajewski et al.* [1993] belongs to the second category of physical-statistical techniques. They employed a stochastic rainfall model to generate a two-dimensional (2-D) true rainfall field. Given the rainfall rates, the study proposed a technique to obtain parameters of corresponding DSD, which were then transformed into the radar reflectivities. The DSD parameters were assumed to be independent in time and space. They imposed error fields (which are function of radar parameters) onto the reflectivity fields and applied a *Z-R* transformation to obtain radar-rainfall fields. *Anagnostou and Krajewski* [1997] extended the work of *Krajewski et al.* [1993] to the 3-D space by linking 2-D rainfall fields with the cloud-top height models. The generated radar observables included the effects of DSD variability, partial beam filling, vertical and horizontal gradients in reflectivity, attenuation, and random measurement errors. *Haase and Crewell* [2000] developed a framework to generate 3-D radar-rainfall observations using the output from high-resolution mesoscale weather forecast model. These generators, based on the physics of radar observation process, were reasonably successful in producing a realistic *Z-R* scatter. However, the main limiting factors in these models are the ad hoc assumptions regarding the error structure. Recent literature has seen error models proposed based on strong empirical evidence. Some of these error models were discussed in section 3. We now proceed to generators based on such empirically based error models.

4.3. Empirically Based Techniques

Germann et al. [2006c, 2009] developed an ensemble generator based on the uncertainty model discussed in section 3.4. The generator employed LU factorization using Cholesky decomposition, where the covariance matrix is decomposed into a product of lower triangular matrix $\mathbf{L}$ and an upper triangular matrix $\mathbf{U} = \mathbf{L}^{\mathrm{T}}$, to obtain spatially correlated error fields. The LU factorization approach allows full representation of the spatial dependence of the radar-rainfall errors, a critical aspect in a mountainous region such as Switzerland. They also proposed to use a modified form of the Cholesky algorithm or singular value decomposition technique to avoid numerical instability when the covariance matrix is close to singular. Assuming, in a first step, temporal independence of the errors, the generator obtained an ensemble of Gaussian fields with zero mean and covariance structure given by equation (7). The temporal structure of the errors is then introduced by second-order autoregressive filtering. According to the AR(2) filtering technique, the error field δ_t at time step t is obtained as follows.

$$\delta_t^{'} = Ly_t - a_1\delta_{t-1}^{'} - a_2\delta_{t-2}^{'} \qquad (10)$$

$$\delta_t = \mu + \upsilon\delta_t^{'}, \qquad (11)$$

where a_1 and a_2 are the AR(2) model parameters, Ly_t is the zero mean spatially correlated field at time t, and υ is the variance rescaling factor. The model parameters a_1 and a_2 are functions of lag-1 and lag-2 autocorrelations in time. The error fields were then imposed on the log-transformed radar-rainfall field to obtain an ensemble of rainfall fields Φ_t.

$$10\log[\Phi_t] = 10\log[R_{r_t}] + \delta_t. \qquad (12)$$

Figure 5 shows sample realizations from the ensemble generated using *Germann et al.* [2009]. Adopting logarithmic error definition as in equation (5), *Llort et al.* [2008] characterized the errors for a 64×64 km^2 radar domain in the Catalunya area, Spain. The radar-rainfall products were available with a spatial resolution of 1 km and every 10 min. The errors were found to closely follow a Gaussian distribution. The mean and the standard deviation of the error distribution were estimated, and the spatial dependence of the errors was characterized in terms of power spectrum. The information was used to generate Gaussian perturbation fields with the desired mean and error standard deviation. The error correlations were induced through the power-law filtering. The ensemble rainfall fields were then obtained by imposing error fields over the radar-rainfall fields as in equation (12).

Villarini et al. [2009b] developed a generator based on the PED model to obtain an ensemble of probable areal rainfall fields conditional on the radar-rainfall field. The generator involves the following steps: (1) the radar-rainfall field is rescaled with the overall bias factor B_o, (2) the B_o-corrected

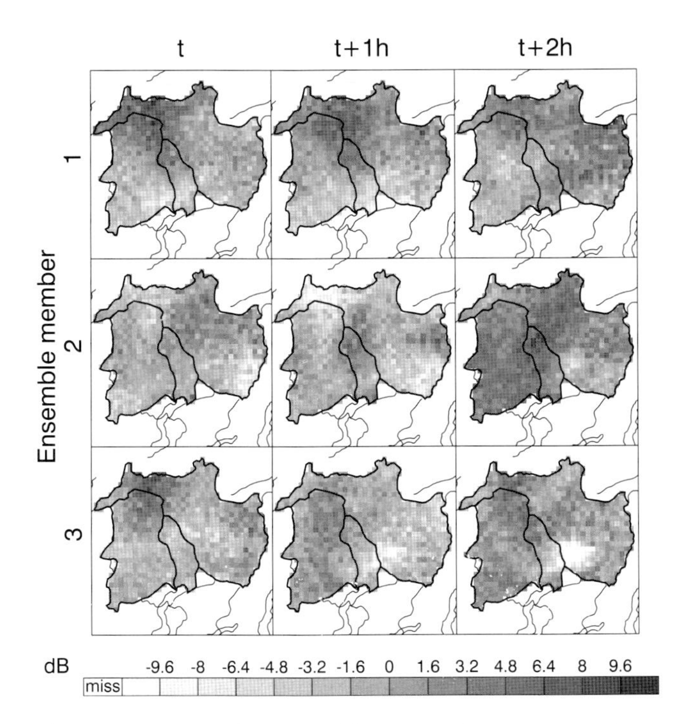

Figure 5. Sample realizations of three ensemble members at three consecutive time steps from radar-rainfall ensemble generated using REAL generator. Figure 5 illustrates the spatial and temporal correlation of radar-rainfall residual errors in an Alpine catchment of size 2800 km^2. Adapted from the work of *Germann et al.* [2009], reprinted with permission.

field R_{br} is then transformed to the deterministic field $h(\cdot)$ using a power law relation (equation (2)), and (3) an ensemble of spatially correlated Gaussian random error fields with a mean of 1, and standard deviation that depends on the radar estimate (equation (3)) is generated and imposed on the $h(\cdot)$ field from step 2. The information on the overall and conditional biases in steps 1 and 2 and the random error characteristics (standard deviation and spatial correlation) in step 3 are obtained from the PED error model [*Ciach et al.*, 2007] and depend on the season, time scale, and spatial zone. Similar to the study of *Germann et al.* [2006c, 2009], the study employed LU factorization approach to obtain spatially correlated fields. By repeating the aforementioned steps for each time step in an event, one can generate an ensemble of probable rainfall events that are consistent with the radar-rainfall error structure. One main limitation of the generator is that the temporal dependence of radar-rainfall errors is not explicitly incorporated in the generator. The resulting probable rainfall events are not temporally independent as the time dependence in the deterministic component will propagate into the ensemble. However, one does not have full control over the temporal evolution of the errors. Because of the Gaussian probability distribution, the error fields from step 3 may contain some negative values (particularly for smaller values of R_R, where the standard deviation of random error increases in a hyperbolic manner), which are eventually set to zero.

Aghakouchak et al. [2010] employed the error model described in section 3.5 to generate an ensemble of rainfall fields. The parameters A and B in equation (8) were used to convert the radar reflectivity field into the rainfall field. Then, the conditional and purely random Gaussian error fields with zero mean and standard deviations of σ_1 and σ_2, respectively, were imposed on the radar-rainfall field to obtain an ensemble of rainfall fields. For the pixels containing rain gauges, the errors were generated such that the total error is similar to the observed error value. As mentioned in section 3.5, the generated error fields are independent in space and time.

5. UNCERTAINTY PROPAGATION

Several recent studies have propagated the uncertainties in radar-rainfall products through hydrological models and assessed the impact of the rainfall estimation errors on the modeled fluxes [e.g., *Borga*, 2002; *Sharif et al.*, 2002; *Carpenter and Georgakakos*, 2004; *Gabellani et al.*, 2007; *Schröter et al.*, 2008; *Germann et al.*, 2009]. *Borga et al.* [2000] evaluated the impact of the mean field bias and range-dependent errors in the C band radar-rainfall estimates on the accuracy of streamflow predictions for two mountainous basins of size 77 and 116 km^2, respectively. The radar-rainfall estimates and the basin-averaged rain gauge and radar-rainfall estimates were used as input to the semidistributed hydrologic model. The results showed that the hydrologic model magnified the rainfall estimation errors, and the streamflow prediction accuracy depended on the distance of the basin from the radar. However, the accuracy improved significantly when the mean field bias and range-dependent error corrections were applied to the raw radar-rainfall estimates [*Borga et al.*, 2000]. In a similar study, *Borga* [2002] analyzed the impact of the mean field bias, errors due to nonuniform VPR, and the anomalous propagation on the streamflows simulated for a 135-km^2 Brue catchment in the United Kingdom. The high-density rain gauge network in the basin allowed them to better characterize the radar-rainfall errors. The streamflows obtained from radar-based rainfall estimates were compared with those from the rain gauge-based rainfall measurements and observed streamflows for a 2.5-year period. The study reported that the streamflow simulation efficiency increased up to 30% with the use of corrected radar estimates. In addition, the study reported that the residual errors in the corrected radar-rainfall estimates lowered the efficiency by 10%. They also noted that the effect of nonuniform VPR is more dominant than the anomalous propagation. Continuing along the same lines, *Borga et al.* [2006] investigated the relative roles of hydrologic model input errors and model structure errors using the GLUE framework described briefly in section 4. The results showed that the hydrologic model parameters were significantly affected by the radar errors. The parameter sensitivity to radar uncertainties increases rapidly with the radar scans, and beam geometry, thus restricting the transferability or regionalization of parameters to shorter ranges (<70 km for a C band radar). The study also suggested that the runoff simulation efficiency is strongly dependent on the spatial structure of the radar-rainfall errors [*Borga et al.*, 2006].

Sharif et al. [2002] employed a physically based simulation framework to study the effects of radar-watershed orientation, radar-watershed range, and radar-rainfall estimation errors on the streamflow predictions for 21 km^2 Goodwin Creek watershed in Mississippi. They used a numerical weather prediction model called Advanced Regional Prediction System [e.g., *Xue et al.*, 2000] to simulate a well-documented convective storm. A virtual radar was placed at different orientations and ranges from the watershed, and radar observations of the storm were obtained using a 3-D radiative transfer model. The study demonstrated that the range and orientation errors are amplified as they propagated through the hydrologic model. To study the effects of radar-rainfall estimation errors, they imposed a Gaussian noise of zero mean and standard deviation of 1 dBZ on reflectivities obtained by the radar simulator. However, the results suggested that the random errors have minor impact compared to range/orientation effects [*Sharif et al.*, 2002]. *Carpenter and Georgakakos* [2004] carried out a Monte Carlo experiment to investigate the effect of uncertainties in the hydrologic model parameters and radar-rainfall estimates on the ensemble flow predictions for five watersheds in the central United States. The uncertainty in rainfall estimates was approximated in the form of two probability distributions: (1) a uniform distribution and (2) an exponential distribution. The errors were assumed to be uncorrelated in space and time. The main result from the study was that the effect of errors decreases with the increase in the spatial scale. They found that the uncertainty in ensemble predictions decreased with the logarithm of increasing drainage area across a range of scales (~50–2000 km^2).

Gabellani et al. [2007] analyzed the impact of uncertainty in the space-time structure of rainfall events on the runoff predictions from a semidistributed hydrologic model for two mountainous basins of size 840 and 3415 km^2, respectively. They generated an ensemble of rainfall events with approximately the same volume (up to 1% accuracy), duration, and space-time power spectra, but with different space-time distribution of intensities. The ensemble runoff predictions from the hydrologic model displayed significant sensitivity to the localization of the rainfall peaks for both basins. The study also showed that the flow predictions are sensitive to the uncertainty in the estimation of power spectra. The studies discussed so far in this section have approximated the uncertainty in rainfall using arbitrary models. However, there are studies that assessed the impact of radar-rainfall errors through empirically estimated error models.

For the 21-km^2 Goodwin Creek watershed in Mississippi, *Habib et al.* [2008] showed that the systematic errors in radar-rainfall fields were amplified as they propagated through the hydrologic model. To study the impact of random errors, they generated an ensemble of rainfall fields using an empirically driven radar-rainfall error model (described in section 3.3) and used them as input to the distributed hydrological model. They noted that for most of the time, the ensemble streamflow predictions enclosed those obtained from the bias-corrected radar-rainfall and reference rainfall fields. The study also reported that the uncertainty in the ensemble streamflow predictions is sensitive to the radar-rainfall random error spatial correlations. *Germann et al.* [2009] generated radar-rainfall ensembles using the model described in section 3.4 and set up a real-time experiment coupling the rainfall ensembles with the semidistributed hydrologic model for a 2800-km^2 area in Southern Alps. They found the accuracy of the radar ensemble-driven and rain gauge-driven streamflow predictions to be comparable. The study also reported that the generator was bias-free and did not overstate the uncertainty that was already present in the deterministic component. The experiment has been in operation since May 2007 and being updated on an hourly basis.

The PED model described in section 3.2 and the corresponding ensemble generator described in the previous section have been used for many applications. *Villarini et al.* [2009b] exploited the Gaussian nature of the random errors in the PED model and developed an analytical framework for generating the probability of exceedance maps of areal rainfall. For each pixel in a given rainfall field, they obtained probability that the pixel-averaged rainfall exceeds a particular value R_{thresh}. In addition to pixel-scale probability maps, they also extended the analytical expression to obtain exceedance probabilities for a group of pixels. The error model and the analytical results were employed to evaluate the TRMM satellite rainfall products [*Villarini et al.*, 2009a] and to assess the uncertainties in the Flash Flood Guidance System [*Villarini et al.*, 2010]. *Villarini and Krajewski* [2009b] employed the model to assess the impact of radar-rainfall uncertainties on the spatial scaling properties of 15 rainfall events over Oklahoma, United States. They reported that the errors in radar-rainfall products significantly bias the estimated scaling properties. *Mandapaka et al.* [2010] analyzed the impact of radar-rainfall errors on the estimates of the spatial structure of 10 events over Kansas, United States, and concluded that the overall effect of the errors is to smooth the radar-rainfall fields at smaller scales.

6. OPEN QUESTIONS

Besides recent advances in modeling the radar-rainfall residual errors, and evaluating the impact of the errors in various hydrometeorological applications, there are several open questions in the field of radar-rainfall uncertainty analysis and propagation. Questions such as the appropriate definition of the radar-rainfall error, dependence of errors on rainfall intensity, type, and space-time scales, interdependence of source-specific errors, and adequate size of the ensemble to sample full range of uncertainty need to be addressed. Here we outline some important questions and possible future research directions.

6.1. Additive Versus Multiplicative Error Structure

We have seen in sections 2 and 3 that some studies defined the error as the difference between the radar-rainfall estimate and the corresponding ground reference, while others have adopted a multiplicative definition for error (defined as the ratio of radar-rainfall estimate to the ground reference). There is no general consensus on the definition of the errors in the radar-rainfall uncertainty analyses. In a recent study, *Villarini and Krajewski* [2010b] showed that the statistical properties of the errors and their sensitivity to different error sources depend on the definition of the error. From a physical point of view, error sources such as the uncertainties in various terms of the radar equation [*Skolnik*, 2008, equation (19.12)], attenuation by wet radome and strong rain, and fluctuations in *Z-R* relation are multiplicative in nature. Therefore, it is logical to employ multiplicative error definition (sometimes in logarithmic scale) as evidenced in many modeling studies [e.g., *Anagnostou et al.*, 1999; *Jordan et al.*, 2003; *Gabella et al.*, 2005; *Ciach et al.*, 2007; *Habib et al.*, 2008; *Germann et al.*, 2009]. However, one main limitation of the multiplicative framework is that it is restricted to nonzero rainfall values, and zero-rainfall values require special treatment. While the multiplicative framework is logical for understanding the radar-rainfall uncertainties from different sources, an additive framework may be easy to interpret from an end-user perspective. This is still an open area of research, which needs to be addressed.

6.2. Error Distribution

In many of the radar-rainfall evaluation studies, the errors were shown to closely follow Gaussian or lognormal distribution depending on the definition of the error [e.g., *Ciach et al.*, 2007; *Habib et al.*, 2008; *Germann et al.*, 2009; *Aghakouchak et al.*, 2010]. This is a convenient result especially for generating the ensembles as well-established techniques exist to produce correlated Gaussian and lognormal fields. *Villarini and Krajewski* [2010b] reported that distribution of errors varied with time scale. While the errors followed a Gaussian distribution at scales of 1 h and longer, there was a significant departure from the Gaussian model at subhourly scales. *Kirstetter et al.* [2010] suggested double exponential probability distribution for the errors. What do we do when errors are not Gaussian? Quantile transformation is a possible option, but it is not a trivial task to generate fields with desired probability distribution at the same time preserving the error correlations.

6.3. Dependence Among the Individual Errors

Information regarding the dependence among the errors from each source of radar-rainfall uncertainty is still lacking. While a lot of work has been done in identifying each source of radar-rainfall uncertainty and in characterizing the corresponding errors, their interdependence is still a subject of active research. For example, qualitatively, we can say that the errors due to random variability in VPR and those due to the variability in *Z-R* are related due to their dependence on the underlying meteorological conditions. Section 2.3 discussed a model by *Berenguer and Zawadzki* [2008], which characterized the relation between VPR and *Z-R* errors in terms of correlation matrix. But how are the VPR and Z-R errors related to other error sources such as beam blockage? Are the errors due to signal attenuation related to those from other sources? It is important to address these questions to use the source-specific error modeling approach to the full potential.

6.4. Nonstationarity of the Errors

Due to the very nature of the sources of radar-rainfall uncertainties, there is a significant dependence of errors on such factors as distance from the radar, radar visibility, and meteorological conditions. The studies discussed in sections 2 and 3 did provide enough evidence that the errors are nonstationary in nature. Therefore, it is important that the realizations in the ensemble reflect the observed nonstationarity. *Villarini et al.* [2009b] used the conditional statistics given by *Ciach et al.* [2007] to interpolate the error structure to ungauged pixels, while *Germann et al.* [2009] used a linear interpolation scheme. Moreover, presence of high-quality strategically placed dense rain gauge networks would help in better characterizing the nonstationarity of error structure.

6.5. Statistical Structure of Radar-Rainfall Versus Ensemble

The main purpose of the ensemble is to represent the random residual uncertainties, and therefore, all the systematic errors in the radar-rainfall fields should be removed before imposing the stochastic component. By removing systematic errors, the ensemble generator is bias-free in terms of rainfall volume. It is important to conserve the volume especially while assessing how the random errors propagate through the hydrologic models and impact streamflow predictions. However, conserving the rainfall structure in terms of extremes and higher-order statistics is a challenging task. Statistics such as variance, when not conserved, would lead to misleading interpretations of the effect of random errors. *Mandapaka et al.* [2010] compared several statistics (power spectra, scaling properties) of the bias-corrected radar-rainfall field and the corresponding ensemble (produced using *Villarini et al.* [2009b]) for 10 warm season rainfall events over Wichita, Kansas. Figure 6 shows the power spectra for 1 of those 10 events. It can be noticed that the power in the ensemble departs systematically from that of radar-rainfall field, as we move toward smaller scales. *Mandapaka et al.* [2010] interpreted this systematic departure as the effect of random errors on the spatial characterization of rainfall events. However, the departure could also be due to the manner in which the random errors were characterized by the error model. The importance of preserving the higher-order error statistics by the ensemble and their effect on the hydrologic simulations need to be studied in more detail.

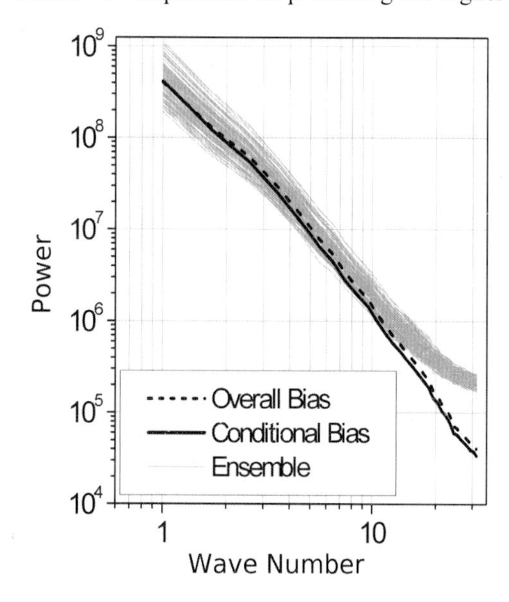

Figure 6. Effect of radar-rainfall random errors on the spatial characterization of the rainfall events. Figure 6 compares the power spectra of the rainfall ensemble with those of the radar-rainfall fields after accounting for overall and conditional biases. The ensemble realizations are produced using the generator developed by *Villarini et al.* [2009b].

6.6. Temporal Dependence in Ensemble

It is clear from the literature on radar-rainfall evaluation that the errors are dependent on time [e.g., *Ciach et al.*, 2007; *Habib et al.*, 2008; *Germann et al.*, 2009; *Kirstetter et al.*, 2010]. However, as mentioned by *Villarini et al.* [2009b], introducing temporal dependence of errors in the ensemble is not a trivial task especially in the presence of nonstationarity. *Germann et al.* [2009] included the temporal evolution of errors in their ensemble generator by assuming that the parameters in their autoregressive model are invariant in space. As an alternative, the study also proposed to use LU factorization approach for both space and time, which allows full flexibility regarding the error covariance.

7. CLOSING REMARKS

Evaluation of radar-rainfall products has progressed from simulation studies based on ad hoc assumptions to empirically based studies. The chapter discussed important developments that took place in the last 10 years in the area of radar-rainfall uncertainty analysis and propagation. We focused mainly on the error models based on strong empirical evidence and radar-rainfall ensemble generators. Two different approaches toward modeling of errors were presented: one based on identifying and characterizing each source of radar-rainfall uncertainty and the other based on modeling the total error between the radar estimate and the corresponding ground reference. Identification of relevant error sources and characterizing the complex interactions among each source of error is not a trivial task and is the main limitation for adopting the former approach. Recent studies such as the works of *Lee et al.* [2007] and *Berenguer and Zawadzki* [2008] made considerable progress in modeling the VPR and *Z-R* related uncertainties. However, much research needs to be done to characterize other major sources of errors for different space and time scales, climate, and radar systems. With the availability of high-quality dense rain gauge networks in different parts of the world, many recent studies employed the second approach of modeling the total residual error in radar-rainfall products. Irrespective of the selected approach, radar-rainfall evaluation studies in recent years improved our understanding of the error characteristics for different geographical and meteorological conditions across a range of space and time scales. The chapter also summarized recent studies that employed error models together with radar measurements to generate an ensemble of equiprobable rainfall fields. The ensemble generation is an elegant and straightforward way of propagating the uncertainty through hydrometeorologic models. However, generation of nonstationary, non-Gaussian, space-time correlated error fields is a challenging task. Another open question is how to estimate the error structure in real time for any x and t. Computational requirements should also be considered while designing generators, especially when the purpose is to create an ensemble in real time.

Acknowledgments. We would like to acknowledge all the participants of the COST-731 and IMPRINTS joint workshop on radar ensemble quantitative precipitation estimation and forecasting, which took place in Stresa, Italy, in February 2010. We also wish to thank Marco Gabella and Gabriele Villarini for their helpful comments.

REFERENCES

Aghakouchak, A., E. Habib, and A. Bardossy (2010), Modeling radar rainfall estimation uncertainties: Random error model, *J. Hydrol. Eng.*, *15*, 265–274.

Ahnert, P. R., W. F. Krajewski, and E. R. Johnson (1986), Kalman filter estimation of radar-rainfall field bias, paper presented at 23rd Conference on Radar Meteorology, Am. Meteorol. Soc., Snowmass, Colo.

Anagnostou, E. N., and W. F. Krajewski (1997), Simulation of radar reflectivity fields: Algorithm formulation and evaluation, *Water Resour. Res.*, *33*, 1419–1428.

Anagnostou, E. N., W. F. Krajewski, D.-J. Seo, and E. R. Johnson (1998), Mean-field rainfall bias studies for WSR-88D, *J. Hydrol. Eng.*, *3*, 149–159.

Anagnostou, E. N., W. F. Krajewski, and J. A. Smith (1999), Uncertainty quantification of mean-areal radar-rainfall estimates, *J. Atmos. Oceanic Technol.*, *16*, 206–215.

Austin, P. M. (1987), Relation between measured radar reflectivity and surface rainfall, *Mon. Weather Rev.*, *115*, 1053–1071.

Bell, T. L. (1987), A space-time stochastic model of rainfall for satellite remote sensing studies, *J. Geophys. Res.*, *92*, 9631–9643.

Bell, V. A., and R. J. Moore (1998), A grid-based distributed flood forecasting model for use with weather radar data. 2. Case studies, *Hydrol. Earth Syst. Sci.*, *2*, 278–283.

Bellon, A., G. Lee, and I. Zawadzki (2005), Error statistics of VPR corrections in stratiform precipitation, *J. Appl. Meteorol.*, *44*, 998–1015.

Berenguer, M., and I. Zawadzki (2008), A study of the error covariance matrix of radar rainfall estimates in stratiform rain, *Weather Forecast.*, *23*, 1085–1101.

Beven, K., and J. Freer (2001), Equifinality, data assimilation, and uncertainty estimation in mechanic modeling of complex environmental systems using the GLUE methodology, *J. Hydrol.*, *249, 11–29.*

Borga, M. (2002), Accuracy of radar-rainfall estimates for streamflow simulation, *J. Hydrol.*, *267*, 26–39.

Borga, M., and F. Tonelli (2000), Adjustment of range-dependent bias in radar rainfall estimates, *Phys. Chem. Earth, Part B*, *25*, 909–914.

Borga, M., E. N. Anagnostou, and E. Frank (2000), On the use of real-time radar rainfall estimates for flood prediction in mountainous basins, *J. Geophys. Res.*, *105*(D2), 2269–2280.

Borga, M., S. D. Esposti, and D. Norbiato (2006), Influence of errors in radar rainfall estimates on hydrological modeling prediction uncertainty, *Water Resour. Res.*, *42*, W08409, doi:10.1029/2005WR004559.

Bowler, N. E., C. E. Pierce, and A. W. Seed (2006), STEPS: A probabilistic precipitation forecasting scheme which merges an extrapolation nowcast with downscaled NWP, *Q. J. R. Meteorol. Soc.*, *132*, 2127–2155.

Carpenter, T., and K. P. Georgakakos (2004), Impact of parametric and radar rainfall uncertainties on the ensemble streamflow simulations of a distributed hydrologic model, *J. Hydrol.*, *298*, 202–221.

Chumchean, S., A. Sharma, and A. Seed (2003), Radar rainfall error variance and its impact on radar rainfall calibration, *Phys. Chem. Earth*, *28*, 27–39.

Chumchean, S., A. Seed, and A. Sharma (2006a), Correcting of real-time radar rainfall bias using a Kalman filtering approach, *J. Hydrol.*, *317*, 123–137.

Chumchean, S., A. Sharma, and A. Seed (2006b), An integrated approach to error correction for real-time radar-rainfall estimation, *J. Atmos. Oceanic Technol.*, *23*, 67–79.

Ciach, G. J., and W. F. Krajewski (1999a), On the estimation of radar rainfall error variance, *Adv. Water Resour.*, *22*, 585–595.

Ciach, G. J., and W. F. Krajewski (1999b), Radar-rain gauge comparisons under observational uncertainties, *J. Appl. Meteorol.*, *38*, 1519–1525.

Ciach, G. J., M. L. Morrissey, and W. F. Krajewski (2000), Conditional bias in radar rainfall estimation, *J. Appl. Meteorol.*, *39*, 1941–1946.

Ciach, G. J., E. Habib, and W. F. Krajewski (2003), Zero-covariance hypothesis in the Error Variance Separation method of radar rainfall verification, *Adv. Water Resour.*, *26*, 673–680.

Ciach, G. J., W. F. Krajewski, and G. Villarini (2007), Product-error-driven uncertainty model for probabilistic quantitative precipitation estimation with NEXRAD data, *J. Hydrometeorol.*, *7*, 1325–1347.

Collier, C. G. (2002), Developments in radar and remote-sensing methods for measuring and forecasting rainfall, *Philos. Trans. R. Soc. London, Ser. A*, *360*(1796), 1345–1361.

Collier, C. G., P. R. Larke, and B. R. May (1983), A weather radar correction procedure for real-time estimation of surface rainfall, *Q. J. R. Meteorol. Soc.*, *109*, 589–608.

Crane, R. K. (1990), Space-time structure of rain rate fields, *J. Geophys. Res.*, *95*(D3), 2011–2020.

Fabry, F., G. L. Austin, and D. Tees (1992), The accuracy of rainfall estimates by radar as a function of range, *Q. J. R. Meteorol. Soc.*, *118*, 435–453.

Fan, Y., E. E. Wood, M. L. Baeck, and J. A. Smith (1996), Fractional coverage of rainfall over a grid: Analyses of NEXRAD data over the southern Plains, *Water Resour. Res.*, *32*, 2787–2802.

Fulton, R. A., J. P. Breidenbach, D. J. Seo, D. A. Miller, and T. O'Bannon (1998), The WSR-88D rainfall algorithm, *Weather Forecast.*, *13*, 377–395.

Gabella, M. (2004), Improving operational measurement of precipitation using radar in mountainous terrain—Part I: Methods, *IEEE Geosci. Remote Sens. Lett.*, *1*, 78–83.

Gabella, M., M. Bollinger, U. Germann, and G. Perona (2005), Large sample evaluation of cumulative rainfall amounts in the Alps using a network of three radars, *Atmos. Res.*, *77*, 256–268.

Gabellani, S., G. Boni, L. Ferraris, J. von Hardenberg, and A. Provenzale (2007), Propagation of uncertainties from rainfall to runoff: A case study with a stochastic rainfall generator, *Adv. Water Resour.*, *30*, 2061–2071.

Gebremichael, M., W. F. Krajewski, M. L. Morrissey, D. Langerud, G. J. Huffman, and R. Adler (2003), Error uncertainty analysis of GPCP monthly rainfall products: A data based simulation study, *J. Appl. Meteorol.*, *42*, 1837–1848.

Germann, U., and J. Joss (2001), Variograms of radar reflectivity to describe the spatial continuity of Alpine precipitation, *J. Appl. Meteorol.*, *40*, 1042–1059.

Germann, U., I. Zawadzki, and B. Turner (2006a), Predictability of precipitation from continental radar images. Part IV: Limits to prediction, *J. Atmos. Sci.*, *63*, 2092–2108.

Germann, U., G. Galli, M. Boscacci, and M. Bolliger (2006b), Radar precipitation measurement in a mountainous region, *Q. J. R. Meteorol. Soc.*, *132*, 1669–1692.

Germann, U., M. Berenguer, D. Sempere-Torres, and G. Salvade (2006c), Ensemble radar precipitation estimation—A new topic on the radar horizon, in *Proceedings of Fourth European Conference on Radar in Meteorology and Hydrology (ERAD 2006)*, pp. 559–562, Barcelona, Spain.

Germann, U., M. Berenguer, D. Sempere-Torres, and M. Zappa (2009), REAL—Ensemble radar precipitation estimation for hydrology in a mountainous region, *Q. J. R. Meteorol. Soc.*, *135*, 445–456.

Haase, G., and S. Crewell (2000), Simulation of radar reflectivities using a mesoscale weather forecast model, *Water Resour. Res.*, *36*, 2221–2231.

Habib, E., and W. F. Krajewski (2002), Uncertainty analysis of the TRMM ground validation radar-rainfall products: Application to the TEFLUN-B field campaign, *J. Appl. Meteorol.*, *41*, 558–572.

Habib, E., A. V. Aduvala, and E. A. Meselhe (2008), Analysis of radar-rainfall error characteristics and implications for streamflow simulation uncertainty, *Hydrol. Sci. J.*, *53*, 568–587.

Harrison, D. L., S. J. Driscoll, and M. Kitchen (2000), Improving precipitation estimates from weather radar using quality control and correction techniques, *Meteorol. Appl.*, *6*, 135–144.

Hicks, N. S., J. A. Smith, A. J. Miller, and P. A. Nelson (2005), Catastrophic flooding from an orographic thunderstorm in the central Appalachians, *Water Resour. Res.*, *41*, W12428, doi:10.1029/2005WR004129.

Javier, J. R. N., J. A. Smith, M. L. Baeck, and G. Villarini (2010), Flash flooding in the Philadelphia metropolitan region, *J. Hydrol. Eng.*, *15*, 29–38.

Jordan, P., A. Seed, and G. Austin (2000), Sampling errors in radar estimates of rainfall, *J. Geophys. Res.*, *105*(D2), 2247–2257.

Jordan, P. W., A. W. Seed, and P. E. Wienmann (2003), A stochastic model of radar measurement errors in rainfall accumulations at catchment scale, *J. Hydrometeorol.*, *4*, 841–855.

Joss, J., and U. Germann (2000), Solutions and problems when applying qualitative and quantitative information from weather radar, *Phys. Chem. Earth, Part B*, *25*, 837–841.

Joss, J., and R. Lee (1995), The application of radar-gauge comparisons to operational precipitation profile corrections, *J. Appl. Meteorol.*, *34*, 2269–2280.

Kitchen, M., and R. M. Blackall (1992), Representativeness errors in comparisons between radar and gauge measurements of rainfall, *J. Hydrol.*, *132*, 13–33.

Kitchen, M., R. Brown, and A. G. Davies (1994), Real-time correction of weather radar data for the effects of bright band, range and orographic growth in widespread precipitation, *Q. J. R. Meteorol. Soc.*, *120*, 1231–1254.

Kirstetter, P. E., G. Delrieu, B. Boudevillain, and C. Obled (2010), Toward an error model for radar quantitative precipitation estimation in the Cévennes-Vivarais region, France, *J. Hydrol.*, in press.

Krajewski, W. F., and K. P. Georgakakos (1985), Synthesis of radar rainfall data, *Water Resour. Res.*, *21*, 764–768.

Krajewski, W. F., and J. A. Smith (2002), Radar hydrology—Rainfall estimation, *Adv. Water Resour.*, *25*, 1387–1394.

Krajewski, W. F., R. Raghavan, and V. Chandrasekhar (1993), Physically based simulation of radar rainfall data using a space-time rainfall model, *J. Appl. Meteorol.*, *32*, 268–283.

Krajewski, W. F., G. J. Ciach, J. R. McCollum, and C. Bacotiu (2000), Initial validation of the global precipitation climatology project monthly rainfall over the United States, *J. Appl. Meteorol.*, *39*, 1071–1086.

Krajewski, W. F., G. Villarini, and J. A. Smith (2010), Radar-rainfall uncertainties: Where are we after thirty years of effort?, *Bull. Am. Meteorol. Soc.*, *91*(1), 87–94.

Kumar, P., and E. Foufoula-Georgiou (1994), Characterizing multiscale variability of zero intermittency in spatial rainfall, *J. Appl. Meteorol.*, *33*, 1516–1525.

Lee, G. W., A. Seed, and I. Zawadzki (2007), Modeling the variability of drop size distributions in space and time, *J. Appl. Meteorol. Climatol.*, *46*, 742–756.

Li, P. W., and E. S. T. Lai (2004a), Applications of radar-based nowcasting techniques for mesoscale weather forecasting in Hong Kong, *Meteorol. Appl.*, *11*, 253–264.

Li, P. W., and E. S. T. Lai (2004b), Short-range quantitative precipitation forecasting in Hong Kong, *J. Hydrol.*, *288*, 189–209.

Llort, X., C. Velasco-Forero, J. Roca-Sancho, and D. Sempere-Torres (2008), Characterization of uncertainty in radar-based precipitation estimates and ensemble generation, in *Proceedings of the Fifth European Conference on Radar in Meteorology and Hydrology* [CD-ROM], Finn. Meteorol. Soc., Helsinki.

Mandapaka, P. V., W. F. Krajewski, R. Mantilla, and V. K. Gupta (2009a), Dissecting the effects of rainfall variability on the statistical structure of peak flows, *Adv. Water Resour.*, *32*, 1508–1525.

Mandapaka, P. V., W. F. Krajewski, G. J. Ciach, G. Villarini, and J. A. Smith (2009b), Estimation of radar-rainfall error spatial correlation, *Adv. Water Resour.*, *32*, 1020–1030.

Mandapaka, P. V., G. Villarini, B.-C. Seo, and W. F. Krajewski (2010), Effect of radar-rainfall uncertainties on the spatial characterization of rainfall events, *J. Geophys. Res.*, *115*, D17110, doi:10.1029/2009JD013366.

Marshall, J. S., and W. M. Palmer (1948), The distribution of raindrops with size, *J. Meteorol.*, *5*, 165–166.

Morin, E., Y. Enzel, U. Shamir, and R. Garti (2001), The characteristic time scale for basin hydrological response using radar data, *J. Hydrol.*, *252*, 85–99.

Morin, E., D. C. Goodrich, R. A. Maddox, X. G. Gao, H. V. Gupta, and S. Sorooshian (2006), Spatial patterns in thunderstorm rainfall events and their coupling with watershed hydrological response, *Adv. Water Resour.*, *29*, 843–860.

Onof, C., and H. S. Wheater (1996), Analysis of the spatial coverage of British rainfall fields, *J. Hydrol.*, *176*, 97–113.

Pegram, G. G. S., and A. N. Clothier (2001), High resolution space-time modelling of rainfall: the "String of Beads" model, *J. Hydrol.*, *241*, 26–41.

Pessoa, M. L., R. L. Bras, and E. R. Williams (1993), Use of weather radar for flood forecasting in the Sieve River basin: A sensitivity analysis, *J. Appl. Meteorol.*, *32*, 462–475.

Rosenfeld, D., D. B. Wolff, and D. Atlas (1993), General probability matched relations between radar reflectivity and rain rate, *J. Appl. Meteorol.*, *32*, 50–72.

Rouault, P., K. Schroeder, E. Pawlowsky-Reusing, and E. Reimer (2008), Consideration of online rainfall measurement and nowcasting for RTC of the combined sewage system, *Water Sci. Technol.*, *57*, 1799–1804.

Sanchez-Diezma, R., I. Zawadzki, and D. Semperre-Torres (2000), Identification of the bright band through the analysis of volumetric radar data, *J. Geophys. Res.*, *105*(D2), 2225–2236.

Schröter, K., X. Llort, C. Velasco-Forero, D. Muschalla, M. Ostrowski, and D. Sempere-Torres (2008), Accounting for uncertain radar rainfall estimates in distributed hydrological modeling, paper presented at International Symposium on Weather Radar and Hydrology, Grenoble, France

Seed, A. W., R. Srikanthan, and M. Menabde (1999), A space and time model for design storm rainfall, *J. Geophys. Res.*, *104*(D24), 31,623–31,630.

Semperre-Torres, D., C. Coral, J. Raso, and P. Malgrat (1999), Use of weather radar for combined sewer overflows monitoring and control, *J. Environ. Eng.*, *125*, 372–380.

Seo, D.-J., and J. P. Breidenbach (2002), Real-time correction of spatially nonuniform bias in radar rainfall data using rain gauge measurements, *J. Hydrometeorol.*, *3*, 93–111.

Seo, D.-J., J. P. Breidenbach, and E. R. Johnson (1999), Real-time estimation of mean field bias in radar rainfall data, *J. Hydrol.*, *223*, 131–147.

Seo, D.-J., J. P. Breidenbach, R. A. Fulton, D. A. Miller, and T. O'Bannon (2000), Real-time adjustment of range-dependent bias in WSR-88D rainfall data due to nonuniform vertical profile of reflectivity, *J. Hydrometeorol.*, *1*, 222–240.

Seo, D.-J., A. Seed, and G. Delrieu (2010), Radar and multisensor rainfall estimation for hydrologic applications, in *Rainfall: State of the Science*, *Geophys. Monogr. Ser.*, doi: 10.1029/2010GM000952, this volume.

Sharif, H. O., F. L. Ogden, W. F. Krajewski, and M. Xue (2002), Numerical simulations of radar-rainfall error propagation, *Water Resour. Res.*, *38*(8), 1140, doi:10.1029/2001WR000525.

Skolnik, M. I. (2008), *Radar Handbook*, 3rd ed., 1328 pp., Mc-Graw Hill, New York.

Smith, J. A., and W. F. Krajewski (1991), Estimation of the mean field bias of radar rainfall estimates, *J. Appl. Meteorol.*, *30*, 397–412.

Smith, J. A., and W. F. Krajewski (1993), A modeling study of rainfall rate reflectivity relationships, *Water Resour. Res.*, *29*, 2505–2514.

Smith, J. A., M. L. Baeck, and M. Steiner (1996a), Catastrophic rainfall from an upslope thunderstorm in the central Appalachians: The Rapidan storm of June 27, 1995, *Water Resour. Res.*, *32*, 3099–3113.

Smith, J. A., D.-J. Seo, M. L. Baeck, and M. D. Hudlow (1996b), An intercomparison study of NEXRAD precipitation estimates, *Water Resour. Res.*, *32*, 2035–2045.

Smith, M. B., V. I. Koren, Z. Zhang, S. M. Reed, J. J. Pan, and F. Moreda (2004), Runoff response to spatial variability in precipitation: An analysis of observed data, *J. Hydrol.*, *298*, 267–286.

Steiner, M., R. A. Houze, Jr., and S. E. Yuter (1996), Climatalogical characterization of three-dimensional storm structure from operational radar and rain gauge data, *J. Appl. Meteorol.*, *34*, 1978–2007.

Szturc, J., K. Ośródka, A. Jurczyk, and L. Jelonek (2008), Concept of dealing with uncertainty in radar-based data for hydrological purpose, *Nat. Hazards Earth Syst. Sci.*, *8*, 267–279.

Tadesse, A., and E. A. Anagnostou (2005), A statistical approach to ground radar-rainfall estimation, *J. Atmos. Oceanic Technol.*, *22*, 1720–1732.

Tessier, Y., S. Lovejoy, and D. Schertzer (1993), Universal multifractals in rain and clouds: Theory and observations, *J. Appl. Meteorol.*, *32*, 223–250.

Venugopal, V., E. Foufoula-Georgiou, and V. Sapozhnikov (1999), Evidence of dynamic scaling in space-time rainfall, *J. Geophys. Res.*, *104*(D24), 31,599–31,610.

Vignal, B., G. Galli, J. Joss, and U. Germann (2000), Three methods to determine profiles of reflectivity from volumetric radar data to correct precipitation estimates, *J. Appl. Meteorol.*, *39*, 1715–1726.

Villarini, G., and W. F. Krajewski (2008), Empirically-based modeling of spatial sampling uncertainties associated with rainfall measurements by rain gauges, *Adv. Water Resour.*, *31*, 1015–1023.

Villarini, G., and W. F. Krajewski (2009a), Empirically based modeling of uncertainties in radar rainfall estimates for a C-band radar at different time scales, *Q. J. R. Meteorol. Soc.*, *135*, 1424–1438.

Villarini, G., and W. F. Krajewski (2009b), Inference of spatial scaling properties of rainfall: Impact of radar-rainfall estimation uncertainties, *IEEE Geosci. Remote Sens. Lett.*, *6*, 812–815.

Villarini, G., and W. F. Krajewski (2010a), Review of the different sources of uncertainty in single-polarization radar-based estimates of rainfall, *Surv. Geophys.*, *31*, 107–129.

Villarini, G., and W. F. Krajewski (2010b), Sensitivity studies of the models of radar-rainfall uncertainties, *J. Appl. Meteorol. Climatol.*, *49*, 288–308.

Villarini, G., F. Serinaldi, and W. F. Krajewski (2008), Modeling radar-rainfall estimation uncertainties using parametric and non-parametric approaches, *Adv. Water Resour.*, *31*, 1674–1686.

Villarini, G., W. F. Krajewski, and J. A. Smith (2009a), New paradigm for statistical validation of satellite precipitation estimates: Application to a large sample of the TMPA 0.25° 3-hourly estimates over Oklahoma, *J. Geophys. Res.*, *114*, D12106, doi:10.1029/2008JD011475.

Villarini, G., W. F. Krajewski, G. J. Ciach, and D. L. Zimmerman (2009b), Product-error-driven generator of probable rainfall conditioned on WSR-88D precipitation estimates, *Water Resour. Res.*, *45*, W01404, doi:10.1029/2008WR006946.

Villarini, G., W. F. Krajewski, A. A. Ntelekos, K. P. Georgakakos, and J. A. Smith (2010), Towards probabilistic forecasting of flash floods: The combined effects of uncertainty in radar-rainfall and flash flood guidance, *J. Hydrol*, in press.

Wesson, S. M., and G. G. S. Pegram (2006), Improved radar rainfall estimation at ground level, *Nat. Hazards Earth Syst. Sci.*, *6*, 1–20.

Wilson, J. W., and E. A. Brandes (1979), Radar measurement of rainfall—A summary, *Bull. Am. Meteorol. Soc.*, *60*, 1048–1058.

Wojcik, R., D. McLaughlin, A. G. Konings, and D. Entekhabi (2009), Conditioning stochastic rainfall replicates on remote sensing data, *IEEE Trans. Geosci. Remote Sens.*, *47*, 2436–2449.

Wyss, J., E. R. Williams, and R. L. Bras (1990), Hydrologic modeling of New England river basins using radar rainfall data, *J. Geophys. Res.*, *95*(D3), 2143–2152.

Xue, M., K. K. Droegemeier, and V. Wong (2000), The Advanced Regional Prediction System (ARPS)—A multiscale nonhydrostatic atmospheric simulation and prediction tool, part I, Model dynamics and verification, *Meteorol. Atmos. Phys.*, *75*, 161–193.

Young, C. B., B. R. Nelson, A. A. Bradley, J. A. Smith, C. D. Peters-Lidard, A. Kruger, and M. L. Baeck (1999), An evaluation of NEXRAD precipitation estimates in complex terrain, *J. Geophys. Res.*, *104*(D16), 19,691–19,703.

Zawadzki, I. (1973), The loss of information due to finite sample volume in radar-measured reflectivity, *J. Appl. Meteorol.*, *12*, 683–687.

Zawadzki, I. (1982), The quantitative interpretation of weather radar measurements, *Atmos. Ocean*, *20*, 158–180.

Zhang, Y., T. Adams, and J. V. Bonta (2007), Sub-pixel scale rainfall variability and the effects on separation of radar and gauge rainfall errors, *J. Hydrometeorol.*, *8*, 1348–1363.

Zhang, Y., J. A. Smith, A. A. Ntelekos, M. L. Baeck, W. F. Krajewski, and F. Moshary (2009), Structure and evolution of precipitation along a cold front in the Northeastern United States, *J. Hydrometeorol.*, *10*, 1243–1256.

U. Germann and P. V. Mandapaka, MeteoSwiss, Locarno CH-6605, Switzerland. (pradeep.mandapaka@meteoswiss.ch)

Framework for Satellite Rainfall Product Evaluation

Mekonnen Gebremichael

Department of Civil and Environmental Engineering, University of Connecticut, Storrs, Connecticut, USA

This chapter focuses on three main issues: (1) standard framework for evaluating satellite rainfall products using ground reference rainfall data, (2) use of streamflow observations for evaluating satellite rainfall products through hydrological models, and (3) differences in performances of various satellite rainfall products. A standard framework for satellite rainfall product evaluation is presented. Three classes of evaluation metrics, each containing specific evaluation metrics, with varying levels of sophistication and information are identified. The most comprehensive information of the estimation error can be achieved by attaching the full distribution of estimation error to each deterministic satellite rainfall estimate. This requires a conditional error model that has the capability to generate the full distribution of estimation error for any given satellite rainfall estimate. An overview of existing and emerging error models is given. As far as the use of streamflow observations for evaluation of satellite rainfall products is concerned, the impact of different calibration strategies (rain gauge versus satellite inputs) is discussed. Comparisons of the performances of different satellite rainfall algorithms are discussed with the goal of answering questions like the following: Is it better to use satellite-gauge rainfall products or just satellite-only products? Is it better to use microwave-based products or IR-based products?

1. INTRODUCTION

All satellite rainfall estimates are unavoidably imperfect due to various kinds of errors involved in the estimation process. Information on the magnitude of the estimation error provides quantitative confidence, which is necessary to users for better decision making and to estimation algorithm developers for eventual improvement in the accuracy of the satellite rainfall estimates. The quantification of the satellite rainfall estimation error has been the subject of several studies [e.g., *Hong et al.*, 2004; *Gebremichael et al.*, 2005; *Gottschalck et al.*, 2005; *Brown*, 2006; *Ebert et al.*, 2007; *Tian et al.*, 2007; *Zeweldi and Gebremichael*, 2009a, 2009b; *Bitew and Gebremichael*, 2010; *Hirpa et al.*, 2010; *Dinku et al.*, 2010; *Sapiano et al.*, 2010]. The possible consequence of these errors in hydrologic simulations has also been explored by some [e.g., *Yilmaz et al.*, 2005; *Su et al.*, 2008; *Artan et al.*, 2007; *Tobin and Bennett*, 2009]. The studies clearly reveal that the estimation errors vary with geographical region, rain intensity, and estimation algorithm. Quantification of the errors in different regions of the world and characterization of their dependence on rain intensity is therefore critically important. Efforts are on the increase to establish the so-called "ground validation sites" in different regions of the world in order to quantify errors in satellite rainfall estimates [*Smith et al.*, 2007]. The purposes of this chapter are to recommend standard strategies for quantifying errors in satellite rainfall estimates, to provide a state-of-the-art knowledge on error models, and to highlight some of the general findings emerging from past and ongoing error studies.

Rainfall: State of the Science
Geophysical Monograph Series 191

10.1029/2010GM000974

2. ART OF EVALUATION

How can we quantify estimation errors in satellite rainfall estimates? Since the "true rainfall value" is unattainable, the typical approach is the routine comparison of satellite rainfall estimates to ground-based reference rainfall estimates that have a much smaller estimation error.

2.1. Basic Definitions

The definitions that are used throughout this chapter are shown in Table 1.

Table 1. Basic Definitions

	Definition
True rainfall value	amount of rain water that has fallen on the ground over a specified area and time period
Satellite rainfall estimate (S)	estimate of the true rainfall based on satellite data corresponding to the same area and period
Ground reference rainfall estimate (R)	estimate of the true rainfall based on data from a network of rain gauges or ground-based radar corresponding to the same area and period, where R has a much lower estimation error than S
Satellite rainfall estimation error (ε)	difference between S and R
Evaluation of satellite rainfall	quantification of ε

2.2. Ground Reference Rainfall Estimates

Ground reference rainfall data must have much smaller (by an order of magnitude or so) estimation errors to be able to evaluate satellite rainfall estimates. The main sources of ground reference rainfall data are rain gauge observations and radar rainfall estimates. Each has its own sources of errors, and the resulting error varies depending on several factors. The adequacy of each must be assessed prior to its usage in satellite rainfall estimate evaluation.

2.2.1. Rain gauge observations. Rain gauges provide direct, therefore accurate, measurements of rainfall but at a point scale. Extrapolation of point measurements to areal values, commensurate with the spatial resolution of satellite rainfall products, introduces estimation errors. The basic relation between the point rain gauge measurements $G(x)$ and the rainfall averaged G_A over area A can be written as

$$G_A = \frac{1}{A}\int_A G(x)\mathrm{d}x^2,$$

where x is the location vector. G_A is generally estimated as a linear combination of the rain gauge measurements:

$$G_A = \sum_{(i=1)}^{N} C_i G_i,$$

where $\{C_i\}$ is the set of weights associated with each layout. The standard deviation of the point-to-area extrapolation error can be expressed as [e.g., *Ciach and Krajewski*, 1999]:

$$\sigma_g^2\left(1 - \frac{2}{A}\int_A \rho(x_g, x)\mathrm{d}x^2 + \frac{1}{A^2}\int_A\int_A \rho(x,y)\mathrm{d}x^2\mathrm{d}y^2\right),$$

where x_g is the rain gauge location within the satellite grid, and $\rho(x,y)$ is the spatial correlation function. Therefore, the magnitude of the point-to-area extrapolation error depends on the number and layout of rain gauges and the spatial variability of rainfall within the area. The point-to-area extrapolation error is the major source of error in rain gauge-based ground rainfall estimates, and this error needs to be sufficiently small to use the data for evaluating satellite rainfall products.

2.2.2. Radar rainfall estimates. Radar rainfall estimates have the advantage of providing gridded rainfall estimates at high space-time resolution. The disadvantage is that they themselves are indirect estimates of rainfall and are prone to a number of error sources (detailed information on the error sources is given in reviews of *Krajewski and Smith* [2002], *Krajewski et al.* [2010], *Villarini and Krajewski* [2010]). Careful quality control of the data and bias correction using rain gauge observations can correct much of the errors in radar rainfall estimates.

3. EVALUATION METRICS

As discussed above, the estimation errors in satellite rainfall estimates can be approximated as

$$\varepsilon = S - R.$$

This results in distributions of ε, which needs to be communicated in a meaningful manner. A number of evaluation metrics exist on different aspects of ε with varying levels of sophistication. In this section, we provide an overview of the evaluation metrics that we recommend for the satellite rainfall users community. The evaluation metrics can be grouped into the following three classes: categorical metrics, unconditional metrics, and conditional metrics.

3.1. Categorical Metrics

Categorical metrics are useful to evaluate binary cases. They are typically used to evaluate the ability of the satellite rainfall estimates to accurately report the presence or absence of rainfall. The 2×2 contingency table (see Table 2) will be referenced in the definition of the categorical metrics. The entries represent the number of cases that fall in each class.

The key categorical metrics are [*Wilks*, 1995] probability of detection, false alarm rate, and Heidke Skill Score.

3.1.1. Probability of detection (POD). The POD measures the fraction of the number of actually "rainy" events that were correctly reported as "rainy events" by the satellite rainfall estimates. POD varies from 0 to 1.

$$\mathrm{POD} = \frac{a}{a+c}.$$

3.1.2. False alarm rate (FAR). The FAR measures the fraction of the number of actually "nonrainy" events that were incorrectly reported as "rainy events" by the satellite estimates. FAR varies from 0 to 1.

$$\mathrm{FAR} = \frac{b}{a+b}.$$

3.1.3. Heidke Skill Score (HSS). The HSS measures the fraction of the total number of cases accurately reproduced by the satellite estimates as "rainy' or "nonrainy" events relative to the fraction of correct random estimations. The HSS varies from -1 (perfect negative skill) to 1 (perfect skill), with 0 representing no skill relative to chance.

$$\mathrm{HSS} = \frac{2(ad - bc)}{(a+c)(c+d)+(a+b)(b+d)}.$$

3.2. Unconditional Metrics

Unconditional metrics measure how the satellite rainfall estimates agree with the reference rainfall on average. These metrics provide single values that can be used to compare different satellite rainfall products and to get an overall performance of the satellite products. These metrics ignore the dependence of the satellite rainfall estimation accuracy on the magnitude of the rainfall estimate and are therefore not appropriate to understand the estimation uncertainty associated with each rainfall estimate.

The key unconditional metrics are basic descriptive statistics, bias, correlation, root-mean-square error, and mean absolute error.

Table 2. Contingency Table Referenced in the Definition of the Categorical Metrics[a]

	Reference Rainfall (r)		
Satellite Estimate (s)	Yes	No	Outcome
Yes	a	b	$a+b$
No	c	d	$c+d$
Outcome	$a+c$	$b+d$	$n = a+b+c+d$

[a]The entries represent the number of cases that fall in each class.

3.2.1. Basic descriptive statistics. Basic descriptive statistics of satellite rainfall estimate and the reference rainfall, such as, mean or median (which measures central tendency), variance or standard deviation (which measures spread) provide an initial "feel" of the two data sets being compared.

3.2.2. Bias (also known as overall bias or unconditional bias). Bias measures the agreement between the means of the satellite rainfall estimate and the reference rainfall. Bias can be expressed in two forms:

$$\text{Additive Bias }(s,r) = E[s] - E[r]$$

$$\text{Multiplicative Bias }(s,r) = \frac{E[s]}{E[r]},$$

where $E[\ ]$ is the expectation operator. The ideal additive bias is 0, and the ideal multiplicative bias is 1. Care must be taken when using the multiplicative bias, as it could lead to misinterpretation for very small magnitudes of reference rainfall.

3.2.3. Correlation ρ. Here ρ measures the strength of the linear relationship between the satellite estimates and the reference rainfall:

$$\rho(s,r) = \frac{\mathrm{Cov}[s,r]}{\sqrt{V[s]}.\sqrt{V[r]}},$$

where $V[\ ]$ and $\mathrm{Cov}[\ ]$ are the variance and covariance operators, respectively; ρ varies from -1 (perfectly negative correlation) to 1 (perfectly positive correlation), with 0 indicating no correlation; ρ is not affected by bias, and so it provides additional independent information. Evaluation statistics that involve second-moment orders, such as correlation, place undue weight on the outliers or extreme events, and therefore, care must be taken when calculating correlation. Correlation is very much dependent upon the characteristics of the rainfall produced by both the estimate and reference data sets; for example, estimates that overestimate rain area have different correlations than those that underestimate rain area. This can be attributed to the zero-rain and magnitude of the values in the calculations; this, in turn, is dependent upon the spatial/temporal resolution. The coefficient of determination ρ^2 measures the variance of reference rainfall explained by the satellite estimate.

3.2.4. Root-mean-square error. Root-mean-square error (RMSE) measures the average degree of agreement between individual satellite estimates and reference rainfall:

$$\mathrm{RMSE}(s,r) = \sqrt{E[s-r]^2}.$$

RMSE can be decomposed as

$$\mathrm{RMSE}(s,r) = \sqrt{(\text{additive bias }(s,r))^2 + V[s] + V[r] - 2\sqrt{V[s]}.\sqrt{V[r]}\rho(s,r)}.$$

Note that RMSE is affected by both bias and correlation. Some report bias-adjusted RMSE, where bias is first removed prior to the calculation of RMSE. The RMSE values are sensitive to outliers as they involve second-moment orders.

3.2.5. Mean absolute error. Like RMSE, mean absolute error (MAE) measures the average degree of agreement between individual satellite estimates and reference rainfall:

$$\mathrm{MAE}(s,r) = E|s-r|,$$

where | | is the absolute value operator. Taking the absolute value of the error term rather than its square removes the bias toward outliers; therefore, MAE is less sensitive to extreme values than RMSE. On the other hand, RMSE is generally amenable to more in-depth mathematical or statistical analyses than MAE.

3.3. Conditional Metrics

Conditional metrics refer to those that vary with each satellite rainfall estimate. Since the estimation errors depend on the magnitude of the rainfall [e.g., see *Gebremichael et al.*, 2005], these metrics are more meaningful and useful to users for decision making. We recommend the following conditional metrics: a number of quantile values or error bounds, such as (5%, 95%), (15%, 85%), (25%, 75%), (35%, 65%), for each satellite rainfall estimate.

Such multiple quantiles or error bounds allow users to directly use the estimation uncertainty associated with each satellite rainfall estimate in their applications.

4. ERROR MODELS

High-resolution satellite rainfall estimates provide alternative source of rainfall data for hydrological applications especially in regions where adequate ground-based instruments are unavailable. However, errors in rainfall estimates compromise the reliability of operational applications and reduce our ability to identify other sources of error thereby slowing down scientific advancement [*Kuczera et al.*, 2006]. Therefore, quantitative information on the full distribution of these errors is fundamental to success in satellite rainfall-runoff modeling. This information enables users to take appropriate actions on risk and water management options according to their own risk assessment and cost/benefit analysis. Despite the importance of estimation errors, operational satellite rainfall products are still deterministic and lack any estimate of their uncertainty. What is needed is a paradigm shift toward converting deterministic satellite rainfall estimates to probabilistic form by overlaying an estimated error distribution around the deterministic rainfall estimate. This requires the use of appropriate error model, which is currently missing for high-resolution satellite rainfall products. Only little work exists on satellite rainfall error models, and most of them are not suitable for high-resolution products. Here we provide an overview of existing and emerging error models, focusing only on the conditional distribution of S given $R = r$.

4.1. Power Law Model of Gebremichael and Krajewski [2004] and Steiner et al. [2003]

Gebremichael and Krajewski [2004] and *Steiner et al.* [2003] developed, using different approaches, the following power law model for temporal sampling errors in satellite rainfall estimates:

$$\frac{\sigma_{es}}{\bar{R}} = a\left(\frac{1}{L}\right)^b\left(\frac{\Delta t}{T}\right)^c \bar{R}^d,$$

where σ_{es} is the standard deviation of sampling error, $\bar{R}$ is large-scale averaged reference rainfall, L is the spatial scale, T is the temporal scale, $\Delta t = \frac{T}{N}$ is the satellite sampling frequency assuming that the satellite makes N visits over area A during the period of T, and a, b, c, and d are regression parameters. The above model was derived from simulated data sets assuming perfect instrument, perfect algorithm, complete coverage of area by satellite, and fixed sampling intervals. These assumptions are, however, far from reality.

Hong et al. [2004] adapted the above formula to total estimation (sampling and algorithm) error as

$$\sigma_e = a\left(\frac{1}{L}\right)^b \left(\frac{\Delta t}{T}\right)^c S^d,$$

where σ_e is the standard deviation of estimation error. Use of σ_e to generate ensemble of realizations of observed rainfall given satellite rainfall estimates as well as interpretation of σ_e require knowledge about the probability distribution function of the estimation error. Assuming that errors in PERSIANN-CCS [*Hong et al.*, 2004] satellite rainfall estimates (spatial resolution: 0.04° × 0.04°, temporal resolution: 1 h) are log-normally distributed, *Moradkhani et al.* [2006] and *Moradkhani and Meskele* [2010] assessed the impact of these errors on lumped streamflow simulation accuracy. However, it was shown by *Gebremichael and Krajewski* [2004] that it is unrealistic to assume log-normal distribution for errors especially at such high spatial and temporal resolutions.

4.2. Semiparametric Conditional Model of Yan and Gebremichael [2009]

Yan and Gebremichael [2009] developed a semiparametric conditional model $f(r|s)$ that generates possible realizations of observed rainfall values given satellite rainfall estimates. The model, developed for coarse resolution (monthly) satellite rainfall estimates, consists of two components: a conditional gamma density given each S and a smooth functional relationship. The gamma conditional model is

$$f(r|s;\alpha_s,\beta_s) = \frac{1}{\Gamma(\alpha_s)\beta_s^{\alpha_s}} s^{\alpha_s-1} e^{-s/\beta_s},$$

where the parameters α_s (shape parameter) and β_s (scale parameter) are smooth functions of s over S. The two parameters determine the conditional mean μ_s and variance σ_s^2 of the gamma density: $\mu_s = \alpha_s\beta_s$ and $\sigma_s^2 = \alpha_s\beta_s^2$. For ease of interpretation and usage of existing *R* software package, the density is reparameterized in terms of location parameter μ_s and a dispersion parameter $ø_s$ through

$$\alpha_s = \frac{1}{ø_s^2} \quad \text{and} \quad \beta_s = \mu_s ø_s^2$$

such that the mean of the distribution is μ_s and the variance is $ø_s^2\mu_s^2$. This parameterization is often used in generalized linear models in which the variance is a function of the mean. The parameters are modeled by smooth functions of s:

$$\mu_s = B_1^T(s)\theta,$$

$$ø_s = B_2^T(s)\omega,$$

where $B_i(s)$, $i = 1,2$ are predefined cubic spline basis evaluated at s, and θ and ω are basis coefficients to be estimated. B_1 and B_2 are allowed to be different such that different complexities of μ_s and $ø_s$ are possible. The spline bases are defined with fixed knots at quantiles of the observed S data. The degree of freedom of a spline basis is the number of basis which determines the flexibility of the model.

4.3. Nonparametric Conditional Model

We found that the above semiparametric model is not suitable for high-resolution satellite rainfall products. In an ongoing work, we have developed the following new nonparametric model for satellite rainfall estimation error using CMORPH [*Joyce et al.*, 2004] satellite rainfall estimates (0.25°×0.25°, 3-hourly) and rain gauge-adjusted ground-based radar rainfall values as reference rainfall. Because rainfall can be zero with strictly positive probability, we model the conditional distribution as the mixture of a positive continuous distribution and a point mass at 0, for any given value of $S = s$. In particular, our conditional distribution of R given $S = s$ is

$$R = \begin{cases} 0 & \text{with probability} \quad p_s, \\ z & \text{with probability} \quad 1-p_s, \end{cases}$$

where Z is a positive continuous variable with density $f(r|s)$. The conditional distribution is characterized by $p_s = P(R = r|S = s)$ and density function $f(r|s)$. Suppose the observed data are (R_i,S_i), $i = 1,\ldots,n$. Depending on $s = 0$ or not, we have different estimation approaches for p_s and $f(r|s)$.

4.3.1. Case 1: $s = 0$. In this case, p_0 is simply the success rate of a Bernoulli variable and is estimated as

$$\hat{p}_0 = \frac{\sum_{i=1}^n I(S_i = 0, R_i = 0)}{\sum_{i=1}^n I(S_i = 0)},$$

where $I(.)$ is the indicator function.

To estimate $f(r|0)$, we use a univariate kernel density based on only those observations with $S_i = 0$ and $R_i > 0$. In particular,

$$\hat{f}_h(r|0) = \frac{\sum_{i=1}^n I(S_i = 0) K\left(\frac{r - R_i}{h}\right)}{h\sum_{i=1}^n I(S_i = 0)},$$

where K is some kernel, and h is the bandwidth that controls the smoothness of the kernel density estimate. A commonly used kernel is the density of $N(0,1)$.

4.3.2. Case 2: $s > 0$. When $s > 0$, we model p_s as a smooth function of s. For example, one may use basis splines $B_1, \ldots, B_q$ to model p_s:

$$p_s = \sum_{j=1}^{q} \theta_i B_i(s),$$

where $(\theta_1,..,\theta_q)$ are coefficients of basis. Define binary variable $Z_i = I(R_i = 0)$. These coefficients can be estimated by fitting a logistic model for Z_i based on a subset of data $\{(Z_i,S_i): S_i > 0\}$.

Density $f(r|s)$ is estimated with data $\{(R_i,S_i):S_i > 0, R_i > 0\}$. We use the conditional density estimation of *Hyndman et al.* [1996], with implementation available in R package hdrcde. This approach estimates $f(r|s)$ as

$$\hat{f}(r|s) = \frac{\hat{g}(s,r)}{\hat{h}(s)},$$

where $\hat{g}(s,r)$ is a bivariate kernel density of the joint distribution of (S,R) for $S > 0$ and $R > 0$,

$$\hat{g}(s,r) = \frac{\sum_{i=0}^{n} I(S_i > 0, R_i > 0) K\left(\frac{s - S_i}{a}\right) K\left(\frac{r - R_i}{a}\right)}{ab\sum_{i=1}^{n} I(S_i > 0, R_i > 0)},$$

$\hat{h}(s)$ is a univariate kernel density of $S > 0$,

$$\hat{h}(s) = \frac{\sum_{i=1}^{n} I(S_i > 0) K\left(\frac{s - S_i}{b}\right)}{b\sum_{i=1}^{n} I(S_i > 0)},$$

and a and b are bandwidth parameters. In practice, this estimation is done for a grid of s values that covers the observed range of S.

4.3.3. Sampling from model. It is of practical interest to sample from the estimated conditional distribution. For a given CMORPH satellite rainfall estimate s, the following algorithms summarize how to sample for possible realizations of the reference rainfall R given $S = s$.

1. Compute $\hat{p}_s$.
2. Generate u from the uniform distribution $U(0, 1)$.
3. If $u < \hat{p}_s$, $R = 0$; else generate R from $\hat{f}(r|s)$.

Because the conditional density $f(r|s)$ is nonparametric, sampling is not as straightforward as for parametric density. The estimate $\hat{f}(r|s)$ may not even integrate to 1. This is a problem of sampling from a density with unknown normalizing constant. As long as the shape of the estimated density is available, one can use the general purpose adaptive rejection Metropolis sampling [*Gilks et al.*, 1995]. Implementation of the algorithm is available in R package HI [*Petris and Tardella*, 2006]. Random numbers generated from this approach are, in general, autocorrelated due to the Metropolis step. Nevertheless, the sample is still a valid sample. Further, one can always thin the sample to reduce autocorrelation.

5. HYDROLOGICAL EVALUATION

In gauged watersheds where ground-based measurements of rainfall are not available at commensurate scales, satellite rainfall products can be evaluated in terms of their ability to serve as input in hydrological modeling for streamflow simulation. The procedure consists of inputting satellite rainfall estimates in hydrologic models and then comparing the model simulations to the observations. The difference between the simulations and the observations gives total errors resulting from a combination of satellite rainfall estimates, hydrologic model, and model parameter estimates. Therefore, care must be taken when interpreting the results and selecting the models and fixing parameter estimates. One issue that arises in this context is whether to use rain gauge calibration parameter values or satellite rainfall calibration parameter values. Studies have indicated that the performance of satellite-based model simulations increases when the model is calibrated with satellite data than with rain gauge data [*Yilmaz et al.*, 2005; *Artan et al.*, 2007].

In our ongoing work, we have evaluated the ability of four widely known high-resolution satellite rainfall products, namely, CMORPH [*Joyce et al.*, 2004], TMPA 3B42 (3B42 for short) [*Huffman et al.*, 2007], TMPA 3B42RT (3B42RT for short) [*Huffman et al.*, 2007], and PERSIANN [*Sorooshian et al.*, 2000] to serve as input into the physics-based semidistributed hydrologic model Swat and Water Assessment Tool (SWAT) [*Arnold et al.*, 1998] for streamflow simulation in the Gilgel Abay watershed (area: 1656 km^2; location: 36°48′E–37°24′E, 10°56′N–11°23′N) in the highlands of Ethiopia. Plate 1 compares the performance of the satellite-based SWAT streamflow simulations when the model is calibrated with rain gauge data and when the model is calibrated with corresponding satellite data. Both the Nash-Sutcliffe efficiency and percent bias (Bias, %) performance statistics improve when the model is calibrated with satellite data than with rain gauge data, for all satellite rainfall products considered.

Let us now compare the SWAT model parameter values obtained from the various rainfall inputs focusing on the most sensitive parameter that affects the overland flow: curve number (CN). CN determines the fraction of rainfall that becomes direct runoff. The calibrated watershed-average CN value for Gilgel Abay is 50 (with rain gauge input data), 57 (with CMORPH or 3B42RT), 65 (with PERSIANN), and 67 (with 3B42). This sequence of increasing CN is in agreement with the sequence of increasing the degree of rainfall underestimation by the products (results not shown here). In other words, the satellite rainfall products that have the largest negative bias in rainfall result in higher CN values. Therefore,

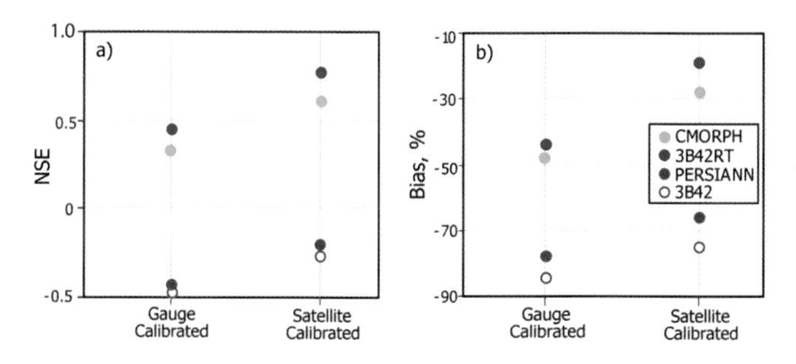

Plate 1. Evaluation of satellite-based SWAT streamflow simulation performance for Gilgel Abay watershed in Ethiopia for various sources of satellite rainfall inputs (CMORPH, TMPA 3B42RT, PERSIANN, and TMPA 3B42) and for two calibration strategies (model calibrated with rain gauge data and model calibrated with satellite data) using statistics (a) Nash-Sutcliffe efficiency (NSE) and (b) bias.

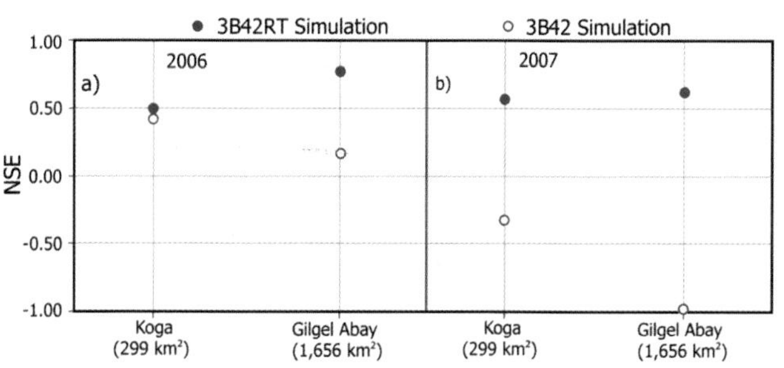

Plate 2. Comparison of the performance statistics (NSE) of TMPA 3B42 and TMPA 3B42RT SWAT daily streamflow simulations in year (left) 2006 and (right) 2007 for Koga (299 km^2) and Gilgel Abay (1656 km^2) watersheds in Ethiopia where the rain gauge network is sparse.

improvement in the satellite streamflow simulations, when the model is calibrated with satellite data than with rain gauge data, is mainly due to adjustments on some of the sensitive parameters that compensate for some of the errors in satellite rainfall estimates. This means that improvements in streamflow simulations come at the expense of estimation accuracy of other water balance components such as groundwater. So, while the use of model calibrated with satellite data rather than rain gauge data on the basis of streamflow improves the satellite-based streamflow simulation accuracy, it deteriorates the satellite-based simulation accuracy of other water balance components such as groundwater.

6. SPECIAL ISSUES: WHICH SATELLITE RAINFALL PRODUCTS ARE BETTER?

The growing availability of high-resolution satellite rainfall products is making them an alternative source of rainfall data for rainfall-runoff modeling, especially in regions where ground-based rainfall measuring instruments are lacking. The concept behind the high-resolution satellite rainfall algorithms is to combine information from the more accurate (but infrequent) microwave (MW) with the more frequent (but indirect) IR to take advantage of the complementary strengths. The combination has been done in a variety of ways. The

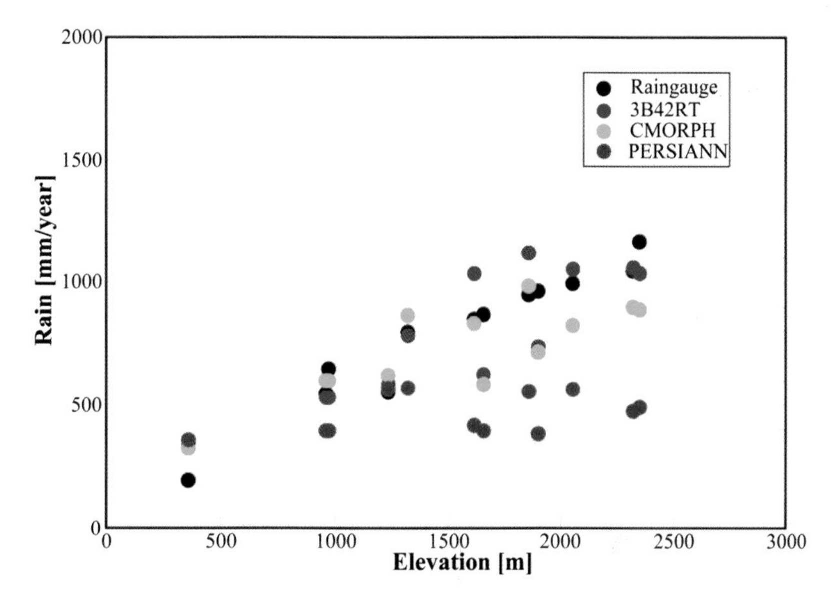

Plate 3. Five-year (2003 through 2007) mean annual rainfall derived from satellite products and rain gauge measurements as a function of elevation only at 0.25° satellite pixels that contain rain gauges.

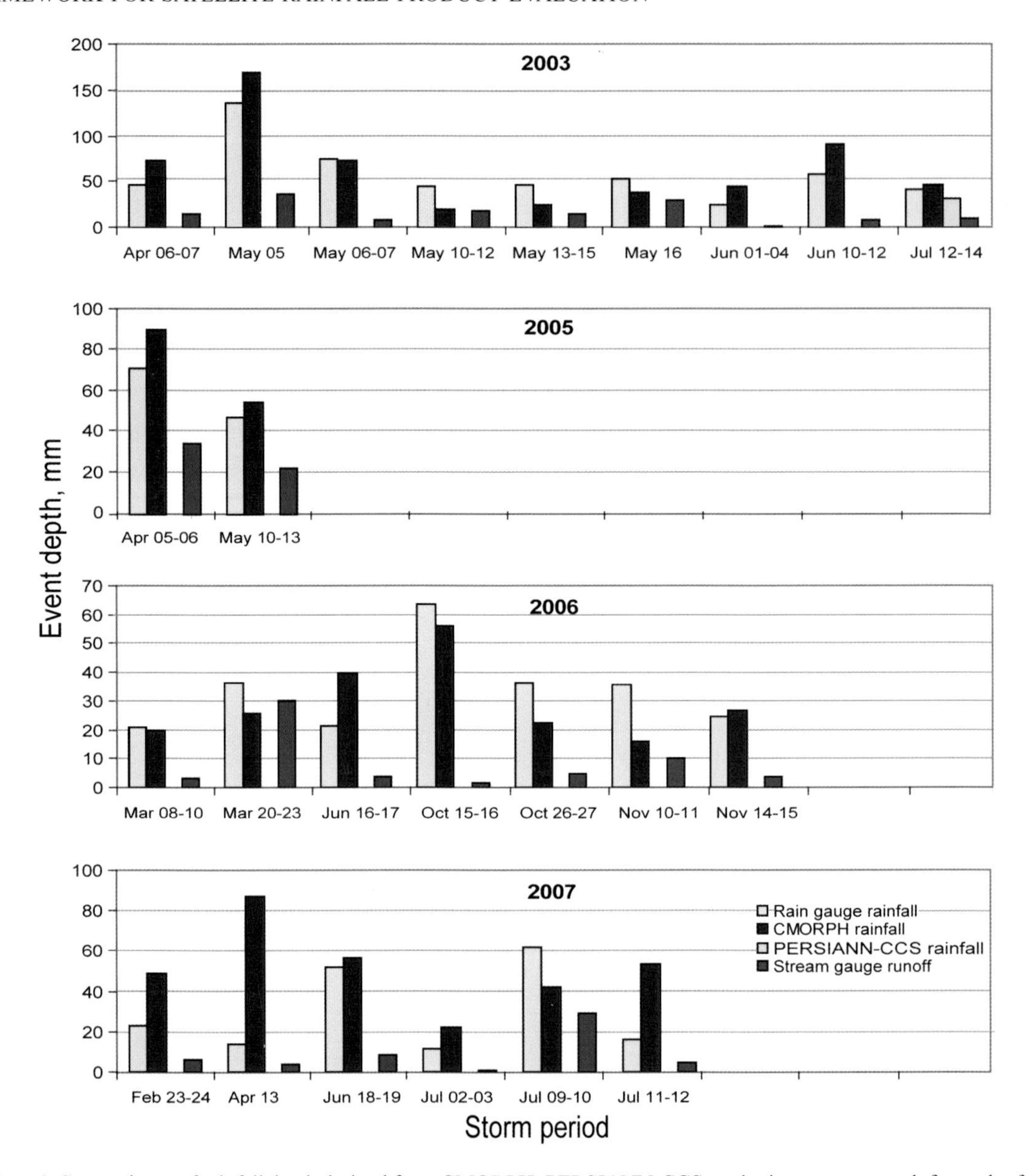

Plate 4. Comparisons of rainfall depth derived from CMORPH, PERSIANN-CCS, and rain gauge network for each of the 24 large storm events in Walnut Gulch watershed. Also shown are runoff depths from a stream gauge located at the outlet of the watershed.

Tropical Rainfall Measuring Mission (TRMM) Multi-Satellite Precipitation Analysis (TMPA) method [*Huffman et al.*, 2007] uses MW data to calibrate the IR-derived estimates and creates estimates that contain MW-derived rainfall estimates when and where MW data are available and the calibrated IR estimates where MW data are not available. The CMORPH method [*Joyce et al.*, 2004] obtains the rainfall estimates from MW data but uses a tracking approach in which IR data are used only to derive a cloud motion field that is subsequently used to propagate raining pixels. The PERSIANN method [*Sorooshian et al.*, 2000] uses a neural network approach to derive relationships between IR and MW data, which are applied to the IR data to generate rainfall estimates. So while both CMORPH and TMPA products rely primarily on MW data for rainfall estimates, PERSIANN relies primarily on IR data. The resolutions of these products are 0.25° and 3 hourly, although finer resolutions are also available for CMORPH and PERSIANN. CMORPH and PERSIANN products are available only post-real time. The TMPA products are available in two versions: real-time version (3B42RT) and post-real-time research version (3B42). The main difference between the two versions is the use of monthly rain gauge data for bias adjustment in the post-real-time research product. The 3B42 products are released 10–15 days after the end of each month, and the 3B42RT are released about 9 h after overpass. Given a variety of satellite rainfall products, an important question users often face is which product(s) to use. Below, we will focus on only two aspects of the products.

6.1. Satellite-Gauge Versus Satellite-Only Products

The conventional notion that the combination of rain gauge data with satellite rainfall estimates performs better than satellite-only estimates has led to the incorporation of rain gauge data into global satellite rainfall products. Recently, *Habib et al.* [2009] compared the performances of 3B42 (that contains rain gauge data) and 3B42RT (that does not contain rain gauge data) rainfall estimates during six heavy rainfall events in Louisiana, United States, by using rainfall data from a dense rain gauge network and weather radar as a ground reference. They reported that the 3B42 product has better performance than the 3B42RT. Nonetheless, there is a concern that the improved accuracy of 3B42 over 3B42RT is attributable to the reasonably dense rain gauge network of the United States and that similar level of improvement may not be achieved in many parts of the world where dense rain gauge networks do not exist.

In an ongoing work, we have used 3B42 and 3B42RT rainfall inputs, separately, as input into the SWAT hydrological model for two adjoining watersheds with drainage areas of 299 km^2 (Koga) and 1656 km^2 (Gilgel Abay) in the highlands of Ethiopia where the rain gauge network is sparse, and compared the daily streamflow simulations to the observations. Results turn the conventional notion on its head: the satellite-only TMPA 3B42RT products are found to be much better than the satellite-gauge TMPA 3B42 products in terms of their ability in reproducing daily streamflow (Plate 2). Incorporating rain gauge data in satellite rainfall products has the undesirable consequence of deteriorating the quality of the satellite rainfall products in this region. Apparently, the use of rain gauge information from sparsely distributed network is introducing additional error in the satellite rainfall products. This conclusion may or may not hold in other regions, but it certainly shows that the conventional notion is not necessarily valid everywhere.

6.2. MW- Versus IR-Based Products

Satellite rainfall algorithms could be grouped into two categories: those that use primarily MW data (e.g., CMORPH) and those that use primarily IR data (e.g., PERSIANN). There are also other algorithms in between, such as TMPA 3B42RT, which are MW-based when MW data are available, and IR-based when MW data are not available. The conventional notion is that MW-based algorithms have better performance than the IR-based. *Bitew and Gebremichael* [2010] deployed a dense network of rain gauges over a 25-km^2 region in Ethiopia and compared the performances of CMORPH to those of PERSIANN-CCS (a variant of PERSIANN but at finer, 0.04°, resolution) [*Hong et al.*, 2004]. They reported better performance for CMORPH over PERSIANN-CCS. *Hirpa et al.* [2010] compared the performances of PERSIANN, CMORPH, and 3B42RT over a 110,000-km^2 mountainous watershed in Ethiopia. Plate 3, adapted from *Hirpa et al.* [2010], presents a 5-year mean annual rainfall derived from rain gauge measurements and satellite rainfall estimates as a function of elevation, only at pixels that contain rain gauges. It is clearly shown that PERSIANN significantly underestimates rainfall in the high-elevation areas (1400 m above sea level to 2400 m). *Hong et al.* [2004] also reported the setback of PERSIANN in missing rain events in the mountainous region of Mexico.

In an ongoing work, we have evaluated the performances of CMORPH and PERSIANN-CCS in the Walnut Gulch experimental watershed using dense rain gauge rainfall observations and streamflow observations. The watershed has the following properties: drainage area of 21.4 km^2, location in Mississippi, elevation ranging from 71 to 127 m, and climate characterized by long and hot humid summers, with mean annual precipitation of 1450 mm. We selected 24 largest storm events (i.e., events with peak flow rates greater than 0.5 m^3 s^{-1}) during a 4-year period (2003, 2005–2007; we excluded 2004 due to stream gauge malfunctioning) and compared event depths of rainfall and streamflow (Plate 4). While CMORPH saw all these 24 storms, PERSIANN-CCS missed all of them except one. Based on these discussions, we conclude that PERSIANN and its variants (i.e., PERSIANN-CCS) are unreliable sources of rainfall data. Furthermore, this indicates that the IR-based rainfall algorithms deserve additional scrutiny.

7. CONCLUSIONS

This chapter has presented a standard framework for the purpose of evaluating satellite rainfall products. Such a standard framework would facilitate comparison of evaluations carried out by various researchers and ensure that the users' needs are met. Three classes of evaluation metrics, each containing specific evaluation metrics, with varying levels of sophistication and information have been identified as measures of satellite rainfall estimation error that meet the needs of all users. The most comprehensive information of the estimation error can be achieved by attaching the full distribution of estimation error to each deterministic satellite rainfall estimate. This requires a conditional error model that has the capability to generate the full distribution of estimation error for any given satellite rainfall estimate. Although such error model is currently lacking for satellite rainfall products, we have provided an emerging model that may have the potential to overcome the problem.

The chapter also discusses issues associated with the use of streamflow observations to evaluate satellite rainfall products through hydrological models. Significant improvement in satellite-based streamflow simulation is obtained when the model is calibrated with satellite data than with rain gauge data. The improvement comes at the expense of estimation accuracy of other water balance components such as the groundwater. Therefore, caution must be exercised when using satellite-based simulations of water balance components such as groundwater, when the model is calibrated with satellite data on the basis of observed streamflow.

Finally, the chapter highlights some special issues emerging from past and current satellite rainfall evaluation studies. As far as the comparison between satellite-gauge and satellite-only rainfall products is concerned, there is no clear winner. However, the conventional notion that says the satellite-gauge products are better than the satellite-only products is not necessarily true, especially in rain gauge sparse regions. As far as the comparison between MW- and IR-based rainfall algorithms is concerned, the MW-based algorithms have better performance. In particular, the IR-based PERSIANN (or its variants) satellite rainfall products cannot be taken as reliable sources of rainfall data.

Acknowledgments. Support for this study came from two NASA grants (Award NNX08AR31G and NNX10AG77G) to the University of Connecticut. This work has benefited from the contributions of my graduate students Menberu Bitew, Feyera Hirpa, and Dawit Zeweldi.

REFERENCES

Arnold, J. G., R. Srinivasan, R. S. Muttiah, and P. M. Allen (1998), Large-area hydrologic modeling and assessment: Part I. Model development, *J. Am. Water Resour. Assoc.*, *34*(1), 73–89, doi:10.1111/j.1752-1688.1998.tb05961.x.

Artan, G., H. Gadain, J. Smith, K. Asante, C. Bandaragoda, and J. Verdin (2007), Adequacy of satellite-derived rainfall data for streamflow modeling, *Nat. Hazards*, *43*, 167–185, doi:10.1007/s11069-007-9121-6.

Bitew, M. M., and M. Gebremichael (2010), Evaluation through independent measurements: Complex terrain and humid tropical region in Ethiopia, in *Satellite Rainfall Applications for Surface Hydrology*, edited by M. Gebremichael, and F. Hossain, pp. 205–214, doi:10.1007/978-90-481-2915-7_12, Springer, Netherlands.

Brown, J. E. M. (2006), An analysis of the performance of hybrid infrared and microwave satellite precipitation algorithms over India and adjacent regions, *Remote Sens. Environ.*, *101*, 63–81.

Ciach, G. J., and W. F. Krajewski (1999), On the estimation of radar rainfall error variance, *Adv. Water Resour.*, *22*(6), 585–595.

Dinku, T., S. J. Connor, and P. Ceccato (2010), Comparison of CMORPH and TRMM-3B42 over mountainous regions of Africa and South America, in *Satellite Rainfall Applications for Surface Hydrology*, edited by Dinku, T., and F. Hossain, pp. 193–204, doi:10.1007/978-90-481-2915-7_11, Springer, Netherlands.

Ebert, E. E., J. E. Janowiak, and C. Kidd (2007), Comparison of near-real-time precipitation estimates from satellite observations and numerical models, *Bull. Am. Meteorol. Soc.*, *88*(1), 47–64.

Gebremichael, M., and W. F. Krajewski (2004), Characterization of the temporal sampling error in space-time-averaged rainfall estimates from satellites, *J. Geophys. Res.*, *109*, D11110, doi:10.1029/2004JD004509.

Gebremichael, M., W. F. Krajewski, M. Morrissey, G. Huffman, and R. Adler (2005), A detailed evaluation of GPCP one-degree daily rainfall estimates over the Mississippi River Basin, *J. Appl. Meteorol.*, *44*(5), 665–681.

Gilks, W. R., N. G. Best, and K. K. C. Tan (1995), Adaptive rejection Metropolis sampling, *Appl. Stat.*, *44*, 455–472.

Gottschalck, J., J. Meng, M. Rodell, and P. Houser (2005), Analysis of multiple precipitation products and preliminary assessment of their impact on global land data assimilation system land surface states, *J. Hydrometeorol.*, *6*, 573–598.

Habib, E., A. Henschke, and R. F. Adler (2009), Evaluation of TMPA satellite-based research and real-time rainfall estimates during six tropical-related heavy rainfall events over Louisiana, USA, *Atmos. Res.*, *94*(3), 373–388, doi:10.1016/j.atmosres.2009.06.015.

Hirpa, F. A., M. Gebremichael, and T. Hopson (2010), Evaluation of high resolution satellite precipitation products over very complex terrain in Ethiopia, *J. Appl. Meteorol. Climatol.*, *49*, 1044–1051, doi:10.1175/2009JAMC2298.1.

Hong, Y., K. L. Hsu, S. Sorooshian, and X. Gao (2004), Precipitation estimation from remotely sensed imagery using an artificial neural network cloud classification system, *J. Appl. Meteorol.*, *43*, 1834–1852.

Huffman, G. J., R. F. Adler, D. T. Bolvin, G. Gu, E. J. Nelkin, K. P. Bowman, Y. Hong, E. F. Stocker, and D. B. Wolff (2007), The TRMM Multisatellite Precipitation Analysis (TMPA): Quasi-global, multilayer, combined-sensor, precipitation estimates at fine scale, *J. Hydrometeorol.*, *8*, 38–55.

Hyndman, R. J., D. M. Bashtannyk, and G. K. Grunwald (1996), Estimating and visualizing conditional densities, *J. Comput. Graph. Stat.*, *5*, 315–336.

Joyce, R. J., J. E. Janowiak, P. A. Arkin, and P. Xie (2004), CMORPH: A method that produces global precipitation estimates from passive microwave and infrared data at high spatial and temporal resolution, *J. Hydrometeorol.*, *5*, 487–503.

Krajewski, W. F., and J. A. Smith (2002), Radar hydrology: Rainfall estimation, *Adv. Water Resour.*, *25*, 1387–1394.

Krajewski, W. F., G. Villarini, and J. A. Smith (2010), Radar-rainfall uncertainties: Where are we after thirty years of efforts?, *Bull. Am. Meteorol. Soc.*, *91*(1), 87–94.

Kuczera, G., D. Kavetski, S. Franks, and M. Thyer (2006), Towards a Bayesian total error analysis of conceptual rainfall-runoff models: Characterising model error using storm dependent parameters, *J. Hydrol.*, *331*, 161–177.

Moradkhani, H., and T. T. Meskele (2010), Probabilistic assessment of the satellite rainfall retrieval error translation to hydrologic response, in *Satellite Rainfall Applications for Surface Hydrology*, edited by Moradkhani, H., and F. Hossain, pp. 229–242, doi:10.1007/978-90-481-2915-7_14, Springer, Netherlands.

Moradkhani, H., K. Hsu, Y. Hong, and S. Sorooshian (2006), Investigating the impact of remotely sensed precipitation and hydrologic model uncertainties on the ensemble streamflow forecasting, *Geophys. Res. Lett.*, *33*, L12401, doi:10.1029/2006GL026855.

Petris, G., and L. Tardella (2006), Transdimensional Markov Chain Monte Carlo using hyperplane inflation in locally nested spaces, *Technical Rep. 4*, Dipartimento di Statistica, Probabilitàe Statistiche Applicate, Università di Roma "La Sapienza", Rome, Italy.

Sapiano, M. R. P., J. E. Janowiak, W. Shi, R. W. Higgins, and V. B. S. Silva (2010), Regional evaluation through independent precipitation measurements: USA, in *in: Satellite Rainfall Applications for Surface Hydrology*, edited by Sapiano, M. R. P., and F. Hossain, pp. 169–191, doi:10.1007/978-90-481-2915-7_10, Springer, Netherlands.

Smith, E. A., et al. (2007), International Global Precipitation Measurement (GPM) Program and Mission: An Overview, in *Measuring Precipitation From Space*, edited by Smith, E. A., P. Bauer, and J. Turk, pp. 611–653, Springer, Dordrecht, Netherlands.

Sorooshian, S., K. Hsu, X. Gao, H. V. Gupta, B. Imam, and D. Braithwaite (2000), Evaluation of PERSIANN system satellite-based estimates of tropical rainfall, *Bull. Am. Meteorol. Soc.*, *81*, 2035–2046.

Steiner, M., T. L. Bell, Y. Zhang, and E. F. Wood (2003), Comparison of two methods for estimating the sampling-related uncertainty of satellite rainfall averages based on a large radar dataset, *J. Clim.*, *16*, 3759–3778.

Su, F., Y. Hong, and D. P. Lettenmaier (2008), Evaluation of TRMM Multisatellite Precipitation Analysis (TMPA) and its utility in hydrologic prediction in the La Plata Basin, *J. Hydrometeorol.*, *9*, 622–640.

Tian, Y., C. D. Peters-Lidard, B. J. Chaudhury, and M. Garcia (2007), Multitemporal analysis of TRMM-based satellite precipitation products for land data assimilation applications, *J. Hydrometeorol.*, *8*, 1165–1183, doi:10.1175/2007JHM859.1.

Tobin, K. J., and M. E. Bennett (2009), Using SWAT to model streamflow in two river basins with ground and satellite precipitation data, *J. Am. Water Resour. Assoc.*, *45*(1), 253–271, doi:10.1111/j.1752-1688.2008.00276.x.

Villarini, G., and W. F. Krajewski (2010), Review of the different sources of uncertainty in radar-based estimates of rainfall, *Surv. Geophys.*, *31*, 107–129.

Wilks, D. S. (1995), *Statistical Methods in the Atmospheric Science*, 465 pp., Academic, San Diego, Calif.

Yan, J., and M. Gebremichael (2009), Estimating actual rainfall from satellite rainfall products, *Atmos. Res.*, *92*(4), 481–488, doi:10.1016/j.atmosres.2009.02.004.

Yilmaz, K. K., T. S. Hogue, K.-L. Hsu, S. Sorooshian, H. V. Gupta, and T. Wagener (2005), Intercomparison of rain gauge, radar, and satellite-based precipitation estimates with emphasis on hydrologic forecasting, *J. Hydrometeorol.*, *6*, 497–517.

Zeweldi, D. A., and M. Gebremichael (2009a), Evaluation of CMORPH precipitation products at fine space–time scales, *J. Hydrometeorol.*, *10*(1), 300–307, doi:10.1175/2008JHM1041.1.

Zeweldi, D. A., and M. Gebremichael (2009b), Sub-daily scale validation of satellite-based high-resolution rainfall products, *Atmos. Res.*, *92*(4), 427–433, doi:10.1016/j.atmosres.2009.01.001.

M. Gebremichael, Department of Civil and Environmental Engineering, University of Connecticut, 261 Glenbrook Road, Unit 2037, Storrs, CT 06269-2037, USA. (mekonnen@engr.uconn.edu)

AGU Category Index

Aerosols and particles, 29
Climate and interannual variability, 189
Climate impacts, 215
Cloud physics and chemistry, 29, 49
Clouds and aerosols, 49
Clouds and cloud feedbacks, 49
Computational hydrology, 171
Estimation and forecasting, 79
Extreme events, 159, 171
General or miscellaneous, 7, 29
Global climate models, 189
Hydrology, 61, 171
Hydrometeorology, vii, 1, 7, 79, 105, 127, 159
Measurement and standards, 61
Modeling, 171
Particle precipitation, 189
Precipitation, vii, 1, 7, 49, 61, 79, 105, 127 159, 215, 247 265
Precipitation-radar, vii, 1, 7, 79, 105, 247
Remote sensing, vii, 1, 105, 127, 247, 265
Spatial analysis and representation, 189
Stochastic hydrology, 215
Streamflow, 265
Troposphere: composition and chemistry, 29
Uncertainty assessment, 61, 247, 265
Water cycles, 127
Water supply, 215

Index

Note: Page numbers with italicized *f* and *t* refer to figures and tables

A

active microwave methods, 140–141
additive bias, 267
AHB10 error model, 253*t*, 254–255
Akaike information criterion (AIC), 177
Algorithm Intercomparison Programme (AIP), 147
All-Weather Precipitation Accumulation Gauge (AWPAG), 62
alternating renewal process-based models, 220*t*
annual rainfall generation, 217–218
ANUSPLIN software, 191
Archimedean copulas, 183
areal reduction factor, 163
Arkansas-Red Basin River Forecast Center (ABRFC), 94
Atmospheric InfraRed Sounder (AIRS), 133
Automatic Surface Observing System (ASOS), 62
auxiliary predictors, 203–206
average received power, 80
Azimuthal Equidistant projection system, 201

B

Bayesian information criterion, 177
Beard-Chuang equilibrium, 109
Best Linear Unbiased Predictor (BLUP) model, 191
bias, in satellite rainfall estimates, 267
Bilogora case study, 200–206
  spatial prediction using auxiliary predictors, 203–206
  spatial prediction using geostatistics, 201–203
bivariate copulas, 183*t*
bivariate frequency analysis, 182–183
Boltzmann transport equation, 31
breakup, collision-induced, 43, 50–53
breakup efficiency, 50
BZ08 error model, 249, 253*t*

C

calibration errors, 63–65
canonical random multiplicative model, 229–230
CASA Integrated Project 1 (IP1) network, 121–122
categorical metrics, 267
Clayton copulas, 184*f*
climate change, 230–235
Climate Prediction Center Merged Analysis of Precipitation (CMAP), 142
Climate Prediction Center Morphing (CMORPH), 138*f*, 142, 270, 271*f*, 272–273
cloud profiling radar (CPR), 133
cloud water path (CWP), 135–136
Cloud-Aerosol Lidar with Orthogonal Polarization, 133
cloud-resolving model (CRM), 146
Clouds-Aerosols-Precipitation Satellite Analysis Tool (CAPSAT), 134
CloudSat, 133, 141
cluster-based point processes, 220*t*
coagulation, 34–39
  breakup, 43
  collision kernels, 43–45
  definitions of specific processes, 34–36
  integral rates of total number and mass densities, 38–39
  integrals of spectral rates, 37–38
  spectral rates, 35–36
coalescence efficiency, 50
collision efficiency, 32–33, 51
collision kernels, 43–45
collisional interaction, 32–34
collision-induced breakup, 43, 50–53
complete-duration series (CDS), 171
conditional metrics, 268
Consiglio Nazionale delle Ricerche, 206
constant altitude plan position indicator (CAPPI), 249
convective-stratiform separation algorithm (CSSA), 85
Cooperative Institute for Climate Studies (CICS), 143
copolar correlation coefficient, 110
copulas, 3, 167. *See also* Archimedean copulas; bivariate copulas; Clayton copulas
correlation, 267–268
CPC Morphing Technique (CMORPH), 196
Croatia, 201
CSU-CHILL radar, 121*f*
CSU-HIDRO algorithm, 119–120
CSU-ICE algorithm, 112, 117–118, 119
curve number, 270

D

daily rainfall generation, 218–227. *See also* rainfall generation
  alternatives aimed at reducing overdispersion, 221*t*
  alternatives for stochastic generation, 220*t*
  on fine spatial grid, 226–227
  model nesting at multiple time scales, 223–225
  modified Markov model, 223
  multisite alternatives, 225*t*
  ROG-RAG model, 221–223
  stochastic disaggregation of rainfall, 227
  transition probability matrix, 219–223
Defense Meteorological Satellite Program (DMSP), 133
delta-change downscaling, 232*t*
dielectric factor, for water, 11
differential reflectivity, 11, 109
digital terrain model (DTM), 82
disdrometers, 69–74. *See also* rain gauges
  basic definitions, 69–70
  measurement errors, 73–74
    instrumental uncertainty, 73
    observational uncertainty, 74
    sampling uncertainty, 73–74
  types of, 70–72
    2D video disdrometer, 70–71
    impact disdrometer, 70
    Joss-Waldvogel disdrometer, 276
    optical disdrometer, 70–72
    radar-based disdrometer, 72–73
downscaling, 232–235. *See also* rainfall generation
  analog, 233
  delta-change approach, 232*t*
  dynamic vs. statistical, 231*t*
  GLIMCLIM method, 233–234
  MMM-KDE method, 234
  nonhomogenous hidden Markov model, 234
  regression-based approaches, 232*t*
  scaling, 232*t*, 233
  statistical, 233
  statistical downscaling model, 234–235
  stochastic, 233
  weather classification-based approaches, 232*t*
  weather generators, 232*t*
drop eccentricity, 11
drop oscillation, 12–16
drop shape, 10–12
  classification of, 11
  gravity models, 12
  perturbation models, 12
  physical factors, 11
drop size distribution (DSD), 49–56
  collision-induced breakup, 50–53
  defined, 69
  drop shape and, 10
  measurements of, 53–54
  observed vs. modeled, 55–56
  overview, 49–50
  parametric forms, 108
  radar rainfall estimation and, 81
  representation of, 54–55
  stationary, 51–53
drop towers, 21
dual-polarization radar, 105–122. *See also* radar rainfall estimation
  future research in, 119–122
  overview, 105–106
  parameters, 108–110
    copolar correlation coefficient, 110
    differential reflectivity, 109
    linear depolarization ratio, 110
    specific differential phase, 109–110
  rainfall estimation with, 110–119
    classification, 111–112
    optimal rainfall estimators, 117
    physical approach, 113
    preprocessing, 110–111
    quantification, 112–119
    single measurement estimators, 113–114
    statistical approach, 112–113
    three measurement estimators, 115–117
    two measurement estimators, 114–115
  rainfall measurements, 107–108
  weather radar rainfall estimation, 106–107
duration. *See* complete-duration series; intensity-duration-frequency curves
dynamic downscaling, 231*t*

E

effective velocity, 31
electrostatic forces, 11, 18–20
ensemble generation, 255–257. *See also* error models
  empirically based techniques, 256–257
  physical-statistical techniques, 256
  statistical structure, 260
  statistical techniques, 255–256
  temporal dependence, 260
equilibrium distribution, 51
error models, 249–255, 268–270
  AHB10, 254–255
  for first- and second-order moments, 250–251
  HAM08, 252–254

error models (*continued*)
KDBO10 error, 255
nonparametric conditional model, 269–270
power law model, 268–269
product error-driven, 251–252
REAL, 253–254
sampling from model, 270
semiparametric conditional model, 269
source-specific, 248–249
BZ08, 249
JSA00, 248–249
JSW03, 249
error variance separation (EVS), 250
estimation, 2–3
EUMETSAT, 130
evaporation losses, rain gauge, 65
external aerodynamic pressure, 11
extreme rainfall events, frequency analysis of, 171–185
bivariate frequency analysis, 182–183
case study, 180–182
data requirements, 171–173
data validation, 171–172
goodness-of-fit tests, 177
information criteria, 177–178
Kendall test for stationarity, 173
Mann-Whitney test for homogeneity and stationarity, 173
nonstationarity, 178–180
nonstationarity GEV model, 178–179
nonstationarity model selection, 179–180
nonstationarity of hydrological series, 178
parameter estimators for model with covariates, 179
notation, 184–185
overview, 171
parameter estimating methods, 176–177
plotting positions, 177
probability plots, 177
statistical criteria and tests, 172
statistical distributions, 173–176
generalized extreme value, 174
harmonic distribution, 175
three-parameter LN distribution, 175
Wald and Wolfowitz test for independence, 172–173

F

false alarm rate (FAR), 267
FAOCLIM climate database, 195
Fast Fourier Transform (FFT), 95
fields package, 194
Flash Flood Monitoring and Production System (FFMP), 82
floating gauges, 62
fluid dynamics, raindrop, 8–10
free-falling drop studies, 21–22
frequency analysis, 171–185
bivariate frequency analysis, 182–183
case study, 180–182
data requirements, 171–173
data validation, 171–172
of extreme rainfall events, 184
goodness-of-fit tests, 177
information criteria, 177–178
Kendall test for stationarity, 173
Mann-Whitney test for homogeneity and stationarity, 173
nonstationarity, 178–180
notation, 184–185
overview, 171
parameter estimating methods, 176–177
plotting positions, 177
probability plots, 177
statistical criteria and tests, 172
statistical distributions, 173–176

G

gamma distribution, 108
gauge-only analysis, 94–95
general circulation models (GCMs), 230–235
analog, 233
atmospheric, 231
delta-change approach, 232*t*
dynamic vs. statistical, 231*t*
GLIMCLIM method, 233–234
MMM-KDE method, 234
nonhomogenous hidden Markov model, 234
oceanic, 231
regression-based approaches, 232*t*
scaling, 232*t*, 233
statistical, 233
statistical downscaling model, 234–235
stochastic, 233
weather classification-based approaches, 232*t*
weather generators, 232*t*
generalized extreme value (GEV) distribution, 174
generalized linear model for daily climate (GLIMCLIM) method, 233–234
generalized linear models (GLMs), 225*t*
generalized maximum likelihood (GML), 176
geometric collision kernel, 44
geoR package, 194
geostationary satellites, 130–134
geostatistics, 201–203
GLO distribution, 176

Global History Climatology Network (GHCN), 194–195
Global Mapping of Precipitation (GSMaP), 196–197
Global Observing System (GOS), 131*f*
Global Precipitation and Climatology Project (GPCP), 142, 148
Global Precipitation Climatology Centre, 195
Global Precipitation Climatology Project (GPCP), 196
Global Precipitation Measurement (GPM), 139
Global Summary of Day (GSOD), 195
Goddard Institute for Space Studies (GISS), 146
Goddard profiling technique (GPROF), 137
GOES Multi-Spectral Rainfall Algorithm, 135
goodness-of-fit tests, 177
gravity models, of raindrop shape, 12
ground reference rainfall estimates, 266, 266*t*
ground-based radar images, 199–200
gstat package, 194

H

HAM08 error model, 252–254, 253*t*
Heidke Skill Score (HSS), 267
HIB distribution, 175
higher-order Markov chain models, 220*t*
"hybrid-order" Markov models, 220*t*
HYDRO-35, 163
Hydrologic Automated Data System (HADS), 67
hydrologic forecasting, 79–98
  multisensor rainfall estimation, 89–98
  radar rainfall estimation, 79–89
hydrologic frequency analysis (HFA), 171–185
  bivariate frequency analysis, 182–183
  case study, 180–182
  data requirements, 171–173
  data validation, 171–172
  of extreme rainfall events, 184
  goodness-of-fit tests, 177
  information criteria, 177–178
  Kendall test for stationarity, 173
  Mann-Whitney test for homogeneity and stationarity, 173
  nonstationarity, 178–180
    nonstationarity GEV model, 178–179
    nonstationarity model selection, 179–180
    nonstationarity of hydrological series, 178
    parameter estimators for model with covariates, 179
  notation, 184–185
  overview, 171
  parameter estimating methods, 176–177
  plotting positions, 177
  probability plots, 177
  statistical criteria and tests, 172
  statistical distributions, 173–176
    generalized extreme value, 174
    GLO distribution, 176
    harmonic distribution, 175
    HIB distribution, 175
    Pearson type 3 distribution, 175–176
    three-parameter LN distribution, 175
  Wald and Wolfowitz test for independence, 172–173
Hydrometerological Automated Data System (HADS), 93

I

impact disdrometer, 70
intamap package, 194, 207–208
intensity-duration-frequency (IDF) curves, 159–167
  annual maxima-centered approach, 162
  annual vs. partial duration series, 164
  data adjustments/corrections, 164–165
  equations for, 163–164
  frequency analysis, 165
  geographically fixed relationships, 162
  historical development of, 159–161
  HYDRO-35, 163
  multivariate analyses, 166–167
  overview, 3, 159
  precipitation frequency data server, 165
  from raw data, 164–165
  scaling, 166
  seasonality, 166
  spatial precipitation frequency, 161–163
  spatial smoothing, 165
  storm-centered relationships, 162
  technical papers, 160–161
internal circulation, 11
internal hydrostatic pressure, 11
International Precipitation Working Group (IPWG), 138*f*

J

Jacobi determinant, 30
Joint Polarization Experiment (JPOLE), 119
Joss-Waldvogel disdrometer, 276
JSA00 error model, 248–249, 253*t*
JSW03 error model, 249, 253*t*

K

KDBO10 error model, 253*t*, 255
Kendall test, 173

kinetic coagulation equation, 34
kriging, 192
  conventional, 203
  regression, 206, 211*f*
  transGaussain, 208
kriging with external drift (KED), 96, 192

L

Lagrangian model, 142
Legendre functions, 12, 24
levitated drop studies, 21
"light precipitation" error type, 68
lightning imaging sensor (LIS), 133
linear depolarization ratio (LDR), 110
local bias (LB) correction, 94
low Earth orbiting (LEO) satellites, 130–134
Lower Mississippi River Forecast Center (LMRFC), 67
low-frequency variability, 216, 218, 223, 237
low-order Markov chain models, 220*t*

M

Mahane Kahane Pierre function, 230
Mann-Whitney test, 173
Markov chain models, 220*t*
Marshall-Palmer spectrum, 10
maximum likelihood (ML) method, 176
mean absolute error (MAE), 268
mean field bias correction, 90–93
measurement, 2–3
Mesoscale Model (MM5), 139
Meteostart Second Generation (MSG) satellites, 130
MeteoSwiss QC system, 67
microcanonical cascades, 230
microphysics, 1–2, 87
microwave remote sensing, 195–196
MMM-KDE method, 234
Moderate Resolution Imaging Spectroradiometer (MODIS), 197
modified Markov model, 223
morphodynamics, raindrop, 7–24
  dimensionless parameters, 8–9
  drop oscillation, 12–16
  drop shape, 10–12
  electrostatic effects, 18–20
  experimental techniques, 21–23
    2D video disdrometer, 22
    free-falling drop studies, 21–22
    levitated drop studies, 21
    PMS device, 22
  fluid dynamics, 8–10
  overview, 7–8
  terminal velocity, 16–18, 23–24
multifractal simulation, 220*t*
multiplicative bias, 267
Multisensor Precipitation Estimator (MPE), 67, 89–90
multisensor precipitation reanalysis (MPR), 97
multisensor rainfall estimation, 89–98, 142–143. *See also* radar rainfall estimation
  delineation of effective radar coverage, 90
  detection bias in radar rainfall estimates, 96
  gauge-only analysis, 94–95
  local bias correction, 94
  mean field bias correction, 90–93
  mosaicking estimates from multiple radars, 93–94
  overview, 89–90
  parameter estimation, 96–97
  radar-gauge analysis, 95–96
  reanalysis, 97–98
  scale of analysis, 96
multisite Markov models, 225*t*
multisite subdaily rainfall generation, 230
multistate Markov models, 220*t*

N

National Climatic Data (NCDC), 195
National Oceanographic and Atmospheric Administration (NOAA), 161
National Severe Storms Laboratory (NSSL), 67, 118
National Weather Service (NWS), 160–161
National Weather Service (NWS) Surveillance Radar, 106
Navier-Stokes equation, 8
Next Generation Weather Radar (NEXRAD), 107, 139
Nimrod system, 82
noncollecting gauges, 62–63
nonhomogenous hidden Markov (NHMM), 234
nonparametric conditional model, 269–270
nonparametric multisite models, 225*t*
nonparametric precipitation amounts model, 220*t*
nonparametric stochastic disaggregation model, 228*t*
nonrecording storage gauges, 62
numerical weather prediction (NWP), 82

O

optical disdrometer, 70–72
optical gauges, 62–63
oscillations, raindrop, 12–16

P

parametric precipitation amounts model, 220*t*
parametric stochastic disaggregation model, 228*t*
particle fall speed, 107
Particle Measurement Systems (PMS), 53
passive microwave methods, 136–140
Pearson type 3 distribution, 175–176
PERSIANN, 141, 196, 269, 270, 271*f*, 272–273
PERSIANN-CCS, 196
PERSIANN-Multi-Spectral Analysis (PERSIANN-MSA), 141
perturbation models, of raindrop shape, 12
Poisson cluster process, 228, 228*t*
polarization-corrected temperature (PCT), 140
power law model, 268–269
Precipitation Estimation from Remotely Sensed Information using Artificial Neural Networks. *See* PERSIANN
Precipitation Frequency Data Server (PFDS), 165
Precipitation Intercomparison Projects (PIP), 146–147
Precipitation-elevation Regressions on Independent Slopes Model (PRISM), 90
probability of detection (POD), 267
probability plots, 177
product error-driven (PED) model, 251–252, 258
Program for the Evaluation of High-Resolution Precipitation Products (PEHRPP), 147
PRO-OBS-5, 196

Q

quality control, 61, 106. *See also* dual-polarization radar
  UK Met Office, 4
quantitative precipitation estimation, 127, 134–144
  active microwave methods, 140–141
  multisensor techniques, 142–146
  passive microwave methods, 136–140
  snowfall retrievals, 143–144
  VIS/IR methods, 134–136

R

R+OSGeo software, 194
radar rainfall estimates, 266
radar rainfall estimation, 79–89. *See also* multisensor rainfall estimation
  applications of, 82
  applied science, 106
  basic science, 106
  detection bias in, 96
  dual-polarization, 110–119
    classification, 111–112
    optimal rainfall estimators, 117
    physical approach, 113
    preprocessing, 110–111
    quantification, 112–119
    single measurement estimators, 113–114
    statistical approach, 112–113
    three measurement estimators, 115–117
    two measurement estimators, 114–115
  ensemble generation, 255–257
  equations for, 80–81
  error models, 248–255
  error sources, 80–81
  hardware and beam propagation, 87–89
  maps for cool season event in complex terrain, 86*f*
  microphysics of, 87
  overview, 80–81
  quantitative hydrologic applications, 84*f*
  reflectivity morphology, 82–87
  sampling geometry, 82–87
  uncertainty propagation, 257–259
radar reflectivity, 69
radar signal wavelength, 11
radar-based disdrometer, 72–73
radar-gauge analysis, 95–96
radar-rainfall error
  additive vs. multiplicative error structure, 259
  dependence among individual errors, 259
  distribution of, 259
  nonstationarity of errors, 259
  statistical structure, 260
rain gauges, 61–69. *See also* disdrometers
  errors, 63–66
    bad location, 66
    calibration-related, 63–65
    classification of, 63*f*
    local random errors, 65–66
    local systematic, 63
    malfunctioning problems, 66
    wetting, evaporation, and splashing losses, 65
    wind effects, 63–64
  network, 106, 114, 121*f*
  observations, 266
  quality control of measurements, 66–68
    automated real-time, 67
    historical data, 67–69
    multisensor observations, 67
  types of, 61–63
    floating gauges, 62
    noncollecting gauges, 62–63
    nonrecording storage gauges, 62
    optical gauges, 62–63
    recording gauges, 62

rain gauges (*continued*)
types of (*continued*)
tipping bucket gauges, 62
weighing gauges, 62
raindrop
breakup, 43
chord axis ratio, 10
collision kernels, 43–45
collisional interaction of, 32–34
electrostatic forces, 18–22
morphodynamics, 7–24
oscillation, 12–16
self-collection of droplets, 35
shape, 10–12
spectra, 29–45
terminal velocity, 16–18, 23–24
raindrop size distributions (RSDs), 49–56
collision-induced breakup, 50–53
defined, 69
drop shape and, 10
measurements of, 53–54
observed vs. modeled, 55–56
overview, 49–50
parametric forms, 108
radar rainfall estimation and, 81
representation of, 54–55
stationary, 51–53
rainfall
dual-polarization radar estimation, 110–119
extreme, 171–185
frequency analysis, 171–185
intensity, 64, 69
measurement and estimation, 2–3
microphysics, 1–2
multisensor estimation, 89–98
quantitative precipitation estimation, 134–144
radar estimation, 79–89
satellite estimation, 265–274
spatial prediction, 189–212
statistical analyses, 3–4
rainfall generation, 215–238
for climate change conditions, 230–237
daily, 218–227
downscaling techniques, 232–235
general circulation models, 230–235
analog, 233
delta-change approach, 232*t*
dynamic vs. statistical, 231*t*
GLIMCLIM method, 233–234
MMM-KDE method, 234
nonhomogenous hidden Markov model, 234
regression-based approaches, 232*t*
scaling, 232*t*, 233
statistical, 233
statistical downscaling model, 234–235
stochastic, 233
weather classification-based approaches, 232*t*
weather generators, 232*t*
overview, 215–216
seasonal to annual, 217–218
stochastic, 216–230
subdaily, 227–230
Randburg station, 180–182
random cascade models, 225*t*
random multiplicative cascades, 228–229
Rayleigh conditions, 110
REAL error model, 253–254, 253*t*
recording gauges, 62
reflectivity, differential, 11
reflectivity factor, 80–81, 107
refractive index, 11
regression-based downscaling, 232*t*
regression-kriging, 206, 211*f*
reshuffling approach-based models, 225*t*
Reynolds number, 8–9
Reynolds transport theorem, 31
ROG-RAG model, 221–223
root-mean-square error (RMSE), 268

S

SAGA GIS software, 194
satellite imagery, 195–199
satellite rainfall estimates, 265–274
defined, 266*t*
definition of, 266*t*
error, 268–270
defined, 266*t*
nonparametric conditional model, 269–270
power law model, 268–269
sampling from model, 270
semiparametric conditional model, 269
evaluation metrics, 266–268
categorical metrics, 267
conditional metrics, 268
unconditional metrics, 267–268
ground reference rainfall estimates, 266
hydrological evaluation, 270–271
overview, 265
products, 271–273
infrared-based, 273
microwave-based, 273

satellite rainfall estimates (*continued*)
products (*continued*)
satellite-gauge, 273
satellite-only, 273
radar rainfall estimates, 266
rain gauge observations, 266
web sites, 132–133*t*
satellites, 127–150
abbreviations and definitions, 128–130*t*
applications, 144–146
climate studies, 146
hydrology, 144–145
latent heat evaluation, 145–146
process studies, 145
water cycle, 144–145
geostationary, 130–134
low Earth orbiting, 130–134
overview, 127–130
quantitative precipitation estimation, 134–144
scale invariance theory, 228*t*
scaling, 232*t*, 233
Schwarz method, 177
seasonal rainfall generation, 217–218
self-collection of cloud droplets, 35–36
semiparametric conditional model, 269
SEVIRI PRO-OBS-5, 196
shape, raindrop, 10–12
classification of, 11
gravity models, 12
perturbation models, 12
physical factors, 11
single-optimal estimation, 95
Smoluchowski equation, 34
snowfall retrievals, 143–144
Soil Conservation Service, 160
Southeast River Forecast Center (SERFC), 67
Southern Oscillation Index (SOI), 180–182
spatial interpolation, 190
spatial precipitation frequency, 161–163
spatial prediction, 189–212
atmospheric model, 190
case studies
Bilogora, 200–206
Italy, 206–211
data sources, 194–200
ground-based radar images, 199–200
rainfall measurements, 194–195
satellite imagery, 195–199
estimation error, 191
software, 194
specific properties of rainfall data, 192–194
splines with tension model, 190–191
stochastic model, 190–191
specific differential phase, 109–110
spectra, raindrop, 29–45
breakup processes, 43
coagulation, 34–39
definitions of specific processes, 34–36
integral rates of total number and mass densities, 38–39
integrals of spectral rates, 37–38
spectral rates, 35–36
collision kernels, 43–45
collisional interaction of drops, 32–34
general definitions and relations, 29–32
numerical case study, 39–43
overview, 29
spectral number density function, 30
Spinning Enhanced Visible and Infrared Instrument (SEVIRI), 195
splashing losses, rain gauge, 65
Spring Enhanced Visible and InfraRed Imager (SEVIRI), 136
stationary distribution, 51–53
statistical analyses, 3–4
statistical downscaling, 231*t*, 233
statistical downscaling model (SDSM), 234–235
stochastic collection equation, 34
stochastic rainfall generation, 216–230
canonical RMC model, 229–230
for climate change conditions, 230–235
daily, 218–227
extension to multisite generation, 225–226
on fine spatial grid, 226–227
general circulation models, 230–235
analog, 233
delta-change approach, 232*t*
dynamic vs. statistical, 231*t*
GLIMCLIM method, 233–234
MMM-KDE method, 234
nonhomogenous hidden Markov model, 234
regression-based approaches, 232*t*
scaling, 232*t*, 233
statistical, 233
statistical downscaling model, 234–235
stochastic, 233
weather classification-based approaches, 232*t*
weather generators, 232*t*
microcanonical cascades, 230
model nesting at multiple time scales, 223–225
Poisson cluster process-based models, 228
random multiplicative cascades, 228–229
seasonal to annual, 217–218
stochastic disaggregation of rainfall, 227
subdaily, 227–230
transition probability matrix, 219–223

Strouhal number, 8–9
subdaily rainfall generation, 227–230. *See also* rainfall generation
  canonical RMC model, 229–230
  microcanonical cascades, 230
  models, 228*t*
  multisite, 230
  Poisson cluster process-based models, 228
  random multiplicative cascades, 228–229
surface tension, 11

T

terminal velocity, 16–18, 23–24
three-parameter LN distribution, 175
time series models, 220*t*
tipping bucket gauges, 62
TMPA 3B42, 270, 271*f*, 273
TMPA 3B42RT, 270, 271*f*
TMPA 3B42T, 273
total mass density, 30
total number density, 30
transGaussain kriging, 208
transition probability matrix (TPM), 219–223
transmitter power, 80
TRMM (Tropical Rainfall Measuring Mission), 133, 134, 140–141, 145, 148
TRMM Microwave Imager, 133
TRMM Multi-Satellite Precipitation Analysis (TMPA), 142, 196
TRMM Precipitation Radar, 87–89
TRMM-Large-Scale Biosphere-Atmosphere Experiment (TRMM-LBA), 117
true rainfall estimate, 266*t*
true stochastic coalescence equation (TSCE), 34
turbulent collision kernel, 45
two-dimensional video disdrometer (2DVD), 22, 70–71, 73

U

uncertainty propagation, 257–259
unconditional bias, 267
unconditional metrics, 267–268
universal kriging, 192
U.S. Climate Reference Network (USCRN), 62

V

VDIAZ spectrofluviometer, 22
velocity, effective, 31
vertical profile of reflectivity (VPR), 81, 85
Visible and InfraRed Scanner (VIRS), 133
VIS/IR methods, 134–136

W

wakes, classification of, 10*t*
Wald and Wolfowitz test for independence, 172–173
water cycle, 144–145
Water Vapor Strong Lines, 139
Weather Bureau, 160–161
Weather Bureau Air Force Number (WBAN), 200
weather classification-based downscaling, 232*t*
weather generators, 232*t*
weather radar rainfall estimation, 106–107
weather state-based hidden Markov models, 225*t*
Weather Surveillance Radars-1988 Doppler (WSR-88D) radars, 2
Weber number, 8–9
weighing gauges, 62
wet spell-based precipitation amount models, 220*t*
wetting losses, rain gauge, 65
wind effects, on rain gauges, 63–64
World Meteorological Organization (WMO), 130*t*
WSR-88D, 81, 82*f*, 88
WSR-88D radars, 2

X

X band network, 121–122

Z

"zero precipitation" error type, 68
zero-inflated Poisson (ZIP) model, 192–193
Z-R relations, 90, 106, 249